D1433998

INDUSTRIAL MINERALS AND THEIR USES

INDUSTRIAL MINERALS AND THEIR USES

A Handbook and Formulary

Edited by

Peter A. Ciullo

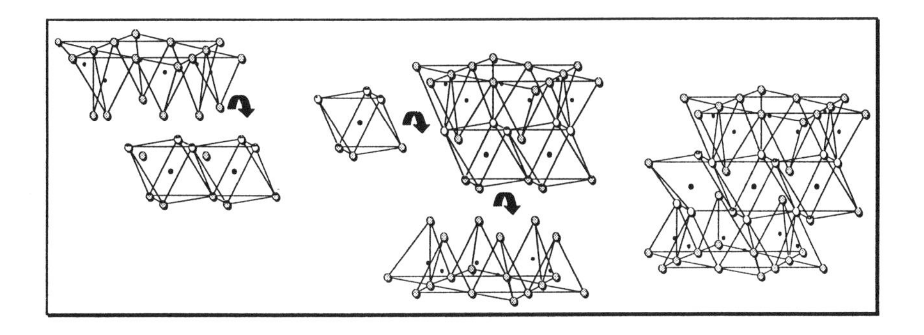

NOYES PUBLICATIONS
Westwood, New Jersey, U.S.A.

Library of Congress Catalog Card Number: 96-29173
ISBN: 0-8155-1408-5
Printed in the United States

Published in the United States of America by
Noyes Publication
369 Fairview Ave.
Westwood, New Jersey 07675

10 9 8 7 6 5 4 3 2 1

Library of Congress Cataloging-in-Publication Data

Industrial minerals and their uses : a handbook and formulary / edited by Peter A. Ciullo.
p. cm.
Includes index.
ISBN 0-8155-1408-5
1. Inorganic compounds--Industrial applications--Handbooks. manuals, etc. I. Ciullo, Peter A., 1954- .
TP200.I52 1996
661--dc20 96-29173
CIP

PREFACE

The technical orientation of most formulators and compounders is chemistry, not mineralogy. They may have a natural grasp of their chemical ingredients, but many lack training or background in the minerals they use as “fillers”. The reference works on formulating technology for the various minerals-consuming industries likewise often treat the mineral additives in a cursory fashion, if at all. My primary purpose in compiling *Industrial Minerals and Their Uses* has been to provide product development professionals – novice and seasoned – with a better understanding of their mineral raw materials. My hope is that through this understanding they can develop their skills in matching the most appropriate minerals to their applications while gaining an appreciation of both the common ground and differences in approach they have with counterparts in industries other than their own.

Industrial Minerals and Their Uses accordingly offers a concise profile of the structure, properties and uses of eighteen of the most commonly employed industrial minerals, plus a comprehensive overview of how and why these minerals are used in eight consuming industries. Paints and coatings, paper, rubber, adhesives and sealants, and plastics technology are reviewed as major beneficiaries of the use of minerals as functional additives. Chapters on pharmaceuticals and pesticides are included as a contrast in perspective regarding the selection and use of mineral additives, while the chapter on ceramics and glass is offered as an introduction to the use of minerals as primary raw materials or reactants, with chemicals relegated to the role of additives.

While formulators and compounders are the main audience for *Industrial Minerals and Their Uses*, the producers and marketers of the industrial minerals themselves will undoubtedly find this book a valuable resource for identifying potential new markets for current products, and for discovering opportunities for the development of new ones. It is, in fact, because the industrial minerals producers have been so successful in tailoring the particle size, shape, surface area, and surface properties of their raw materials that the classification “filler”, although still used generically, is now an anachronism and generally misapplied.

In preparing *Industrial Minerals and Their Uses*, I have been very fortunate in obtaining the aid of minerals and formulating experts whose knowledge and experience extend well beyond my own. They have generously contributed their time and expertise in the form of several chapters in this book. For this I am deeply grateful. I must also sincerely thank Frank Alsobrook of Alsobrook & Co., Inc. and Dr. Slim Thompson, mineralogist emeritus at R.T. Vanderbilt Co., Inc. and my minerals mentor, for their much appreciated editorial attention to chapters one and two. My gratitude extends as well to R.T. Vanderbilt Co. for providing the opportunity over the past twenty years to learn about and contribute to the industrial minerals and their uses, and in particular to Bob Ohm, editor of the *Vanderbilt Rubber Handbook* for alerting me to the unsuspected (i.e., nerve-wracking) challenges of compiling a book of this nature, and most especially for not discouraging me from doing it anyway.

Above all, my thanks and my love to my wife Claudia, and to Marissa and Adam, my children, for their unwavering support and understanding in this and all my seemingly neverending book projects.

Peter A. Ciullo

NOTICE

CONTENTS

ONE

SILICATE STRUCTURES

Peter A. Ciullo
R.T. Vanderbilt Company, Inc.
Norwalk, CT

Technologists charged with using industrial minerals typically draw their expertise from disciplines other than mineralogy. They may have a strong practical understanding of chemical raw materials but often lack an appreciation of minerals beyond their obvious effects on product properties and cost. A mineral's name and chemical formula are admittedly seldom enlightening. A grasp of mineral crystal architecture can therefore provide at least a foundation for using industrial minerals as constituents in many products. Conjuring up images of mineral structures will perhaps unleash long-suppressed memories of space groups, unit cells, and planes of symmetry. These topics of Inorganic Chemistry 101 seem of little relevance when the job at hand is to improve the heat deflection temperature of polypropylene or to ensure the durability of a bridge coating. Picturing the structures of common industrial silicate minerals can, nevertheless, at least provide insight into their common features and subtle differences and how these are reflected in the properties and uses described in the following chapter.

In simplest terms, the silicate minerals can be considered inorganic polymers based on two basic "monomer" structures. These are the tetrahedron of Figure 1 and the octahedron of Figure 2. Many of the silicates can be pictured as the configurations made by joining of such tetrahedra and octahedra to themselves and to each other in three dimensions. These involve the sharing of corners, edges, and faces in numerous conformations. The possible geometric permutations are further modified by chemical substitutions within the structure, which usually depend on how well a metal ion will fit among close-packed oxygen ions. This is largely a matter of relative ionic radii. Given an O^{2-} ionic radius of 1.40 angstroms, the preferred (most stable) coordination of cations common in industrial silicate minerals has been calculated and expressed in terms of ionic radius ratio.

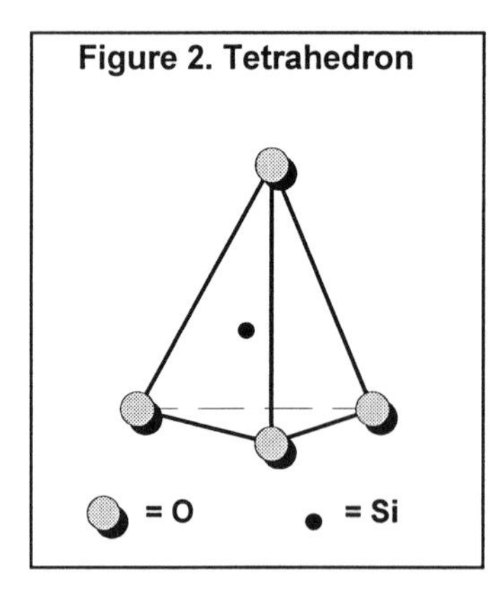

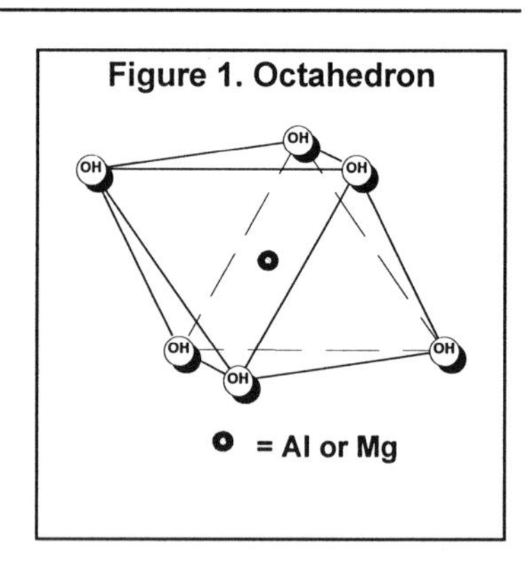

Tetrahedral, four-fold coordination is theoretically preferred when the radius ratio of metal cation to oxygen ion is in the range 0.225 to 0.414; for octahedral, six-fold coordination, this range is 0.414 to 0.732; for cubic, eight-fold coordination, it is 0.732 to 1.000. In nature, these ranges overlap to some extent, and the mineral lattice will distort to a limited degree to accommodate ions that are not a perfect fit. Aluminum, for example, is found in both tetrahedral and octahedral coordination. The following table lists the atomic radii of common metals found in silicate minerals, along with their ratio compared to O^{2-} and their coordination number.

Ion	Radius, A	R_M:R_O	Coordination No.
Si^{4+}	0.39	0.278	4
Al^{3+}	0.51	0.364	4
Al^{3+}	0.51	0.364	6
Fe^{3+}	0.64	0.475	6
Mg^{2+}	0.66	0.471	6
Li^{+}	0.68	0.486	6
Fe^{2+}	0.74	0.529	6
Na^{+}	0.97	0.693	8
Ca^{2+}	0.99	0.707	8
K^{+}	1.33	0.950	8-12

A mineral's unit cell formula or structural representation will usually reflect the theoretical composition or one with the most common substitutions. As the table above suggests, however, like-size cations can and do substitute for the theoretical components in nature. Chemical purity of industrial minerals is a concern when it adversely affects color or when the mineral is being used at least in part for its chemical constituents, as in ceramics.

⌘⌘⌘

Mineralogical purity of industrial minerals is a factor distinct from chemical purity for some end uses. Commercially exploitable ore deposits are rarely monomineralic. A substantial component of the value-added cost of many industrial mineral products is the expense incurred by the producer in reducing mineral impurities by screening, air classification, washing, flotation, centrifuging, magnetic separation, heavy media separation, electrostatic separation, or various combinations of these. Conversely, some rocks such as nepheline syenite derive commercial value from their component mineral properties. The reasons why certain minerals coexist in ore deposits are beyond the intent of this chapter. An understanding of the structural relationships among these minerals, however, can help to explain this coexistence. Perhaps more interesting to those who use industrial minerals are the structural features common to minerals that are otherwise mutually exclusive in use.

Quartz

The fundamental structural unit of industrial silicate minerals is the silica tetrahedron. Quartz is just a densely packed arrangement of these tetrahedra, as depicted in Figure 3.

Figure 3. Quartz

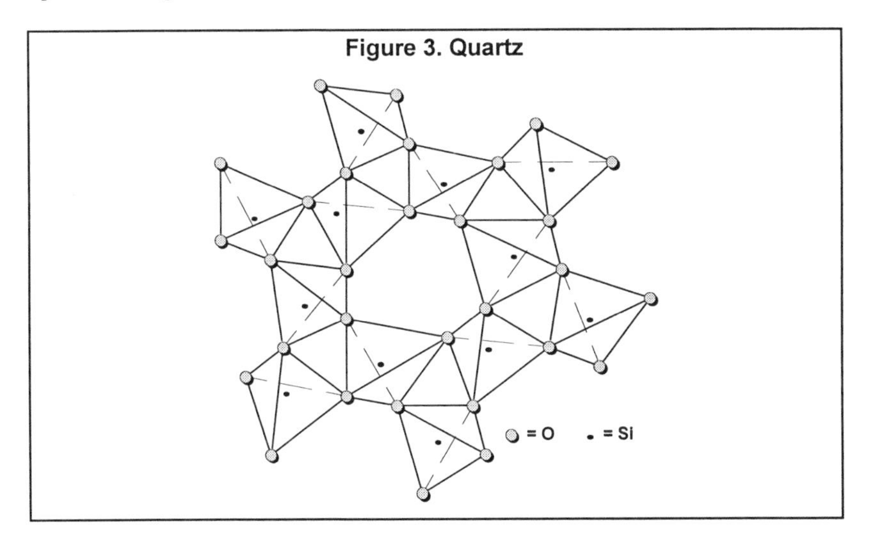

Extended in three dimensions, this structure provides the characteristic hardness and inertness of quartz. The different forms of crystalline silica – most commonly quartz, cristobalite, and tridymite – differ mainly in the relative orientation of adjacent tetrahedra and the shape of voids created within a given plane.

⌘⌘⌘

Feldspar

Similar dense-packed tetrahedra characterize the crystal framework of feldspar minerals, as depicted in Figure 4.

Figure 4. Feldspar

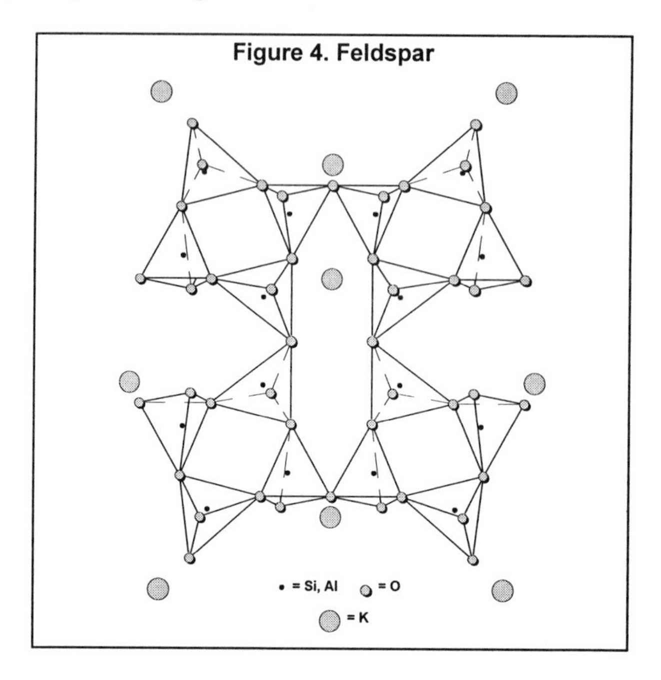

This figure shows a single layer viewed perpendicular to its plane. The framework is extended by rotating each successive layer 90°. Feldspars usually have one of every four Si^{4+} substituted with Al^{3+}. The resulting charge imbalance is compensated by sodium, and/or potassium ions. Some feldspars have half their silicon replaced by aluminum with calcium balancing the framework charge. Feldspars are nearly as hard as quartz and are exploited for their chemistry in glassmaking and ceramics, since the aluminum content improves chemical and physical stability while its alkali content provides fluxing action.

Wollastonite

Among the industrial minerals high terahedra density and consequent hardness are also found in chain silicates. Wollastonite is characterized by the repeating, twisted, three-tetrahedra unit depicted in Figure 5.

The chains formed by these silica tetrahedra are connected by calcium in octahedral coordination. Because of this chain structure wollastonite can occur as acicular crystals, in some cases of macroscopic dimensions. This acicular particle shape is important in certain uses as a functional mineral filler.

⌘⌘⌘

Figure 5. Wollastonite

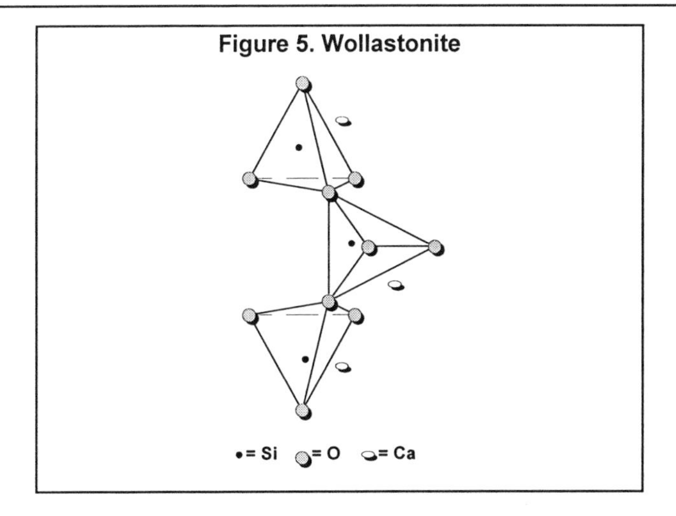

Phyllosilicates

Silica tetrahedra also can join into rings, as depicted in Figure 6.

Figure 6. Silica Ring

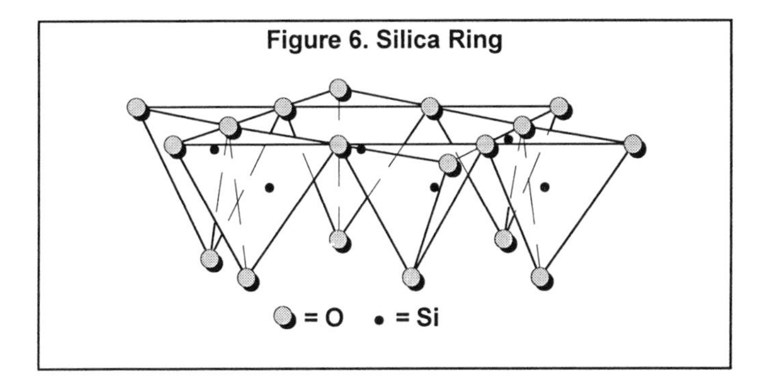

Phyllosilicates are characterized in part by an indefinitely extended sheet of rings, with three of the tetrahedral oxygens shared and the fourth (apical) oxygen in each case pointing in the same direction, as illustrated in Figure 7.

Another characteristic of most phyllosilicate minerals is the presence of an hydroxyl group central to the apical oxygens. This configuration is achieved through bonding of the silica sheet to a continuous sheet of octahedra, with each octahedron tilted onto one of its triangular sides. These octahedra, shown in Figure 2, most often contain either Mg^{2+} or Al^{3+}. When the metal cation is trivalent, as with aluminum, charge balancing requires only two of every three octahedral positions to be filled. This structure, that of the mineral gibbsite, is called dioctahedral. For divalent cations such as magnesium, all octahedral positions must be filled for charge balancing. This structure, that of the mineral brucite, is called trioctahedral. Phyllosilicates accordingly are often differentiated as dioctahedral or trioctahedral based on octahedral occupancy.

Figure 7. The Silica Sheet

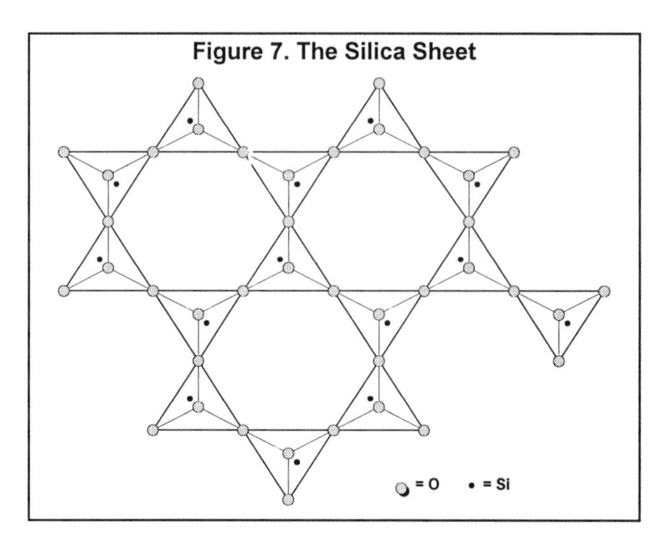

Kaolinite – When a layer of silica rings is joined to a layer of alumina octahedra through shared oxygens, as shown in Figure 8, the mineral kaolinite is formed. Kaolinite is the sole or dominant constituent of what is known as kaolin clay or simply kaolin.

Figure 8. Kaolinite

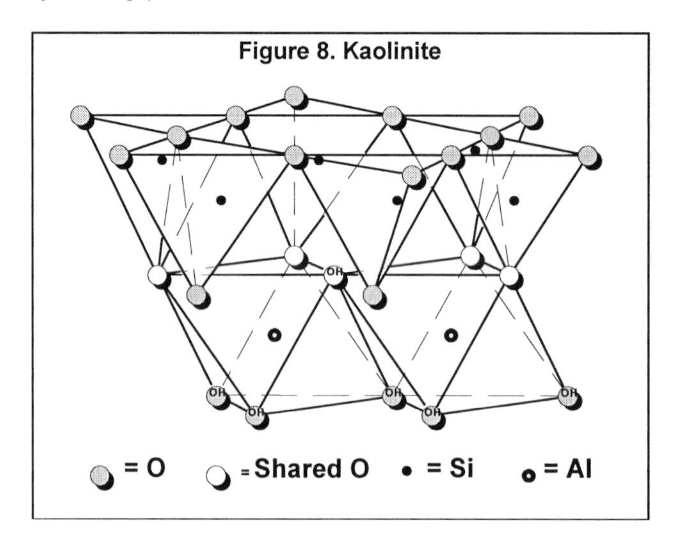

⌘⌘⌘

Kaolin may be considered the prototypical phyllosilicate in that its sheet structure results in platy or flake-shaped particles that occur as overlapping, separable layers. Because an individual kaolin particle has an oxygen surface on one side and an hydroxyl surface on the other, it is strongly hydrogen bonded to the laminae above and below it. These particles stack together in such a way that under magnification they look like sheaves of paper and are often called “books”. It is difficult to delaminate kaolin books into individual platelets, although this is done commercially. Compared to the silica, feldspar, and chain silicate structures, kaolin, and phyllosilicates in general, are relatively soft and lower in specific gravity.

Pyrophyllite – If the kaolin structure is bound through shared oxygens to a layer of silica rings on its alumina side, the pyrophyllite structure of Figure 9 results. Because both faces of a pyrophyllite platelet are composed of silica oxygens, interlaminar bonding is by relatively weak van der Waals forces. Pure pyrophyllite is therefore soft with talc-like slipperiness, because its laminae will slide past each other or separate fairly easily.

Figure 9. Pyrophyllite

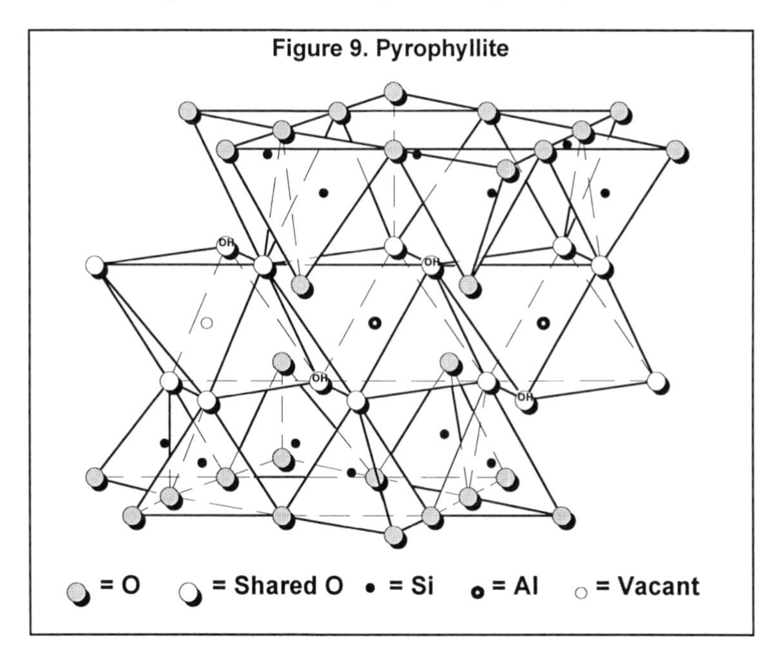

Serpentines – If a magnesia octahedral layer rather than an alumina layer is joined to one sheet of silica rings, two minerals of the serpentine group result. These differ markedly from each other and from the analogous aluminum-based kaolinite. One is the mineral antigorite, whose sheet structure does not directly correspond to that of kaolinite. This is because brucite does not fit quite as well to the silica sheet as does gibbsite. This minor mismatch is compensated by a slight stretching of the apical silica oxygens so that they can form a common oxygen link with the magnesium-based octahedral layer. This stretching results in a bending of the entire structure. Antigorite is laminar because its tetrahedral silica layer is continuous, although it periodically rotates 180°, preventing continuity of the octahedral layer. The face of an antigorite platelet is therefore corrugated, as pictured schematically in Figure 10.

Figure 10. Antigorite Corrugation

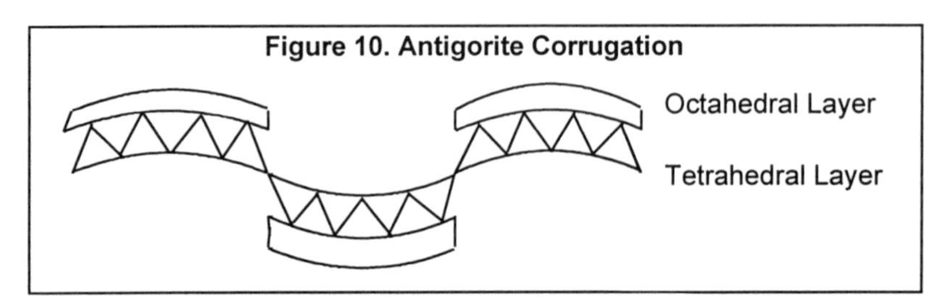

When both the octahedral and tetrahedral sheets are continuous (no rotation of the silica layer), the brucite-silica mismatch causes a continuous bending into long tubes. This results in the asbestos mineral chrysotile. Chemically, kaolinite and chrysotile differ only in their octahedral cation. This relative subtlety, however, explains the difference between their respective microscopically laminar and macroscopically fibrous morphologies.

Talc – If a sheet of silica rings is attached to the magnesia side of chrysotile, the bending tendencies on either side of the octahedral layer negate each other. The mineral structure remains planar, and the laminar trioctahedral analogue of pyrophyllite results. This is the talc structure shown in Figure 11.

As with pyrophyllite, individual talc laminae are held together by weak van der Waals forces. Sliding and delamination are relatively easy, giving talc its characteristic soft, slippery feel.

Tremolitic talc is a related industrial mineral that, despite its classification as a talc product, is actually a natural mineral blend with tremolite as the major component and talc as a minor component. As such, tremolitic talc has properties and uses that depend primarily on its tremolite content. The hardness and prismatic shape of tremolite crystals are derived from a structure analagous to that of wollastonite. While wollastonite is comprised of single chains of silica tetrahedra, tremolite is comprised of double chains, as depicted in Figure 12.

⌘⌘⌘

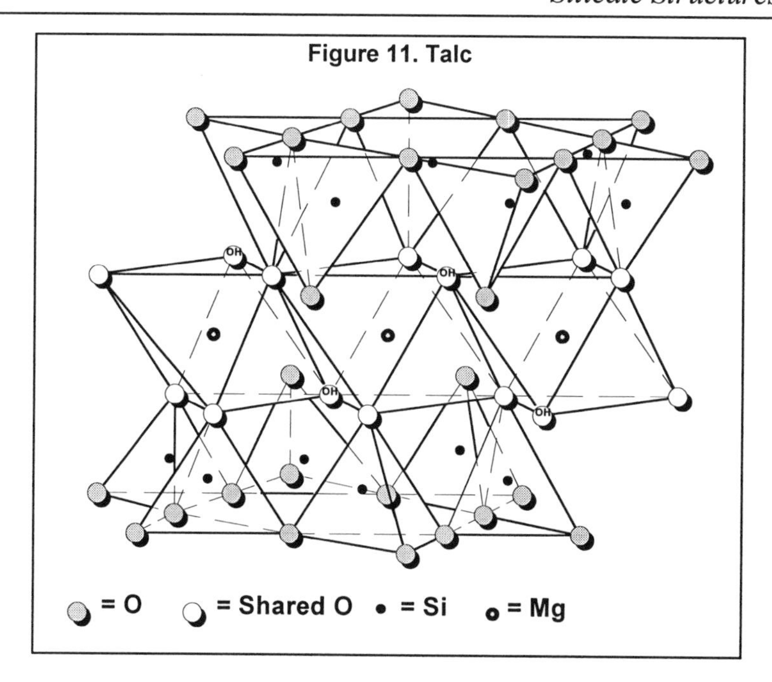

These double silica chains form the hexagonal rings common to phyllosilicates, but they extend in one direction instead of two. While the single wollastonite chains are joined by octahedrally coordinated calcium, the double chains of tremolite are joined by octahedrally coordinated magnesium between apical oxygens and by calcium on the opposite side. The schematic view of this structure shown in Figure 13 also suggests why hard prismatic tremolite can coexist with soft laminar talc in the same ore body.

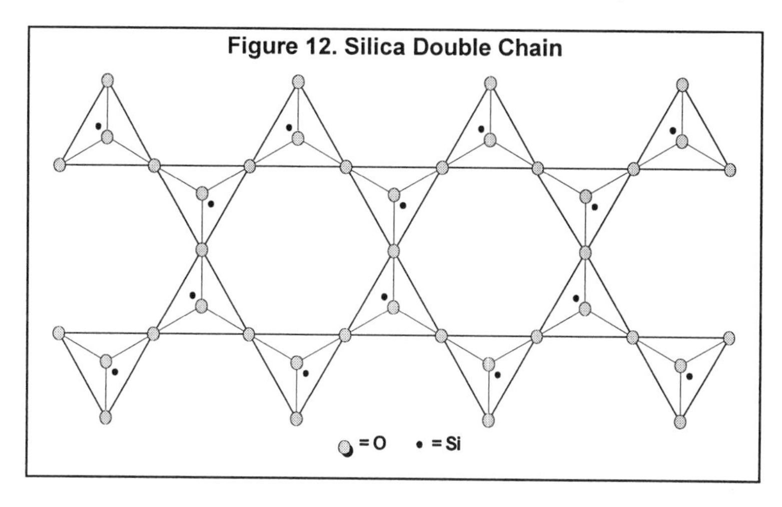

⌘⌘⌘

This is because tremolite can be viewed as offset strips of talc strongly linked back-to-back by calcium ions. A simple analogy would be to compare the tremolite structure to a brick wall, with the talc strips represented by the bricks and the calcium ions by the mortar. The structure is dense, rigid, and of high structural integrity. Talc, on the other hand, might be viewed as a stack of ceramic tiles. With little effort, one tile can be pushed across or removed from an adjacent tile in the stack.

Figure 13. Tremolite

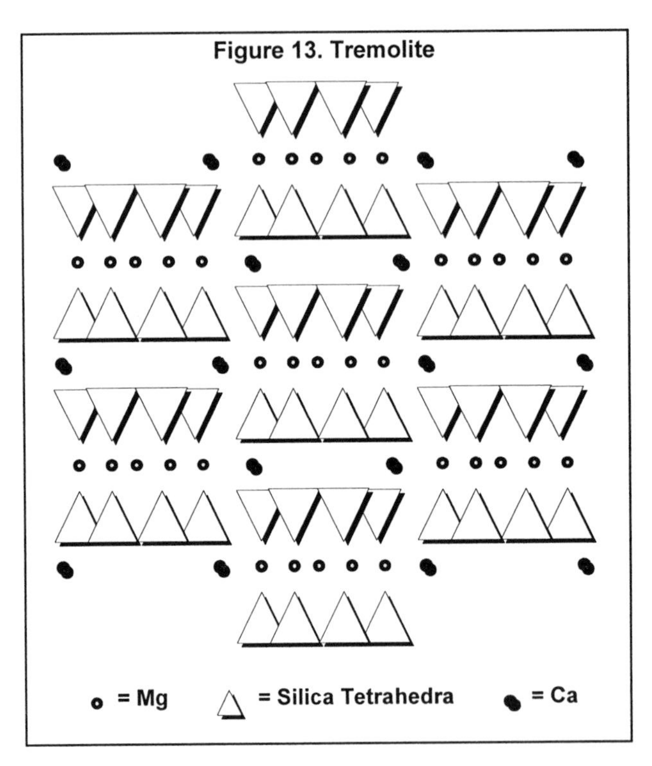

Hormite clay – Hormite clays are trioctahedral chain silicate minerals having certain structural features in common with both tremolite and antigorite, although their properties are quite unlike either. As in the case of antigorite, silica sheets are continuous but periodically inverted. Since hormites have a silica layer on both sides of the octahedral layer, silica sheet inversions limit the width of the octahedral sheet, leaving it to grow in just one direction. The result is talc-like strips resembling tremolite. However, these strips are joined by shared tetrahedral oxygens at the lines of inversion. This creates channels that are filled with water, as depicted schematically in Figure 14. Removal of this water confers highly absorptive properties.

⌘⌘⌘

This structure accounts for the high surface area and acicular particle shape of the commercial hormites – palygorskite (attapulgite) and sepiolite. Sepiolite is the high-magnesia end member containing minor substitution of Al^{3+} and/or Fe^{3+} for octahedral Mg^{2+} and tetrahedral Si^{4+}. Palygorskite exhibits higher substitution, principally aluminum for magnesium. The charge imbalance arising from these substitutions is compensated by exchangeable alkaline and alkaline earth cations. Palygorskite and sepiolite differ in the number of octahedral sites per unit cell.

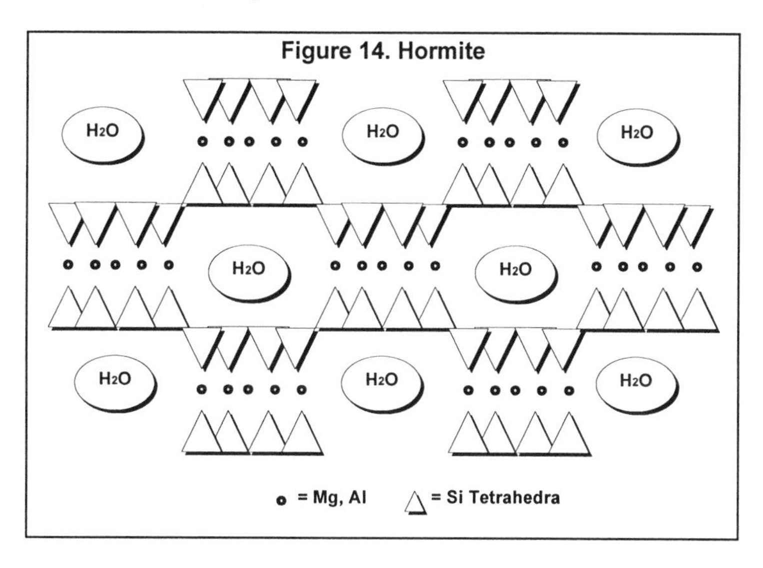

In addition to their use as absorbents, hormite clays are used as rheological agents. When dispersed in water, their needle-like particles deagglomerate in proportion to the amount of energy applied and form a random colloidal lattice.

Chlorite – Chlorite is an accessory mineral in some talc ores. It is laminar and composed of alternating talc and brucite sheets. The chlorite structure depicted in Figure 15 includes the upper silica layer of an adjoining platelet. Unlike talc, chlorite accommodates appreciable substitution of both tetrahedral and octahedral cations. Up to half of the tetrahedral Si^{4+} and up to one third of the octahedral Mg^{2+} may be replaced by Al^{3+}. Fe^{2+} and Fe^{3+} both commonly substitute for part of the Mg^{2+} as well. The charge imbalance from tetrahedral substitution is generally balanced by octahedral substitution either in the talc structure or the brucite structure. Hydroxyl-bearing brucite sheets between the talc sheets allow for hydrogen bonding and a corresponding increase in delamination difficulty.

⌘⌘⌘

Figure 15. Chlorite

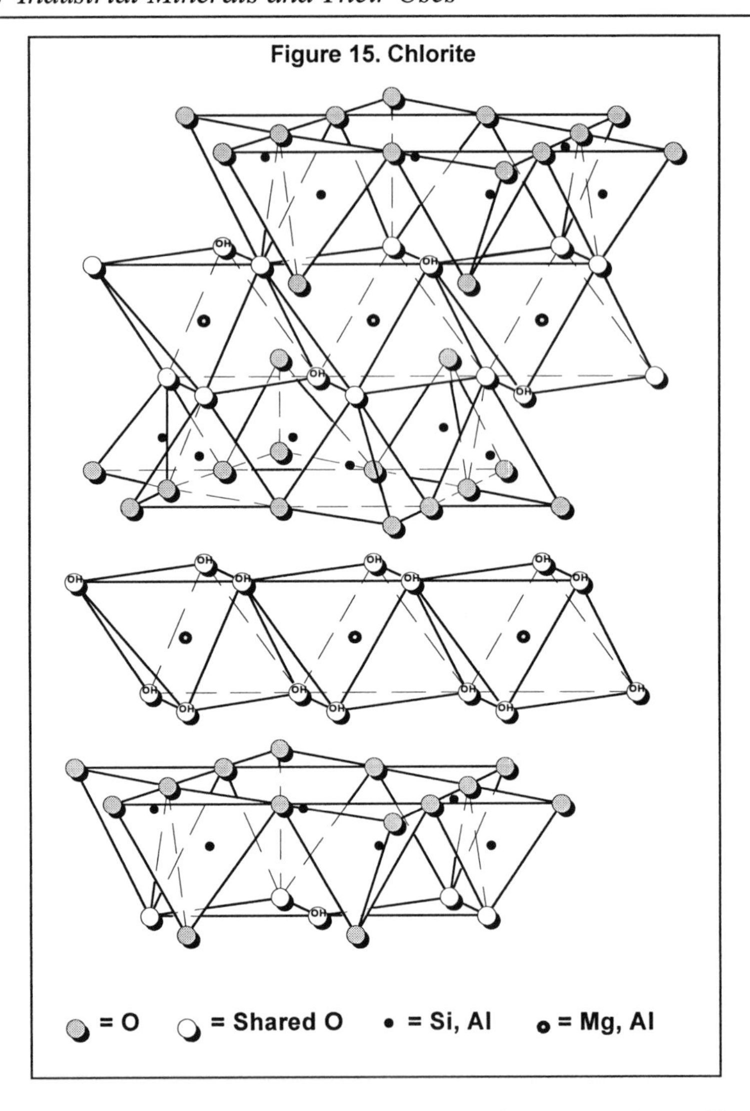

Vermiculite – The basic talc structure also typifies vermiculite, as illustrated in Figure 16. Vermiculite differs from talc primarily in its substitution of Al^{3+} for tetrahedral Si^{4+} and the presence of two oriented layers of water between individual laminae. Limited substitution of octahedral Mg^{2+} by Fe^{3+} and Al^{3+} also occurs. The charge imbalance arising primarily from tetrahedral substiutions is compensated by cations, usually Mg^{2+}, between interlaminar water layers. Because these cations are not structural components, they can be exchanged with other charge-balancing cations under the proper conditions.

⌘⌘⌘

As a consequence, vermiculite has the greatest cation exchange capacity among all of the phyllosilicates, at 100 to 260 meq/100 g.

The water-Mg^{2+}-water structure has nearly the same height, at approximately 5 angstoms, as does a brucite sheet. The talc-like laminae of chlorite and vermiculite are therefore separated by about the same distance, although the interlaminar structure of vermiculite is less rigid and usually less regular. Vermiculite is nearly as soft as talc, but delamination is prevented by the attraction of opposing laminae to exchangeable cations plus the simultaneous hydrogen bonding of oriented water to laminae faces while forming hydration shells around these cations. When heated rapidly to high temperature, however, interlaminar water volatilizes and pushes the talc-like layers apart. The result is low-density, high-porosity, concertina-shaped particles.

Figure 16. Vermiculite

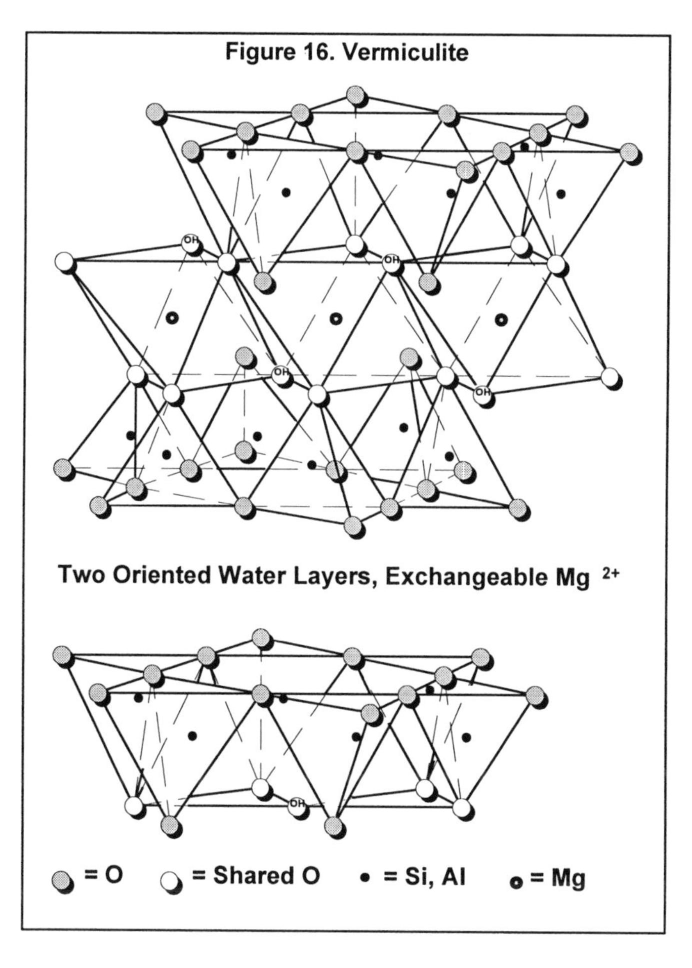

⌘⌘⌘

Mica – Most vermiculite has been altered from biotite, a trioctahedral mica containing substantial substitution of Fe^{2+} for octahedral Mg^{2+}. Biotite itself is not produced as a commercial industrial mineral. Phlogopite, a trioctahedral mica with less octahedral substitution, is available commercially, but dioctahedral muscovite is the most commonly used mica. All micas have either a talc or pyrophyllite structure and accordingly are characterized by platy or flake-shaped particles. For both phlogopite and muscovite there is some replacement of OH^- with F^-, and about one of four tetrahedral Si^{4+} is replaced by Al^{3+}. The resulting charge imbalance is compensated most often by K^+ located central to the opposing hexagonal openings in the silica sheets of adjacent platelets. There is usually little or no water between mica plates. Mica is well known for its ready delamination, even in the form of large sheets. This is due to the relatively weak bonding effect of the univalent counterion. The muscovite structure is depicted in Figure 17.

Figure 17. Muscovite

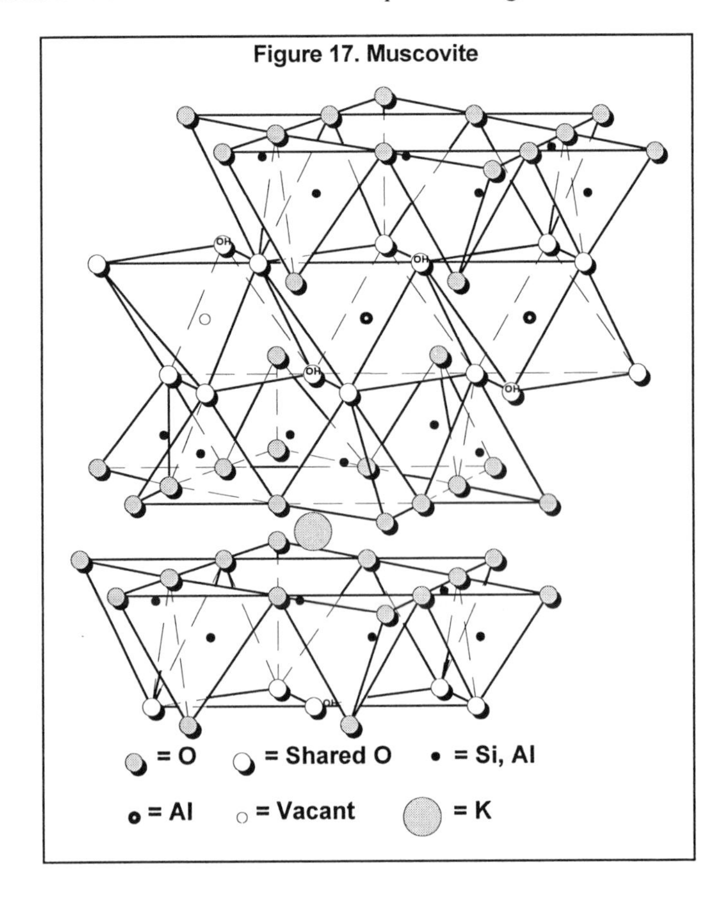

⌘⌘⌘

Smectite clay – Like mica, smectite clay (commonly called bentonite) has either a pyrophyllite or talc structure. Montmorillonite, a common high-aluminum smectite, can be characterized by the pyrophyllite crystal structure with a small amount of octahedral Al^{3+} replaced by Mg^{2+}. The resulting charge imbalance is compensated by exchangeable cations, usually Na^{+} or Ca^{2+}, between the laminae. In addition to these counterions, oriented water, similar to that in vermiculite, occupies the interlaminar space. When Ca^{2+} is the exchangeable cation, there are two water layers, as in vermiculite; when Na^{+} is the counterion, there is usually just one water layer. Figure 18 shows the montmorillonite structure.

Figure 17. Montmorillonite

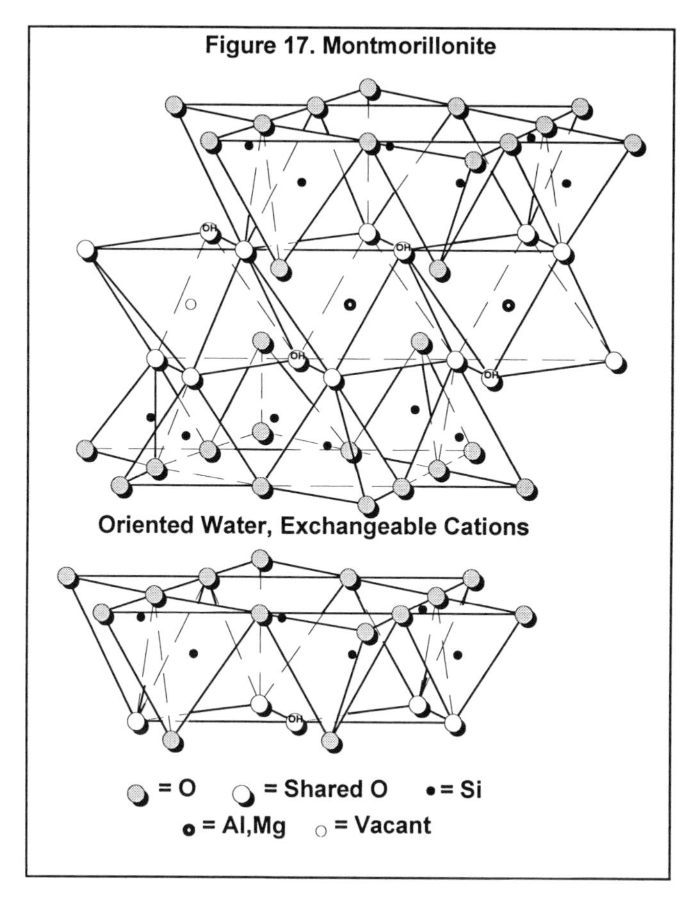

Unlike vermiculite, the smectite crystal structure accommodates additional interlaminar water layers, due at least in part to its lower counterion density. This allows for hydraulic delamination. On immersion in water, sodium smectites incorporate enough additional water layers to overcome weak

⌘⌘⌘

lamina-lamina attractions so that particles ultimately separate. Sodium smectites are therefore used as rheology control agents because of the colloidal structure their delaminated particles form in water. Calcium smectites also swell through interlaminar water absorption, but will not proceed to complete delamination due to the greater bonding effect of their divalent cations. Smectites can also absorb polar liquids other than water and will accomodate organic cations in exchange for their native counterions. This enables them to be used as absorbents and as rheological agents in nonaqueous systems.

Saponite, a high magnesium smectite, is similar in structure to talc but with limited substitution of tetrahedral Si^{4+} by Al^{3+}, while hectorite has the talc structure but with limited substitution of Li^{+} for octahedral Mg^{2+} and F^{-} for OH^{-}. As with montmorillonite, the resulting charge imbalance is compensated by Na^{+} or Ca^{2+} residing with oriented water in the interlaminar spaces. Saponite and hectorite have swelling, ion exchange, and absorbent properties similar to those of montmorillonite.

BIBLIOGRAPHY

Carr, D.D. (ed.), 1994, *Industrial Minerals and Rocks*, 6th Ed., Society for Mining, Metallurgy, and Exploration, Inc., Littleton, CO

Deer, W.A., Howie, R.A., Zussman, J., 1978, *Rock-Forming Minerals*, Vols. 1-5, Wiley, New York

Grim, R.E., 1953, *Clay Mineralogy*, McGraw-Hill, New York

Grim, R.E., 1962, *Applied Clay Mineralogy*, McGraw-Hill, New York

Grimshaw, R.W., 1980, *The Chemistry and Physics of Clays*, 4th Ed., Wiley, New York

Hurlbut, C.S., Klein, C., 1977, *Manual of Mineralogy*, 19th Ed., Wiley, New York

Newman, A.C.D. (ed.), 1987, *Chemistry of Clays and Clay Minerals*, Wiley, New York

van Olphen, H., 1977, *An Introduction to Clay Colloid Chemistry*, Wiley, New York

⌘⌘⌘

TWO

THE INDUSTRIAL MINERALS

Peter A. Ciullo
R.T. Vanderbilt Company, Inc.
Norwalk, CT

This guide to the industrial minerals is designed as a convenient reference for those who use these materials in the design, formulation or manufacture of finished goods. Each monograph includes the mineral's basic properties, the general types or grades of the mineral available, and the products in which it is commonly used. The industrial minerals today also include a number of synthetic materials; these are described in the appropriate sections when they are analogues or derivatives of a natural mineral.

The Industrial Minerals

ASBESTOS

Chrysotile $Mg_3Si_2O_5(OH)_4$
Refractive Index: 1.53-1.56
Specific Gravity: 2.5-2.6
Mohs Hardness: 2.5-4

Asbestos is a generic term applied to six minerals that occur in nature as strong, flexible, heat-resistant fibers. Nearly all (>98%) commercial asbestos is the mineral chrysotile. Chrysotile is differentiated from the other five asbestos minerals by its tubular serpentine rather than ribbon-like amphibole structure, its generally greater fiber flexibility and strength, its lower heat resistance, its greater surface area and positive surface charge, its lower refractive index, and its greater susceptibility to decomposition by strong acids. Of the ampibole asbestos minerals, only two – amosite, a magnesium iron silicate, and crocidolite, a sodium iron silicate – are produced in commercial quantities. The other three – anthophyllite asbestos, tremolite asbestos, and actinolite asbestos – are rare asbestiform varieties of the nonasbestiform prismatic minerals anthophyllite, tremolite, and actinolite. It is the rarity of these three asbestiform varieties that has precluded the assignment of distinct mineral names and unduly caused the prismatic analogues to become identified with asbestos. The nonasbestos forms of chrysotile, amosite, and crocidolite are sufficiently common to have earned separate mineral identities as antigorite, cummingtonite-grunnerite, and riebeckite, respectively. Asbestos ore is typically processsed in highly automated operations designed to minimize worker exposure to mineral dust. The ore is crushed, dried, screened, milled, and air separated to produce a variety of grades.

TYPES

The two major world producers, the former Soviet Union and Canada, designate several major grades of asbestos, with further subdivisions within each grade. Grades are based on fiber length, strength, color, and purity, plus intended application. The following grades are based on Canadian standards.

Spinning fiber – The cleanest and longest fibers, to >12mm, are reserved for producing woven asbestos textiles.

Asbestos cement fiber – This is the longest fiber grade that is <12mm.

Paper/shingle fiber – This is essentially <5 mm (-4 mesh) fiber, with shingle fiber being generally shorter than paper fiber.

⌘⌘⌘

Shorts/floats – These are the shortest fibers, with most shorts or all floats <2mm (-10 mesh).

Crudes – This is crushed ore containing staple fibers >10mm. Crudes are sold to customers who process them into fibers for their own purposes.

USES

Approximately 3.5 million metric tons of asbestos are produced annually. Major producers are the former Soviet Union (60%) and Canada (17%). Production and use in the United States is very minor due to health and liability concerns, although California hosts a short fiber chrysotile deposit considered to be the largest single mineral ore body in the world. Major asbestos applications worldwide are asbestos cement, friction products, roofing, insulation, flooring, plastics, and gaskets.

Asbestos cement – In asbestos cement pipe asbestos provides good drainage and high green strength during manufacture, plus high pipe tensile strength, impact strength, heat resistance, and alkali resistance. In asbestos cement sheets it provides high flexural stregth as well.

Friction products – Paper and shingle fibers are used in molded clutch plates and disk brake pads, while short and float fibers are used in brake linings. Clutch plates are also made from open-weave asbestos cloth impregnated with resin. In all cases, asbestos is used for its durability, heat and moisture resistance, low thermal conductivity, and high strength.

Roofing – Short, float, and shingle fiber are used in asphalt shingles and roofing felts and in asphalt-based roof coatings to provide dimensional stability and flexibility, to enhance crack resistance and weatherability, and to control rheology (coatings).

Insulating products – Textiles for heat-resistant protective clothing are woven from spinning fiber, but most asbestos insulation products are in the form of paper, paperboard, millboard, and mat from paper-grade fiber. Asbestos provides flexibility, dimensional stability, tear resistance, heat resistance, chemical resistance, moisture resistance, low thermal conductivity, and high electrical resistivity. Products include pipe wrap, thermal insulation in appliances, and electrical and heat insulation in electronics.

Flooring – Short fiber is used in vinyl tile to provide flexibility, resilience, durability, fire resistance, and dimensional stability. Short fiber is also coated

⌘⌘⌘

with rubber latex and formed into paper used as backing for vinyl sheet flooring.

Plastics – Abrasion-free asbestos is used to thicken and reinforce thermosets, providing heat, tear, and electrical resistance, low heat deformation, high strength, and stiffness. Short and float fibers are used as fillers; mat, felt, paper and cloth are impregnated with resin to form laminates.

Gaskets – Abrasion-free asbestos cement- and paper-grade fibers are used in rubber-based gaskets and packing to provide resilience, plus resistance to heat, tear, and chemical attack. Densified latex-asbestos paper is also used to make gaskets.

Other uses – Short and float fibers are used in textured paints, drywall joint cements, caulking compounds, automotive undercoatings, and asphalt paving mixes for high traffic areas.

⌘⌘⌘

BARITE

Barite $BaSO_4$
Refractive Index: 1.64-1.65
Specific Gravity: 4.5
Mohs Hardness: 3-3.5

The commercial significance of barite is related almost entirely to its high specific gravity. Most processed barite (90%) is used as a weighting agent in well drillling fluids. Its physical and chemical properties assume more importance for its filler applications and for its use as a source of barium. Common mineral impurities in barite ores are quartz, carbonate minerals, sulfide minerals, and clay. Processing of barite depends upon ore purity and the nature of associated minerals. Drilling grades often require only crushing, grinding, screening, and milling. An intermediate washing step may be employed to achieve required minimal specific gravity and $BaSO_4$ content. Applications requiring high brightness and high chemical or mineralogical purity may necessitate flotation, bleaching with sulfuric acid, and wet grinding. Barite is also known as barytes and heavy spar. It is used as a source of bariuim in a number of barium compounds, including blanc fixe, a high-purity precipitated barium sulfate that is used in place of processed barite in certain demanding applications.

TYPES
Other than drilling fluid products, which account for most of production, barite is differentiated according to its chemical and filler uses.

Drilling grade – Barite for well drilling fluids is typically a 200 mesh product with specific gravity of at least 4.2 (i.e., >90% $BaSO_4$). Color is not critical, but water-soluble alkaline earth metals are controlled so as not to interfere with drilling fluid rheology.

Glass grade – Glass-grade barite is generally -30+140 mesh, with 96 to 98% $BaSO_4$, <2.5% SiO_2, and <0.15% Fe_2O_3. Iron and slica content may be further restricted for specific uses, and there may be limits on TiO_2 and Al_2O_3.

Chemical grade – Barite for barium chemicals is -16mm+0.84mm (20 mesh) and contains at least 95% $BaSO_4$, <1% $SrSO_4$, <1% combined iron oxides, and no more than a trace of fluorine.

Filler grade – Filler uses for barite generally require high brightness, high purity, and fine particle size, usually -325 mesh or finer. Purity is typically

⌘⌘⌘

>95% $BaSO_4$ and <0.1% Fe_2O_3, with no more than 0.5% moisture. The highest quality filler grades are made by flotation, followed by wet grinding, bleaching with sulfuric acid, washing, drying, and milling.

Blanc fixe – Blanc fixe is precipitated barium sulfate for uses where higher brightness and purity and finer particle sizes are required than are generally available with barite. The precursor to blanc fixe is common to most barium compounds made from barite. Crushed barite is first roasted with coke in a rotary kiln at about 1200°C. This reduces the barium sulfate to barium sulfide in the form called black ash. The hot black ash is quenched in water and countercurrent leached to produce a barium sulfide solution. Blanc fixe is produced by treating this solution with sodium sulfate to precipitate ultrafine barium sulfate. This is then filtered, washed, milled, and dried.

Lithopone – Lithopone production starts with the same process used for blanc fixe, except that zinc sulfate is used in place of sodium sulfate. The intimate mixture of barium sulfate and zinc sulfide that precipitates is filtered, washed, dried, calcined, water quenched, wet ground, and dried. The result is a white mixture of barium sulfate, zinc sulfide, and zinc oxide. Lithopone was one of the first fine white pigments for industry but is now rarely used.

USES

Annual production of barite worldwide is approximately 5.4 million metric tons, dominated by China (33%), India (11%), and Morocco (8%). Ground barite for well drilling fluids accounts for 90% of all production. The balance is used in the manufacture of barium chemicals and glass and in filler applications.

Well drilling fluids – Drilling fluids are designed to cool the drill bit, lubricate the drill stem, seal the walls of the well hole, remove cuttings, and confine high oil and gas pressures by the hydrostatic head of the fluid column. A high specific gravity fluid is required to maintain sufficient hydrostatic pressure to control hydrocarbon release and prevent gushers and fires. Barite is uniquely suited as the weighting agent because it is heavy, chemically inert, and nonabrasive. The deeper the hole the more barite is used, because hydrocarbon pressure rises strongly with depth below about 2100 meters. In most drilling fluids barite is the major ingredient by weight percent.

Glass – In glassmaking barite saves fuel by reducing the heat-insulating froth on the melt surface. It also acts as an oxidizer and decolorizer, making the glass more workable. It reduces seeds and annealing time and improves glass toughness, brilliance, and clarity.

⌘⌘⌘

Coatings – Paints and primers represent the largest use for filler-grade barite. High-brightness micronized barite is used as an extender to provide the weight that customers equate with quality and because of its low binder demand, which allows high loadings. Blanc fixe is used where a finer particle size is needed for denser packing of the paint film, as in premium metal primers, and to provide resistance to corrosion by acids and alkalis. Despite their high brightness barite and blanc fixe have poor hiding and tinting strength because they are close to the refractive index of binders. They function instead as extenders and spacers, keeping the pigment particles separated and uniformly disseminated to optimize light scattering.

Polymers – Finely ground barite is used in rubber, where its weight, inertness, isometric particle shape, and low binder demand are advantageous. It has little effect on cure, hardness, stiffness, or aging. It is used in acid-resistant compounds, in white sidewalls for tires, and in floor mats. Blanc fixe fine enough to be semireinforcing is used to provide the same compound softness and resilience as barite but better tensile strength and tear resistance. Barite is used in PVC and polyurethane foam backings for carpeting and sheet flooring because of its ability to form dense coatings due to its high specific gravity and its ability to be used at high loadings.

Other uses – Because barium sulfate is insoluble and opaque to X-rays, blanc fixe meeeting pharmacopeia specifications is used as an indicator in medical X-ray photography. Natural barite is used in concrete for the construction of facilities handling nuclear materials because it absorbs gamma radiation. Micronized white barite and blanc fixe are used as fillers and extenders, primarily to add weight, in bristolboard, playing cards, and heavy printing papers. Blanc fixe is used in the base coat of photographic papers to supply an inert substrate for the silver halide emulsion coat. Finely ground (-325 mesh) barite is used as an inert filler in brake linings and clutch plates.

⌘⌘⌘

CALCIUM CARBONATE

Calcite $CaCO_3$
Refractive Index: 1.66-1.74
Specific Gravity: 2.71
Mohs Hardness: 3

Aragonite $CaCO_3$
Refractive Index: 1.68-1.69
Specific Gravity: 2.95
Mohs Hardness: 3.5-4

The mineral calcite is the major or sole constituent of most commercial calcium carbonate products. These include natural limestone, marble, and chalk, plus most precipitated calcium carbonate. Aragonite is a metastable polymorph of calcite that typically has an acicular crystal shape. Natural aragonite products are less common, but precipitated varieties are available. Many calcium carbonate deposits are the remains of the shells and skeletons of ancient sea life. The color, purity, density, and crystal morphology depend upon the influence of waves and water currents before burial, and upon temperature, pressure, and tectonic activity after burial. The most common mineral impurities are quartz and clay. The most common substitutes for calcium are other divalent cations, such as magnesium, strontium, and barium, although the amount of substitution is usually no more than a few percent. The exception is magnesium, which can substantially replace calcium to form the mineral dolomite, $CaMg(CO_3)_2$. Calcium carbonate rocks, a common constituent of the Earth's surface, range from high-calcium limestones containing >95% calcite to dolostones containing 90% or more dolomite.

Chalk is a fine-grained, white, friable, high-purity limestone. Marble is a dense, hard, low porosity stone composed substantially or solely of calcium carbonate. It is formed by the action of heat and pressure on buried limestone beds. Marble used for ground calcium carbonate products is chosen for color and purity. Marble for decorative and dimension stone (blocks, fascia slabs, tile) is selected for its characteristic shadings or veining, caused by minor mineral impurities, and for its ability to take a polish. Travertine is a banded, dense calcium carbonate also used as decorative and dimension stone. It is formed by rapid chemical precipitation around natural hot springs. A similar material is precipitated from cold water solutions as stalagmites and stalactites. This is known as onyx marble (true onyx is banded quartz), Mexican onyx, Algerian onyx, and oriental alabaster (true alabaster is dense gypsum).

Aragonite sand comprises extensive marine deposits off the south Florida coast. It is recovered by suction dredging and after drying and screening

⌘⌘⌘

grades about 96% calcium carbonate. Most is used locally in cement manufacture.

TYPES

Filler uses for calcium carbonate generally require white color and a high degree of mineralogical purity, plus control of particle size and shape, surface area, and liquid absorptivity. Natural calcium carbonate fillers are generally called ground limestone or ground calcium carbonate but may also be sold as ground chalk, ground marble, or whiting. The synthetic alternatives are known as precipitated calcium carbonate, or PCC.

Ground calcium carbonate – Natural calcium carbonate ores high in chemical and mineralogical purity are wet or dry ground to a wide range of products. Dry-ground calcium carbonates, comprising nominal 200 to 325 mesh products, are among the least expensive white fillers available. They are simply ground from ore but may also be beneficiated by air separation. Wet-ground products are produced in finer particle size ranges and may be beneficiated by washing or flotation. They are informally classified by particle size as fine ground (FG; 3 to 12 micrometers median, 44 micrometers top), and ultrafine ground (UFG; 0.7 to 2 micrometers median, 10 micrometers top). There is some overlap between these classifications from one producer to another. As dry grinding technology advances, dry-ground products in fineness ranges previously associated with wet-ground grades are becoming more common. Wet processed FG and UFG products are of necessity more expensive than dry-ground products due to the cost of drying and in some cases remilling to break up agglomerates of ultrafine particles. Wet-ground fine and ultrafine products are also sold in 75% solids slurry form for high-volume paint and paper applications and in stearic acid- and stearate-treated forms for use in polymers.

Precipitated calcium carbonate – Precipitated calcium carbonate (PCC) is produced for applications requiring any combination of higher brightnesss, smaller particle size, greater surface area, lower abrasivity, and higher purity than is generally available from ground natural products. In the US PCC is most commonly made by the carbonation process. Limestone controlled for coloring oxides (e.g. of Mn and Fe) is calcined to calcium oxide and carbon dioxide. The calcium oxide (burnt lime) is then slaked with water to form calcium hydroxide (milk of lime). The carbon dioxide liberated on calcining is then reintroduced to precipitate calcium carbonate. Manipulation of process variables determines particle size and shape, surface area, and whether the product is isomorphous calcite or acicular aragonite. PCC products are also made by the lime-soda process, where milk of lime is reacted with sodium carbonate to form a calcium carbonate precipitate and a sodium hydroxide solution. This process is used by commercial alkali manufacturers to make a relatively coarse PCC as a byproduct of sodium hydroxide recovery. A third production route is to react milk of lime with ammonium chloride, forming ammonia gas and a calcium chloride solution. This solution is purified and reacted with sodium carbonate to form a calcium carbonate precipitate and a sodium chloride solution. This process is the simplest of the three, but to be economical it is usually carried out in a satellite facility adjacent to a Solvay-process soda ash plant. Although still common elsewhere, the Solvay process became obsolete in the US in 1986. PCC products are typically offered as fine (0.7 micrometer median) and ultrafine (0.07 micrometer median) grades, with and without stearate surface treatments.

USES

The major filler uses of calcium carbonate, both natural and PCC, are paper, paint, adhesives and sealants, and polymers. Filler uses account for only about 1% of the 700 to 800 million metric tons of calcium carbonate produced in the United States annually. Production is overwhelmingly dominated by commodity, low-value crushed stone, mainly for civil engineering uses and as aggregate for concrete and asphalt.

Paper – In alkaline papermaking, calcium carbonate is used as a paper filler and coating. Both uses require high brightness, high purity, small particle size, and lack of abrasion. Precipitated products generally retain a performance edge over the best ultrafine wet-ground grades. Commercial PCC products are at a disadvantage, however, in their high cost as dry products and in their difficulty in forming the high solids (usually 75%) slurries, due to their extremely small particle size, that large paper mills prefer. In Europe, where alkaline papermaking has been more common, an acceptable balance of performance and price has been met with high-quality ground chalk and marble. In the US PCC is preferentially used because the more recent and ongoing conversions to alkaline papermaking have beeen accompanied by the establishment of satellite PCC production facilities adjacent to paper mills. Byproducts of the pulping process are diverted to economically produce PCC, which is pumped to the paper mill in slurry form. About half these US satellite PCC plants produce enough slurry to also supply smaller mills where the construction of a full-scale PCC plant may not be justified. In Europe the establishment of satellite PCC plants is just now gaining popularity. Whether ground natural or precipitated, calcium carbonate is used as a paper filler and in coatings to provide opacity, high brightness, and improved printability due to its good ink receptivity.

⌘⌘⌘

Coatings – Gound natural calcium carbonate is the most widely used white pigment in paints because it is available at relatively low cost and has high brightness for TiO_2 extension, high purity, low abrasivity, and resistance to weathering. The coarsest grades are used at loadings up to 50% in textured paints. Fine and ultrafine grades, including PCC, are used in a wide variety of other decorative and protective coatings. PCC products, with generaly higher brightness, provide better TiO_2 extension. PCC and ultrafine wet-ground grades contribute to rheology and stability and provide good dry hide and gloss retention. Slurries of 75% ultrafine natural calcium carbonate are sold for convenient use in the large-scale manufacture of water-based coatings. Stearate-treated natural and PCC products are used in nonaqueous coatings.

Adhesives and sealants – Calcium carbonate is the most widely used filler in adhesives and sealants. The coarsest grades of ground natural products are used at high loadings in drywall joint cements and in ready-mix adhesives for heavy wall tile. Somewhat finer and generally off-color natural products are used at high loadings in oil-based putties. Finely ground white grades are used as a general-purpose filler in most types of adhesives, sealants, and gap fillers for their balance of low binder demand and narrow particle size distribution. This allows economically high loadings without adversely affecting flow. High performance polymer-based adhesives and sealants use stearate-coated PCC and ultrafine natural products to control flow and slump on application and to provide low modulus with good tear and tensile properties in the cured state.

Polymers – Ground natural calcium carbonate is the most commonly used filler in plastics due to its low cost, low abrasion, low oil absorption, low moisture, high brightness, and easy dispersion with conventional mixing equipment. These attributes account for its widespread use in elastomers as well, where it can be used at very high loadings with little loss of compound softness, elongation, or resilience. PCC products, particularly stearate-coated grades, are used as functional fillers in rubber and plastics. In plastics they are used to improve mar and impact resistance, surface gloss, weatherability, shrinkage control, cold flow properties, low and high temperature properties, and dielectric properties, and to reduce plasticizer migration and crazing of molded parts. PCC is most widely used in rigid and flexible PVC and PVC plastisols. Fine and ultrafine natural products are also commonly used in PVC as well as in polypropylene and in polyester molding compounds. Lesser quantities are used in polyurethane foam, epoxies, and phenolic resins. In rubber stearate-coated ultrafine PCC is used for its low moisture absorption, good dispersion, and good elastomer-filler contact. This enables it to perform as a semireinforcing filler, imparting good tensile strength, tear resistance, resilience, abrasion resistance, and flex crack resistance. Dry-ground calcium carbonate is the most common filler in carpet backings. It is used to add weight and body and to extend the binder, usually latex, which secures the loop of tufted carpeting. It is also used as the filler in secondary backing and in separate underlay (the foam pad), both of which are usually urethane foam.

Other uses – Calcium carbonate meeting pharmacopeia requirements is used as a therapeutic source in anatacids and calcium supplements and as a tableting excipient. Fine particle size products are used as a fire extinguisher foam filler, as an abrasive in household cleaners, as a flux in welding rod coatings, as a diluent in agricultural pesticide dusts, and as a dusting agent to mitigate the explosion potential of coal dust generated during underground mining. Substantial quantites of crushed and ground limestone are also used in the manufacture of portland cement, lime, glass, and metallurgical fluxes, as well as in flue gas desulfurization processes and as a soil amendment.

⌘⌘⌘

DIATOMITE

Diatomite SiO_2
Refractive Index: 1.40–1.43 / calcined 1.43–1.47
Specific Gravity: 2.0–2.1 / calcined 2.1–2.3
Mohs Hardness: 4.5–5 / calcined 5.5–6

Diatomite is comprised of the sedimentary remains of diatoms, a class of unicellular algae found in both fresh and sea water. It is their microscopic skeletons, composed primarily of amorphous opaline silica. These skeletons take a variety of forms – spheres, disks, wheels, needles, ladders – but all are characterized by a porous, lace-like structure. Individual skeletons, called frustules, can range from less than 1 micrometer to 1000 micrometers; more typically, they are in the 10 to 150 micrometer range. Diatomite products generally assay 86 to 94% silica, with typical impurities being clay, quartz, and feldspar. The chemistry and morphology of diatomite products provide the high porosity, high surface area, low bulk density, chemical inertness, and mild abrasivity that account for their commercial utility. Ores are processed by drying, gentle milling to preserve the skeleton structure, and air classification. Much of the diatomite sold is further procesed by calcining and size clasification to control surface area, porosity, and inertness for major filtration applications. Diatomite is also known as diatomaceous earth and (in Europe) kieselguhr. In the past it was also known as tripoli powder and tripolite.

TYPES
Diatomite products are generally differentiated by process and particle size.

Natural diatomite – This is diatomite ore which has been gently crushed and milled to retain the frustule shape and then screened and air classified to remove impurities and to segregate products into coarse fractions for filtration applications and fine fractions for filler uses.

Calcined diatomite – Also known as straight-calcined, this is diatomite which has been calcined at between 870° and 1100°C in a rotary kiln. This process burns off organic matter, converts some of the opaline silica to cristobalite, shrinks, hardens, and reduces the fine structure of individual particles, and forms agglomerates or clusters of particles through fusion. The overall effect is to decrease surface area but to increase bulk density and void volume due to the nature of agglomerate packing. Calcination generally turns the white to off-white natural diatomite pink from iron oxidation. The calcined diatomite is carefully milled, screened, and air classified to various size fractions, primarily for filtration uses.

⌘⌘⌘

Flux-calcined diatomite – This is diatomite that has been calcined at about 1200°C with sodium carbonate or sodium chloride as a fluxing agent. The flux converts iron oxides to a colorless glassy phase, resulting in a white product, and produces greater agglomeration. Temperature, time, and flux content determine agglomerate size distribution, which is generally higher overall than for straight-calcined products. Flux-calcined diatomite is milled, screened, and air classified, with coarse fractions sold for filtration uses and fine fractions as white fillers. Flux-calcined products provide maximum void volume, which can exceed 90%, and consequent high absorptivity.

USES

Annual worldwide production of diatomite is approximately 1.6 million metric tons, dominated by the United States with about 37% of this total. Other major producers are France, the former Soviet republics, and Spain. Production in Romania and the former Soviet republics is almost exclusively of relatively low-value material for construction products. Uses elsewhere are dominated by filtration applications (80% in the US, 60% worldwide). The only other major use is as a filler.

Filtration – Diatomite is widely used as a filter aid for the separation of suspended solids from liquids. It is used as a precoat on the filter medium or dispersed in the liquid to be filtered; these two uses are often employed together. The precoat traps solids from the liquid and prevents them from blocking the filter medium. The dispersed aid adds to the precoat as filtration proceeds, so that the surface of the precoat does not build an occlusive layer of solids. The choice of diatomite grade for a particular filtration application depends upon the balance desired between filtrate clarity and flow rate. Natural diatomite provides the best clarity but the slowest rate. Flux-calcined products, with the highest level of agglomeration, provide less clarity but faster rates. Straight-calcined products provide intermediate results. Diatomite filtration is used in the clarification or processing of beer, wine, sugar liquors, swimming pool water, fruit juices, vegetable oils, water, pharmaceuticals, effluent wastes, and phosphoric acid.

Fillers – White, flux-calcined 325 mesh and finer diatomite is used as a functional filler, primarily in paint. The diatomite particles roughen the paint film to provide flatting and improved adhesion of subsequent coats. They also improve film toughness and durability, while their porosity helps control vapor permeability for reduction of blistering and peeling. Natural and flux-calcined products are used in certain specialty papers as a lightweight bulking agent, as a drainage aid, as an opacity builder, and as a fiber dispersion aid. In polyolefin films fine particle size flux-calcined products are used as

⌘⌘⌘

antiblocking agents by projecting through the film surface and providing mechanical separation of film layers. Natural, straight-calcined, and flux-calcined diatomites are used as processing aids (absorbents) in high-oil, highly loaded rubber compounds and as both processing aids and semireinforcing fillers in mechanical goods.

Absorbents – Since calcined and flux-calcined diatomites can absorb 2 to 3 times their weight of liquids while remaining free flowing, they are used in pet litter and floor sweeping compounds, as pitch absorbers in paper manufacture, as pesticide carriers, as anticaking agents, and in the storage and transport of hazardous liquids such as phosphoric and sulfuric acids. Absorbent grades are controlled in particle size, dust free, and physically stable when wet.

Other uses – Diatomite is used as a nonscratching abrasive in polishes and cleansers, in concrete, mortars, grouts, plasters, and stucco for improved plastic and cured properties, in asphalt products to decrease cracking caused by rapid temperature changes, in catalyst carriers, in chromatographic supports, and as a silica source in the production of synthetic calcium and magnesium silicates.

⌘⌘⌘

FELDSPAR

Orthoclase $KAlSi_3O_8$
Refractive Index: 1.52-1.54
Specific Gravity: 2.57
Mohs Hardness: 6

Microcline $KAlSi_3O_8$
Refractive Index: 1.52-1.54
Specific Gravity: 2.54-2.57
Mohs Hardness: 6

Albite $NaAlSi_3O_8$
Refractive Index: 1.53-1.54
Specific Gravity: 2.62
Mohs Hardness: 6

Anorthite $CaAl_2Si_2O_8$
Refractive Index: 1.58-1.59
Specific Gravity: 2.76
Mohs Hardness: 6

Feldspars are the most widespread mineral group, comprising approximately 60% of the Earth's crust. Feldspar minerals are sodium, potassium, and calcium aluminosilicates. Their ores are commonly associated with quartz and mica and may also contain spodumene (a lithium aluminosilicate), kaolin, garnet, or iron minerals depending upon the type of deposit. Most commercial feldspars are produced by flotation and magnetic separation followed by milling. Air classification is used for the finest grades. Products typically contain quartz – low levels in high grade feldspars, high levels in feldspar/quartz mixtures sold for glassmaking. Feldspar ores and products generally contain mixtures of feldspar minerals rather than one mineralogically distinct feldspar type.

TYPES

Feldspars occur as solid solutions between their principal end member (Na, K, or Ca) compositions. Commercial products in most cases are employed for their chemistry (aluminum and alkali content) and are generally characterized accordingly. Certain products are more broadly described in geological terms.

Alkaline feldspar – This is feldspar with chemistry ranging between the potassium and sodium end members.

⌘⌘⌘

Plagioclase feldspar – This is feldspar with chemistry ranging between the sodium and calcium end members.

Potash spar – Potash spar is principally orthoclase or microcline and contains at least 10% K_2O.

Soda spar – Soda spar is mostly albite and contains at least 7% Na_2O.

Calcium spar – This is principally anorthite.

Aplite – Aplite is a grainy granitic (containing quartz and alkali feldspar) rock containing a substantial proportion of plagioclase. It is also called lime-soda feldspar. Low-iron aplite is sold in the US for glassmaking.

Feldspathic sand – This is natural silica sand containing 10-35% feldspar, usually potash spar or soda spar. This sand is described, in decreasing feldspar content, as river sand, dune sand, and beach sand.

See also *NEPHELINE SYENITE.*

USES

Of the approximately 5 million metric tons of feldspar products produced annually, Italy is the leading supplier with about 29%, followed by the US with about 12%. Other major producers are Japan (aplite), France, Germany, Korea, and Thailand. The two major applications for feldspar worldwide are glass and ceramics, which rely more on feldspar chemistry than mineralogy. In the US, 54% is used in glass and 44% in ceramics.

Glass – Feldspar is used primarily in container glass, followed by flat glass and glass fiber. Glass-grade products are typically coarsely ground, 20-40 mesh, and contain 4-6% K_2O, 5-7% Na_2O, about 19% Al_2O_3, and less than 0.1% Fe_2O_3. Feldspar is used mainly as a source of alumina, which improves both the workability of the glass melt and the chemical and physical stability of the finished product. It also provides the alkaline oxides (Na_2O, K_2O) that provide fluxing in partial substitution for calcium oxide, which improves chemical resistance, and for more expensive soda ash. Feldspar is also a source of silica. Soda spars are favored over potash spars in most cases. Low-iron aplite is used as a lower cost alternative. In areas where inexpensive feldspathic sands are abundant, they may be used in preference to processed feldspar. Feldspar beneficiated free of all impurities except quartz is sold for glassmaking as well. Feldspar typically makes up about 5-6% by weight of a batch of container glass.

⌘⌘⌘

Ceramics – Feldspars are used in sanitaryware, tiles, electrical porcelain, tableware, and glazes. Finely ground products, 120 mesh and finer, with a high K:Na ratio are generally preferred as fluxes in ceramic bodies to lower firing temperatures and allow faster firing schedules. Fine, 200 mesh, potash and soda spars are used in glazes as a source of alumina and water-insoluble alkali. Their function in glazes is similar to that in glass. Usage levels range from less than 10% in certain wall tiles to as much as 60% in some floor tiles. A glaze may contain 30-50% feldspar.

Other uses – Feldspar is used as a flux in welding electrodes, and beneficiated 325 mesh soda spar finds limited use as a filler in plastics, rubber, adhesives, and coatings. In filler applications it offers low vehicle demand, high dry brightness with low tint strength, and resistance to abrasion and chemical degradation. In coatings it also provides good film durability and high resistance to chalking and frosting.

⌘⌘⌘

GYPSUM

Gypsum $CaSO_4 \cdot 2H_2O$
Refractive Index: 1.52-1.53
Specific Gravity: 2.32
Mohs Hardness: 2

Gypsum is hydrated calcium sulfate formed by the evaporation of brines. Commercial deposits invariably contain some anhydrite ($CaSO_4$), the anhydrous form of calcium sulfate, and may also contain clay minerals, silica, limestone, dolomite, and soluble chloride and sulfate impurities. Gypsum is beneficiated solely by selective mining and screening. Products are typically white to off-white and range from 80 to 95% purity. The principal commercial use of gypsum is based upon the ability of calcium sulfate to readily take on and give up water of crystallization. The dihydrate ($CaSO_4 \cdot 2H_2O$), as mined, is converted to the hemihydrate form ($CaSO_4 \cdot {}^1/_2H_2O$) by partial calcination. When hemihydrate is then mixed with water, it hardens as it returns to the dihydrate form.

TYPES

Most of the gypsum products sold worldwide are produced as low cost ingredients for construction products. Although gypsum is commonly used as the collective term for all types of gypsum-derived products, the various forms differ in properties and use.

Gypsum – This is ground gypsum ore of at least 80% and more typically greater than 90% purity, with anhydrite as the principal impurity.

Stucco – This is the name used in the wallboard industry for the hemihydrate formed by partially calcining gypsum at 350°F. Also known as plaster of paris, this is the main product in North America because of its widespread use in wallboard and plasters.

Selenite – This is gypsum that forms in large, transparent crystals. Its commercial availability is limited.

Synthetic (byproduct) gypsum – Gypsum is produced as a byproduct of flue gas desufurization at coal-fired electric power plants, and is known as FGD gypsum. Chemical gypsum is generated as a byproduct during the production of phosphoric acid (phosphogypsum), hydrofluoric acid (fluorogypsum), and sulfate-process TiO_2 (titanogypsum). Until recently this gypsum was landfilled as waste. In most cases it is an unattractive alternative to natural sources

⌘⌘⌘

because of impurities, crystal morphology, and/or price. It is disposed of as wet filter cake, but must incur the expense of drying to compete with natural gypsum. As environmental concerns result in restricted landfill access and increasing output of FGD gypsum, more widespread commercial use of byproduct gypsum may occur.

USES

Annual worldwide production of natural gypsum is approximately 100 million metric tons. The leading producer is the US (15%), followed by China, Iran, and Canada. Most of this production is used in the hemihydrate form, particularly in economically developed countries. In the US 75% is used in this form. Annual worldwide production of byproduct gypsum is estimated in the tens of millions of tons. FGD gypsum, at approximately 7 million tons per year, is the most likely candidate for expanded use in traditional gypsum applications. The main uses of gypsum products are in wallboard, portland cement, and soil conditioners.

Wallboard – The use of gypsum-based wallboard, once common only in the US and Canada, has spread to virtually all developed countries. The gypsum core provides a strong, fire-resistant, and inexpensive construction material. The -100 mesh stucco (hemihydrate) form is used. Japan is the largest user of synthetic gypsum products in wallboard, because it lacks natural gypsum resources. The other major source of synthetic (FGD-based) product for this market is Germany, which serves all of Europe. Most of current US wallboard production is based on natural sources.

Portland cement – Uncalcined gypsum crushed to $-1^1/_2$ $+^3/_8$ inch is used with portland cement to retard the setting time of concrete. This is the primary market in developing countries.

Soil conditioners – Gypsum rock ground to -100 mesh is used in treating alkaline, saline, and clayey soils, and as a source of sulfur. In the US its main application is in the cultivation of peanuts.

Other uses – The hemihydrate form is used in a wide variety of construction and industrial plasters. Minor amounts of purified gypsum are used in glassmaking and as a white filler (terra alba) in paint and paper. "Soluble anhydrite", made by dehydrating gypsum into porous, highly absorbent granules, is used as a desiccant.

⌘⌘⌘

HORMITE

Palygorskite (Attapulgite) $(Mg, Al)_5Si_8O_{20}(OH)_2(OH_2)_4 \cdot 4H_2O$
Sepiolite $Mg_8Si_{12}O_{30}(OH)_4(OH_2)_4 \cdot 8H_2O$
Refractive Index: 1.52
Specific Gravity: 2.0
Mohs Hardness: 2-2.5

The hormites are clay minerals with a chain instead of a sheet structure. The most common commercial varieties are acicular or needle-like palygorskite, also known as attapulgite in the US, and sepiolite. Like the sheet silicates (see *SMECTITE*), the central octahedral layer is situated between silica layers. The palygorskite unit cell has five octahedral sites and alternate pairs of silica tetrahedra inverted. The sepiolite unit cell has eight octahedral sites and alternate sets of four silica tetrahedra inverted. Because of the configuration of the silica layers, the central octahedral layer can grow only in one direction, giving a characteristic high surface area and acicular morphology. The inversion of silica tetrahedra also creates channels in the chain structure that are filled with water. This is both absorbed free water, also called zeolitic water, and the water of crystallization, which completes the coordination of the octehadral cations at the edges of the internal layer. If the water is removed, the resulting product has highly absorptive properties.

Commercial sepiolite typically has longer and more flexible needles than palygorskite. While the term attapulgite implies short (<2 micrometers) and low aspect ratio (<10:1) needles, long-needle palygorskite is produced, primarily in Ukraine. Sepiolite is the high-magnesia end member, with minor substitution by Al^{3+} and/or Fe^{3+} for octahedral Mg^{2+} and tetrahedral Si^{4+}. Palygorskite has higher substitution, principally aluminum for magnesium. Attapulgite has a Mg:Al ratio of approximately 1.5. The charge imbalance arising from these substitutions is compensated by exchangeable alkaline and alkaline earth cations. The most common impurities in commercial hormites are smectite, silica, and carbonate minerals.

In addition to their high absorptive capacity hormite clays offer useful thickening properties. When dispersed in water they do not swell, as smectites do. Instead, their needle-like colloidal particles deagglomerate in proportion to the amount of shear applied and form a random colloidal lattice. This loosely cohesive structure thickens the water and imparts thixotropy, pseudoplasticity, and yield value. Because of their mechanically-based dispersion and colloidal structure building, hormite clays are largely insensitive to the types and levels of acids, bases, and salts dissolved in the aqueous systems in which they are used. Since their dispersion is mechanically rather than ionically driven, as with smectites, they can be used in nonaqueous applications in much the same

⌘⌘⌘

way as smectite-based organoclays. While smectite organoclays must be preformed before dispersion in an oil or solvent, hormites can be dispersed in hydrophilic form and then reacted *in situ* with the appropriate quaternary ammonium compound or amine.

The processing of hormite clays involves crushing, extruding, drying, milling, screening, and air classifying, as required, to produce a range of products from partially calcined, coarse absorbent granules to fine colloidal powders for thickening applications.

TYPES

In addition to the mineralogical distinction between palygorskite (attapulgite) and sepiolite, hormite clay products are generally classified by function and water content.

Gelling clay – Products for rheology control are typically dried to 12-16% free moisture and micronized. For use in well drilling fluids the American Petroleum Institute (API) sets specifications for moisture, viscosity, and yield (barrels of mud per ton of clay) that cover both smectite and hormite clays. The viscosity of the latter, however, is determined with a NaCl solution. Gelling grades may be extruded before drying to facilitate separation of the acicular particles. Drilling mud grades may have 1-2% MgO added to improve viscosity in salt water. Small tonages are water washed to remove impurities for pharmaceutical and cosmetic applications. Because the preparation of aqueous hormite dispersions is energy intensive, concentrated predispersed suspensions are sold as a convenience to major consuming industries.

RVM clay – These products for absorbent applications are dried at about 200°C to remove zeolitic water and thereby leave the internal channels of the clay open to penetration by gases and polar liquids. RVM (regular volatile matter) grades may be extruded during processing to improve absorption capacity and are produced in a wide range of sizes from coarse granules to fine powders.

LVM clay – These products (low volatile matter) are partially calcined at about 500°C to remove both zeolitic water and water of crystallization. This improves absorption capacity for oils, greases, and low-polarity organic compounds.

Fullers earth – This is a generic term that derives from an early use, the cleansing and decolorizing of woolens by fullers. It implies no specific mineralogy and in the US can encompass both hormite and smectite clays. In

⌘⌘⌘

the United Kingdom fullers earth is smectite clay. The term is generally used to indicate high absorptive capacity and the ability to decolorize oils and fats.

USES

In the US attapulgite is available in gelling and absorbent grades from extensive reserves in Georgia and Florida, but American sepiolite is rare. The reverse is true in Europe, where extensive sepiolite deposits exist in Spain, and palygorskite is rare. Annual worldwide production of palygorskite is approximately 1.2 million metric tons, more than 80% of which comes from the United States. Annual production of sepiolite is approximately 1 million metric tons, dominated 90% by Spain. The uses for palygorskite are essentially the same as for sepiolite in countries where the latter is more readily available. Worldwide, the major application is pet waste absorbents. Other major uses are oil and grease absorbents, agricultural products, drilling fluids, coatings, and construction products.

Pet litter – Cat litter is the single largest application for hormite clays, typically RVM grades. The clay granules absorb both liquid waste and ammonia odor.

Floor absorbents – Both RVM and LVM products are sold as oil and grease absorbents for industrial and commercial workplaces (e.g. metalworking and automotive shops) and as absorbents in slaughterhouses, butcher shops, and animal barns.

Agricultural products – Gelling-grade clays are used to stabilize fertilizer suspensions, animal feed suspensions, and pesticide suspension concentrates. They are preferred to smectites in these applications because of their tolerance of and compatibility with high solute levels. Absorbent grades are used as pesticide carriers for dry applications and as binders and nutrient carriers in animal feeds.

Drilling fluids – Gelling grades meeting API specifications are produced for salt water drilling fluids since they are insensitive to electrolyte levels that will flocculate bentonite-based fluids. Because of their acicular shape, however, they are inferior to bentonite in controlling fluid loss through the bore hole walls.

Coatings and construction products – Gelling-grade clays are used as thickeners/stabilizers in paints, asphaltic coatings, adhesives, mastics, mortars, and cements. There is also limited use as a partial asbestos replacement in wallboard and certain cements, where long-needle products are preferred.

⌘⌘⌘

Other uses – Hormite clays are also used as thickeners in grease, polyesters, and vinyl plastisols, odor absorbents, bleaching (decolorizing) agents for petroleum, mineral, and vegetable oils, filter aids, anticaking agents, pharmaceutical absorbents, acicular fillers in rubber (sepiolite), catalyst supports (sepiolite), and cosmetic/pharmaceutical thickeners and emulsion stabilizers.

⌘⌘⌘

KAOLIN

Kaolinite $Al_2Si_2O_5(OH)_4$
Refractive Index: 1.55-1.57 / calcined: 1.62
Specific Gravity: 2.58 / partially calcined: 2.50 / fully calcined: 2.63
Mohs Hardness: 2 / partially calcined: 4-6 / fully calcined: 6-8

Commercial grades of kaolin are composed primarily of the mineral kaolinite, a sheet silicate, and may contain greater or lesser quantities of related sheet silicates (mica, illite, chlorite, smectite) and quartz. An individual kaolinite particle has the shape of an hexagonal plate. In nature these plates occur in stacks or "books" that exhibit varying degrees of stacking regularity. Kaolin is hydrophilic (readily water dispersible); for nonaqueous applications matrix compatibility can be improved by surface treatment.

TYPES

Descriptive classifications of kaolin products are many and in certain cases specific to particular markets. The most common designations are as follows.

Primary (residual) kaolin – In reference to geologic origin, the clay occurs in the deposit where it formed.

Sedimentary (secondary) kaolin – In reference to geologic origin, the clay has been eroded and transported from its site of formation and deposited at a distant location. The world's major kaolin belt, the 250 miles between Aiken, SC and Eufala, AL, consists of sedimentary kaolin.

Hard clay – In the rubber industry hard clay is very fine-grained, relatively poorly crystallized kaolin. It is used as a reinforcing filler in rubber, where it provides high modulus, high tensile strength, good abrasion resistance, and stiff (hard) uncured compounds.

Soft clay – In the rubber industry soft clay is coarser, better crystallized kaolin. It has low reinforcing effect in rubber, where it provides lower modulus, tensile strength, and abrasion resistance and softer uncured compounds than does hard clay.

China clay – This is a ceramics industry term synonymous with what is today called simply kaolin. China clay is substantially pure white or near-white kaolinite characterized by low plasticity, low green strength, and good fired whiteness.

⌘⌘⌘

Ball clay – In the ceramics industry ball clay is a highly plastic, fine-grained sedimentary clay containing 70+% kaolinite. Ball clay is characterized by the presence of organic matter, high green strength, and light fired color.

Fire clay – In the ceramics industry fire clay is a refractory (heat resistant), high-kaolinite content clay often found in association with coal beds. Used in refractories or to raise the vitrification temperature of ceramics, the clay's fired color ranges from buff to gray.

Flint clay – A highly refractory hard rock, flint clay is composed principally of well-ordered kaolinite and is low in iron and fluxing compounds.

Airfloat clay – This is dry-ground kaolin that has been air separated to remove impurities and control the particle size distribution.

Water-washed clay – Water-washed kaolin has been slurried in water and centrifuged or hydrocycloned to remove impurities and produce specific particle size fractions. The refined slurry is either dewatered (to reduce soluble impurities) and dried, or concentrated to 70% solids and sold in slurry form. Water-washed clays are often treated to improve brightness. This includes chemical bleaching and/or high-intensity magnetic separation to remove iron and titanium impurities.

Delaminated clay – The coarse clay fraction from water washing is attrition milled to break down the kaolinite stacks into thin, wide individual plates. This improves brightness and opacity.

Calcined clay – The kaolin, usually water-washed soft clay, is calcined to either partially or totally remove surface hydroxyl groups. Calcining increases brightness, opacity, oil absorption, and hardness (i.e., abrasivity).

Surface-treated clay – This is processed kaolin that has been surface modified (e.g., with stearates or silanes) to improve compatibility with and performance in organic matrices.

USES

The United States is the leading supplier of kaolin clays, with about 40% of world production. Second largest is the United Kingdom with about 12%. US annual production capacity of 10 million metric tons is comprised of 40% water-washed, 15% airfloated, 13% calcined , and 10% delaminated clay. The major worldwide use for kaolin is in paper. In the US 40% of kaolin consumption is for coating and filling paper. Much lower but still substantial

⌘⌘⌘

amounts are used in ceramics, refractories, paint, polymers (plastics, rubber, adhesives), and the production of fiberglass.

Paper coatings – High-brightness, low abrasion water-washed and delaminated kaolins are used in coatings for acid-sized paper to improve brilliance, gloss, smoothness, and ink receptivity.

Paper fillers – High-brightness, low-abrasion airfloated, water-washed and delaminated kaolins are used as pulp extenders in acid-sized paper, where they improve opacity, smoothness and ink receptivity. Partially calcined clay is used as a TiO_2 extender.

Ceramics – Whitewares use a combination of china clay (usually airfloat) and ball clay selected for consistent chemical composition so that firing and vitrifying characteristics of the body are controlled and color and translucency of the fired ware are satisfactory. The ratios used are chosen to optimize green strength, plasticity, and casting behavior.

Refractories – In North America certain foundry refractories contain flint clay to give a dense, strong product able to withstand much higher temperatures than ordinary clay-based refractories. In Europe calcined ball clay (chamotte) is used in place of flint clay. China clay is used in cordierite-based kiln furniture and insulating firebrick.

Coatings – The principal use of kaolin in coatings is as a TiO_2 extender. Partially calcined grades generally provide the best extension, durability, and dry hide. Water-washed and delaminated clays are used in water-based coatings to control gloss (coarser = flatter, finer = glossier), film integrity, durability, scrub resistance, covering power, suspension ability, flow, and leveling.

Plastics – The largest single use of kaolin in plastics is for calcined kaolin in PVC wire insulation to improve electrical resistivity. Calcined kaolin is also used in agricultural polyethylene films to improve infrared absorption characteristics and in engineering resins, in both silane-treated and untreated forms, for improved physical properties and heat deflection. Both airfloat and water-washed, fine-particle size kaolins are used in thermosets to provide a smooth surface finish, reduced cracking, warping and crazing, and to obscure fiber reinforcement patterns. Delaminated clays improve thermoplastic physical properties, including enhanced impact resistance when surface treated.

⌘⌘⌘

Rubber – About 80% of the kaolin for rubber is airfloated hard clay. Water-washed and delaminated clays are used for further improved color, physical propeties, and abrasion resistance. Calcined and surface-treated clays are used for improved electrical properties and ease of extrusion.

Adhesives and sealants – Kaolin is used to control flow, penetration, and specific adhesion on application, and adhesive strength, tear strength, tensile strength, and elongation after cure. For aqueous systems the choice of airfloated vs. water-washed clay is dictated by cost, color, abrasion, and rheological properties pre-cure and reinforcing properties post-cure. Stearate-coated clays are used for improved compatibility in nonaqueous systems.

Fiberglass – Low-iron, low-alkali, low-moisture, low-cost, airfloated kaolin is used in the manufacture of continuous filament fiberglass.

Other Uses – Kaolin clays are used in the manufacture of aluminum chemicals, bricks, cements, cosmetics, pharmaceuticals, animal feeds, fertilizers, catalysts, wallboard, printing ink, linoleum, flexible tile, pesticides, and roofing granules.

⌘⌘⌘

MICA

Muscovite $KAl_2(AlSi_3)O_{10}(OH,F)_2$
Refractive Index:1.58-1.62
Specific Gravity: 2.76-2.88
Mohs Hardness: 2-2.5

Phlogopite $KMg_3(AlSi_3)O_{10}(OH,F)_2$
Refractive Index: 1.56-1.64
Specific Gravity: 2.78-2.85
Mohs Hardness: 2.5-3

Micas are sheet silicates historically significant for their ability to be split into large, thin sheets that are uniquely useful for their electrical, thermal, and mechanical properties. They have high electrical and thermal insulating properties; they are resistant to chemical attack; they can be split into transparent or optically flat films; and they can be cut or stamped to shape. Most mica used today, however, is in ground form, although the mineral's platy nature and inertness are still primary attributes. The mica of commerce is principally muscovite. Muscovite products range in hue from colorless to pale green or ruby. Minor amounts of sericite, a fine-grained form of muscovite, are also sold. Phlogopite is the only other mica of commercial significance. It ranges in color from pale yellow to light brown. Phlogopite has superior thermal stability but is not as commonly available as muscovite.

TYPES
Mica products often are clasified in terms of to their traditional production in sheet form as opposed to the more recent filler forms.

Sheet mica – Sheet mica consists of "books" of mica laminae and is classified according to the thickness, purity, and maximum usable area that can be cut or stamped. These criteria dictate the subcategories, in order of decreasing thickness, of block, thins, film, and splittings. Film mica is split from block and thins. Splittings are very thin but have relatively small usable areas, so they are used mostly to fabricate built-up mica. The ASTM (American Society for Testing and Materials) designates 13 quality groups for sheet mica based on color and visible imperfections and 12 additional groups based on maximum usable rectangle. Although its overall volume is minor compared to ground mica, sheet mica is still widely used in the electrical and electronics industry for its combination of thermal and electrical insulating properties and high mechanical strength.

⌘⌘⌘

Built-up mica – Also known as micanite, built-up mica is a fabricated sheet of desired thickness made by overlapping irregularly shaped splittings and then binding them together with an organic or inorganic binder, heat, and pressure. Some products are reinforced with a paper, fiberglass, or textile backing. Built-up mica is produced as an alternative to natural sheet mica for electrical insulation applications.

Mica paper – Mica paper is a fabricated alternative to natural sheet and built-up mica products. Scrap mica is delaminated through a combination of thermal, chemical, and mechanical treatments. The mica pulp is then processed on a papermaking machine into a continuous, homogeneous sheet of uniform thickness.

Scrap mica – The term "scrap" is a holdover from the days when sheet was the major mica product. It is basically mica that is insufficient in size or quality to qualify as sheet. Scrap is produced at mines and at sheet mica factories. Scrap mica can be a byproduct of sheet mica mining or of the recovery of other minerals or the sole product of a mine. Factory scrap is the trimmings from sheet mica. Scrap is the source for ground mica products, which today account for more than 95% of all mica sold worldwide.

Flake mica – Flake mica is essentially nonfactory-generated scrap mica. It is recovered by flotation from mica ores, from which quartz and feldspar are also generally separated and recovered, or as a floated byproduct during the beneficiation of feldspar, kaolin, or lithium-bearing ores.

Wet-ground mica – Flake mica concentrates from flotation are ground wet in mills designed to delaminate as well as to grind. Such mills provide products having a higher aspect ratio, sheen, and slip compared to dry-ground mica. Trimmings from good-quality sheet mica are also used as feed for wet-grinding plants.

Dry-ground mica – Flake mica flotation concentrates are at least partially dried and then ground in mills appropriate for the particle size desired. Coarse-milled products (>100 mesh) are processed with hammer mills and screens or air separators. Fine-ground products, -100 mesh to -325 mesh in particle size, are processed in fluid energy mills, usually with superheated air.

Micronized mica – This is dry-ground mica milled to -20 or -10 micrometers in fluid energy mills using superheated steam. Some is calcined for use in cosmetics.

⌘⌘⌘

USES

The United States is the leading supplier of mica, with about 46% of the world's production. The US produces only ground muscovite. The only significant production of phlogopite is in Canada and Finland, which together account for about 15% of total worldwide mica output. India produces 80% of the world's sheet mica. In the US fully half the mica produced is used in wallboard joint cements. Other major end uses are coatings, plastics, and well drilling fluids.

Joint cements – Fine dry-ground (fluid energy-milled) muscovite is used in drywall joint compounds, where it contributes to consistency and workability, smooth surface finish, and resistance to shrinkage and cracking.

Coatings – Fine-ground, -325 mesh and micronized mica grades are used in paint as a pigment extender and for dry film reinforcement. The inert, platy mica improves suspension stability, controls film checking, chalking, shrinkage, and blistering, improves resistance to weathering, chemicals, and water penetration, and improves adhesion to most surfaces. Coarser grinds are used in textured paints, and wet-ground mica is used in high quality exterior house paints. High aspect ratio grades are preferred for porous surface sealers to seal pores, control penetration, and reduce sagging and film cracking. Automotive paints use high aspect ratio mica to achieve a metallic effect either as is, or after conversion to pearlescent pigments by surface coating with metal oxides.

Plastics – Finely ground, -325 mesh and micronized micas are used in plastics to improve electrical, thermal, and insulating properties. Mica is considered the most effective mineral for reducing warpage and increasing stiffness and heat deflection temperature in plastics. In general, mica reinforces crystalline better than amorphous polymers. Best results are obtained with nonpolar polymers when mica is pretreated with a coupling agent to improve wetting. Mica is used in both thermoplastics and thermosets. Its largest single use is in polyolefins, even though it requires stabilizers to prevent degradation of polypropylene. Both muscovite and phlogopite micas are used in plastics, with high aspect ratio grades preferred for their superior reinforcement properties.

Drilling fluids – Coarse, hammermilled (+10 mesh) mica is used in water-based oilwell drilling fluids to prevent fluid loss into porous rock formations. The coarse mica flakes bridge openings and seal porous sections of the drill hole against loss of circulation. Mica's platy nature also aids in the suspension of drilling fluid solids and cuttings.

⌘⌘⌘

Other uses – Ground mica is used as an asbestos substitute in certain thermal boards, brake linings, gaskets, and cement pipes, as a filler and nonstick surface coating for roll roofing and asphalt shingles, as a mold lubricant and release agent in the manufacture of tires and other molded rubber goods, as a flux coating on welding rods, and as a pearlescent pigment in wallpapers.

⌘⌘⌘

NEPHELINE SYENITE

Nepheline — $(Na,K)AlSiO_4$
Refractive Index: 1.53-1.55
Specific Gravity: 2.57
Mohs Hardness: 5.5-6

Microcline — $KAlSi_3O_8$
Refractive Index: 1.52-1.54
Specific Gravity: 2.54-2.57
Mohs Hardness: 6

Albite — $NaAlSi_3O_8$
Refractive Index: 1.53-1.54
Specific Gravity: 2.62
Mohs Hardness: 6

Nepheline syenite is a rock composed of soda and potash feldspars (see *FELDSPAR*) and nepheline. Nepheline has the theoretical chemical composition $NaAlSiO_4$, but potassium invariably substitutes for a portion of the sodium. The amount of potassium in natural nephelines ranges from 3 to 12% K_2O by weight. Nepheline is related to the structure of high-trydimite (a high-temperature form of silica), with Al^{3+} replacing Si^{4+} in half the tetrahedra. The charge imbalance is compensated by Na^+ and K^+. Unlike feldspars, the sodium is exchangeable for H^+, making nepheline less acid stable. Nepheline will gelatinize in HCl, while feldspar will not. Nepheline can form only in a geologic environment deficient in silica. Commercial nepheline syenite deposits and the products made therefrom are consequently free of crystalline silica impurities. These deposits are likewise exploited for their high mineral brightness and low level of dark mineral impurities. Commercial products are made by crushing, dry magnetic separation, and milling. Fine particle size grades are produced by air classification. In North America typical nepheline syenite is approximately 25% nepheline, 20% microcline feldspar, and 55% albite feldspar.

TYPES

Nepheline syenite is characterized primarily by chemistry and particle size.

Glass grade – Coarsely ground nepheline syenite, typically -40+200 mesh with <0.1% Fe_2O3, >23% Al_2O_3, and >14% total Na_2O+K_2O, is used in glassmaking. Refractory minerals must be absent. A higher iron grade containing 0.35% Fe_2O_3 maximum is used in amber glass and glass fiber.

⌘⌘⌘

Ceramic grade – Grades for ceramics are finely ground (typically 200, 270, or 400 mesh), controlled for PCE (pyrometric cone equivalent), free of dark impurities, and white-firing without specking.

Filler grade – Filler grades are finely ground (325 mesh to 1250 mesh) and have a high brightness (>93) and low vehicle demand.

Aggregate grade – Off-color, gray and blue nepheline syenite is produced in Arkansas for use as a construction aggregate and as roofing granules.

See also *FELDSPAR*.

USES

Of the approximately 4 million metric tons of nepheline syenite produced annually, Russia is the largest source, accounting for more than 75%. Russian output is captive, however, and is devoted almost entirely to the making of portland cement and to the production of alumina and aluminum due to the relative local scarcity of bauxite. Russian products are generally too high in iron for use in major North American and European applications – glassmaking and ceramics. These markets are served instead by products from Canada and Norway. Of the approximately 900,000 metric tons annually produced outside of Russia, Canada supplies about 63% and Norway 37%. Glassmaking is the major use, consuming 70% of Canadian and 80% of Norwegian production. Both sources supply about 15% of their output for ceramics and the balance for filler applications.

Glass – Container glass is the single largest application for nepheline syenite, followed by flat glass and glass fiber. Nepheline syenite is used as a source of alumina, which improves both the workability of the glass melt and the chemical and physical stability of the finished product. It also contributes alkaline oxides (Na_2O, K_2O) that provide fluxing in partial substitution for more expensive soda ash. Nepheline syenite is a source of noncrystalline silica as well. It competes with feldspar and aplite by offering a higher alumina and alkali content per unit weight.

Ceramics – Nepheline syenite is used in sanitaryware, tiles, electrical porcelain, tableware, and glazes as an alternative to feldspar. It acts as a flux in ceramic bodies to lower firing temperatures and allow faster firing schedules. It is used in preference to feldspar when its chemistry and/or price warrants or when crystalline silica is an overriding issue.

⌘⌘⌘

Coatings – Although nepheline syenite's acid solubility makes it less resistant than feldspar to frosting, it does offer similarly low vehicle demand, high brightness with low tint strength, abrasion resistance, and good film durability. It is, in fact, more widely used than feldspar in coatings, due at least in part to its freedom from crystalline silica. Coatings consume about 10% of North American production.

Other uses – Nepheline syenite is also used as a filler in plastics, rubber, and adhesives because of its low vehicle demand, high dry brightness with low tint strength, nearly lamellar shape, and resistance to abrasion and chemical degradation. It is used as a sandblasting medium due to its absence of crystalline silica.

⌘⌘⌘

PERLITE

Perlite Volcanic Glass

Refractive Index: 1.48-1.49
Specific Gravity: 2.5-2.6
Mohs Hardness: 5.5-7.0

Perlite is a hydrated volcanic glass composed chiefly of amorphous silica with 12-18% aluminum oxide, lesser amounts of the oxides of potassium and sodium, and minor amounts of iron, magnesium, calcium, and titanium. It is because of the 2-5% water contained in perlite that this rock is commercially valuable. Perlite as mined is distinguished by its concentric fractures, a result of rapid cooling of the molten volcanic glass, which gives rise to an onion-skin appearance. Upon heating above 870°C the contained water, as both silanol and molecular water, expands the perlite grains, forming minute bubbles within the glass matrix. The result is low-density particles with cellular interiors. These particles are used for their acoustical and thermal insulating properties, their chemical inertness and physical resilience, their fire resistance, their water retention ability, and their low bulk density.

TYPES

Because most of the perlite used commercially is in its expanded form, products are differentiated according to whether they have been expanded.

Processed perlite – Since the objective of nearly all perlite production is the expanded form, the principal product of perlite mining is processed perlite. The ore is crushed and ground and then classified to remove as many nonexpandable impurities as possible. The size of expanded perlite particles is largely determined by the particle sizes of the processed perlite from which they were made. The last step in the production of processed perlite, therefore, is its segregation by screening or air classification into more than a half dozen size classifications. These range from -200 mesh to +12 mesh. Processed perlite is supplied primarily to regional plants that carry out the thermal expansion and local marketing.

Expanded perlite – Expanded perlite has a very low bulk density, so expansion of the processed material is caried out in regionally located plants in order to minimize the otherwise high cost of transportation to market per unit weight of product. The perlite is expanded by quick exposure in furnaces to temperatures of 870° to 1100°C. This allows simultaneous softening of the glass matrix and volatilization of the contained water. The perlite expands in

⌘⌘⌘

volume up to 20 times and decreases in bulk density by up to 90%. The size and quality of the processed perlite feed and the design and operation of the furnace are controlled to maximize the production of desired particle size, density, and strength. Insufficient expansion will produce smaller, denser particles while overexpansion can cause the crude particles to shatter into fines. After expansion, the perlite is air classified to remove any unexpanded particles and fines. Depending upon the intended end use, the expanded product may be further size classified, surface treated, or milled.

Milled perlite (filter aid) – Some expanded perlite is milled to -100 mesh and sold as a filteraid, especially for use in rotary precoat filtration. Careful milling is required, because oversize particles will inhibit the formation of filtercake while excessive fines will slow the filtration rate.

USES

In a world market of approximately 1.8 million metric tons, the leading perlite producers are the former Soviet republics of Armenia and Ukraine (35%), the United States (25%), and Greece (12%). In the US the market for perlite is dominated by formed insulation products (58%), followed by filter aid, horticultural products, and filler applications.

Formed insulation products – Medium- to fine-grained expanded perlite is incorporated into pipe insulation, insulation boards, acoustical ceiling tiles, precast concrete floor and roofing tiles, and precast roof decks. The perlite contributes to reduced weight, acoustical and thermal insulation, and fire resistance.

Filter aid – Expanded and milled perlite is used as a filteraid in the processing of foods such as sugar, oils, beer, and fruit juices and in the filtration of chemical products and industrial effluents.

Horticultural products – The low density and high absorptive capacity of expanded perlite is useful in soil conditioning. Expanded products provide a permeable soil structure with good ventilation and water retention and are a source of aluminum, potassium, and sodium nutrients. In nursery potting soils they provide these same benefits and serve as a lightweight substrate that facilitates transplanting. Expanded perlite also is used as a rooting medium, as a growing substrate in hydroponics, as a fertilizer carrier, as an additive in animal feeds, where it is a carrier of liquid additives, a binder, an anticaking agent, and a digestive aid, and as an additive to manures to decrease their odor and improve their fertilizer value.

⌘⌘⌘

Filler – Expanded perlite is used as a filler in textured paints and in auto underbody plastisol coatings to reduce weight and improve sound insulation. It is used in syntactic foams and in automotive polyester molding compounds and body fillers to reduce weight and improve sanding characteristics. Its spherical shape allows more perlite to be incorporated, compared to irregularly shaped fillers, because there is less effect on melt or plastisol viscosity.

Other uses – Expanded perlite is used as an aggregate in plasters, wallboard, lightweight concrete, and lightweight fire-resistant coatings for structural steel and concrete. It is used as cryogenic insulation in storage vessels for liquified gases, and with silicone treatment as loose-fill insulation in building cavities. Unexpanded perlite is used as an abrasive, a chemical source of silica, a foundry slag coagulant, and to cover molten metal prior to casting, such that the heat of the metal expands the mineral into an insulating layer.

⌘⌘⌘

PYROPHYLLITE

Pyrophyllite $Al_2Si_4O_{10}(OH)_2$
Refractive Index: 1.59-1.60
Specific Gravity: 2.8
Mohs Hardness: 1.5

Pyrophyllite in pure or near-pure form is rare in nature and in commerce. It can be viewed as an aluminum analogue of talc, and a pure, platy specimen would share many of talc's physical characteristics. In addition to the platy morphology, pyrophyllite occurs in nature as massive aggregates of small crystals and as large needle-like crystals. In Japan and Korea, where pyrophyllite is abundant, beneficiated forms have been used in applications traditionally reserved elsewhere for kaolin and talc. The requirements for competing in a world economy, however, have more recently moved Pacific Rim countries toward wider use of imported kaolins and talcs. In other producing countries a relatively low value is assigned to pyrophyllite, so that many natural mineral mixtures (some even devoid of this mineral) are sold under its name, generally at low prices. In many cases the lack of purity has been used to advantage. The major current uses actually exploit the unique and desirable features of the mineral blends that constitute most commercial grades. The minerals most commonly associated with commercial pyrophyllite are the related sheet silicates (mica, kaolinite, chlorite, smectite, illite), plus diaspore, andalusite, and quartz.

TYPES
Since the pyrophyllite market is dominated by ceramics and refractories uses, the products available are best characterized in terms of mineralogy and resultant market suitability.

Refractory grade – For refractory uses the product must be low in alkalis (<1%), which necessitates a low mica content. Alkalis reduce the melting point of pyrophyllite. Content of fluxes (Fe_2O_3, FeO, TiO_2) should be <1%.

Ceramic grade – For ceramics the preferred product is high in alkalis, so a higher mica content is advantageous. The lower melting point caused by a high alkali content allows for faster firing. A low coloring oxides content is required for whiteware applications.

Agricultural grade – For use as a carrier for the active ingredients in pesticide dusts, pyrophyllite's neutral pH, inertness and nonhygroscopic nature

⌘⌘⌘

are natural attributes. Beyond this, market acceptance is dictated by performance, which is affected by fineness and bulking values.

Filler grade – In the US quartz-induced abrasion restricts the use of pyrophyllite in polymers, although the fine platy nature of the pyrophyllite and mica components can be expected to contribute to physical properties in a manner similar to kaolin or talc. In paints a fine grind and good color are required for pigment extension, while a relatively coarse grind is needed for reinforcing high-build coatings. For wallboard joint cements grind, color, and water absorption are the key properties.

USES

Japan and Korea account for 80-85% of the world's annual production of about 2 million metric tons. US pyrophyllite, 3% of the world's total, is all dry ground and more or less abrasive depending on the particle size distribution of its quartz content. As such, it is generally overlooked in filler applications in favor of kaolin or talc. In the US the ceramics and refractories industries consume about 75% of the pyrophyllite produced. Pesticides take 10% of production, and filler uses (paint, joint cements, mastics) take 3-5%.

Refractories – Low-alkali pyrophyllite is used in refractories for its permanent expansion on firing, excellent reheat stability, low hot load deformation, good hot creep resistance, low reversible thermal expansion, low thermal conductivity, and high resistance to corrosion by molten metals and basic slags. Refractory grades are used in metal pouring refractories, kiln car furniture, insulating firebrick, ramming and gunning mixes, foundry mold coatings, and refractory mortars. The largest use is in ladle bricks.

Ceramics – High-alkali products are used in whitewares for their contribution to increased firing strength in vitreous bodies, improved thermal shock resistance from their low coefficient of thermal expansion, the ability to produce bodies with little or no shrinkage and reduced warpage, and the ability to produce low moisture expansion bodies with excellent craze resistance. Ceramic grades are used in tile, sanitaryware, pottery, chinaware, electrical porcelain, and glazes. The largest use is in wall tile, where the mineral contributes to increased fired strength and faster firing.

Pesticides – Pyrophyllite is used as a carrier in insecticide dusts because it is nonhygroscopic, fluffy, neutral in pH, and compatible with both acid and alkaline active ingredients. In passing through the blowers of dusting machinery it picks up an electrostatic charge, which causes its attraction to the underside of leaves as well as the exposed upper plant surfaces.

⌘⌘⌘

Coatings – Coatings use a finely ground, high-brightness pyrophyllite for low-cost pigment extension. The product's platy nature, from its sheet silicate structure, promotes good dispersion by inhibitting pigment settling, helps film dry, and increases resistance to film cracking. Worldwide, paint is the largest filler market. In the US coatings are a minor market, generally utilizing relatively coarse grades for imparting mud-crack resistance to high-build coatings (textured paints, block fillers) and checking/cracking/frosting resistance to exterior latex paints.

Joint cements and mastics – Finely ground paint-grade pyrophyllite is used in wallboard joint cements and in mastics to control rheology and provide reinforcement. For joint cements in particular, water absorption is an important property because it affects the critical cement properties of consistency, spreading, shrinkage on drying, and resistance to cracking.

Other uses – In the US the mostly platy nature, low moisture content and inertness of pyrophyllite products qualify them as low-cost alternatives to kaolin and talc for other filler applications that can accommodate higher abrasivity.

⌘⌘⌘

SILICA

Quartz SiO_2
Refractive Index: 1.54
Specific Gravity: 2.65
Mohs Hardness: 7

Synthetic (amorphous) Silica SiO_2
Refractive Index: 1.45
Specific Gravity: 2-2.3
Mohs Hardness: 5-6

Natural silica products are crystalline and are classified for most of their uses under the term "sand and gravel". Finely ground natural silica with high mineralogical and chemical purity is nevertheless produced in substantial quantities. It is generally characterized as off-white to white and abrasive, with low suface area, low binder demand, and low cost. Amorphous forms of natural silica exist as diatomaceous earth (see *Diatomite*), volcanic glass (see *Perlite*), and noncommercial opaline minerals. Synthetic silicas are generally characterized as amorphous and white, with high purity, a high surface area, high liquid absorption, and ultrafine particle size. These include silica gel and precipitated silica, also known as hydrated silicas, and fumed silica, also known as pyrogenic or anhydrous silica. Hydrated silicas are made in aqueous media, which results in a relatively high density of surface silanol (-Si-OH) groups. They therefore absorb water readily and are typically sold with 5 to 6% free moisture. The pyrogenic process used to produce fumed silica leaves most surface silanols condensed to siloxane bridges (-Si-O-Si-), so that silanol surface density is only about 25% of that of hydrated silicas. There is consequently less tendency to absorb water, and the products typically contain less than 2% free moisture.

TYPES

The primary differentiation among silica products is their natural vs. synthetic origin, which translates to a division between crystalline and amorphous forms.

Ground silica – Ground silica, also known as ground quartz and silica flour, is produced by grinding high-purity quartz, quartzite, sandstone, or silica sand to finer than 200 mesh. Air separation is used as required to remove kaolin, mica, feldspar, or calcite impurities. Silica sand recovered as a byproduct during the flotation of other minerals is also used as a source. Ground silica for filler uses typically assays >99% SiO_2 and offers high brightness, low moisture, chemical

⌘⌘⌘

inertness, relatively low surface area, and the low liquid absorption that allows high loading levels. Because of its hardness and the creation of sharp, angular particles upon grinding, most ground silica is too abrasive for use in certain plastics compounding machinery.

Novaculite – This form of ground silica is microcrystalline quartz that is milled to low-moisture, high-purity (>99% SiO_2), plate-like or disc-like particles. Brightness is generally lower than for other forms of ground silica, but particle shape, lower binder demand, lower abrasivity, and availability in a range of particle size distributions (to as small as 2 micrometers average) has earned novaculite a place in filler applications, particularly in paint.

Tripoli – This is a general term for a friable or powdery rock composed substantially of microcrystalline quartz. Commercial tripoli is ground mainly for use as an abrasive agent in buffing and polishing compounds, since particle edges tend to be less sharp than for most other forms of ground silica.

Flint – This is a relatively high silica (≥85% SiO_2) rock composed of microcrystalline quartz, with calcite as the common major impurity. Flint is characteristically gray or black and consequently is not a source of ground silica for most applications. The exception is ceramics, particularly in Europe, where low-iron ground flint is used as a silica source in whitewares because it fires white. In Europe silex is a commercial term for flint.

Silica sand – Silica sand can be broadly defined as a natural quartz sand that is -4 mesh and +200 mesh. Silica sand has more rounded grains than are obtained by grinding high-silica rocks to an equivalent size range, although grain angularity varies from source to source. Products are commonly named for their intended use – e.g., glass sand, foundry sand, blasting sand, and hydraulic fracturing sand. Each term indicates compliance with requirements for minimum SiO_2, maximum chemical impurities, particle size distribution, and grain shape for that application. Most uses require at least 95% SiO_2 and have limits on the Al_2O_3 and Fe_2O_3 content.

Precipitated silica – Precipitated silica is made by the controlled neutralization of a sodium silicate solution under alkaline conditions using either sulfuric or carbonic acid (as CO_2 + HCl). This is carried out in a heated stirred reactor. The rate and order of addition of reactants, reactant concentration, and reaction temperature are varied to manipulate particle size and particle structure. Precipitation produces a low-solids slurry of amorphous silica particles within the byproduct salt solution. The discrete primary silica particles that initially form fuse into aggregates, which in turn form loose agglomerates. The precipitate is filtered, washed, and dried (usually spray dried) to a neutral pH product. The dry silica is milled to reduce average agglomerate size and is then frequently compacted to balance bulk density, freedom from dustiness, and dispersibility. The particle size conventionally reported is that of primary particles (10 to 30 nanometers), although it is the aggregates (30 to 150 nanometers) that are the actual functional particles. BET surface area is therefore frequently preferred for classifying the various grades. Products are also classified according to structure, which is a function of aggregate size, shape, and pore volume. For convenience, structure has been related to oil absorption, with five structure levels designated. These range from VLS (very low structure) at less than 75ml/100g to VHS (very high structure) at greater than 200ml/100g. In elastomer compounding, which uses approximately 75% of the precipitated silica produced, the silica is commonly treated with silane coupling agents *in situ* to promote better matrix compatibility and improved compound properties. Silica grades pretreated with silane are also available.

Silica gel – Silica gel is made by reacting a sodium silicate solution with sulfuric acid under acid conditions. Silica properties are manipulated by controlling the reaction rate, order of reactant addition, reactant concentration, reaction temperature, and mixing conditions. The silicate-acid reaction first forms a hydrosol, which is aged to a transparent rigid gel. This gel is broken into small lumps, washed to remove residual salts, and dried. When the gel's water is removed slowly, the colloidal silica structure contracts and a dry silica xerogel results. When the water is removed quickly, so that the colloidal silica structure is preserved, a dry silica aerogel with lower density and higher pore volume results. Aerogels are neutral in pH, while xerogels are usually processed to produce an acid pH (4 to 5) in water. The dried xerogel or aerogel is milled to a controlled particle size distribution, with the average particle size measurement reflecting aggregate rather than primary particles. Surface-treated grades are available for improved organic matrix compatibility. Major applications for silica gel depend on its small particle size for rheology control and its porosity for low particle density and high liquid absorption.

Fumed silica – Fumed silica is prepared by the hydrolysis of silicon tetrachloride in a flame of hydrogen and oxygen. Particle size and surface area are controlled by varying the ratio of the three reactants. The primary silica particles are round and 7 to 40 nanometers. These primary particles collide and fuse into branched or chain-like clusters. These aggregates, the functional particles, average about 1 micrometer, and in turn form agglomerates. In aqueous dispersions fumed silica provides a pH of appproximately 4. The very low bulk density of the agglomerates is increased by compaction with vacuum

⌘⌘⌘

deaeration before packaging. Fumed silica is sold in a hydrophilic form as produced and in a hydrophobic form with silane surface treatment.

USES

To a large extent natural and synthetic silicas have little in common other than their gross chemistry. Applications reflect this. The United States produces about 25 milllion metric tons of silica annually under the general description "industrial sand and gravel". Of this total about 40% is glass sand, 20% is foundry sand, 7% is abrasive sand, 6% is hydraulic fracturing sand, and 6% is gravel. Specialty silica, consisting of ground natural and synthetic silicas, accounts for 2 to 4% of total silica output. For silica sand the major use is as an inexpensive source of SiO_2. This encompassses the production of nearly all types of glass, plus ferrosilicon, silicon metal, silicon carbide, and metallurgical fluxes. Each of these applications has its own requirements for particle size distribution, minimum SiO_2, and maximum Fe and Al, as well as control over other chemical and mineralogical impurities. Other major silica sand uses have similar requirements, plus additional specifications on particle shape. Fracturing sand, for example, must penetrate and maintain openings through which oil or gas can flow to a well. It must consist of well-rounded grains to allow ease of placement and maximum permeability. Fractured angular grains with sharp edges are obviously prefered for most abrasive sand uses. The major uses of specialty silicas are in ceramics, coatings, rubber, plastics, adhesives, and sealants.

Ceramics – Low-iron ground silica, typically -200 mesh, and calcined silica are used in whiteware formulations to facilitate drying of the body, to control expansion characteristics and compatibility between the body and glaze to prevent crazing, and to provide whiteness and acid resistance. Ground silica is also used as a source of SiO_2 in glazes and enamels.

Coatings – Ground quartz and novaculite are used as extender pigments because of their low binder demand, which allows high loadings. The platy shape of novaculite imparts additional mar, wear, and weather resistance. Synthetic silicas are used in coatings to provide flatting, mar resistance, and abrasion resistance. In certain specialty coatings they are used as well for rheology control and as suspension aids.

Rubber – Finer (<0.025 micrometer) precipitated silica is the only fully reinforcing alternative to carbon black for general rubber compounding. Most of the precipitated silicas used in rubber are reinforcing grades rather than the coarser extending grades. Precipitated silica is used in compounds designed to be translucent or colored, and in general compounding to promote abrasion resistance, cut growth resistance, tear strength, elastomer-to-textile adhesion, and resistance to heat aging. It is often compounded with silane coupling agents to improve matrix compatibility. Because of its particular attributes, precipitated silica is used in radial ply passenger tires, off-road tires, and footwear. Fumed silica is the conventional reinforcing filler in silicone elastomers, although more recently it has been supplemented with precipitated silica and finely ground natural silica that extend its reinforcing properties at lower cost. Synthetic silicas have an index of refraction close to that of silicone rubber, so that transparent compounds can be formulated.

Plastics – Synthetic silicas, particularly fumed silica, are used as thixotropes in unsaturated polyester resins and gel coats and in epoxy resins. Fumed and precipitated silicas are used as thixotropes in PVC plastisols. Synthetic silicas are also used as antiblocking and antislip agents by temporarily absorbing plasticizers that can cause tack and by providing an imperceptible surface roughness. They are used as matting or flatting agents and as plate-out agents in highly plasticized compounds. Precipitated silica is used as a reinforcing filler in thermoset EVA and to provide controlled porosity in polyethylene battery separators. Finely ground natural silica is used in thermosets to provide dimensional stabiliy, improved thermal conductivity, and good electrical insulation properties at low cost.

Adhesives and sealants – Synthetic silicas are used in adhesives, caulks, and sealants to control flow and sag, improve bond strength, and provide reinforcement. Moisture-sensitive products, such as one-part RTV silicones, use fumed silica preferentially. Finely ground natural silica is used for its low moisture content, low cost, low binder demand, and its ability to improve tensile strength without affecting flexibility or durability.

Other uses – Synthetic silicas are used as thickeners in printing inks, as carriers for liquid and active ingredients, and as anticaking and moisture control agents. Fumed and precipitated silicas are used in high-temperature silicone greases. Silica gel is used as a thickening and polishing agent in dentifrices and as the abrasive agent in silicon wafer polishing compounds.

⌘⌘⌘

SMECTITE

Montmorillonite $(Al,Mg)_2Si_4O_{10}(OH)_2$
Hectorite $(Mg,Li)_3Si_4O_{10}(OH,F)_2$
Saponite $Mg_3(Si,Al)_4O_{10}(OH)_2$
Refractive Index: 1.50-1.64
Specific Gravity: 2.5
Mohs Hardness: 1-1.5

Smectites are water swellable clays that have a platy structure. Smectite is the mineralogical term for a group of clays, which includes montmorillonite, hectorite, and saponite. Most smectites are more commonly known under the geological term bentonite. By convention, bentonite is understood to be an ore or product with a substantial smectite content. The range of possible chemical variations in the basic smectite trilayer lattice starts with montmorillonite, the high-aluminum end member. Montmorillonite is composed of a central alumina octahedral layer sandwiched between tetrahedral silica layers. This is identical to the dioctahedral pyrophyllite structure except for small substitutions of Mg^{2+} for Al^{3+} in octahedral positions and Al^{3+} for Si^{4+} in tetrahedral positions. The resulting charge imbalance is compensated by exchangeable alkali and alkaline earth cations, which contribute to the ability of the clay to swell. When the exchangeable cations are predominately sodium, the individual platelets separate to produce a colloidal structure in water.

At the other end of the smectite series are the high-magnesium members, hectorite and saponite. These clays possess a talc structure, with a trioctahedral magnesia layer sandwiched between the silica layers. Swellability results from minor substitution of aluminum for silicon in saponite or lithium for magnesium in hectorite. As with montmorillonite, the type of exchangeable cation determines the degree of swelling. Hectorite also has partial substitution of lattice hydroxyls by fluorine.

Each macroscopic smectite particle is composed of thousands of submicroscopic platelets stacked in sandwich fashion with exchangeable cations and oriented water between each. The platelet faces carry a negative charge from lattice substitutions, while edges have a slight positive charge from broken bonds and cation adsorption. When smectite and water are mixed, water penetrates the area between the platelets forcing them farther apart. With calcium as the major exchange ion the platelets will swell in this fashion but have limited ability to completely delaminate. When sodium is the predominant exchange ion, the platelets separate farther apart and the exchange ions begin to diffuse away from the platelet faces. Further penetration of water betwen the platelets then proceeds in an osmotic manner until they are completely separated. The presence of dissolved substances in the water will prolong hydration time by inhibiting this osmotic swelling. Once the smectite platelets are separated, the weakly positive platelet edges are attracted to the negatively charged platelet faces. The resulting three dimensional structure, often referred to as a "house of cards", imparts thixotropy, pseudoplasticity, and yield value.

Most smectities are processed by drying, crushing, and milling to a 200 mesh powder, with mineralogical purity determined by ore selection. Granular grades are produced for absorbent uses. A relatively small quantity of white or light colored smectite is beneficiated by hydroclassification to produce products of sufficient mineralogical, chemical, and microbiological purity for pharmaceutical, cosmetic, and the more demanding industrial uses. The impurities most commonly associated with commercial smectites are silica, feldspar, zeolites, and carbonate minerals.

TYPES

There are a number of descriptive terms for smectite products based on geographic source, exchangeable cations, production process, and end use application.

Sodium bentonite – Sodium bentonite is composed substantially of smectite, usually montmorillonite, with sodium as the major exchangeable cation. As such it is water swellable and will hydrate to form the characteristic colloidal structure.

Calcium bentonite – Calcium bentonite is composed substantially of smectite, usually montmorillonite, with calcium as the major exchangeable cation.

Ion-exchanged bentonite – This is calcium bentonite that has been dry blended or hydrotreated with sodium carbonate to increase the availability of exchangeable sodium ions sufficiently to enable the clay to behave like sodium bentonite. This is also called sodium-exchanged bentonite.

Acid-activated clay – This is calcium bentonite that has been treated with sulfuric or hydrochloric acid to exchange calcium with hydrogen ions, to dissolve carbonate impurities, to leach some tetrahedral aluminum and octahedral magnesium, aluminum, and iron, and to delaminate the edges of the stacks of clay platelets. The overall effect is to increase porosity and surface area and, thereby, the clay's absorptive properties. This type of bentonite is used primarily to decolorize and deodorize petroleum and edible oils.

⌘⌘⌘

Bleaching clay – Although certain natural smectite and hormite clays are used to decolorize oils, the term bleaching clay most commonly refers to acid-activated bentonite as described above.

Absorbent clay – This is usually calcium bentonite that has been partially calcined at 200° to 550°C to remove interlayer water and develop absorbency.

Wyoming bentonite – Also known as Western bentonite and swelling bentonite, this is generally a high mineralogical purity, tan to green sodium bentonite with good thickening properties. It is actually mined in Wyoming, South Dakota, and Montana. This term is sometimes used generically to denote a sodium bentonite with good thickening properties.

Southern bentonite – This is calcium bentonite mined in Texas, Mississippi, and Alabama. It is also known as nonswelling bentonite.

White bentonite – This is bentonite that is naturally light enough in color for use in color-sensitive applications such as pharmaceuticals, cosmetics, ceramics, laundry products, and the paper industry.

Hydroclassified smectite – This is white sodium montmorillonite, saponite, or hectorite that is dispersed in water to colloidal dimensions, beneficiated by centrifuges or hydroclones to remove mineral impurities, and dried. Some products are based on white calcium smectite that is exchanged with sodium carbonate prior to centrifugal beneficiation. Water processed smectites, individually and in blends, are produced for uses where high mineralogical and bacterial purity, control of whiteness and heavy metals, and greater performance efficiency justify their higher price.

Synthetic smectite – Smectite is synthesized in commercial quantities from component oxides at high temperature and pressure. The synthetics, most commonly hectorite, are sold for applications where a premium price is justified by their purity, consistency, and transparent or translucent colorless dispersions.

Organoclay – Organophilic smectite is most often produced by reacting water-purified clay in an aqueous dispersion with quaternary ammonium compounds, then filtering and drying. These products are designed to perform in oils and solvents in much the same way that their untreated counterparts perform in water.

⌘⌘⌘

Pillared clay – Pillared clay is purified smectite, usually montmorillonite, treated in an aqueous dispersion with large metal hydroxide complexes (e.g., aluminum chlorohydrate), dried, partially calcined, and milled. The metal complexes form durable "pillars" between clay platelets, maintaining a fixed, well defined separation between them. Because of their size- and shape-selective absorption properties, pillared clays share certain molecular sieve applications with zeolites.

USES

Annual worldwide production of smectite clays is approximately 8.5 million metric tons. The leading producers are the United States with about 40%, followed by the former Soviet Union, Greece, and Japan. The largest worldwide application is well drilling fluids. In the US the major end uses are iron ore pelletizing, drilling fluids, and foundry sand bonding. Other significant applications are oil bleaching, absorbents, and water impedance.

Ore pelletizing – Finely ground taconite ore is bonded into pellets one inch or more in diameter using about 0.5% sodium bentonite. The pellets are then used as blast furnace feed. The bentonite provides both green and dry strength to facilitate forming, drying, and handling of the pellets.

Drilling fluids – Natural sodium bentonite and sodium-exchanged bentonite meeting American Petroleum Institute specifications for moisture, viscosity, water loss, wet screen residue, and yield (barrels of mud per ton of clay) are the most commonly used smectites for freshwater drilling fluids. Their primary functions are to lubricate the drill bit, seal the walls of the hole against fluid loss, and carry cuttings to the surface for removal. Hectorite is used in some instances for its higher efficiency and better heat stability in deep holes. Brine-based muds, or those used to drill through saltwater formations, use palygorskite. Mineral oil or synthetic oil-based drilling fluids, usually water-in-oil emulsions, use organoclays.

Foundry sand – Sodium bentonite, as well as some ion-exchanged and calcium bentonites, are used to provide bonding strength and plasticity to the molding sands used in metal casting. The choice of bentonite is based on a balance of green, dry, and hot strengths. Sodium bentonite generally provides superior dry and hot strength, whereas calcium bentonite provides better green strength.

Oil bleaching – Acid-activated and certain natural smectites are used to decolorize and in some cases to deodorize, dehydrate, neutralize, and filter mineral, vegetable, and animal oils. Products are qualified according to procedures prescribed by the American Oil Chemists Society. The method of preparation determines the suitability of a given clay for a specific oil.

Absorbents – Granular absorbent-grade (partially calcined) calcium bentonite is used as a floor absorbent, particularly for oily materials, and in pet litters. Uncalcined sodium bentonite is used as the agglomerant in scoopable cat litter.

Water impedance – Once swollen by contact with water, a layer of smectite will prevent further passage of water. Smectites, particularly sodium bentonite, are therefore widely used to prevent seepage loss from ponds, ditches, reservoirs, and waste disposal areas, to line and waterproof tunnels and the below-grade walls of residential and commercial buildings, and to seal cracks and fissures in rocks and concrete.

Other uses – Smectite clay is used to clarify water, wine, beer, and other consumable liquids because it will adsorb suspended impurities and facilitate their removal by filtration or flocculation. Hydroclassified sodium smectite is used to thicken and stabilize water-based coatings and cleaning products as well as cosmetic and pharmaceutical emulsions and suspensions. Organoclays are used similarly in solvent- or oil-based products and also to thicken oils into nonmelting greases. Finely ground high purity smectite is used as a binder for animal feed and to provide plasticity and green strength in ceramics.

⌘⌘⌘

TALC

Talc $Mg_3Si_4O_{10}(OH)_2$
Refractive Index: 1.59-1.60
Specific Gravity: 2.75
Mohs Hardness: 1

Commercial talc is composed primarily of the mineral talc, a sheet silicate, but may contain related sheet silicates such as chlorite and serpentine, plus prismatic tremolite, anthophyllite, and carbonates such as magnesite, dolomite, and calcite. Talc particles are characteristically platy in morphology and are oleophilic/hydrophobic; they are wetted by oil instead of water. Talc's reinforcing and pigmenting properties, together with good color, make it desirable as a functional filler in both aqueous and nonaqueous applications. For nonaqueous uses its naturally good matrix compatibility can be further enhanced by surface treatment.

TYPES

Talcs are clasified by mineralogy, morphology, and geographic source. Talc products are processed using various combinations of dry grinding, air separation and flotation depending upon the quality of the crude ore and the properties required for intended applications.

Platy talc – This is distinctly lamellar, soft talc, typically of >90% purity naturally or through beneficiation, which is used in cosmetic, pharmaceutical, and reinforcing filler applications.

Steatite – This originally was a mineralogical name applied to pure talc. Today it refers to high-purity, dense, very fine-grained talc that can be machined. On firing, it has good electrical insulating properties and is used in the manufacture of electrical porcelain.

Soapstone – This is typically a less pure form of steatite that can be carved, sawed, drilled, or machined. Because of its chemical resistance, dense nature, and refractory qualities, it is fabricated into shaped products such as sinks and stoves.

Tremolitic talc – This is fine-grained "hard" talc that usually contains <50% talc but major quantities of hard, prismatic tremolite and fine, platy serpentine. It may also contain a minor amount of prismatic anthophyllite and traces of carbonates and quartz. It lacks the distinctly lamellar, soft, hydrophobic characteristics normally associated with talc and is consequently excluded

⌘⌘⌘

from certain traditional talc applications. Its atypical properties, however, are used to advantage in ceramics and paint.

New York talc – This is generally considered synonymous with tremolitic talc as described above.

Vermont talc – Most crude Vermont talc is characterized by a significant content, 20-30%, of magnesite. The ore is beneficiated to produce products containing various amounts of residual magnesite and occasionally other carbonates, mainly for filler applications. High-purity cosmetic- and pharmaceutical-grade floated talc has been produced in Vermont as well.

Montana talc – Montana talc is known for its naturally high purity and brightness. The principal impurities are small amounts of chlorite, dolomite, and magnesite. The high natural purity makes it possible to produce products for a wide range of uses with little or no beneficiation.

Texas talc – Crude Texas talc is usually gray or black, due to organic materials, and contains quartz and dolomite impurities. It can be processed to filler grades having acceptable brightness, but its principal application is in ceramics. This talc fires to an acceptable color and is suitable for use in fast-fired bodies.

Canadian talc – Although the crude talc in Canada varies considerably in purity and color from one deposit to another, most of the material imported into the US is floated, high-brightness talc for filler and pitch adsorption applications. Some tremolitic talc is available in Canada.

Italian talc – Italian talc has the reputation as being the world's purest. Virtually all the Italian talc imported into the US is a particularly high-purity, soft, cosmetic and pharmaceutical grade.

Chinese talc – The main talc-producing region in China is the Haicheng district of Liaoning Province. Imports into the US are sometimes simply called Haicheng talc, although this includes talc of varying quality from a number of different mines. Chinese talc is known for its high purity and brightness, and filler uses account for most of its consumption in the United States.

USES

The United States, with 1.1 million metric tons of annual output, is the world's second largest producer of talc after China. Other major producing countries are France, Finland, and Australia. Imports into the US from these countries are generally of high-purity platy talc. The single largest use in the US is in ceramics (30-33%), followed by paint, paper, roofing, plastics, cosmetics, and pharmaceuticals.

Ceramics – Talc for ceramic applications is low in iron and carbonates, uniform in chemical composition and fired shrinkage, hard and fine grained, controlled in particle size distribution, and white or near-white firing. Talc's high fusion point and fluxing action enable lower firing temperatures and quicker firing schedules to be used in the production of wall tiles, sanitaryware, vitreous china and cordierite bodies (catalytic converter substrates and electrical insulators). Tremolitic talc is prefered for wall tile production, because bodies can be dry pressed without lamination and because its contribution to high uniform thermal expansion and low moisture expansion prevent crazing of glazed products.

Coatings – The principal use for talc in coatings is as a TiO_2 extender. It also contributes to suspension stability, flatting and sheen control, chemical resistance, leveling, film integrity, and weatherability. Architectural paints generally use -325 mesh talcs, whereas industrial paints use micronized grades. Because it is prismatic and lower in oil absorption, tremolitic talc provides easier dispersion, higher loading levels, less flatting, and better dry hide than does platy talc. It also provides better durability in exterior and traffic paints.

Paper – In the US high-purity, micronized platy talc is the preferred material for pitch adsorption. Its low abrasion and ability to preferentially wet oily materials in the presence of water are unique among mineral alternatives. Pitch adsorbers account for more than 70% of the talc sold to the US paper industry. The balance is high-purity, high-brightness platy talc for paper filling and coating. This talc is used for TiO_2 extension and for improved gloss, opacity, brightness, and ink holdout. A minor use is in the deinking of recycled paper. In Europe, where it can compete more effectively with kaolin, approximately 70% of the talc used by the paper industy is in filler applications, and pitch control accounts for less than 20%.

Roofing – Low-cost, off-color platy talc is used in roofing felts to add weight for lower cost per ft^2, to provide UV protection, to prevent oil penetration and migration, and to increase fire resistance.

Plastics – Platy talc is used for reinforcing and/or filling both thermosets and thermoplastics, although principally the latter. Talc is used in thermoplastics to control melt flow, reduce creep in molded parts, increase molding cycles,

⌘⌘⌘

increase heat deflection temperature, and improve dimensional stability. The single largest use is in polypropylene to increase both stiffness and resistance to high temperature creep. The main requirements for plastics-grade talc are low iron, low moisture, and low abrasion. The color required depends on the finished plastic. Black polypropylene is less demanding than white or pigmented polypropylene used in furniture, appliances and automobiles.

Cosmetics and pharmaceuticals – Only high-brightness platy talc of exceptional purity is used in cosmetics and pharmaceuticals. The talc must have good lubricity or "slip" and acceptable fragrance retention and moisture absorption. It must also adhere to strict limits on acid-soluble substances, loss on ignition, microbial content, chloride, iron, and heavy metals. In cosmetics talc is used in makeup and dusting powders. In pharmaceuticals it is used in tablet manufacture as a filler and lubricant/glidant, and as a reinforcing agent in film coatings.

Other uses – Talc is used for flow control in animal feeds, as a dusting agent for rubber, as an anticaking agent for fertilizers, as a carrier for insecticides, as a wicking preventer in automotive undercoatings, as a filler in carpet and textile backings, as a filler in wallboard joint compounds and grouts, and as a functional filler in adhesives and sealants.

⌘⌘⌘

VERMICULITE

Vermiculite $Mg_{0.3}(Mg,Fe)_3(Al,Si)_4O_{10}(OH)_4 \cdot 8H_2O$

Refractive Index: 1.55-1.58
Specific Gravity: 2.1-2.8
Mohs Hardness: 1.5-2.8

Vermiculite is a platy sheet silicate that is similar in appearance to the mica from which it was altered and similar in ion exchange properties to the trioctahedral smectites which it resembles in structure. Most commercial vermiculite was formed by the alteration of biotite or iron-bearing phlogopite micas. An individual vermiculite particle is composed of an octahedral magnesia layer sandwiched between two tetrahedral silica layers. There can be some substitution of iron and aluminum for magnesium in the octahedral layer. There is always some substitution of aluminum for silicon in the tetrahedral layer, creating a charge imbalance. Vermiculite layers are separated by two oriented water layers, between which magnesium cations reside to balance the charge on the mineral lattice. Because of the separation of platelets by the oriented water, the magnesium ion is not fixed but exchangeable for other cations. This ion exchange capacity is exploited in certain industrial applications. Unlike smectite, vermiculite will not readily incorporate additional layers of water between platelets and will not, therefore, completely delaminate in aqueous systems. Most commercial vermiculite is heat treated to volatilize its interlayer water and produce an expanded (exfoliated) product. Upon heating, the flat, macroscopic "books" of vermiculite flakes expand into elongated, concertina-like particles. Expanded vermiculite posseses absorptive, insulating, and ion exchange properties that serve as the basis for its industrial uses.

TYPES

Because more than 90% of vermiculite used is in the low-density, expanded form, commercial vermiculite is differentiated according to whether or not it has been exfoliated.

Vermiculite concentrate – Since the objective of nearly all vermiculite production is the expanded form, the principal product of vermiculite mining is a vermiculite concentrate. Vermiculite ores typically contain only about 35% vermiculite. The balance may consist of various proportions of mica, chlorite, and other sheet silicates. The ore is beneficiated to vermiculite flake concentrates of at least 90% purity by either dry or wet processes. The former include screening and air separation, while the latter may employ either froth

⌘⌘⌘

flotation, hydroclassification, or heavy media separation as required. Vermiculite concentrates are supplied to regional plants which carry out the thermal exfoliation that results in the expanded products. Vermiculite concentrates are screened and sold under various grade designations based on relatively narrow particle size distributions (e.g., Grade No.1 is 4x8 mesh and Grade No. 2 is 8x16 mesh). Flakes that are less than 60 mesh are too fine for exfoliation and are discarded during beneficiation.

Expanded vermiculite – Expanded vermiculite has a very low bulk density, so exfoliation is carried out in regionally located plants in order to minimize the high cost of transportation to customers per unit weight of product. Vermiculite concentrates are expanded in furnaces by quick exposure (0.25–8 seconds) to 900°C or more. The heat rapidly turns the interlayer water to steam, which pushes the individual vermiculite laminae apart. The vermiculite increases in volume by 8-20 times and decreases in bulk density by 85-90%. The expanded particles are 90% entrapped air by volume and thereby posses excellent insulation properties. The expanded material is separated from nonexpanding impurities by air classification and is either bagged directly or ground to sizes suitable for various applications. If properly exfoliated, expanded vermiculite retains 90-95% of its ion exchange capacity.

Chemically exfoliated vermiculite – Vermiculite can be chemically delaminated by treating an aqueous dispersion alternately with a brine solution and an n-butylammonium solution. The mineral's exchangeable magnesium is sequentially replaced by sodium and the alkylammonium cation. This increases the interlayer separation sufficiently to allow the osmotic entry of water. The mineral thus expands to form an aqueous gel. This gel is used to coat glass fibers and fiberglass fabrics, thereby doubling their effective working temperature.

USES

In a world market of approximately 600,000 metric tons, the United States is second to South Africa in vermiculite production, with 30% vs 40%, but leads in consumption by far with 50%. The US market for expanded vermiculite is divided one-third each between agricultural products and ready-mix plaster and cement premixes. The balance is used in insulation products of all types and as a lightweight concrete aggregate.

Agricultural products – In decreasing order of volume consumed, the major agricultural uses of vermiculite are as a fertilizer carrier, in horticultural products, and as a soil conditioner. Expanded vermiculite will absorb fertilizer and trace nutrient liquids, and the resulting free-flowing particles provide

⌘⌘⌘

controlled, slow release of these liquids to the soil. The low density, high absorption, and ion exchange properties of expanded vermiculite find use in potting, nursery, and farm soils. Exfoliated vermiculite provides a permeable soil structure having good aeration and water retention. It also controls the release of nutrients to the plant, and in nurseries it ensures a lightweight soil that facilitates the transplanting of large nursery stock. Expanded vermiculite is also used as a growing substrate in hydroponics, as a carrier of liquid additives, binder, anticaking agent, and digestive aid in animal feeds, and as an additive to manures to decrease their odor and improve their fertilizer value.

Plaster and cement premixes – Expanded vermiculite is used in ready-mix plasters and cements to impart low density, high thermal insulation, and low thermal conductivity. These mixes are sprayed on as lightweight, fire-resistant coatings for structural steel and concrete. The laminar nature of the mineral also protects against cracking and spalling of these coatings when subjected to temperature extremes and mechanical shock. Vermiculite is also used in fire-protection boards at loading levels of up to 90%. These boards are used as cladding for structural steel and in residential and commercial buildings as more aesthetic alternatives to sprays.

Insulation – Vermiculite is used in thermal and acoustical insulation boards (e.g., ceiling tiles), lightweight insulating blocks, refractory blocks, and fire bricks. Coarse expanded vermiculite is used as loosefill insulation in interior building walls and, with a silicone coating to minimize moisture absorption, in exterior walls.

Lightweight aggregate – Coarse, up to 8mm, expanded vermiculite is used as an aggregate in ready-mix and preformed concrete to reduce weight and improve insulation properties. Of the alternatives (including pumice, pumicite, and perlite), vermiculite provides the lightest weight but physically weakest concrete.

Other uses–Expanded vermiculite is used as an absorbent for industrial spill containment and cleanup, in ground form as an alternative to asbestos for brake linings, in refractory mold releases, and as a water filtration medium in aquaculture. Unexpanded vermiculite is used to cover molten metal prior to casting, where the heat of the metal expands the mineral into an insulating layer.

⌘⌘⌘

WOLLASTONITE

Wollastonite $CaSiO_3$
Refractive Index: 1.62-1.65
Specific Gravity: 2.92
Mohs Hardness: 4.5

Commercial grades of wollastonite are typically high in purity because most ores must be beneficiated by wet processsing, high-intensity magnetic separation, and/or heavy media separation to remove accessory minerals. The minerals most commonly found associated with wollastonite are calcite (calcium carbonate), diopside (calcium magnesium silicate), and garnet (calcium aluminum silicate). Wollastonite is hard, white, and alkaline (pH 9.8). It is exploited for its chemistry as a source of CaO and SiO_2, and its low ignition loss, low oil absorption, very low moisture absorption, and acicular morphology. For polymer applications matrix compatibility is improved by surface treatment.

TYPES

The natural morphology of wollastonite is acicular because its crystals grow longer than they are wide. This gives wollastonite particles a needle-like shape. The uses for commercial products are dictated by the length of individual needles in the ore and the extent to which this shape is preserved during grinding of the finished products. The acicular nature of wollastonite is measured in terms of aspect ratio – the ratio of particle length to width.

Wollastonite powder – This is milled wollastonite that has a low aspect ratio (3:1 - 5:1). It is produced from naturally low-aspect ratio ores and from high-aspect ratio ores that have been ground in a way that breaks the needles widthwise. Despite their relatively low aspect ratios, powder grades can provide better reinforcing properties than prismatic and even platy minerals in some applications.

Acicular wollastonite – This is produced from ore containing a suitably high percentage of long needles. The ore is milled and air separated in such a way that very fine, needle-like particles are preserved and recovered. Acicular grades typically have aspect ratios of 15:1 to 20:1.

Surface-treated wollastonite – Both powder and acicular forms of wollastonite are readily available with silane, organosilicone, and titanate treatments.

⌘⌘⌘

USES

The United States is the world's leading supplier of wollastonite, with 34% of total annual production of approximately 365,000 metric tons. China is a close second at 33%. The single largest application is in ceramics, which consumes 40-45% of production. Significant uses also are found in plastics, asbestos substitution, metallurgical powders, and paint.

Ceramics – Wall tile is the principal ceramic application for wollastonite, which promotes low shrinkage, low warpage, good strength, and permeability for fast firing with minimal gas evolution. Grades with low ignition loss, high fired brightness, and low coloring oxides of Fe, Ti, and Mn are used. In glazes wollastonite is used as a source of calcium silicate in place of limestone and flint, thereby reducing volatiles and increasing gloss. Wollastonite is also used in small amounts in semivitreous and vitreous bodies to reduce shrinkage and increase strength.

Plastics – Wollastonite is used as a reinforcing filler in plastics because of its low oil and moisture absorption, high brightness and acicularity, and availability with a variety of surface treatments. Acicular wollastonite is a lower cost alternative to short-milled glass fibers for both thermoplastics and thermosets, most notably in BMC polyester and RRIM polyurethane for automotive applications. It is likewise used as an asbestos substitute in phenolic molding compounds and as a reinforcing filler in Nylon 6 and 66. Micronized (-10 micron) wollastonite is used in high-impact and platable nylon compounds. Acid treatment of the plastic surface dissolves the wollastonite and provides a uniform anchor pattern for strong, smooth plating.

Asbestos substitution – High-aspect ratio wollastonite is used in high-temperature insulation board, wallboard, shaped insulation, and roofing tile as a substitute for short-fiber asbestos. It offers thermal stability, reinforcement, good acoustical properties, and relatively light weight in these applications. It is used together with organic and metallic fibers in asbestos-free formulations for friction products such as clutches, brake linings, and brake pistons.

Metallurgical powders – Low ignition loss, low-phosphorous (<0.01%), low-sulfur (<0.01%) wollastonite powder is used as a low-temperature flux in welding and steel casting formulations. In composite flux powders it helps maintain the surface flow of molten steel as it is poured from ladle to tundish, thereby minimizing surface defects.

Coatings – Powder grades are used in coatings as an extender pigment and to provide resistance to flash and early rust. Low oil and water absorption allow

⌘⌘⌘

high loading levels. The acicular nature, even at low aspect ratios, reinforces paint films, providing durability and superior scrub resistance. Finely ground and micronized grades are also used in epoxy powder coatings because they promote smooth flow, water resistance, improved wet adhesion, and good gloss.

Other uses – Wollastonite is used as an energy-conserving alternative to limestone and sand in the production of glass and glass fiber and as a white reinforcing filler in adhesives and sealants, where it can be used at high loadings.

⌘⌘⌘

ZEOLITE

	Channel Size (Angstroms)
Chabazite	3.7 x 4.2
Clinoptilolite	3.9 x 5.4
Erionite	3.6 x 5.2
Faujusite	7.4
Mordenite	2.9 x 5.7
Phillipsite	4.2 x 4.4
	2.8 x 4.8
Type A	4.2
Type X	7.4
Type Y	7.4
ZSM-5	5.4 x 5.6
	5.1 x 5.5

Zeolites are framework silicates based on a three dimensional network of SiO_4 and AlO_4 tetrahedra linked by shared oxygens. Since two aluminum atoms cannot share the same oxygen atom (Lowenstein's Rule), a zeolite unit cell contains at least as many, and usually more, silicon atoms as aluminum atoms. The presence of structural aluminum imparts a negative charge to the zeolite lattice. This is balanced in nature by alkali and alkaline earth cations. Synthetic zeolites are usually formed as the sodium-exchanged species and are subsequently converted to a hydrogen-, alkali-, alkaline earth-, or rare earth-exchanged product as required. The mineral framework contains openings and internal voids or channels of fixed dimensions characteristic of individual varieties. These internal channels are occupied by water that can be reversibly removed, leaving a microporous structure with up to 50% void volume. In general, a higher framework Si:Al ratio results in decreased ion exchange capacity and increased acidity, thermal stability, and hydrophobicity.

TYPES

Zeolites are fundamentally differentiated as being either natural or synthetic. Further classification is based on those specific properties that dictate commercial suitability for the general applications of catalysis, molecular sieving, ion exchange, and adsorption.

Natural zeolites – More than three dozen natural zeolite varieties have been identified, but only clinoptilolite, chabazite, and mordenite are commercially exploited to any significant extent. Natural zeolites are typically crushed and screened or ground and air classified to the required particle size, and some are

⌘⌘⌘

offered in pelletized form. They are available in both hydrated and activated (dehydrated) forms and may be enhanced in H^+ or Na^+ content by washing with acid or NaCl, respectively.

Synthetic zeolites – More than 100 zeolites have been synthesized, but the principal commercial synthetics are Types A, X, Y, and ZSM-5. Type A has a unique structure based on sodalite cages and has a 1:1 Si:Al ratio to maximize ion exchange capacity. It is sold in a sodium form for ion exchange applications or in a calcium form for adsorption applications, since the larger sodium ion restricts internal access. Types X and Y are isostructural with natural faujasite but are tailored in their Si:Al ratios and are exchanged with rare earth cations as required for major catalytic applications. ZSM-5 is based on a unique pentasil structure that allows very high Si:Al ratios. This minimizes ion exchange and hydrophilicity. Despite its low ion exchange capacity, the hydrogen form of ZSM-5 is treated to develop higher catalytic activity than is exhibited by other common synthetics. Its channels are too small for use in petroleum cracking, but they are well suited for catalysis of certain small cyclic hydrocarbons.

USES

Japan is the leading producer of natural zeolites, followed by the former Soviet Union. Of the estimated 1 million metric tons of natural zeolites produced annually worldwide, the United States accounts for only about 2%. The leading producers of synthetic zeolites are the United States, Germany, and Japan. Although the technical properties and performance of natural zeolites and their synthetic counterparts overlap appreciably, in practice their applications are largely different. The higher uniformity, purity, and price of synthetics reserves for them most uses that justify these attributes. The principal uses for natural zeolites are ammonia removal from organic wastes, removal of radioactive cesium^{137} and strontium^{90} from nuclear wastes, odor control, and soil treatment. Synthetic zeolites are used in catalytic cracking at petroleum refineries, for drying, purifying, and separating industrial gas streams, and as detergent builders.

Aquaculture – Zeolites, principally clinoptilolite, are used to remove ammonia from water at fish hatcheries and farms, in aquaria, and in live fish transport systems. This prevents the eutrophication of such systems and thereby enhances fish mortality and growth.

Agricultural products – Zeolites, principally clinoptilolite, are used as vehicles for the controlled release of ammonia in fertilizers and as carriers of insecticides, herbicides, and fungicides. They also are added to contaminated

⌘⌘⌘

soils to scavenge radioactive Cs^{137} and Sr^{90}, plus Pb, Cd, and other toxic heavy metals. Zeolites are used as animal feed supplements to improve nutrient efficiency and growth rates for poultry and swine while reducing manure moisture and odor. Drier, less odoriferous manure can be used as a fertilizer that has slow release properties. Natural zeolites also are used as litter for horse stalls, pig pens, poultry houses, and household pets. Here they control ammonia odors and absorb liquid wastes.

Waste treatment – Clinoptilolite, chabazite, mordenite, and phillipsite remove radioactive Cs^{137} and Sr^{90} from nuclear waste streams and serve as encapsulants for these isotopes to facilitate solid waste disposal. Zeolites, particularly clinoptilolite, are used in treating sewage and industrial waste streams to remove ammonia and heavy metals. Extracted ammonia is subsequently vented, recovered for chemical use, or converted with sulfuric acid to ammonium sulfate fertilizer.

Gas adsorption – Zeolites with appropriate channel dimensions are used for the selective removal of cetain gases from gaseous mixtures. Mordenite and calcium Type A are preferred for adsorbing nitrogen from the air to generate relatively pure oxygen for medical and industrial uses. The latter include oxygenation of pulp and paper mill effluents, waste and sewage treatment streams, and metal smelters. Natural gas and methane generated by animal waste, sanitary landfills, and sewage systems are purified by treatment with synthetic zeolites, principally Type A, to remove H_2O, CO_2, SO_2, and H_2S. Clinoptilolite and mordenite can be used to remove SO_2 from the stack gases at fossil fuel-burning power plants.

Detergents – The sodium form of Type A is used in powdered detergents as a primary sequestrant for calcium. It also removes magnesium, plus the iron and manganese that can cause stains. Its high adsorptivity allows the incorporation of high levels of liquid surfactants, thus enabling the production of highly concentrated, free-flowing products.

Catalysts – The primary application for zeolites in catalysis is for specially stabilized Type Y in petroleum cracking, whereby crude oil is broken down into gasoline and fuel oils. The zeolite content of FCC (fluid cracking catalysts) ranges from 5% to 50% . The ultrastable Type Y zeolite also is used to improve the octane rating of gasoline. ZSM-5 zeolites are used primarily to produce gasoline from methanol by both catalysis and molecular sieving. Methanol is first converted to dimethyl ether, which is further converted by the zeolite to a mixture of hydrocarbons. Organic molecules too large to be suitable for gasoline are trapped within the zeolite structure until they are

⌘⌘⌘

broken down enough to escape. The size- and shape-selectivity of ZSM-5 is likewise used in chemical synthesis for the conversion of toluene to benzene and *p*-xylene. The shape of any *m*- and *o*-xylene initially formed impedes their escape from the zeolite channels. Most remain "trapped" long enough for conversion to the *p* form.

Other uses – Zeolites find additional uses as desiccants, heat storage media, filters for air cleaning, and (in Japan) as paper fillers.

⌘⌘⌘

BIBLIOGRAPHY

Bureau of Mines, 1991, *Minerals Yearbook Volume 1, Metals and Minerals*, United States Department of the Interior, Washington

Bureau of Mines, 1992, *Minerals Yearbook Volume 1, Metals and Minerals*, United States Department of the Interior, Washington

Bureau of Mines, 1993, *Minerals Yearbook Volume 1, Metals and Minerals*, United States Department of the Interior, Washington

Carr, D.D. (ed.), 1994, *Industrial Minerals and Rocks*, 6th Ed., Society for Mining, Metallurgy, and Exploration, Inc., Littleton, CO

Deer, W.A., Howie, R.A., Zussman, J., 1978, *Rock-Forming Minerals*, Vols. 1-5, Wiley, New York

Grim, R.E., 1953, *Clay Mineralogy*, McGraw-Hill, New York

Harben, P.W., Bates, R.L., 1990, *Geology and World Deposits*, Metal Bulletin Plc, London

Hurlbut, C.S., Klein, C., 1977, *Manual of Mineralogy*, 19th Ed., Wiley, New York

Industrial Minerals Magazine, 1985-1995, Metal Bulletin Plc, London

Lewis, P.A. (ed.), 1988, *Pigment Handbook*, Wiley, New York

Milewski, J.V., Katz, H.S., 1987, *Handbook of Fillers for Plastics*, Van Nostrand Reinhold, New York

Milewski, J.V., Katz, H.S., 1987, *Handbook of Reinforcements for Plastics*, Van Nostrand Reinhold, New York

Weiss, N.L. (ed.), 1985, *SME Mineral Processing Handbook*, Society of Mining Engineers, New York

⌘⌘⌘

THREE

MINERAL SURFACE MODIFICATION

Charles Kuhn
Surface Modification Consultant
Hot Springs, AR

Peter A. Ciullo
R.T. Vanderbilt Company, Inc.
Norwalk, CT

Mineral particles are compounded into polymers to constructively alter composite physical properties and to improve economy. Minerals that perform primarily the former function are generically classed as reinforcements, while those that serve the latter are traditionally called fillers. The term *reinforcement* generally implies anisometric particle shape – platy, acicular or fibrous – which allows physical and thermal stress to be transferred to the high tensile mineral. A *filler* was once assumed to be an inexpensive particulate intended to take up space in a polymer matrix, thereby extending the "expensive" polymer without seriously degrading physical properties. The term filler, in this sense, is today nearly obsolete. Most of these materials can now be formulated as nondegrading and often as beneficial to composite properties. This evolution in mineral functionality has been sustained by the use of coupling agents.

Successful compounders of plastics, elastomers, adhesives, sealants, and coatings are skilled in developing the most cost effective mineral-polymer composite that will meet performance requirements and retain critical properties under all expected environmental conditions. This is not an easy job. The interface between mineral and polymer is often one of chemical incompatibility, making difficult even the initial uniform dispersion of mineral particles into liquid or molten polymer. The typically hydrophilic mineral is not easily deagglomerated and wetted by the organic matrix. The mineral particles can be tightly encapsulated within the matrix, but the interface is a collection site for adsorbed water and an impediment to stress transfer. Under a tensile force, the mineral will separate from the matrix. Coupling agents are the means by which mineral and matrix are made compatible and durably bonded.

The concept of bonding an inorganic filler to an organic matrix is not new, having started with fiberglass surface modification more than fifty years ago. Today, glass fiber reinforcement is routinely surface modified to improve dispersion properties, matrix compatibility, composite performance properties, and resistance to deterioration of these properties under long term use conditions. The past two decades have brought a gradual appreciation of the benefits of modifying mineral fillers to similarly enhance their functionality. Acceptance by compounders was initially slowed by the perceived high cost of surface modification, discounting the improvements obtained in mineral-matrix properties. An understanding of the value of surface modified minerals evolved from increased market competition, greater technical sophistication, and new laws and regulations concerning air quality. The paint industry, for example, has been forced to reformulate due to laws limiting the permissible Volatile Organic Content (VOC) that can be released into the atmosphere from each gallon of paint. One way to comply is with high solids coatings. A surface modified filler allows higher loading with little or no viscosity change. The plastics industry has used small quantities of surface modified minerals for years, but now likewise relies on coupled composites for automotive, aerospace, and appliance applications where once only steel was considered.

Modification vs. Treatment

There is sometimes confusion over what constitutes surface modification as opposed to surface treatment. Here, surface modification will mean the durable attachment of a coupling agent to the surface of a mineral particle and to its surrounding organic matrix. The coupling agent cannot be removed from the filler surface by hydrolysis, solvolysis, extraction, or any other mechanism to which the filler-matrix composite may be exposed in use. The coupling agent will hydrophobize the filler surface by exposing organic groups for interaction with the matrix. This interaction establishes a durable coupling agent-polymer bond through chemical reaction or chain entanglement. Surface treatment, on the other hand, is the association of a processing aid with the mineral's surface only. Surface treatments may or may not bond to the filler surface and lack the ability to bond to the organic matrix. Most surface treatments are analogous to the wetting agents used in aqueous systems. One part of the treating agent molecule has an affinity for the mineral surface and the other for the organic medium.

Considerable volumes of natural and precipitated calcium carbonate fillers, for example, are sold with stearic acid treatment. The fatty acid is ionically bound as calcium stearate with the exposed C_{18} chains promoting good wetting of the filler by organics. At best, molecular entanglement of stearate chains and polymer chains may occur in polyolefins. There are certain

⌘⌘⌘

organosilanes and organotitanates, which are otherwise classed as coupling agents, that will bond to filler surfaces but only be wet by, not bound to organic matrices.

Surface treatments facilitate deagglomeration and uniform dispersion of the mineral particles in a liquid organic phase or molten polymer. They thereby reduce viscosity of the blend and allow higher filler loadings. Surface modifiers serve similar processing functions, but also improve impact, tensile, flexural, and dielectric properties in plastics and elastomers; film integrity, substrate adhesion, weathering and service life in coatings; and application and tooling properties, substrate adhesion, cohesive strength, and service life in adhesives and sealants.

There are two common processes whereby coupling agents are applied. One is pretreatment of the mineral and the other is integral addition or the *in situ* method. Pretreatment is the attachment of the coupling agent to the mineral before the mineral is added to the resin system. The advantages of this type of surface modification are reduced compounding time due to faster dispersion of the mineral, and improved physical properties of the finished product. Pretreated fillers also offer the compounder convenience and the assurance that the coupling agent is uniformly distributed on the filler surface. With *in situ* modification, the coupling agent and mineral are added concurrently or consecutively to the resin before other additives are introduced. This method may require up to twice the amount of coupling agent because of its tendency to attach to any available receptive site. The *in situ* method does not generally produce the physical improvements that can be obtained through pretreatment. It can also be more expensive due to the higher amount of coupling agent required. Other disadvantages are the need to remove volatile byproducts from the coupling reaction, prolonged mixing requirements, and the need for other additives to maximize coupling agent performance.

The Coupling Agents

There are several types of coupling agents commercially available, including silanes, titanates, zirconates, zircoaluminates, carboxylated polypropylenes, chromates, chlorinated paraffins, organosilicon chemicals, and reactive cellulosics. Of these, the organosilanes and titanates are the most widely used for modifying mineral surfaces. The other surface modifiers are more narrowly used where they offer specific advantages in handling or performance.

Silanes – The general chemical structure of organosilanes is $RSiX_3$, where X is a hydrolyzable group, such as methoxy, ethoxy, acetoxy, or chloride, and R is a nonhydrolyzable organofunctional group. Most commercial silanes are alkoxy derivates, $RSi(OR')_3$, the most common of which follow.

⌘⌘⌘

Common Commercial Organosilanes

Silane Type	Chemical Name
Amino	Aminopropyltriethoxysilane
Diamino	N-(2-aminoethyl)-3-amino-propyltrimethoxysilane
Methacrylate	3-Methacryloxypropyltrimethoxysilane
Epoxy	3-Glycidoxypropyltrimethoxysilane
Mercapto	3-Mercaptopropyltrimethoxysilane
Vinyl	Vinyltrimethoxysilane Vinyltriethoxysilane
Chloro	3-Chloropropyltriethoxysilane 3-Chloropropyltrimethoxysilane
Alkyl	Methyltrimethoxysilane Methyltriethoxysilane Propyltrimethoxysilane Isobutyltrimethoxysilane Octyltriethoxysilane
Styrylamine	3[2-Vinyl(benxylamino)ethylamino]propyltrimethoxysilane
Silazane	Hexamethyldisilazane
Triamino	Aminoethylaminoethylaminopropyltrimethoxysilane
Phenyl	Phenyltrimethoxysilane

Hydrolysis of an alkoxysilane forms silanetriol and alcohol.

$$RSi(OR')_3 + 2H_2O \rightarrow RSi(OH)_3 + 3R'OH$$

The rate of hydrolysis depends on the nature of the hydrolyzable group. The most rapid hydrolysis occurrs with -Cl, followed by $-OOCCH_3$, $-OCH_3$, $-C_2H_5$, and $-OC_3H_7$. The formation of the respective acids, hydrochloric and acetic, or alcohols as a byproduct of silane hydrolysis necessitates proper venting in use.

The silanetriol will slowly condense to form oligomers and siloxane polymers.

⌘⌘⌘

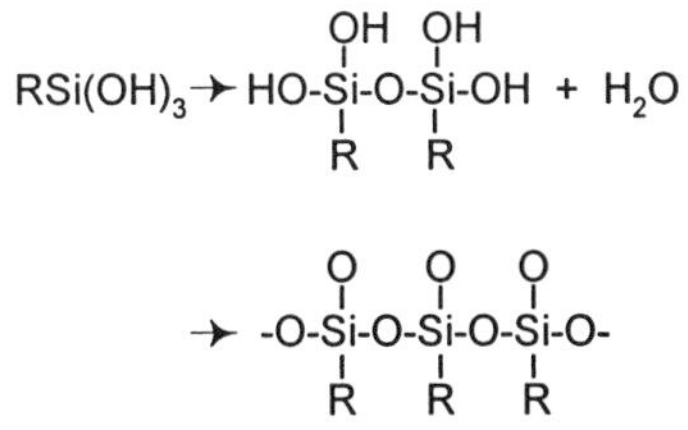

$$RSi(OH)_3 \rightarrow HO\text{-}\underset{R}{\overset{OH}{Si}}\text{-}O\text{-}\underset{R}{\overset{OH}{Si}}\text{-}OH + H_2O$$

$$\rightarrow \text{-}O\text{-}\underset{R}{\overset{O}{Si}}\text{-}O\text{-}\underset{R}{\overset{O}{Si}}\text{-}O\text{-}\underset{R}{\overset{O}{Si}}\text{-}O\text{-}$$

The -Si-OH groups of the silanetriol or oligomer are believed to initially hydrogen bond with -OH groups on the mineral surface. These -OH groups are from aluminols (e.g., on aluminum trihydrate) or silanols. Since the silicas and silicates that are successfuly silane modified generally lack structural silanols, these are ascribed to silanols formed by the reaction of silica surfaces and adsorbed water. This would provide silanol groups in sufficient numbers and distribution to enable uniform silane coating.

The reaction of silanetriol or oligomer with mineral surface -OH may ultimately result in the condensation of siloxane polymer, essentially encapsulating the mineral particle. Silane treatment levels, typically 0.5 to 2.0% on mineral weight, are usually adjusted according to particle size and surface area in order to form a monomolecular layer. This most cost-effectively provides the desired organofunctional surface. If excess silane is present, polymerization results in several molecular layers.

Most commercial silanes are low viscosity liquids with a long shelf life as long as they are kept free of moisture. Exposure to water causes hydrolysis and eventual polymerization, at which point usefulness as a mineral treatment is effectively lost. The exception is aminosilanes, which hydrolyze in water but remain soluble and resistant to polymerization.

Coating or encapsulation of the mineral particle creates a hydrophobic surface from the organofunctional group on the silane. This can range from nonreactive alkyl groups, to reactive epoxy, mercapto, and vinyl groups, among others. The choice of silane organofunctionality is guided by chemical compatibility of silane and matrix, treatment method and conditions, and the particular composite properties targeted for improvement. General recommendations are:

Silane Organofunctional Group	Polymer
vinyl, methacrylate	peroxide cured thermoset polyesters, peroxide cured elastomers, crosslinked polyethylene
methacrylate	polyethylene, polypropylene, polyacrylate

⌘⌘⌘

epoxy	epoxies, phenolics, nylon, thermoplastic polyester
mercapto, polysulfide	sulfur cured elastomers
amino	epoxies, phenolics, urethanes, nylon, thermoplastic polyester, polycarbonate, sulfur cured elastomers

Silane coupling agents were first developed for thermoset composites, with organofunctional groups chosen to participate in the crosslinking reaction – amino for melamine or vinyl for unsaturated polyester, for example. In thermoplastics designed to be unreactive under processing conditions, predicting the most appropriate organofunctional group is more complicated. Azidosilanes were once sold as general purpose coupling agents with the ability to bond to most polymers, so that matching organofunctionality to polymer chemistry was simplified. The general formula of azidosilanes is $X_3SiRSO_2N_3$. The hydrolyzable groups react with siliceous fillers at room temperature. The sulfonyl azide group is stable at room temperature, but decomposes rapidly with heat (e.g., in melt compounding) to form a nitrene and release nitrogen gas. The highly reactive nitrene inserts into the -C-H bonds of the polymer matrix. Azidosilanes were recommended in particular for use in polyolefins, but were withdrawn from the market in the 1980s because of health and safety issues.

Silane-mineral coupling can be performed *in situ* but usually with loss of efficiency, as noted above. The compounder must also make provision for collection of reaction byproducts – water and alcohol or acid. The general recommendation for *in situ* coupling is to add the silane and mineral to the polymer simultaneously with intense mixing (e.g., compounding extruder for thermoplastics, Banbury for elastomers) and before any other additives are introduced. The effectiveness of this method depends on how intimately the three components are mixed, the rate of silane hydrolysis, and how successfully reaction byproducts are prevented from being incorporated into the blend. For successful coupling, the silane must migrate to the mineral surface, hydrolyze, and condense. Hydrolysis and condensation can be catalytically accelerated. At least one supplier offers a product for thermosets consisting of silane and a metallic acid ester catalyst for improved *in situ* coupling.

Pretreatment of the mineral is generally preferred for optimized performance. The conventionally recommended technique is to prepare a solution of 10% to 25% silane in a 90% alcohol/10% water diluent. This is adjusted to pH 4 to 5 with acetic acid, which acts as a hydrolysis catalyst. This composition will facilitate hydrolysis to the silanetriol, while inhibiting

⌘⌘⌘

polymerization for a number of hours. This dilute silanetriol solution is then sprayed onto the well agitated mineral particles to ensure complete and uniform coating. After intimate mixing is complete, the moist mineral is dried under mild forced air conditions to accelerate silane-mineral reaction and polysiloxane formation while evaporating diluent and byproduct volatiles.

When generation of organic volatiles must be minimized, undiluted silane is atomized under dry conditions onto warmed (40° to 70°C) and air fluidized mineral particles. Adsorbed water on the mineral surface is sufficient to hydrolyze the silane. The rate of hydrolysis and silane-mineral reaction are dependent upon the mineral's water content, surface pH, surface temperature, the nature of the hydrolyzable group on the silane, and processing time. Rapid reaction will allow no or minimal post-drying, as long as volatiles are extracted during processing. However, incomplete hydrolysis may cause evolution of volatiles after processing and even after packaging. Aqueous organosilane solutions are now available that have been prehydrolyzed, stabilized against premature polymerization, and stripped of byproduct alcohol so that the user does not have to deal with volatile organics.

Not all minerals are amenable to performance enhancement by silane coupling. On a qualitative scale, the common fillers may be accordingly ranked as follows:

Silane Coupling Effect	Filler
Excellent to Good	Synthetic silicas
	Natural silicas
	Wollastonite
Good to Fair	Mica
	Alumina trihydrate
	Kaolin clays
	Talc
	Barite
	Titanium dioxide
	Iron oxides
Slight to None	Asbestos
	Calcium carbonate
	Carbon black

Titanates – The general chemical structure of organotitanates is $XTiR_3$, where X is a hydrolyzable group and R represents the nonhydrolyzable organofunctional groups. The first commercial titanate coupling agents were monoalkoxy isopropyl and methoxydiglycol derivatives. These bond in theory to filler surfaces by way of a silane-like mechanism of reaction with surface

⌘⌘⌘

protons (-H) to form an ether link (-O-) and liberate alcohol or glycol. They have just one hyrolyzable group, as opposed to a silane's three. This facilitates formation of a monomolecular layer on the substrate and precludes polymerization.

From this start, several approaches have been taken to customizing both the substrate-functional and organofunctional portions of the titanate molecule. This has adapted titanate functionality to a wide range of substrates, a broad cross-section of organic matrices, and a variety of liquid vehicles. Today, titanate coupling agents are available based on the following six generalized structures.

$XTiR_3$

Monoalkoxy

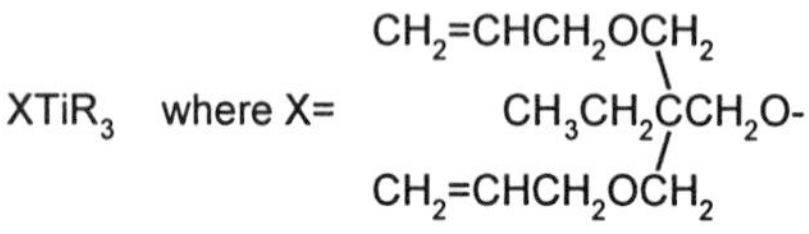

$XTiR_3$ where X= $(CH_2{=}CHCH_2OCH_2)_2C(CH_2CH_3)CH_2O$-

Neoalkoxy

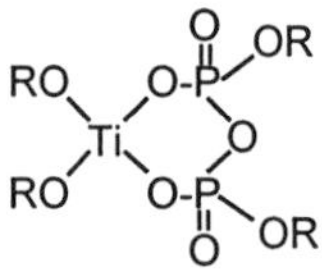

$(RO)_2Ti[OP(O)(OR)O]_2$

Cycloheteroatom

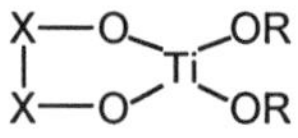

$(XO)_2Ti(OR)_2$

Chelate

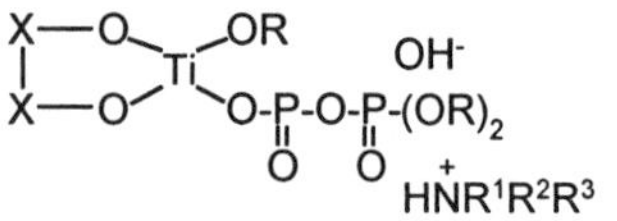

$(XO)_2Ti(OR)[OP(O)OP(O)(OR)_2]$ OH^- $H\overset{+}{N}R^1R^2R^3$

Quat Salt

$X_4Ti{\cdot}(HP(O)OR)_2$

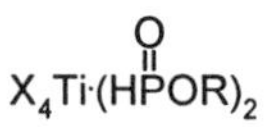

Coordinate

The neoalkoxy and heteroatom types offer successively greater thermal stability for *in situ* addition to molten polymers. The chelates provide better composite stability in wet environments, while the quaternary salt varieties are water soluble. The coordinate type incorporates phosphite functionality.

The two or three organofunctional groups on the various titanate types are not necessarily all the same for a given product. They may include one or all three of the following functionalities in various combinations:

(1.) Reactive groups, typically amino, acrylic or methacrylic, provide polymer-titanate bonding.

⌘⌘⌘

(2.) Long aliphatic and aromatic chains are incorporated to provide polymer compatibility and anchoring through chain entanglement.

(3.) Alkylate, carboxyl, sulfonyl, phenolic, phosphate, pyrophosphate, and phosphite groups are used to tailor the properties of the filler-matrix composite.

The titanates are most often used as *in situ* coupling agents dissolved in a suitable vehicle. Because of the breadth of titanate chemistry commercially available, a water, solvent or oil based diluent can usually be conveniently found. For example, titanate can be added to an elastomer in the process oil before filler addition. The matrix itself can also serve as diluent, as in the case of polyolefins where a neoalkoxytitanate can be blended into the melt flux prior to filler addition.

If the titanate cannot be diluted with a matrix-compatible liquid, the filler can be pretreated by atomizing undiluted titanate into a fluidized bed of filler particles. The neoalkoxy titanates are available in dry powder form as >65% concentrates on an inert carrier. These can be used to pretreat filler by intimate dry blending, although substrate coupling actually occurs *in situ* during compounding.

Since the organotitanates will form a monomolecular layer on filler surfaces, use levels are acccordingly adjusted based on particle size, surface area, oil absorption, and porosity. Typical treatment levels range from 0.5 to 3% based on these factors. Excess titanate can in some cases adversely affect composite properties. With the many variations in titanate chemistry available, substrate receptivity is claimed for a multitude of mineral, metal, glass, ceramic, and even organic surfaces. For coupling to a metallic substrate, the Ti-O-M bond has greater strength and hydrolytic stability than the Si-O-M bond, while the Si-O-Si bond is stronger than Ti-O-Si for coupling to silica surfaces. Titanate coupling has been particularly recommended for those fillers poorly suited to silane modification. The titanate's affinity for varied surfaces has also prompted the general recommendation that all handling and processing equipment be washed with a dilute titanate solution prior to use. This prevents equipment surfaces from appropriating titanate meant for filler-matrix coupling.

Other Coupling Agents – The zirconates, zircoaluminates, and carboxylated polypropylenes are among the other coupling agents used with minerals for specific applications.

Zirconates – Zirconate coupling agents are the direct analogues of the neoalkoxy titanates. They are used to avoid certain specific drawbacks of their titanate counterparts. In thermosets and coatings using peroxide-based cures titanium acts as an oxidizing agent. Peroxide free radicals are eliminated and cure efficiency drops as Ti^{4+} is reduced to Ti^{3+} ($ROOTi^{3+}$) through reaction with the peroxy radical. Tetravalent zirconium is an activator for peroxide and can accelerate peroxide and air-based cures. Titanates can cause discoloration in contact with phenolic antioxidants and chain stops. Zirconates generally are not color body producers in contact with phenols other than nitrophenol. The organozirconates are employed as *in situ* or pretreatment coupling agents using the same methods practiced with the corresponding titanates.

Zircoaluminates – The general chemical structure of the zircoaluminates is

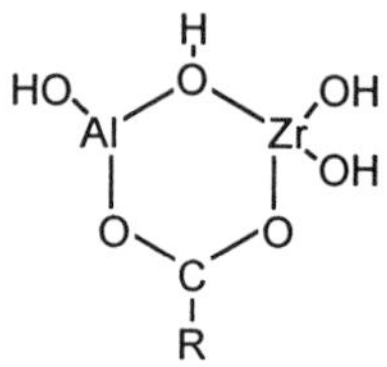

where R is a nonhydrolyzable amine, carboxy, methacryloxy, mercapto, or mercaptoamine organofunctional group. Solvent solutions at less than 30% concentration are stable and resistant to polymerization. The zircolaluminates are usuallly used for *in situ* coupling at 1.5% or more based on polymer weight. A pendant hydroxyl group on each metal atom reacts with hydroxyl groups on mineral surfaces through a combination of hydrogen bonding and covalent bonding with loss of byproduct water. The zircoaluminates are used *in situ* in nylons, polyolefins and thermosets to allow higher filler loadings, lower melt viscosities, high temperature extrusions, and significant impact improvement without loss of tensile or flexural strength. They are also used *in situ* or as a filler pretreatment in elastomers for physical properties improvement.

Carboxylated Polypropylene – Carboxylated polypropylenes are used primarily in polypropylene composites as alternatives to silanes. They are made by grafting an unsaturated carboxylic compound, usually maleic anhydride or acrylic acid, onto the polypropylene backbone. The available grades vary in melt index and the percentage of grafted substituent. The maleated or acrylated polymer is melt compounded with filler and polypropylene resin at levels of 5% to 100% on resin weight, depending upon the properties desired. The assumed coupling mechanism is carboxylate-filler bonding with polypropylene chains extending into the resin. There anchoring can take place through chain entanglement and possibly cocrystallization. Carboxylated polypropylenes also have been used with aminosilane modified fillers for further composite property optimization. The carboxyl groups of the carboxylated polymer can react with the amine group on the silane during melt compounding to form amide linkages.

Modified Mineral Benefits

Surface modification offers many benefits in optimizing mineral-polymer composites. Among these are:

Improved processability – Surface modified minerals deagglomerate easily and disperse well to maximize mineral-matrix contact while promoting lower viscosity. This reduces compounding energy requirements and cost. Table 1, for example, shows the dramatic viscosity change for various fillers in polyester when pretreated with a methacrylate functional silane.

Table 1. Viscosity Reduction in Polyester Resin From Surface Modification of Fillers

	Coupling Agent	
	None	A-174*
Silica, 10 micron	52000 cps	14000 cps
Calcined Kaolin Clay	56000	37000
Hydrous Kaolin Clay	95000	39000
Mica, High Aspect	104000	28000
Talc	58000	30000
Wollastonite	90000	54000
Alumina Trihydrate	65000	48000

*3-methacryloxypropyltrimethoxysilane (OSI Specialty Chemicals)

Higher filler loadings – The mineral-matrix compatibility provided by coupling agents allows higher filler loadings for greater economy without sacrificing composite properties or processability. Table 2 shows the effect of titanate use with a 70% calcium carbonate (2.5 micron) in polypropylene.

Table 2. 70% Calcium Carbonate Filled Polypropylene

	Virgin	No Titanate	Titanate
Polypropylene (Profax 6523)	100	30	70
Calcium Carbonate (Micro White 25)	-	70	70
Isopropyltriisostearoyltitanate (KR TTS)	-	-	2.1
Properties			
Flexural Modulus, psi	$24x10^4$	$53x10^4$	$38x10^4$
Drop Weight Impact, ft-lbs	1.0	1.8	7.5
Mold Shrinkage, mils/in.	14 - 15	10 - 13	11 - 12
Melt Flow, I_2230	5.00	0.07	4.7
Surface Appearance	Good	Poor	Good

(Kenrich Petrochemicals)

⌘⌘⌘

Improved mechanical properties – Optimized filler dispersion and bonding by means of coupling agents offers improvement in tensile and flexural strength; tear, abrasion, impact, and liquid resistance; hardness, elongation, thermal conductivity, and surface appearance; and reduced shrinkage and thermal expansion. Table 3 compares the effects on physical properties of two silanes in three clay-filled elastomers. The results with amino silane vs. chloropropyl silane demonstrates the value of matching organofunctionality to the polymer matrix for the particular desired property modification.

Table 3. Mechanical Properties of Clay-Filled Rubber

	Natural Rubber		
Coupling Agent	**None**	**1% Z-6020**[1]	**1% Z-6076**[2]
300% Modulus, psi	1040	1655	985
Tensile Strength, psi	3435	3925	3690
Elongation, %	585	520	585
Tension Set, %	43	38	43
Compression Set, %	8	6	7
Tear Strength, ppi	127	118	138
Abrasion Resistance	100	107	100
	SBR		
Coupling Agent	**None**	**1% Z-6020**	**1% Z-6076**
300% Modulus, psi	285	400	200
Tensile Strength, psi	1120	1505	1380
Elongation, %	925	885	1015
Tension Set, %	37	35	41
Compression Set, %	13	10	11
Tear Strength, ppi	141	154	150
Abrasion Resistance	100	155	130
	Nitrile		
Coupling Agent	**None**	**1% Z-6020**	**1% Z-6076**
300% Modulus, psi	1230	2125	1355
Tensile Strength, psi	3150	3495	3255
Elongation, %	650	545	665
Tension Set, %	31	21	27
Compression Set, %	7	4	7
Tear Strength, ppi	201	153	176
Abrasion Resistance	100	166	122

[1]N-(2-aminoethyl)-3-aminopropyltrimethoxysilane (Dow Corning)

[2]3-chloropropyltrimethoxysilane (Dow Corning)

⌘⌘⌘

Improved mechanical properties retention – The durability of the coupling agent mediated mineral-polymer bond is essential to the long term retention of composite properties. Because this interface is the destination for absorbed ambient moisture, the mineral-coupling agent bond must be resistant to hydrolysis. Table 4 shows the improvement in flexural strength obtained by the methacrylate functional silane modification of several fillers, plus the superior retention of these improvements under severe hydrolysis conditions. Table 5 likewise compares the properties improvement with two silanes in a phlogopite mica reinforced nylon, together with the retention of those improvements under hydrolysis conditions.

Table 4. Flexural Strength in Polyester Resin: Modified vs. Unmodified Fillers

	Flexural Strength, psi			
			After Immersion[1]	
Coupling Agent	None	A-174	None	A-174[2]
Silica, 10 micron	10300	17700	8600	14600
Calcined Kaolin Clay	11300	16000	8800	14100
Hydrous Kaolin Clay	10100	11000	4500	6500
Mica, High Aspect	2800	6500	2600	3500
Talc	8800	10300	4400	7200
Wollastonite	10400	15600	7700	12500
Alumina Trihydrate	7200	12100	4600	8300

[1]Immersion in boiling water for 8 hours
[2]3-methacryloxypropyltrimethoxysilane (OSI Specialty Chemicals)

Table 5. Mica-Reinforced Nylon 6

	Unfilled	Untreated	0.5% Z-6020[4]	0.5% Z-6076[5]
Flexural strength - Dry	14500	17900	18400	18100
Flexural strength - Wet[2]	7100	10300	12600	10400
Flexural modulus - Dry[3]	3200	14100	14300	14000
Flexural modulus - Wet[2,3]	1240	5340	7350	6750
Tensile strength - Dry	9400	10700	11000	10400
Tensile strength - Wet[2]	7000	6900	8500	8200

[1]psi
[2]16 hours at 50°C in water
[3]$x10^8$
[4]N-(2-aimoethyl)-3-aminopropyltrimethoxysilane (Dow Corning)
[5]3-chloropropyltrimethoxysilane (Dow Corning)

Improved electrical properties – Resistance to hydrolysis and maintenance of a hydrophobic mineral-polymer interface is particularly important to the dielectric properties of composites used in electrical and electronic applications. Table 6 shows the ability of epoxy silane to control degradation of strength and electrical properties of a novaculite (silica) reinforced epoxy.

⌘⌘⌘

Table 6. Silica Reinforced Epoxy Resin[1]

	DRY		WET[2]	
Coupling Agent	**None**	**Z-6040[3]**	**None**	**Z-6040**
Flex Strength, psi	18,800	22,400	14,900	18,500
Volume Resistivity, ohm-cm	2.2×10^{15}	1.1×10^{15}	1.9×10^{12}	6.4×10^{14}
Disipation Factor, x10²	0.0051	0.0046	0.053	0.014

[1] 100 parts D.E.R. 331 epoxy resin, 18 parts Curing Agent Z, 50 parts Malvern Minerals Novacite 1250 (novaculite).

[2] After 4 hours immersion in boiling water.

[3] 1% 3-glycidoxypropyltrimethoxysilane (Dow Corning)

Improved retention of electrical properties – Coupling agents are also used to prevent degradation of electrical properties under harsh environmental conditions. Table 7 compares treated and untreated clay and wollastonite fillers to unfilled thermoplastic polyurethane during one week of water immersion. With surface modification, both minerals are able to maintain or improve upon the electrical properties of the unfilled resin.

Table 7. Electrical Properties of Thermoplastic Urethane Composites[1]

		CLAY		WOLLASTONITE	
	Unfilled	**Untreated**	**A-174[2]**	**Untreated**	**A-174**
Initial					
Dielectric Constant[3]	4.87	5.36	4.70	5.32	4.84
Dissipation Factor[3]	0.024	0.025	0.023	0.024	0.021
1 Day[4]					
Dielectric Constant[3]	5.41	7.56	5.40	9.11	5.67
Dissipation Factor[3]	0.022	0.083	0.022	0.062	0.021
3 Days[4]					
Dielectric Constant[3]	6.11	8.25	5.43	9.61	5.62
Dissipation Factor[3]	0.023	0.093	0.022	0.047	0.027
7 Days[4]					
Dielectric Constant[3]	5.91	8.45	5.49	9.72	5.71
Dissipation Factor[3]	0.023	0.098	0.022	0.047	0.026

[1] Composition, parts by weight: "Roylar" E-9 thermoplastic resin (Uniroyal)-100; filler-33

[2] The filler was treated with A-174 (3-methacryloxypropyltrimethoxysilane; OSI Specialty Chemicals) at 1 wt. % based on filler weight.

[3] Properties measured at 1000 Hz after immersion in water at 50° C.

[4] Immersed in water at 50° C.

⌘⌘⌘

The use of surface modified minerals has been hampered in the past by a perceived cost premium. The actual performance premium and ultimate economy of coupled composites has only more recently been recognized. Lower energy and raw material costs can compensate for the cost of the coupling agent. Higher filler levels can be used in expensive polymers; polyolefins can be used in place of engineering resins; thinner, smaller, and lighter parts can meet or exceed product specifications. Plastic parts can replace metal parts. The replacement cost of plastic components can be significantly deferred when properties are retained longer in use.

BIBLIOGRAPHY

Dannenberg, E.M., 1987, "Product Modifications Important in Fillers", *Elastomerics*, Dec., pp 22-25

Goodman, R., 1995, "Surface Modification of Mineral Fillers", *Industrial Minerals*, Feb., pp 49-55

Griffiths, J.B., 1990, "Minerals as Fillers – Modification to Serve Modern Markets", *Plastics and Rubber Processing and Applications*, Vol. 13, No. 1, pp 3-8

Griffiths, J.B., 1987, "Surface Modified Minerals", *Industrial Minerals*, Oct., pp 23-45

Joslyn, W., 1986, "Optimal Pretreatment of Reinforcement Modifiers with New Generation Silanes", C.H. Kline & Company Conference on Chemically Modified Minerals, October 9, 1986

Katz, H.S., Milewski, J.V., (eds.), 1987, *Handbook of Fillers for Plastics*, Van Nostrand Reinhold, New York

Kuhn, C.R., Johnston, J., 1995, "Surface Modified Minerals–Types of Minerals, Benefits Realized from Their Use and Applications in Industry", SME Annual Meeting, Denver, Colorado, March 6-9, 1995

⌘⌘⌘

NOTES

⌘⌘⌘

FOUR

PAINTS & COATINGS

Frank McGonigle
Whittaker, Clark & Daniels, Inc.
South Plainfield, NJ

Peter A. Ciullo
R.T. Vanderbilt Company, Inc.
Norwalk, CT

The oldest known use of paint dates to approximately 23,000 BC in the cave paintings near Lascaux, France. These drawings were made with pastes of natural iron oxides and manganese oxides ground with crude mortars and pestles and mixed with water. Egg whites, animal fat, or marrow were possible binders. Wood ash, siderite (iron carbonate), and chalk may have been used as well to round out the primitive color palette of black, red, yellow and blends thereof.

The Egyptians are generally credited with advancing the art of decorative paint production and use during the period of 3000 to 600 BC. They used vegetable gums as binders and employed additional mineral colors – blues from azurite and lapis lazuli, green from malachite, white from gypsum – and developed the first synthetic pigment, Egyptian Blue, by heating a blend of malachite, lime, sodium carbonate, and silica to temperatures above 830°C. The Egyptians also compounded the first lake pigments by fixing a red dye from madder root onto ground minerals. They were also the first to use paint for a protective rather than solely decorative purpose, using red lead (lead oxide) in preservative coatings for wood.

The preservative functions of coatings became better developed in the Greek and Rome periods, 600 BC to 400 AD. Varnishes based on drying oils were introduced at this time, although it was not until the Middle Ages that the practice of protecting paintings with an overcoat of varnish became widespread. By the 15th century, artists' command of paintmaking and the manipulation of color had reached a high level of sophistication.

By the dawn of the industrial revolution the demand for paints and varnishes for both decorative and protective uses had spawned an established cottage industry. By the end of the 19th century, the widespread use of iron and steel was accompanied by the need for anticorrosive lead- and zinc-based primers. Metal-based driers had been in use since mid-century and the chemical foundations for the use of nitrocellulose and formaldehyde resins as binders had been laid. Nevertheless, the number of scientifically trained personnel working in the area of paints and coatings up to the First World War was very small.

Following the war, the accelerating pace of technological development, particularly the manufacture of titanium dioxide, developments in alkyd chemistry and the mass production of automobiles (which needed priming and painting) catalyzed scientific research into coatings of all types.

The post World War II era may be characterized, perhaps over-simply, as the gradual and ongoing coversion to solventless coatings technology. The first latex paint was introduced in the late 1940's to only moderate success. Water-based systems have since undergone continuous improvement and are now as durable for most uses as their solvent-based counterparts. Today, powder coatings are widely used as industrial finishes free of both solvent and water.

The market for paints and coatings in the United States is well over one billion gallons annually. The overall U.S. market may be broken down as follows:

	Percentage (Approximate)		
Exterior Paints and Coatings			**24**
Barn and Fence		0.5	
Primers		1.5	
Enamels		3.5	
Trim	0.5		
Deck	3.0		
Misc. Pigmented		5.0	
Exterior House Paint		13.5	
Oil and Alkyd	6.0		
Emulsion	7.5		
Interior House Paints			**58**
Oil and Alkyd		11.0	
Emulsion		38.0	
Misc.		9.0	
Industrial Coatings			**18**

TYPES OF COATINGS

Paints and coatings are typically classified according to functionality or end-use applications. Functionally, coatings fall into three basic categories:

⌘⌘⌘

sealers, primers, and top coats. A sealer is a clear or lightly pigmented base coat designed to inhibit capillary action in a porous substrate, such as masonry or wood. The sealer serves two purposes. It provides a uniform base for the primer, allowing an even film to form, and it prevents the migration of extractives from the substrate into the primer and top coats. The primer is considered the first coat. The purpose of the primer is to insure the adhesion of the coating system to the substrate, to assist in the protection of the substrate, and to provide a uniform base for the top coats. The top coats are the finish coats that provide the final protective barrier and present an aesthetically appealing decorative effect.

In designing a complete coating system, a general rule is to provide increasing flexibility from primer to final top coat. The primer should provide the densest film and the top coat should be the least dense and most flexible layer. Primers traditionally have a much higher pigment content than the top coats. These generalities are valid for the majority of systems in the two broad end-use categories: architectural coatings and industrial coatings.

Architectural Coatings

Architectural coatings are essentially the paints used to protect and decorate the exterior and interior of buildings. They are also called "trade sales paints" when sold through retail outlets. These coatings are most often applied to wood and drywall to control moisture absorption, swelling and degradation, while providing a durable, flexible and decorative film. Interior and exterior paints are available in both oil- and water-based formulations yielding a flat, semi-gloss or gloss finish.

Masonry coatings are applied to concrete, brick, stone and tile, and can include similar materials, such as traffic marking paints. The primary reasons for coating masonry are to prevent the deterioration of the surface, to block the passage of moisture through the masonry, and to improve its appearance. Typical masonry coatings must be resistant to alkalies and alkaline environments, since many masonry substrates contain lime and water-soluble alkaline salts.

Industrial Coatings

Industrial coatings are usually applied to ferrous or non-ferrous metal surfaces. Exposed ferrous surfaces oxidize rapidly forming a non-protective oxide coating. Protective coatings are therefore usually based on primers containing a linseed oil or alkyd binder and anti-corrosion pigments such as red lead, lead chromate, zinc chromate, potassium zinc chromate, and zinc dust. The completed coating system will have a dry thickness of 5 to 10 mils.

⌘⌘⌘

Coil coating refers to the application in a continuous process of a coating to a steel strip. The common coil coatings are based on amine alkyds, epoxies, vinyl alkyds, solution vinyls, thermoset resins, solvent-type and water reducible polyesters, and fluorocarbon resins. The coated metallic sheet is then fabricated into building and construction materials, appliances, commercial equipment and a multitude of other colored metallic products.

Non-ferrous metals may form a protective oxide layer, providing a barrier against further deterioration. The coating of aluminum, magnesium, copper, chromium, cadmium and tin may be required, nevertheless, to protect the surface from deterioration other than oxidation, or a clear coat may be used to preserve the appearance of the virgin metal from oxidation. Zinc surfaces may require a chemical pre-treatment prior to the application of the primer and top coat, depending on the condition of the surface and type of primer used. Lead and lead alloy surfaces are easily coated with a linseed oil based primer and a compatible top coat.

Marine coatings are, by definition, used on marine equipment or ships. They are formulated to protect a variety of substrates from the harshness of the marine environment, retarding corrosion and fouling. In general, a marine coating system consists of a pre-construction primer with corrosion inhibiting properties, a multi-purpose repair primer, a ship bottom coating, antifouling coatings, top side coatings, and deck coatings. The composition and use of these coatings is determined by the environmental conditions in which the vessel is required to operate.

Automotive finishes must provide a high level of protection to the vehicle's substrate and an aesthetically appealing appearance, retaining it's original look for as long as possible. The process for applying a coating system to an automobile body begins with the preparation of the surface by cleaning and degreasing prior to the application of the primer, which must provide excellent adhesion and corrosion resistance. A uniform smooth surface is required as the base for the top coats. The finish or top coats must be compatible with the prime coat to form a final film which can be polished, handled and repaired. The top coat should be free of surface imperfections that could reduce functionality and appearance.

Convertible and Non-Convertible Coatings

Coatings are usually applied as a liquid, after which they form a solid flexible adherent film. There are two additional categories of coatings, based on the nature of this film: convertible and non-convertible. A convertible coating undergoes a chemical change during film formation. This results in the conversion of the polymeric binder into a three dimensional crosslinked network. This film is permanently altered and cannot be reliquified by heat or contact with most solvents. Crosslinking may be achieved through an oxidative mechanism (as with oil and alkyd paints), catalysts (as with urea-formaldehyde coatings), amine curing agents (as with epoxy systems), or simply UV radiation.

Non-convertible coatings form polymeric films through the evaporation of solvent or of an aqueous dispersion medium. These films are typically thermoplastic. Solvent coatings, in simplest terms, are made by dissolving a polymer of suitably high molecular weight at sufficiently low concentration to obtain the appropriate coating viscosity. As the solvent evaporates, the resin concentration increases. Solvent-resin compatibility is essential to keeping the film homogeneous and preventing the resin from precipitating out of solution. As the solvent continues to evaporate, a tacky film forms. Final solvent release determines how long it takes the resulting polymeric film to reach final hardness.

Non-convertible aqueous coatings are typically a latex emulsion consisting of colloidal polymer particles, usually smaller than 0.5 microns, which are stabilized with an emulsifier to prevent flocculation and coalescence. As illustrated in Figure 1, once a latex paint is applied, water begins to evaporate and the polymer colloidal particles become increasingly concentrated. The particles eventually touch and pack together, leaving voids filled with water which must eventually be lost to evaporation or porous absorption of the substrate. Finally the particles completely coalesce and fuse into a continous film.

A number of factors control successful film formation. The polymer must be soft enough and have a sufficiently low glass transition temperature (T_g) to allow fusion. For this reason, many of the latex polymers are copolymers, which are usually softer than homopolymers. Coalescing agents are added as temporary plasticizers. As the water evaporates, they have an increasing softening effect on the polymer particles, helping them to coalesce more easily.

Figure 1. Film Formation From Latex Paint

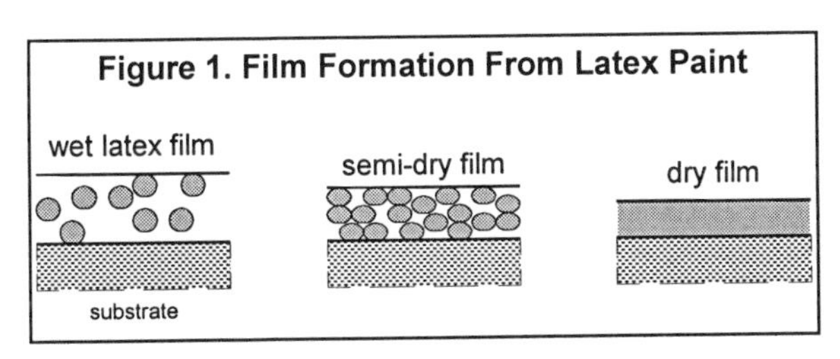

Another important consideration is the temperature at which the latex coating is being applied. If the temperature is too low, the colloidal particles may be too hard to coalesce. The presence of over-size latex polymer particles can also interfere with film formation.

The specific paint formula itself is also important. Too much pigment, for example, can interfere with the smooth formation of a continuous film.

⌘⌘⌘

In fact, any imbalance or excess of paint additives can potentially affect latex film formation.

FORMULATING

A paint formulator must choose from hundreds of different materials in selecting a cost effective combination which will have the desired film-forming and service properties.

Components

A paint formulator chooses raw materials from the following four basic categories.

(1) Binders
(2) Solvents
(3) Pigments
(4) Additives

The binder is the backbone of film formation. The polymeric binder is key to the formation of a continuous film. It literally binds and holds the pigment and extender particles together.

Solvents are needed in liquid coatings to solubilize or disperse the binder. The solvent is lost to evaporation during film formation. Latex emulsions use water as the volatile vehicle.

Pigments are usually added to impart a particular decorative color. They also help protect the binder against UV degradation from outdoor exposure. The minerals added to extend pigments, modify film properties and reduce cost are also commonly refered to as pigments or extender pigments in coatings formulation.

Additives are very broad category including literally hundreds of different chemicals. These ingredients may be used at only a fraction of a percent and usually are not used at levels greater than 5 percent. Even with their relatively low concentration, they have a considerable effect on the paint properties.

Characteristics

Coatings, particularly liquid paints, are conventionally characterized in terms meant to reflect quality and value from the consumer's/users's perspective.

Pounds vs. gallons – Paint formulations are commonly given in both pounds and gallons. Gallons of a given ingredient can be converted to pounds as follows:

⌘⌘⌘

Weight in lbs. = 8.33 x specific gravity x volume (gal.)

where 8.33 is the weight in pounds of one gallon of water at 77°F.

Because paint is generally sold on a volume basis (in gallons) while raw materials are typically purchased on a price per pound basis, paint cost adjustments are usually calculated in cost per gallon.

Non-volatiles content – From a user's perspective, the percent non-volatiles is important, since volatile components are lost after application. While these volatiles are necessary vehicles for applying a coating, they do not have value to the customer in contributing to dry film thickness or total coverage. Therefore, knowing percent non-volatiles can be important to the user in estimating paint coverage at a given dry film thickness.

Some paint formulations will give the percent NV or "dry pounds" for each ingredient. The weight percent non-volatiles is calculated as follows:

$$\text{Wt. \% NV} = \frac{\text{wt. of non-volatiles}}{\text{wt. of formula}} \times 100\%$$

The volume percent non-volatiles is calculated similarly:

$$\text{Vol. \% NV} = \frac{\text{vol. of non-volatiles}}{\text{vol. of formula}} \times 100\%$$

Volatile Organics Content (VOC) – Environmental regulations impose limits on the quantity of volatile organic solvents that a coating may contain. For water-based coatings VOC in pounds/gallon is calculated as:

$$\text{VOC} = \frac{\text{lbs. solvent}}{100 - \text{gals. } H_2O}$$

For solvent coatings:

$$\text{VOC} = \frac{\text{lbs. solvent}}{\text{gals. coating}}$$

VOC in g/L = lbs./gal. x 120

Film coverage – The wet film thickenss is calculated simply as:

$$\text{Wet film thickness} = \frac{\text{volume of applied paint}}{\text{area covered by paint}}$$

⌘⌘⌘

To calculate the dry film thickness, one must use the calculated volume percent non-volatile content to make the correction. The above equation is also used to calculate area that can be covered by a given volume of paint for a given film thickness.

Pigment-to-binder ratio – The pigment-to-binder ratio (P/B) is a calculation which addresses the quality or nature of the solids that remain after film formation.

$$P/B = \frac{\text{wt. of pigment and extenders}}{\text{wt. of total binder solids}}$$

Coatings with P/B values above 4:1, for example, have a relatively small quantity of binder to hold the high quantity of pigment together in the film. Paints with high P/B values are therefore flat in appearance and cannot provide a continuous matrix for effective protection against the elements. On the other hand, paints with very low P/B ratios (<1) are characterized as more durable and possessing a higher gloss (if a flattening agent has not been used).

Pigment volume concentration – While the weight ratio of pigment to binder roughly denotes the quality characteristics of a paint, the volume concentration of the pigments is a more accepted scientific method of predicting how a paint film will perform.

Pigment volume concentration or % PVC is calculated as follows:

$$\%\ PVC = \frac{\text{vol. of pigments}}{\text{vol. of total non-volatiles}} \times 100\%$$

"Volume of pigments" includes the volume of pigments and mineral extenders while "volume of total non-volatiles" includes binders, pigments, additives, extenders, and other non-volatiles.

Another important term is the "critical pigment volume concentration" or CPVC. CPVC is the minimum level of binder required to completely wet the pigment particles and fill the interparticle voids. At pigment volume concentrations above CPVC, there is not enough binder to completely fill the voids between the pigment particles. As a result, the film becomes more porous, permeable, and weak. CPVC usually occurs between 40 to 55% PVC. Figure 2 illustrates the changes in film properties as a function of PVC.

⌘⌘⌘

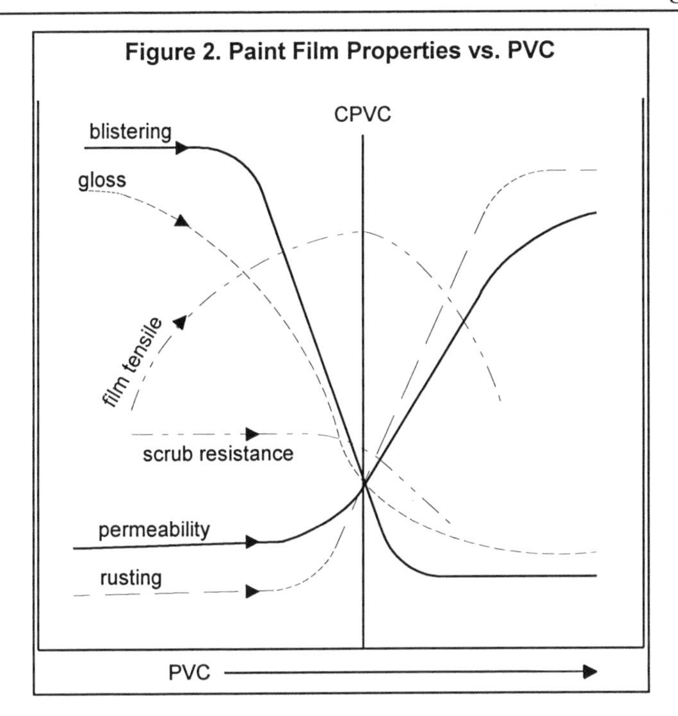

Figure 2. Paint Film Properties vs. PVC

At PVCs above CPVC, the greatly increased film porosity significantly compromises substrate protection. Above CPVC, the pigment particles disrupt the smoothness of the film surface and cause light to be scattered and film gloss to be reduced as illustrated in Figure 3. The film surface becomes very rough with pigment particles, and the film has a very flat appearance.

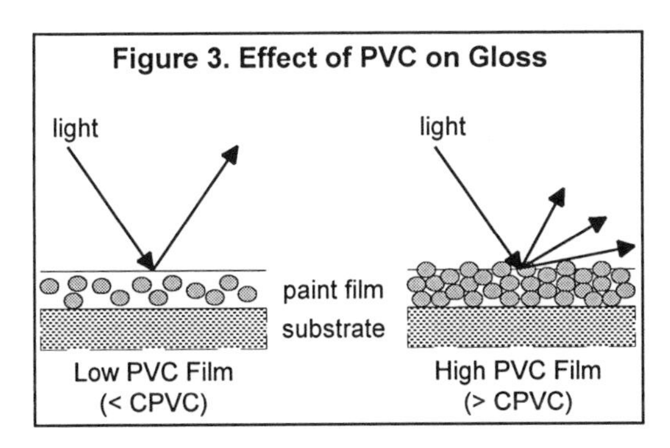

Figure 3. Effect of PVC on Gloss

It is poor practice to formulate a paint at the CPVC. As illustrated in Figure 2, paint properties change very rapidly at this pigment volume concentration such that slight variations plus/minus the CPVC can result in major changes in paint properties. Paints requiring very high gloss, such as

⌘⌘⌘

automobile finishes, must have PVC values well below the CPVC. Conversely, paints that impart a flat appearance, such as some interior house paints, often have relatively high PVC values.

Another consideration regarding PVC is cost. At times, pigments and extenders can be less expensive than binder. Higher PVC paints could then be less costly to formulate than lower PVC paints.

Important Properties

Different coating uses demand diversity in paint properties. An exterior house paint, for instance, must protect the wood substrate principally by controlling the rate of moisture absorption and loss. This helps prevent wood swelling and shrinkage and prevents buckling. A good exterior house paint must not only have good original film flexibility, it must retain that flexibility through severe outdoor weathering. It must resist oxidation and accommodate normal expansion and contractions of the substrate.

Automotive coatings, on the other hand, should be very impermeable to protect the ferrous metal surface from corrosive oxygen, salts and water. The film should also electrically insulate the surface to reduce galvanic action and the flow of electrical current needed to induce corrosion. The protective film should also have high impact resistance to protect against chipping by sand and gravel when driving at high speed. Appearance is also naturally important. The paint must be able to retain a high gloss and have good color stability. For the original equipment market, the speed with which the finish dries is important to maintaining production rates.

Following are the properties a paint formulator may have to consider when developing a new paint formula.

Cost per gallon
Sprayability
Brightness
Sag resistance
Storage stability
Dry film hardness
Dry film durability
Chalking properties
Detergent resistance
Lightfastness
Scrubbability
Dry film adhesion
Salt resistance
Blister resistance
Odor
Ease of application
Hiding power
Hue
Freeze-thaw resistance
Drying time
Dry film flexibility
Dry film weather resistance
Flaking resistance
Chemical resistance
Fire Resistance
Dry film gloss
Dry film impact resistance
Corrosion protection
Ease of manufacture
Toxicity
EPA hydrocarbon emission standards

⌘⌘⌘

The ability to balance these properties to meet the requirements of the specific coating defines the formulator's skill.

RAW MATERIALS

Binders

Binders are the polymeric components in a coating formulation that form the protective film. The binder literally bonds the pigment particles together in the film. If the coating contains no pigments or extenders, then the binder will usually form a glossy and possibly transparent coating.

Drying oils – These natural oils have been used for hundreds of years as film formers and binders in paints. Drying oils are vegetable or fish oils and chemically are triglycerides with the general structure shown in Figure 4.

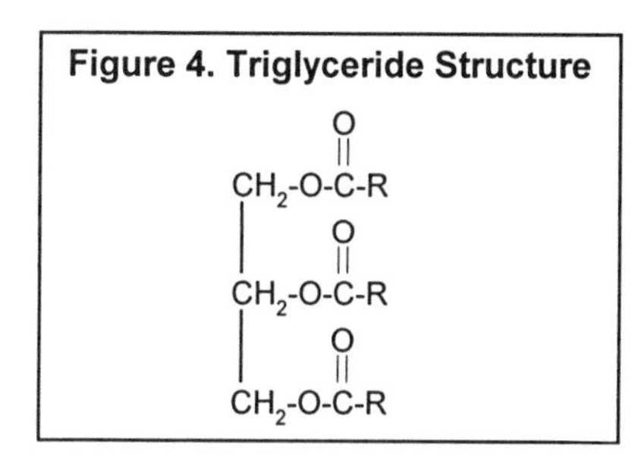

Figure 4. Triglyceride Structure

These triglycerides are based on a blend of saturated, unsaturated, and polyunsaturated fatty acids. Some common examples are shown below.

Caprylic acid $CH_3(CH_2)_6\text{-}COOH$
Capric acid $CH_3(CH_2)_8\text{-}COOH$
Lauric acid $CH_3(CH_2)_{10}\text{-}COOH$
Myristic acid $CH_3(CH_2)_{12}\text{-}COOH$
Palmitic acid $CH_3(CH_2)_{14}\text{-}COOH$
Stearic acid $CH_3(CH_2)_{16}\text{-}COOH$
Oleic acid $CH_3(CH_2)_7CH{=}CH(CH_2)_7\text{-}COOH$
Linoleic acid $CH_3(CH_2)_4CH{=}CHCH_2CH{=}CH(CH_2)_7\text{-}COOH$
Linolenic acid $CH_3CH_2CH{=}CHCH_2CH{=}CHCH_2CH{=}CH(CH_2)_7\text{-}COOH$
Eleostearic acid $CH_3(CH_2)_3CH{=}CHCH{=}CHCH{=}CH(CH_2)_7\text{-}COOH$

Triglyceride oils containing mostly saturated fatty acids will not form films by exposure to the air and are non-drying. They are only used in paints when blended with drying oils. It is the unsaturated fatty acid substituents, particularly the polyunsaturated fatty acids that promote oil "drying" and film formation.

⌘⌘⌘

Drying oils form flexible films through oxidative crosslinks between the polyunsaturated fatty acid groups of different triglyceride molecules. This is based on the formation of hydroperoxides from air exposure, and the subsequent decomposition to free radicals, as in Figure 5.

Figure 5. Drying Oil Oxidation

$$\text{-CH=CH-CH}_2\text{-CH=CH-} + O_2 \longrightarrow \text{-CH=CH-CH(OOH)-CH=CH-}$$

$$\text{-CH=CH-CH(OOH)-CH=CH-} \xrightarrow{\text{drier}} \text{-CH=CH-CH(O}\cdot\text{)-CH=CH-} + \cdot OH$$

$$\cdot OH + \text{-CH=CH-CH}_2\text{-CH=CH-} \longrightarrow \text{-CH=CH-}\dot{C}\text{H-CH=CH-} + HOH$$

The decomposition of the initially formed hydroperoxide is catalyzed by the paint drier. The resulting free radicals, $R^{\cdot}$ and $RO^{\cdot}$, combine to form ether, peroxy and carbon-carbon crosslinks between oil molecules. These crosslinks form a continuous three dimensional matrix.

$RO\cdot + R'O\cdot \longrightarrow$	ROOR'	peroxy crosslink
$RO\cdot + R'\cdot \longrightarrow$	ROR'	ether crosslink
$R\cdot + R'\cdot \longrightarrow$	R-R'	carbon-carbon crosslink

Triunsaturated fatty acids, such as linolenic acid, are much more rapid in forming oxidative crosslinks than diunsaturated fatty acids, such as linoleic acid. Conjugated, triunsaturated fatty acids, such as eleostearic acid, are faster yet.

Linseed oil, tung oil, and fish oil are examples of commonly used drying oils. Linseed oil, from flaxseed, was once widely used by the paint industry. Over fifty percent of its fatty acid component is linolenic acid. The high linolenic acid content tends to cause some yellowing; but in highly pigmented paints this is usually not a problem.

Tung oil is very fast drying because the fatty acid component is 80% eleostearic acid. Because of eleostearic acid's conjugated, triple unsaturation, this oil is faster drying than linseed oil, giving good film hardness and flexibility. Tung oil (also called China wood oil) comes from nuts of the Tung tree indigenous to China.

Fish oil contains both high levels of fast drying acids and non-drying acids (saturated and mono-unsaturated acids). As a result it dries more slowly than linseed oil and gives a softer film because of the plasticizing saturated fatty acids present. Fish oil can impart an undesirable odor.

⌘⌘⌘

Paints based solely on drying oils have been substantially replaced by alkyds, acrylics, and other synthetic polymers. Drying oils provide a softer film than synthetic binders, have a low gloss, and are only used in highly pigmented paints. Even though these paints have relatively poor durability and chemical resistance, they are still used in small volumes for limited applications.

Oleoresinous binders – Oleoresinous binders are used principally in making varnishes. They are produced by "cooking" a given quantity of a natural drying oil, usually tung and/or linseed oil, with a predetermined amount of resin in a large reactor vessel. This process dimerizes and oligomerizes the oil and renders the oil and resin compatible with each other and soluble in a thinning solvent.

The ratio of oil to resin determines many of the final film properties. A higher ratio of oil to resin means a slower drying speed, greater softness, increased flexibility, and reduced gloss. With a lower ratio of oil to resin, there is an increase in drying speed, high film hardness, and increased gloss. The ratio of oil to resin is commonly referred to in the varnish industry as oil length. Oil length is defined as the number of gallons of oil to 100 pounds of resin, as below.

Varnish Type	Oil Length
Short oil	5-15
Medium oil	16-30
Long oil	>30

Many resins are used in making oleoresinous varnishes. These include polyterpene resins, petroleum resins, coumarone-indene resins, gum rosins, wood rosins, tall oil rosins, rosin soaps, rosin esters, novolac phenolic resins and rosin modified phenolic resins.

Most oleoresinous varnishes are made from rosins and phenolic resins. The gum, wood, and tall oil rosins are usually interchangable. The straight rosin is generally not used per se because of its high acidity, brittleness and poor water resistance. Instead, the rosin is used as rosin ester or soap. Rosin esters are obtained by reacting rosin with a polyol such as glycerin. These products have low acid numbers, but a low softening point. The softening point can be increased by using polyols such as pentaerythritol instead of glycerin. These resins can also be hydrogenated or polymerized to increase hardness. Rosin soaps are formed by reacting rosin with an alkali. These products have higher softening points, are harder and more brittle, and provide better gloss.

⌘⌘⌘

In clear coatings, oleoresinuous binders have been largely replaced by polyurethanes with superior hardness, chemical resistance, and water resistance.

Alkyd resins – Alkyd resins are produced by the reaction of a dibasic acid with a polyol and a drying oil. They are called "alkyds" from "al" which means alcohol and "kyd" which stands for the "cid" in acid.

In the early part of this century, experiments were conducted to react dicarboxylic acids with a polyol such as glycerin to form a polyester coating. Useful products were not developed until the 1920s, however, when fatty acids were used to modify these resins to render them soluble in aliphatic solvents and capable of being air dried.

The most common example of alkyd resin formation is from the polycondensation of glycerin, phthalic anhydride, and fatty acid. The reaction between trifunctional glycerin and difunctional anhydride could gel before the polycondensation was complete, but with the addition of a monocarboxylic fatty acid the formation of a somewhat linear polymer is possible. This polymer is soluble in aliphatic solvents if the content of fatty acid is sufficiently high.

Although fatty acids can be used directly, usually less expensive vegetable or marine oils are used to produce these resins in a two stage process. The first stage involves heating the oil and part or all of the glycerin with a suitable catalyst, such as calcium hydroxide, to around 230°C. This results in alcoholysis of the oil, as shown in Figure 6.

Figure 6. Alcoholysis of Triglyceride to Monoglyceride

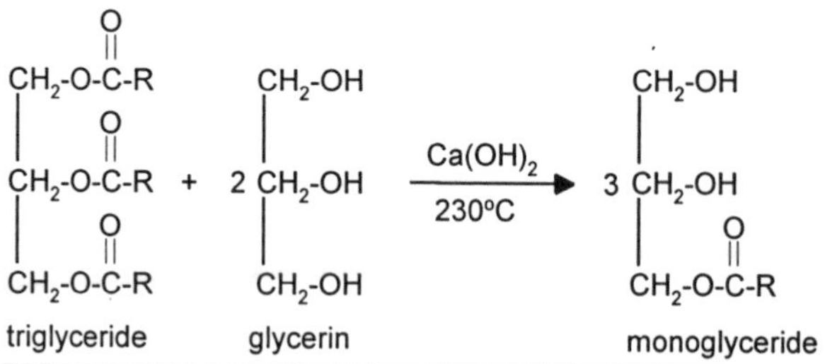

CH_2-O-C(=O)-R | CH_2-O-C(=O)-R | CH_2-O-C(=O)-R (triglyceride) + 2 CH_2-OH | CH_2-OH | CH_2-OH (glycerin) $\xrightarrow[230°C]{Ca(OH)_2}$ 3 CH_2-OH | CH_2-OH | CH_2-O-C(=O)-R (monoglyceride)

In the second stage, the phthalic anhydride reacts with the monoglycerides to produce the alkyd resin. During these reactions, the reactor vessel is purged of oxygen using an inert gas such as nitrogen, since the oils are attacked by oxygen at elevated temperatures and oligomerize.

The variety of alkyd resins used today in the paint industry is seemingly endless because of the wide variety of possible reactants. Although glycerin is the most commonly used polyol, ethylene glycol, pentaerythritol, sorbitol,

⌘⌘⌘

trimethylolethane, trimethylolpropane, diethylene glycol and propylene glycol are used as well.

After glycerin, pentaerythritol is the most commonly employed polyol. Because it is tetrafunctional and very reactive, it is commonly used with other polyols, such as glycerin or ethylene glycol. Pentaerythritol is used to improve water resistance, raise viscosity, increase film hardness, and accelerate drying of the alkyd resin.

Although phthalic anhydride is the most common dibasic acid for producing alkyd resins, isophthalic acid, terephthalic acid, trimellitic anhydride and maleic anhydride are also used. Isophthalic and terephthalic acid are often used in exterior coatings where improved hydrolitic stability is required. Maleic anhydride is used with other polybasic acids as a modifier to increase viscosity and film hardness. Trimellitic anhydride is used in making water soluble alkyds for specific applications.

Soybean oil is widely used to produce alkyd resins because it produces non-yellowing products. When discoloration is less important, linseed oil or tung oil are used for faster drying. Sometimes a non-drying oil, such as cottonseed or coconut is used when the alkyd resin is to serve as a plasticizer rather than a film former.

Oil length also affects alkyd properties. Here oil length may be defined as the number of grams of oil used in making 100 grams of alkyd resin. Short oil alkyds are baked to form a film. Their solubility is limited to aromatic solvents and other non-aliphatic solvents. They are generally restricted to industrial coatings. On the other hand, long oil alkyds are easily air dried and are readily soluble in aliphatic solvents. They are easily applied by brush and are used extensively in architectural coatings. Generally, as oil length increases, viscosity decreases, hardness decreases, flexibility increases and water resistance is reduced.

Alkyd chemistry lends itself to further modification beyond choice of polyol, dibasic acid, and drying oil. Vinyl-modified alkyds, for example, are produced for more durable and quicker drying films, although with some sacrifice in crosslinking rate and consequent development of solvent resistance. Styrene, vinyl toluene, and methyl methacrylate are the most commonly used modifiers. In the presence of a free radical initiator, vinyl polymer will graft onto the alkyd. Tack-free time (i.e. a surface-dry film) may be reduced from 4 to 6 hours for an unmodified alkyd to 1 hour in styrenated form. Acrylics, silicones, phenolic resins, and natural resins are likewise used to tailor film gloss, flexibility, durability, and drying time for certain applications.

Unlike other polymers, alkyds were developed within the paint industry specifically for use in paints. Ninety-five percent of alkyd use is in coating applications. The ability of alkyds to be modified in so many ways account for their use in more different paints than any other binder. Versatility is their primary advantage.

A noted disadvantage is the lack of the alkyd film's chemical resistance. Alkyds, being modified polyesters, do not possess high alkali resistance and are subject to attack by hydrolysis. Conventional alkyds are also dependent on the use of organic solvents, which invoke environmental concerns.

Water-reducible alkyds have been developed to avoid the problem of solvent emissions. The simplest approach has been to emulsify the alkyd, although alkyd emulsion paints do not compete with synthetic latex paints in cost or performance. A more effective approach has been to attach carboxylate groups to the alkyd molecule. Such water-reducible alkyds are made by using low molecular weight stryrene-allyl alcohol polymers as the polyol. The fatty acid (drying oil) esters of these polyols provide rapid drying and rapidly crosslinking films in solvent systems. These alkyds can be maleated by reaction with maleic anydride and then hydrolyzed to carboxylate salt groups with amine and water. The resulting maleated alkyds form stable aquerous dispersions. After the coating is applied, the volatiles, including amine, evaporate and the film crosslinks by autooxidation. The carboxylic acid groups remaining in the crosslinked film, however, generally provide inferior water and alkali resistance compared to the corresponding non-maleated, solvent borne alkyd.

Acrylics – Acrylic binders are widely used in both solvent-based coatings and latex paints for their inherent clarity, color stability and resistance to yellowing.

Thermoplastic acrylics are represented by a number of different polymers, of which poly(methyl methacrylate) is by far the largest volume used. These polymers can be represented by the two general structures in Figure 7.

Figure 7. Acrylic Resins

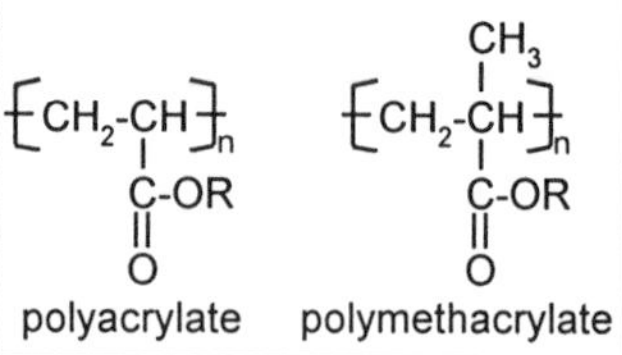

$$\left[CH_2\text{-}\underset{\underset{\underset{O}{\|}}{C\text{-}OR}}{\overset{|}{CH}} \right]_n \qquad \left[CH_2\text{-}\underset{\underset{\underset{O}{\|}}{C\text{-}OR}}{\overset{\overset{CH_3}{|}}{CH}} \right]_n$$

polyacrylate polymethacrylate

Alkyl groups can include methyl, butyl, isobutyl, 2-ethylhexyl, isodecyl and lauryl. As the number of carbon atoms in the alkyl groups is increased, the T_g and brittle point for both types of polymers decreases till about C_8, for poly(alkyl acrylates), and C_{12}, for poly(alky methacrylates), then they rise again. The poly(alkyl methacrylate) has a considerably higher T_g and brittle

⌘⌘⌘

point than the corresponding poly(alkyl acrylate). As a result, poly(methyl methacrylate) is a hard, tough polymer at room temperature while poly(methyl acrylate) is fairly soft. Moreover, poly(ethyl acrylate) is soft and tacky while poly(butyl acrylate) is very soft and tacky.

A combination of methacrylates and acrylates is used on occasion to achieve needed film properties. While poly(methyl methacrylate) is commonly used for film strength and hardness, poly(methyl acrylate) will improve film flexibility. A plasticizing monomer, such as ethyl acrylate or ethylhexyl acrylate is therefore sometimes added with the methyl methacrylate monomer by the polymer manufacturer in an addition-polymerization process to produce a copolymer for use in coating formulations. The softer and more flexible polymer facilitates latex film formation.

Thermosetting acrylics are produced by incorporating functional carboxyl, hydroxyl, or amide groups during free radical polymerization. Acrylamide and acrylic acid are often used as functional monomers for this purpose. The thermosetting acrylics are formulated with other reactive resins, such as epoxies, to provide a crosslinked film on heating. These films are considerably harder, tougher and more chemically resistant than those attainable from thermoplastic acrylics.

Acrylic latex paints dominate the trade sales market for both exterior and interior use because of their balance of film adhesion, durability, flexibility, color stability and alkali resistance, together with resistance to blistering and chalking. Thermoplastic acrylic solvent-based paints are used in high volume applications, like automotive finishes, while thermoset acrylics are used to bake a hard enamel finish on large appliances.

Vinyl resins – In the coatings industry, vinyl resins usually refer to either poly(vinyl chloride) (PVC) or poly(vinyl acetate) (PVAc) which is widely used in interior and exterior latex paints. Produced usually by emulsion polymerization, a PVAc homopolymer is too hard to allow its colloidal latex particles to coalesce well into a continuous film at ambient temperatures. Most PVAc emulsions used in the paint industry are copolymers with a plasticizing monomer such as dibutyl maleate, 2-ethyhexyl acrylate, n-butyl acrylate, dibutyl fumarate, isodecyl acrylate, or ethyl acrylate. By polymerizing under pressure, copolymers of vinyl acetate and ethylene are also produced for latex paints. External plasticizers such as dibutyl phthalate are used as well. All these methods not only soften the polymer to allow the latex particles to coalesce into a continuous film, but also impart the film flexiblity needed in exterior house paints.

PVAc emulsions are less expensive than acrylic emulsions, but impart very good color stability and durability. This has enabled PVAc to penetrate a large portion of the interior and exterior house paint markets. PVAc copolymers are generally superior to acrylics in moisture permeability, which allows houses to "breathe". A disadvantage of PVAc-based latex paints is their relatively poor alkali resistance on masonry compared to acrylics.

As a halogenated polymer, poly(vinyl chloride) is used in specific coatings to form low moisture permeability films. The major use of PVC, usually as a copolymer, is in coatings where corrosion protection is a major function. PVC homopolymer is too poorly soluble in most solvents to be used in solution coatings, and is too hard at room temperataure to allow the formation of a continuous film from emulsion. For fluid coatings, PVC resins are therefore used as copolymers, most often with vinyl acetate. The bulkier acetate group prevents the polymer molecules from packing as tightly and increases polarity thus allowing a much greater range of solvant solubility. A soft monomer like vinyl acetate also reduces resin hardness and allows latex particles to coalesce into a continuous film. Vinyl chloride/vinyl acetate copolymers and other vinyl chloride copolymers are used in solution coating applications, such as container and can linings, and in latex coating applications, such as paper coatings.

High vinyl chloride content PVC copolymers are used in thermoplastic powder coatings, limited in this application form by the deficiencies common to thermoplastic binders. They are difficult to pulverize to small particle size, and their high molecular weight gives high melt viscosity and consequent poor flow and leveling. PVC-based powders are resticted to applications that can accomodate a necessarily thick coating.

Styrene-butadiene binders – Homopolymers of styrene are too hard to enable a continuous film to form from a latex, but a plasticizing monomer can be copolymerized to produce a softer polymer. A ratio of two parts styrene to one part butadiene is commonly used to produce the styrene-butadiene latexes used in coatings. This is nearly the inverse ratio used to produce SBR rubber.

S-B latexes were the first practical latexes to be used in interior architectural applications in the 1950s, but they are not commonly used today. One advantage of S-B emulsion paints is their superior alkali and chemical resistance, greater than either PVAc or acrylic emulsion paints. A major disadvantage of S-B paint is its susceptibility to oxidation because of the unsaturation contained in the butadiene units of the polymer backbone. This oxidation can lead to yellowing and ultimately film embrittlement.

Polyurethane resins – Many solvent-based finishes are based on polyurethane resins because they are characteristically tough yet flexible, and very resistant to abrasion. They also impart good film adhesion, and can be made to harden at ambient temperatures. In the coating industry, any

⌘⌘⌘

polymer containing the urethane linkage of Figure 8 in the backbone, even if that same polymer may also contain many other non-urethane linkages, is termed a polyurethane. Generally, the polyurethane resins involve the reaction of a diisocyanate with a hydroxyl compound.

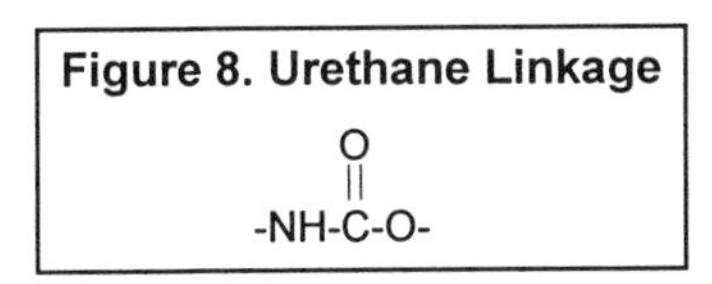

Figure 8. Urethane Linkage

Polyurethane coatings are used as a single-pack system or a two-pack system. There are three basic types of single-pack systems and two of the two-pack system. More recently water-borne urethanes have been developed to reduce solvent emmission.

Oil Modified Type – These finishes are based on a polyurethane-modified alkyd. These are also called "polyurethane-alkyds" or simply "uralkyds", and are prepared in a manner similar to alkyds. In the first stage of the alkyd process, an alcoholysis reaction between a drying oil and polyol, such as glycerin, results in a preponderance of monoglycerides. In the second stage, instead or reacting these monoglycerides with a polycarboxylic acid, they are reacted with a diisocyanate, usually TDI (toluene diisocyanate), as in Figure 9. The ratio of isocyanate to available hydroxyl groups is carefully controlled in a closed system to ensure that there is no unreacted free isocyanate. Thus, the reaction product possesses urethane linkages instead of the ester linkages associated with traditional alkyds.

The percent isocyanate, or conversely oil length, determines the properties of the resulting film. Higher isocyanate content (shorter oil length) reduces film flexibility and impact resistance, increases film hardness and solvent resistance, and reduces drying time. In certain commercial coatings, some of the isocyanate is replaced by phthalic anhydride to make phthalic modified uralkyd. Uralkyds are cured through oxidative drying in the same manner as regular alkyds. Metallic driers can likewise be used to accelerate drying.

Uralkyds are one of the most commonly used polyurethane resin, finding particular application in varnishes for floors, boats and general use. While these resins do not impart the best gloss, they do give excellent abrasion resistance and reportedly better water and alkali resistance than an alkyd.

⌘⌘⌘

Figure 9. Uralkyd Resin Synthesis

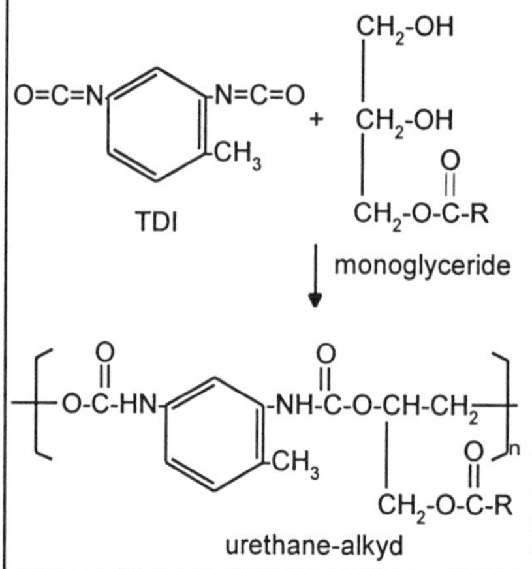

TDI + monoglyceride (CH_2-OH / CH_2-OH / CH_2-O-C(=O)-R) → urethane-alkyd

Moisture Curing Prepolymers – These are single-pack polyurethane systems that cure from exposure to moisture present in the air. Initially a diisocyanate, such as TDI is reacted in a closed system with a high molecular weight polyol, such as castor oil. The TDI reacts with a minimum of three hydroxyl groups per molecule in order to produce trifunctional prepolymer. The prepolymer is cured into a hard film from exposure to moisture in the air. Two isocyanate groups, in the presence of water, bond to form a urea linkage and carbon dioxide gas, as in Figure 10.

During film formation, the carbon dioxide diffuses slowly from the film. Cure rate is greatly dependent on relative humidity. With these polyurethanes, care must be taken that all pigments and mineral additives are free of moisture to avoid reaction before use. This system results in a tough film and is commonly used in floor coatings.

Figure 10. Urea Linkage from Moisture-Cure Urethanes

$$2R\text{-}N{=}C{=}O + H_2O \longrightarrow R\text{-}HN\text{-}\overset{O}{\overset{\|}{C}}\text{-}NH\text{-}R + CO_2$$

Blocked Isocyanates – In this system, polyol is reacted in a closed system with a diisocyanate, such as TDI, to form a structure with isocyanate adducts. These adducts are then blocked by reacting the intermediate with a blocking agent, usually phenol, as in Figure 11.

⌘⌘⌘

Figure 11. Blocked Isocyanate

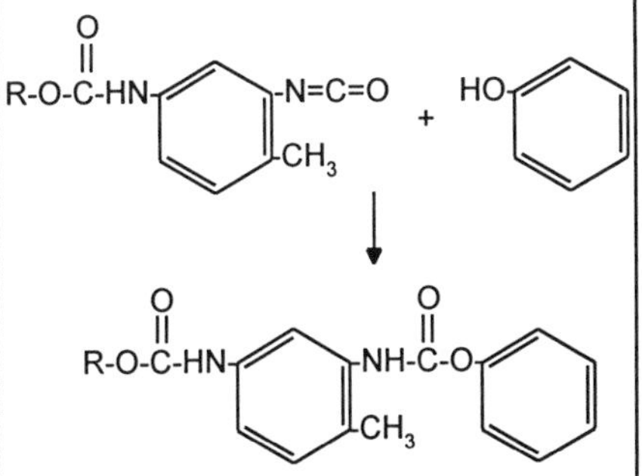

The bond between the phenol and isocyanate is stable up to 150°C, so no reaction will take place while the coating is stored at ambient temperatures. After the coating is applied, it is baked, causing the phenol to split off and allowing the resin to crosslink. Since volatilized phenol is dangerous, the vapors must be safely exhausted from the oven. One common application for this type of resin is in coating electrical wires and parts.

Catalyzed Moisture Cures – This two component system is the same as the one package moisture cure system except a tertiary amine catalyst such as N,N-dimethylethanolamine is added just before application. This shortens cure time and increases the crosslink density of the film for better chemical resistance. As soon as the catalyst is added there is limited pot-life before gelation occurs.

Two-Pack Systems – In these systems, one package contains a low volatility isocyanate, usually a polyisocyanurate (a polymerized TDI), as shown in Figure 12.

Figure 12. Polyisocyanurate

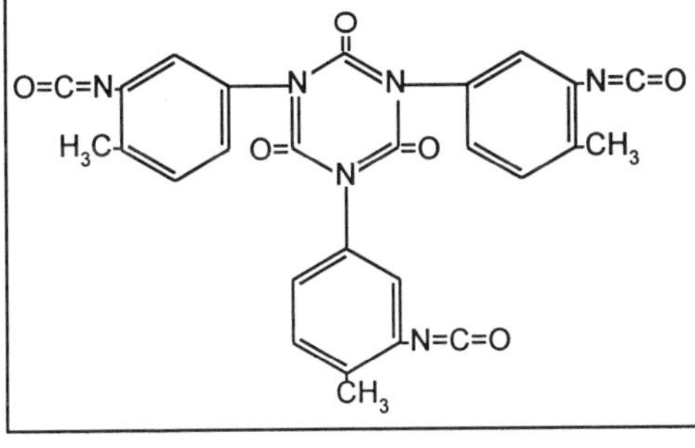

⌘⌘⌘

Another way to prepare a low volatility isocyanate component is to react approximately two moles of TDI with one mole of a low molecular weight polyol such as trimethylolpropane in a closed system. Low volatility is important to minimize worker exposure to isocyanates during application.

The other package consists of the polyol component. The polyol may be either a polyether or saturated polyester resin. Polyethers are generally less expensive but reportedly do not impart as good color retention as polyesters. The distance between the hydroxyl groups on the polyol relates to resultant film flexibility and hardness. If film flexibility is required for greater exterior durability, sufficient space between these hydroxyl groups is needed. On the other hand, if a high crosslink density is desired for greater film harness, the distance between these reactive hydroxyl groups should be less.

When the two packages are mixed, the isocyanate functional groups react with the polyol hydroxyls, as in Figure 13. It is not uncommon for water vapor present to also enter into the reaction resulting in urea crosslinks as well.

Figure 13. Polyurethane from Polyol/Diisocyanate

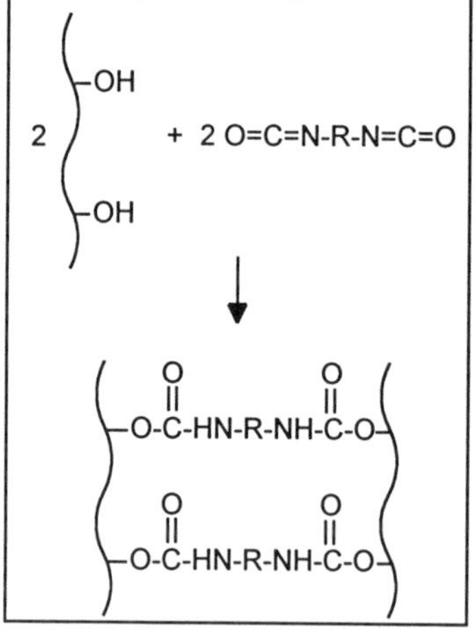

2 { -OH, -OH } + 2 O=C=N-R-N=C=O

↓

{ -O-C(=O)-HN-R-NH-C(=O)-O- }

{ -O-C(=O)-HN-R-NH-C(=O)-O- }

Because of the presence of moisture, usually a slight excess in the stoichiometric ratio of NCO/OH is needed. Care is also required in formulating with a moisture-free solvent.

Two pack systems have a short drying time which has favored numerous products for coating floors, furniture, boats, metals and concrete.

Water-reducible urethanes – In recent years, advances have been made incorporating polyurethane coating attributes into aqueous vehicles. One

⌘⌘⌘

approach is to prepare a hydroxyl-terminated polyurethane (using NCO/OH ratio <1), with 2,2-dimethylolpropionic acid as one of the diols. The carboxylic acid group is sufficiently hindered sterically to minimize amide formation, so that a polyurethane is formed with pendent acid groups. These are neutralized with amine and diluted with water to form stable dispersions. In use they are crosslinked with methylated monomeric melamine-formaldehyde resins.

Another approach is to first prepare an isocyanate-terminated polyurethane in acetone. This is then chain extended with a sulfonate substituted diamine. The acetone solution is diluted with water, and then the acetone is removed by distillation leaving an aqueous polyurethane dispersion stabilized by charge repulsion of sulfonate groups.

Epoxy resins – Approximately half of all the epoxy resins produced in the United States are used in the coatings industry. Epoxy coatings provide superior chemical resistance, high film adhesion strength, and tough but flexible films. On the other hand, epoxies are relatively expensive, slow drying (hardening), and do not provide as high a gloss retention as many other binders.

Many different types of epoxy resins have been used in the coatings industry. The most common are those produced from the reaction of bisphenol A with epichlorohydrin, as illustrated in Figure 14.

Figure 14. Epoxy Resin Formation

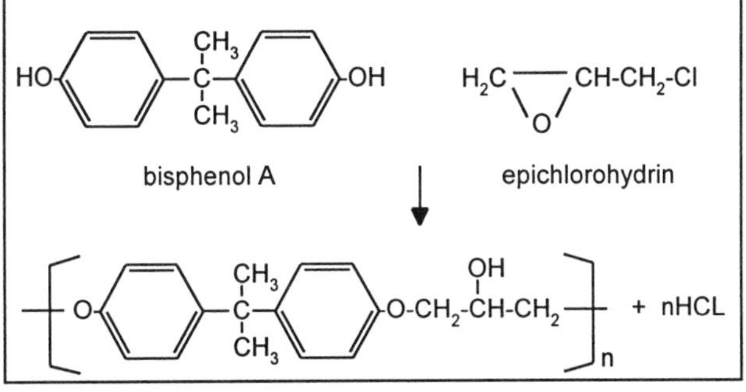

These epoxy resins can be cured in a two pack system using amines, polyamines, polyamides, or acid anhydrides. Aliphatic amines produce faster cures at room temperature than aromatic amines.

An amine combines with an epoxy group in a ring opening reaction, as in Figure 15. Through this type of reaction with a trifunctional amine and diepoxides, a three dimensional crosslinked matrix is obtained. This forms the basis of industrial maintenance coatings and marine coatings, among others. These coatings have exceptional resistance to solvent and chemical attack, but can be slow drying.

Figure 15. Epoxy/Amine Reaction

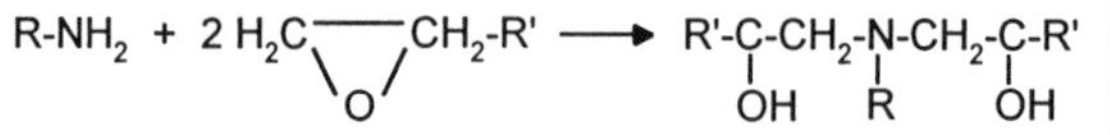

$$R\text{-}NH_2 + 2\,H_2C\underset{O}{—}CH_2\text{-}R' \longrightarrow R'\text{-}\underset{OH}{C}\text{-}CH_2\text{-}\underset{R}{N}\text{-}CH_2\text{-}\underset{OH}{C}\text{-}R'$$

Solvent and chemical resistance is even further improved by blending the epoxy with a phenolic resin and then baking the applied coating. At elevated temperatures, the phenolic resin and epoxy groups react through a ring opening mechanism, shown in Figure 16.

Figure 16. Phenolic/Epoxy Reaction

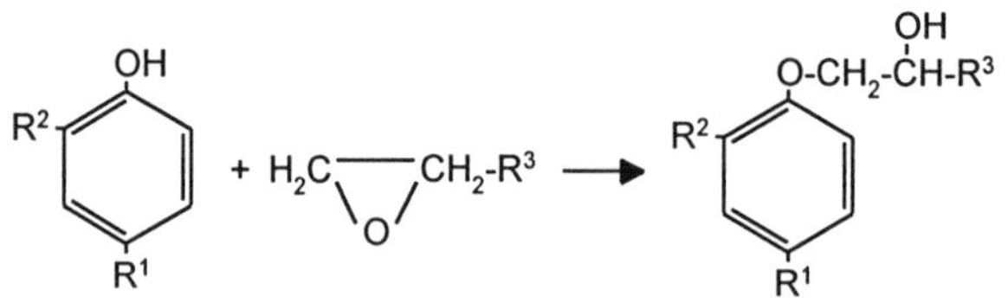

$$(R^1,R^2)C_6H_2\text{-}OH + H_2C\underset{O}{—}CH_2\text{-}R^3 \longrightarrow (R^1,R^2)C_6H_2\text{-}O\text{-}CH_2\text{-}\overset{OH}{C}H\text{-}R^3$$

The secondary hydroxyl groups react further to form a tight, crosslinked network which imparts the optimum solvent and chemical resistance. Flexibility of the film is reduced, however, compared to other epoxies and discoloration can be a problem. If a lighter color is needed, urea-formaldehyde resins are reacted with the epoxy with little loss in chemical and solvent resistance. Epoxy coatings are used to line the interior of cans, drums, tanks and piping.

Epoxy resin esters are formed from reacting an epoxy resin with fatty acids from drying oils in the same type of equipment used to make alkyds. The fatty acids react with both the epoxide and hydroxyl groups of the epoxy resin to form the epoxy ester, as in Figure 17.

Figure 17. Epoxy Ester Formation

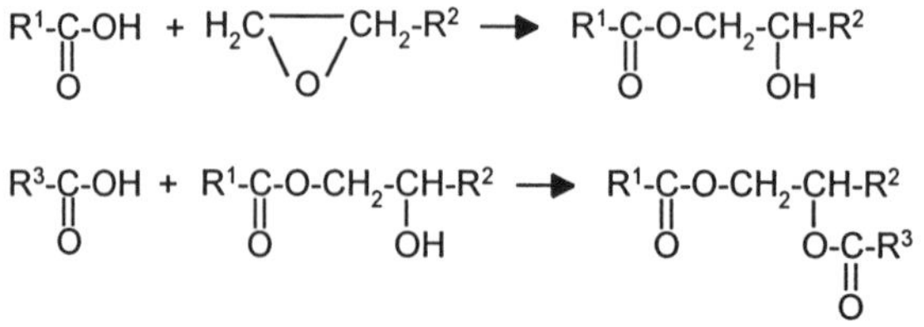

$$R^1\text{-}\underset{\|O}{C}\text{-}OH + H_2C\underset{O}{—}CH_2\text{-}R^2 \longrightarrow R^1\text{-}\underset{\|O}{C}\text{-}O\text{-}CH_2\text{-}\underset{OH}{C}H\text{-}R^2$$

$$R^3\text{-}\underset{\|O}{C}\text{-}OH + R^1\text{-}\underset{\|O}{C}\text{-}O\text{-}CH_2\text{-}\underset{OH}{C}H\text{-}R^2 \longrightarrow R^1\text{-}\underset{\|O}{C}\text{-}O\text{-}CH_2\text{-}\underset{O\text{-}\underset{\|O}{C}\text{-}R^3}{C}H\text{-}R^2$$

⌘⌘⌘

Epoxy resin esters are less expensive than other epoxies because they are partially composed of lower cost fatty acid. They also have superior flexibility. The epoxy esters have poorer solvent and chemical resistance than other epoxy types, but generally display better chemical and alkali resistance and better adhesion than conventional alkyds. As with alkyds, they are usually cured by autooxidative air drying with the aid of metallic driers. These coatings have found applications as industrial and maintenance paints, automotive primers and varnishes. These resins may not compare as well to alkyds in outdoor applications, however, because of a greater tendency to chalk.

Modifications of the above epoxy types are used as well. For example, phthalic anhydride is sometimes used to partially replace fatty acids in producing epoxy resin alkyds. Also short oil length epoxy esters (sometimes based on a non-drying oil), are occasionally modified with an amino resin to yield a coating binder capable of forming a very hard, chemically resistant film. There are also epoxy systems based on modifications of the epoxy resins with acrylic resins or polyisocyanate curing agents.

Amino resins – Amino resins are rarely used as the sole binder in coatings because they form hard, brittle films with only fair adhesion. They are, however, the predominant crosslinking agents for thermosetting coatings. Melamine-formaldehyde (MF) resins are most commonly used for these applications. Amino resins are prepared, in general, by reacting precursor and formaldehyde under alkaline conditions to form methylol groups. Alcohols are then reacted under acid conditions to yield ethers.

MF resins are produced by first reacting melamine and formaldehyde in the presence of aqueous sodium carbonate. The degree of methylolation depends upon the molar ratio of formaldehyde to melamine. Although methylolation will result in a mixture of partially methylolated derivatives, the reaction is usually carried out to optimize the content of either hexa-methylolmelamine or symmetrical trimethylolmelamine, as in Figure 18.

After methylolation, the sodium carbonate is neutralized and the alcohol or alcohol mixture is added together with an acid catalyst, typically nitric acid. Each methylol group and alcohol form the corresponding ether.

The most commonly used alcohols are methanol, butanol, isobutanol and 2-ethylhexanol. Self-condensation of the methylolmelamines also occurs during etherification so that melamine resins typically contain an appreciable quantity of dimers, trimers and higher oligomers.

The first commercial MF resins for coatings were butylated, from butanol or isobutanol with a relatively high level of oligomers. They became quickly popular because of their low cost and their ability to improve the properties of baked alkyd coatings. This class of melamine resin, widely

⌘⌘⌘

used from the 1930s through the 1950s, is characterized by a lower level of methylolation, and consequently lower functionality.

Figure 18. Methylolation of Melamine

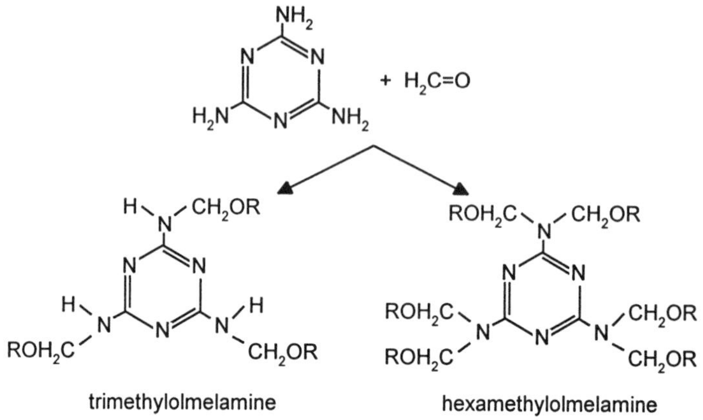

+ $H_2C{=}O$

trimethylolmelamine

hexamethylolmelamine

A class of MF resins with higher functionality and lower oligomer content was developed in the 1940s. These methylated MF resins have grown considerably in use since the 1970s and the shift toward water-based and high solids coatings. The methylated resins contain a high hexamethoxymethylmelamine content and are more suitable in aqueous systems than the older butylated resins, while imparting generally lower viscosity in high solids coating.

Melamine-formaldehyde resins are most often used to crosslink resins having hydroxyl functionality such as alkyds, thermoset acrylics, hydroxy terminated polyesters and bisphenol A epoxy resins. Crosslinking is through transetherification using an acid catalyst. MF resins will also crosslink carboxylic acid functional resins through esterification, although reaction rate is slower than with hydroxyl groups. This is significant in the formation of water-based coatings based on hydroxy-functional binders that have been modified with carboxylic acid groups to improve dispersibility. The curing system for these binders is chosen to ensure a high degree of esterification, so that residual carboxylic acid groups do not increase the water sensitivity of the cured film. MF resins are also increasingly used with binders having both hydroxyl and urethane functionality.

Amino resins analogues to the MF types are produced from glycoluril, acrylamide and methacrylamide (providing greater film flexibility), and benzoguanamine (providing greater alkali resistance), but the next most widely used are urea-formaldehyde (UF) resins. UF resins are generally the least expensive of the amino resins. They are also the most reactive.

⌘⌘⌘

Coatings with UF resins can therefore be formulated for ambient or low heat cures, but the resulting films have relatively poor hydrolytic stability. While the other amino resins are used in metal coatings (automotive finishes, coil coaings, can coatings), UF resins are used in coatings for temperature sensitive substrates, wood finishes for example, where corrosion resistance is unimportant.

Cellulosic binders – Cellulose is a natural anhydroglucose polymer that is insoluble in water and all organic solvents. The cellulose esters, nitrocellulose (actually cellulose nitrate) and cellulose acetobutyrate are solvent-soluble derivatives used in coatings.

Nitrocellulose is made by reacting cellulose with nitric acid. Cellulose hydrolysis and consequent molecular weight reduction are a function of the small amounts of water present during the reaction. Greater hydrolysis results in lower molecular weight, lower solution viscosity and lower film strength. The degree of nitration determines solvent compatibility. Polymers containing approximately 11% nitrogen are alcohol soluble and used in flexographic printing inks. Coatings grades contain about 12% nitrogen and are soluble in esters and ketones and in mixtures of esters and ketones with alcohols and hydrocarbons. Explosive grades contain about 13% nitrogen.

Nitrocellulose lacquers provide thermoplastic films of high gloss, easy repairability, but low durability. They were used for automotive finishes before acrylic lacquers were developed, but today are used principally in wood finishing. Cellulose acetobutyrate (mixed acetate, butyrate esters) has minor use as the binder in lacquers requiring lighter color and better color retention than those based on nitrocellulose. The principal use, however, has been as a component of acrylic-based metallic automotive finishes. The cellulosic contributes to flow control and the orientation of aluminum flake pigment parallel to the film surface for maximum metallic effect. Cellulose acetobutyrate polymers are tailored for solvent solubility and binder compatibility through manipulation of molecular weight and the ratios of acetate to butyrate to unreacted hydroxyl groups.

Extender Pigments

Industrial minerals are used in paints and coatings under the term "extender pigments". Their function as extenders may be interpreted as two-fold. Fine ground minerals have been traditionally used in paints as inexpensive fillers – they extend the film formed by the generally more expensive binder. Current formulation practice more appropriately uses most minerals as functional fillers. They are chosen to benefit coating stability, rheology, and application properties, and to enhance dry film properties. Platy minerals, such as talc and mica, may be used to maintain dispersion stability in storage and improve film durability and tensile properties after application. Very fine ground and ultrafine synthetic (precipitated) forms of silica and silicates may be used to control application rheology and ultimate film gloss.

Very fine particle size (<lμm) minerals are used as pigment extenders in a more literal sense. The pigment that is extended is usually TiO_2, the predominant white opacifying pigment in coatings. When the mineral particle size is comparable to that of the pigment (0.2 to 0.3μm) it can effectively separate the individual TiO_2 particles and provide the optimum pigment spacing for maximum opacifying effect. Pigment extenders are used in this way to reduce the amount of relatively expensive TiO_2 required for a given level of opacification and whiteness.

Calcium carbonate – Calcium carbonates are more widely used in paint and coatings systems on a weight basis than any other mineral filler. These materials are used to reduce the cost of many formulations, while improving both the rheological and mechanical properties of the coating.

In non-aqueous systems, ground limestone products are used extensively in exterior house paints and interior flats and semi-gloss enamels. The higher vehicle demand of the precipitated calcium carbonates limit the use of these materials in non-aqueous systems, although small amounts are used as flow control agents.

In aqueous systems, the hydrophilic nature and low water demand enable the easy incorportion of high levels of calcium carbonate with only minor increases in viscosity. Calcium carbonates are normally used in conjunction with other fillers and titanium dioxide to enhance the dry hiding power and rheological properites of the paint system. The optimum particle size range for ground natural filler grades of calcium carbonate in coatings is 3 to 7μm.

High brightness ultrafine ground and precipitated calcium carbonates are used as TiO_2 extenders and opacifiers in water-based paints. They have a higher binder demand than filler grades, but better dry hide and gloss retention.

Talc – Talc is probably used in a wider variety of coatings than any other single mineral pigment. Talc is normally self-suspending in paint vehicles and assists in keeping other pigments suspended. When settling does occur, it is generally soft and readily redispersed. Talcs are hydrophobic but disperse easily in most paints including aqueous systems. Talcs chalk more readily than most extender pigments, although less than calcium carbonate, and are frequently used in combination with other pigments to provide controlled chalking.

Most paint grade talcs possess both a high specific surface area and a fairly high oil absorption (binder demand). Although these properties limit the amount of talc that can be incorporated into a paint, its presence in moderate amounts contributes to improved rheological properties – antisettling, brushability, leveling, and sag resistance. Certain talcs impart a flatting effect that can be used to control the gloss of enamels. These talcs are generally 10μm grades with top sizes below 25μm. They disperse readily and yield a Hegman Fineness of 5 or above when incorporated on high speed equipment. The finest micronized talcs provide TiO_2 extension, excellent flatting, good low angle sheen and good burnishing resistance. In primers, they exhibit excellent suspension, holdout and sanding properties.

Platy talcs improve the bonding properties, film toughness and general durability of paint films. By their hydrophobic nature they also impart a marked improvemnt in the water and humidity resistance of the film. Tremolitic talcs, because they are prismatic and lower in oil absorption, provide easier dispersion, higher loading levels, less flatting and generally better dry hide than platy talcs. They also provide enhanced durability in exterior architectural paints and in traffic paints.

Talcs impart excellent sanding properties to primers and good scrubability to top coats. Coarse grades of talc are especially suited for wall primers, undercoats and texture paints where residual surface roughness is desirable.

Kaolin clay – The main use of kaolin in coatings is as a TiO_2 extender in water- based architectural paints. Partially calcined clays generally provide the best extension, dry hide, film durability and scrub resistance, and are widely used in latex interior wall paints. Water-washed and delaminated grades also provide these attributes, and contribute to suspension stability, covering power, film integrity, gloss control (finer particle size = glossier film), flow properties and leveling. Delaminated clays are prefered for exterior paints for more controlled chalking and overall durability.

Mica – Water-ground mica is used as an extender and filler in many paint and coating formulations. As a platy mineral, it reinforces the paint film during drying and curing, and as the film ages it reduces the internal stresses from oxidation and from thermal expansion and contraction. Mica increases the film's flexibility and reduces cracking and shrinkage; this results in improved adhesion and better weathering.

Because its flakes align parallel to the substrate, mica is used in exterior coatings to control moisture permeability and to reduce the penetration of light into the film. These coatings are therefore more resistant to fading, weathering and UV degradation, and provide good corrosion control. Mica provides good barrier resistance in primers and roof coatings, and penetration control in sealers for porous surfaces. High aspect ratio mica is used as is or coated with metallic oxides to impart a metallic effect to automotive coatings.

⌘⌘⌘

Mica's relatively low brightness and high binder demand necessitate low loading levels in most decorative coatings.

Diatomite – White, flux calcined diatomaceous silica is used as a functional filler and flatting agent. The varied size and shape and high surface area of the individual particles results in irregular texturing of the coating surface. This reduces the surface gloss and provides improved adhesion of subsequent coats. Flatting efficiency may be adversely affected by overgrinding since this tends to reduce particle irregularities. Diatomite is used in coatings for uniform control of gloss on irregular surfaces, improved sanding properties and increased film vapor permeability. The latter promotes more rapid drying and reduced blistering and peeling. Diatomite, however, reduces resistance of the coating to staining and soiling.

Natural silica – Ground quartz and novaculite are used in a wide variety of industrial and trade sales coatings as inexpensive extenders because their low binder demand allows high loadings. They are especially suited to use in priming systems, where surface hardness and roughness are desirable. The presence of silica in exterior oil-based paints improves the weatherability in a wide range of environments. It also improves the wear resistance of deck enamels and traffic paints. The platy shape of novaculite imparts additional mar, wear, and weather resistance. The use of micronized grades in interior and exterior latex systems increases surface uniformity and provides efficient flatting.

Barite and blanc fixe – Barium sulfate is used in coatings as barite, its natural form, and blanc fixe, the synthetic precipitate. Blanc fixe is used where higher brightness and purity and finer particle size are required. Barium sulfate is used for its low oil absorption, high density, low abrasiveness, chemical inertness, heat resistance, light fastness, low solubility (in water, acids, and alkalies) and its low spectral absorbance over a broad spectral band from the near infrared region through the visible region and into the near ultraviolet.

Finely ground barite and blanc fixe are widely used in the coatings industry, particulary as extenders for both automotive and industrial primers. These primers require materials with high filling power and fine particle size. The low oil absorption of barite and blanc fixe permit high loadings with a minimal effect on rheology. The high filling level results in a smooth non-porous primer surface possessing good sanding properties and allowing excellent gloss retention for subsequent topcoats. The use of these extenders in topcoats improves flow properties, surface hardness, and color stability.

Blanc fixe and finely ground barite are also used as extenders for water-based paints, particularly in wood paints requiring a high level of gloss and minimum color distortion. The high refractive index of these materials allow them to function as a translucent white pigment in latex paints, providing a moderate degree of hiding power, while having a minimal effect on the color of the primary pigment. The chemical resistance of barium sulfate imparts good weathering characteristics to both oil-based and water-based exterior paints.

Wollastonite – Wollastonite has low oil and water absorption, allowing high loadings in coatings. Fine ground wollastonite is used as a pigment extender, and to provide resistance to flash and early rust due to its alkaline pH. The acicular wollastonite particles reinforce paint films, even at the low aspect ratios of fine-ground grades. This contributes to durability and superior scrub resistance. Coarse grades are used in textured paints to impart surface roughness, to aid in peak building, and to inhibit mud-cracking. Finer ground and micronized grades are used in epoxy powder coatings to promote smooth flow, water resistance, improved wet adhesion and good gloss.

Synthetic silica – Precipitated silicas are used in coatings to provide flatting, mar resistance and abrasion resistance. Precipitated and fumed silicas are used as suspending agents and rheology control agents in some specialty coatings. Synthetic silicas are used as flattening agents in clear coatings, as their low refractive index makes them transparent in the film.

Nepheline syenite – Fine-ground, high brightness nepheline syenite is used in exterior architectural paints for its low binder demand (high loading potential), plus its ability to impart abrasion resistance, good tint retention and good film durability through control of checking and cracking. It is also used in interior latex paints for improved scrub resistance.

Synthetic sodium aluminosilicate – The synthetic sodium aluminosilicates (SAS) are produced as high brightness precipitates from the reaction of aluminum sulfate and sodium silicate. Primary particles are only 15 to 25 nm, but these aggregate into functional particles averaging about 5μm. The SAS products are used primarily as TiO_2 extenders in latex paints, and in some cases to improve scrub- and stain resistance, and as flatting agents.

Solvents

In coatings technology, a solvent is generally considered a volatile organic compound used to dissolve the binder resin so that it can be applied in a uniform thin film. This is actually something of an oversimplication, with solvents often classified by function as either active, latent, or diluent. An active solvent will dissolve the binder. A latent solvent will not itself dissolve the binder, but will act as a cosolvent when used with an active solvent. A diluent solvent usually has no solvency for the resin, but can be used to dilute the active solvent without precipitating the binder. Diluent solvents are used to reduce coating viscosity and cost. Water, although it will dissolve certain additives, is considered a vehicle for the dispersed or emulsified binders used in aqueous systems. It is not generally considered a solvent.

The required balance of solvency and evaporation rate guides the selection of solvents for each type of coating. The solvent or solvent blend must keep the binder in solution throughout the drying process in order to optimize subsequent film properties. Evaporation rate must be appropriate for the application method, spraying vs. brushing for example, and in general must be initially relatively quick, to avoid sag, and then more controlled, to allow proper leveling and adhesion.

Coatings solvents are also classified according to chemical nature as hydrocarbons, oxygenated solvents, and "other", which are usually hydrocarbon derivatives.

Hydrocarbon solvents – Hydrocarbons are the most commonly used coatings solvents. They are typically subdivided as principally aliphatic, naphthenic, or aromatic. Most now used are aliphatic, containing mainly iso- and normal paraffins. Some alphatics contain naphthenes (cycloparaffins) and aromatics, which improve solvation power but increase odor.

The most commonly used aliphatic solvent is mineral spirits. Several grades are available based on evaporation rate and solvency. Odorless grades are relatively low in odor because they contain mostly isoparaffins, although a lack of naphthenes and aromatics reduces solvency with many resins. Mineral spirits are widely used in architectural paints because their balance of moderate solvency and moderately slow evaporation impart good brushability, leveling and wet edge. VM&P naphthas are similar in solvency to mineral spirits but evaporate much more quickly. They are used primarily in sprayed coatings.

The aromatic solvents provide stronger solvency but greater odor. The most common are toluene, xylene (mixed isomers), and two high flash point aromatic naphthas. The fast evaporating toluene is used in spray paints and industrial coatings. The mixed xylenes, with moderate evaporation rate, are used in industrial coatings, while the slow evaporating aromatic naphthas are used in baked finishes.

Terpenes, unsaturated cyclic hydrocarbons obtained from pine trees, represent the historically oldest coating solvents, although they are infrequently used today. Turpentine is the most common terpene solvent, although dipentene, with stronger solvency and slower evaporation, is also used. Turpentine offers greater solvency, although in a narrower range, faster evaporation rate, greater odor and higher cost than the mineral spirits which has widely replaced it.

Oxygenated solvents – The principal oxygenated solvents for coatings are ketones, esters, glycol ethers, and alcohols. These are high purity synthetic compounds as opposed to the hydrocarbons, which are naturally derived chemical mixtures. The oxygenated solvents also offer much stronger solvency and partial to complete water miscibility. They are widely used as the active solvents for synthetic binders because of their ability to be blended to provide tailored solvency and evaporation rate.

The ketones are characterized by their strong odor and range of water solubility and evaporation rate. The very fast evaporating acetone is used in sprayed coatings, particularly nitrocellulose and acrylic lacquers, to reduce viscosity and then flash off during application. The fast evaporating methylethyl ketone, and moderate evaporation rate methyl isobutyl ketone are widely used as active solvents in paints and lacquers based on synthetic binders, while the very slow evaporating isophorone is used in baked industrial coatings.

The esters provide solvency nearly equal that of ketones, but with a more pleasant odor. The most common are the alkyl acetates, ranging from the very rapid evaporating methyl acetate to the slow evaporating hexyl acetate. n-Butyl acetate, one of the most widely used ester solvents, has a medium evaporation rate, and is used as a standard reference for expressing the evaporation rate of other solvents. In comparison to its assigned evaporation rate of 100, acetone has an evaporation rate of 1160, and regular mineral spirits has an evaporation rate of 10. Glycol ether acetates are relatively slow evaporating and are used as coalescents in latex paints, and retarder solvents in solvent coatings. In the latter capacity they maintain flow and leveling after the other solvents have evaporated.

The glycol ethers provide both alcohol and ether functionality and are characterized by mild odor, water miscibility, strong solvency and slow evaporation. Ethylene glycol ethers have been widely used in water-based coatings as coupling solvents for water reducible binders, and at low levels in lacquers as retarder solvents. More recently, propylene glycol ethers have replaced ethylene glycol ethers in many applicaitons for health and safety reasons.

Alcohols are used at low levels in water based paints as coupling solvents with glycol ethers and in certain solvent coatings as latent solvents. n-Butanol and denatured ethanol are the most commonly used, with the latter also serving as an active solvent for shellac, poly(vinyl acetate), and several phenolics.

Organic carbonates and furan derivatives are specialty oxygenated solvents. Ethylene carbonate and propylene carbonate are cyclic esters with good solvency, very slow evaporation and very low odor. Tetrahydrofuran, furfuryl alcohol and tetrahydrofurfuryl alcohol are cyclic ethers with particularly strong solvency for vinyl polymers and certain other synthetic binders.

"Other" solvents – N-methyl-2-pyrolidone is a specialty solvent used in water-based paints. It offers strong solvency, water solubility, biodegradability, slow evaporation and low toxicity. 1,1,1-trichloroethane is used as a fast evaporating solvent in some spray coatings because it is not photochemically reactive and is thus not included in VOC content. Health and environmental concerns over chlorinated solvents in general have restricted its use. Supercritical carbon dioxide is used as a diluent solvent in airless spray coating operations for its high solubility in most formulations, its low toxicity, low cost, its viscosity reducing effect, and its behavior on application as a highly volatile solvent.

Pigments

Pigments serve the decorative function of a paint by their effect on the reflection and refraction of light. White pigments reflect all wavelengths of light equally; black pigments absorb all wavelengths equally. A specific colored pigment imparts a desired color, or hue, by selectively absorbing and reflecting different wavelengths of the visible light spectrum.

Pigments differ also in tinting strength (ability to color a white base) and brightness. Different pigments can even appear to have exactly the same shade of color under indoor lighting, but show differences in color in outdoor light. This phenomenon, called metamerism, is due to differing spectrophotometric absorption characteristics between the pigments.

Other important pigment characteristics are lightfastness, bleed resistance and opacity. Lightfastness describes the pigment's resistance to fading or changing color from prolonged light exposure, especially after ultraviolet exposure outdoors. Some pigments are also sensitive to prolonged exposure to heat. Bleed resistance is important when a top coat is applied over a base coat. Some pigments in the base coat may bleed into the applied coat if they are soluble in the top coat's solvents.

Opacity, or hiding power, is a property of particular interest with white pigments such as titanium dioxide. This quality relates to the ability of the pigment particles in the film to prevent light from penetrating through the film to the substrate and back to an observer. It is the pigment's ability in a dried film to obscure any color from the substrate. Hiding power can be measured as the number of square feet covered from one pound of a pigment. If a given paint contains a pigment with good opacity, only one coat may be needed. If opacity is poor, two or more coats may be required.

⌘⌘⌘

A pigment's particle size and refractive index determine its degree of opacity. The smaller the particles, the greater the number of interfaces between the dispersed pigment and binder. This increases light scattering until the particle diameters are about one-half the wavelength of light (0.2 to 0.4 μm). With even smaller particles, scattering power diminishes because the light will go around them.

The greater the difference in refractive index between the binder and pigment, the greater the degree of light scattering and the higher the opacity. If the refractive indexes of the pigment and binder are very similar, little hiding power will result. Flat paints have greater hiding power than high gloss paints because they are typically so highly loaded with pigment that the pigment particles partially interface with the air. Since air has a very low refractive index, higher opacity results.

Chemical reactivity can also be an important consideration in pigment selection. For example, some pigments can react with air pollutants causing discoloration.

There are two broad classes of paint pigments, organic and inorganic. In most cases, there will be several organic and inorganic pigments for the same color group.

White pigments – Many years ago the chief white pigment in paints was basic lead carbonate. Because of toxicity problems associated with lead compounds, this pigment's use has been almost compeltely replaced by other white pigments, mainly titanium dioxide. TiO_2 is an ideal white pigment because it is available with an average particle size close to one-half the wavelength of light, the optimal for light scattering. It has a relatively high refractive index – 2.5 for anatase and 2.7 for the rutile crystallographic form – and is chemically inert. Titanium dioxide is the best general purpose white pigment for hiding power. Rutile is used when the greatest hiding power is needed, while anatase is used when chalking is desired.

Other white pigments are used but all have shortcomings compared to titanium dioxide. Zinc oxide can be chemically reactive with the binder resin and is not as cost effective in hiding power. Lithopone (barium sulfate/zinc sulfide) is generally too costly for high volume use in the United States. Other white minerals, such as talc and calcium carbonate, are not effective compared to titanium dioxide mainly due to insufficiently high refractive index.

Black pigments – Carbon black is the most common black pigment used, although it is one of the most difficult pigments to disperse well. Carbon blacks are insoluble in all solvents but do reportedly have some affinity for aromatic solvents. They have very high tinting strength and opacity. There are three types of carbon blacks available to the paint industry – furnace blacks, thermal black and channel blacks. Furnace blacks are the most commonly used and provide medium jetness (black intensity). Thermal blacks have larger particle size, are more expensive and display a lower degree of jetness. Channel blacks provide the highest degree of jetness and are still used in paints even though they must be imported from abroad.

Black iron oxides are available in a range of jetness and are preferred to carbon blacks in certain applications because of their lower oil absorption and relative ease of dispersion.

Colored pigments – The inorganic colored pigments are generally less expensive and more widely used than their organic counterparts. The inorganics provide excellent heat stability and lightfastness, but are more difficult to keep uniformly suspended due to their relatively high specific gravity. Many of the inorganic pigments are based on lead, chromium, and cadmium, and their use is limited by toxicity concerns. Some representative inorganic pigment types are as follows.

Red (iron oxide): Fe_2O_3
Yellow (hydrated iron oxide): $Fe_2O_3 \cdot H_2O$
Yellow (lead chromate): $PbCrO_4$
Yellow (cadmium sulfide): CdS
Green (chromium oxide): Cr_2O_3
Blue (iron blue): $Fe(NH_4)Fe(CN)_6$
Blue (ultramarine blue): $Na_6Al_6Si_6O_{24}S_4$
Orange (chrome orange): $PbCrO_4 \cdot PbO$
Orange (molybdate orange) $PbCrO_4 \cdot PbMoO_4 \cdot PbSO_4$

A spectrum of hues is obtained through hydration, copreciptation, and blending of pigment compounds in various proportions.

Organic pigments are characterized as generally purer, brighter and higher in tinting strength than the inorganics. The chemical nature of some organic pigments makes them sensitive to heat, solvents, or pH extremes.

Many organic structures absorb electromagnetic radiation in the ultraviolet portion of the spectrum, but not in the lower energy visible portion. Thus the vast majority of organic compounds are clear white powders. In order to obtain a selective absorption in a portion of the visible spectrum, conjugated organic structures consisting of fused aromatic rings, chromophore groups and/or organometallic complexes are used.

Chromophores are those groups within an organic structure that are essential to selective absorption and reflection of specific wavelengths of visible light. They are necessary for the appearance of a color other than white. Chromophore groups are represented by:

⌘⌘⌘

$-N=N-$ $-N=C<$ $-NO_2$

$>C=S$ $>C=O$ $>C=C<$

Auxochromes are substituent groups which modify the color of the basic pigment compound. Auxochrome groups are represented by:

$-NH_2$ $-NHR$ $-NR_2$ $-NO_2$

$-OH$ $-COOH$ $-CH_3$ $-Cl$ $-Br$

The organic pigments can be categorized according to district chemical types as monoazo pigments, diazo pigments, phthalocyanine pigments, quinacridone pigments, acid dye pigments, basic dye pigments, and miscellaneous polycyclic pigments. A description of the chemistry and attributes of these chemical types and the color groups within them is beyond the scope of this chapter. For this, the Pigment Handbook (Peter A. Lewis: Wiley, 1988) is recommended.

Metallic and nacreous pigments – Metal flake pigments are made from aluminum, stainless steel, and bronze. The most widely used are those of aluminum. These are made by milling a mineral spirits suspension of finely divided aluminum metal using a steel ball mill. The resulting thin flakes can be produced in a variety of particle sizes. Aluminum flakes for coatings are produced as leafing and nonleafing. Leafing flakes are surface treated, usually with stearic acid to minimize surface tension. This provides flake orientation at the surface of the coating film. The continuous overlapping of aluminum flakes acts as a barrier to oxygen and water vapor, while providing a bright metallic appearance. Leafing aluminum flake is used in corrosion resistant paints for steel structures, and in roof coatings to improve weatherability and decrease heat absorption.

Nonleafing aluminum flake is not coated and is formulated to be encapsulated within the binder film, but parallel to its surface. It is used most commonly with transparent color pigments in automotive top coats to provide a metallic color that changes shade with the angle of viewing.

Nacreous (pearlescent) pigments are thin transparent platelets of high refractive index which partially transmit and partially reflect light. The pearlescent effect is due to specular reflection from the broad surfaces of the transparent platelets, and the parallel orientation of those platelets at various depths within the binder film. Light transmitted through platelets near the film surface is partially reflected by deeper platelets. The dependence of reflection on viewing angle and the sense of depth created by reflection from many layers produces the pearly luster.

⌘⌘⌘

The first pearl pigment was derived from fish scales as platy crystals of guanine and hypoxanthine. This remained unrivaled in luster for approximately 300 years, until the refinement of crystallization processes for platy basic lead carbonate in the 1950s. Bismuth oxychloride was developed as a nontoxic alternative to lead carbonate, but the most widely used nacreous pigments in coatings today are the coated micas.

White nacreous pigments are made by forming a uniform coating of TiO_2 on mica platelets. The mica serves as a transparent template so that the high refractive index TiO_2 can assume the required platy shape. Pearlescence (white) is optimized when the pigment optical thickness (platelet thickness times refractive index) is in the 100nm (blue white) to 140nm (yellow white) range. Thicker platelets are produced to make interference colors. Light interference is created by interaction of the reflections from the upper and lower surfaces of the platelet. A reflection maximum occurs at the wavelength of light for which these two reflections are in phase. A reflection minimum occurs at the wavelength for which they are exactly out of phase and cancel each other. A transparent, colorless nacreous pigment platelet acts as a filter, separating light into two components – the reflected color and the complementary transmission color.

When coated on a white surface, the nacreous pigment provides a two-tone metallic effect. The reflected color (specular) is seen as a highlight. The background color at nonspecular angles is created by the diffuse reflection (from the white surface) of the transmission color. The complementary reflection/transmission colors as a function of pigment optical thickness range in a continuous spectrum from blue-white/yellow-white at 100nm, to magenta/green at 250nm, to green/red at 370nm.

The color effects of nacreous pigments are supplemented by overcoating the TiO_2-mica platelets with a thin transparent film of a light absorbing colorant. For example, iron oxide for yellow, ferric ferrocyanide for blue, chromium oxide for green, and carmine for red. The combination pigments thus produced can have matched reflectance and absorption color, or these can be two different colors. When the colors are the same, the same color is seen at all viewing angles, with bright (metallic) highlights provided at specular angles. With different colors a two color effect is achieved. Combination interference/absorption pigments are made in the gold to red range by coating Fe_2O_3 directly onto mica.

The TiO_2-coated and Fe_2O_3-coated micas are used to provide metallic, two-tone, and two-color effects in industrial and automotive coatings. They are widely used in the latter with surface treatments designed to improve dispersability and weatherability.

⌘⌘⌘

ADDITIVES

Dispersants and Surfactants

Dispersants are used in water-based coatings specifically to impart an electrical change to pigments and extenders. Agglomeration or flocculation of these solid particles is thereby inhibited through charge repulsion. This retards settling and minimizes the viscosity contribution of particle-particle associations. Dispersants are therefore used to improve suspension stability and to enable the inclusion of a higher level of particulate components without sacrificing rheological properties.

Products sold as dispersants, as opposed to surfactants, are typically sodium or ammonium salts of lignosulfonates, polynaphthalene sulfonates, or relatively short chain polyacrylates. These chemicals have generally little effect on surface tension or interfacial tension, but will align themselves in an aqueous vehicle so that their organic groups loosely attach to and coat the pigment and mineral surfaces, leaving their charged groups protruding into the water. This charged film renders the particles mutually repulsive.

Because dispersants have relatively poor surface activity, they are often used with surfactants. Surfactants reduce surface tension such that the liquid vehicle can effectively wet the pigment and mineral surfaces so that dispersant-particulate association can occur. This is a particular requirement for hydrophobic solids such as talc. Surfactants are used as wetting agents in this regard, and to facilitate coating application by allowing the paint to better wet and spread on the substrate.

Anionic surfactants are typically sodium salts of alkyl sulfates, alkyl sulfonates, alkyl aryl sulfonates and sulfosuccinates. The alkyl and/or aryl group is ideally selected for optimum affinity to the particulate surfaces. Anionic surfactants are often employed to serve simultaneously as dispersants, although it is not uncommon for a paint to contain both surfactants and dispersants for optimum effect.

Nonionic surfactants for aqueous use contain an lipophilic or pigment/mineral compatible group, typically long chain alkyl or alkylaryl group, and a hydrophylic group, typically polyethylene oxide. The inorganophilic groups bond loosely to the pigment and mineral surfaces creating a film of hydrophilic groups to ensure wetting by the aqueous vehicle. Nonionic surfactants can also provide a dispersant effect by providing steric hindrance to particle flocculation.

Dispersion aids for solvent-based paints function primarily as wetting agents, and are typically nonionic. They contain an anchor group, selected for its affinity to the particulates in the paint formulation, and a polymeric or high molecular weight group selected for compatability with the binder and solvent.

⌘⌘⌘

The surfactants used to stabilize binder lattices in latex paints are properly considered emulsifying agents. Their function is analogous to that of wetting agents. An emulsifying agent has an organophilic component that has at least partial solubility in the binder droplets, and a hydrophilic component that allows the droplet to be wet by the aqueous vehicle. Emulsifying agents can be either anionic or nonionic and inhibit droplet coalescence by charge repulsion or steric hindrance. A surfactant-stabilized latex particle is illustrated in Figure 19.

Figure 19. Surfactant-Stabilized Latex

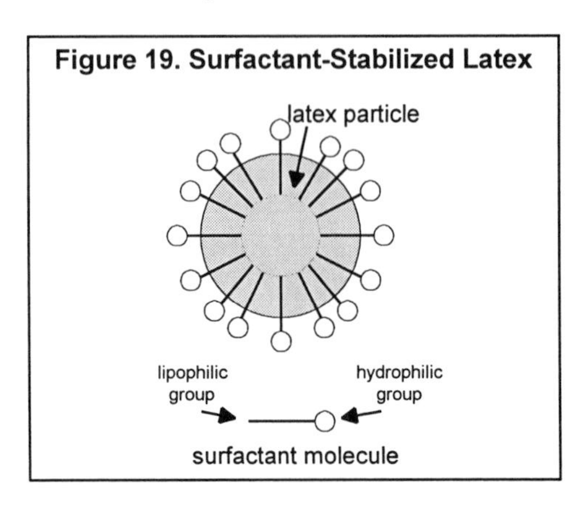

Rheological Agents

Surfactants and dispersants are used to disperse pigment, mineral and latex particles to promote suspension stability, to minimize the viscosity contribution of particle flocculation or agglomeration, and to enable higher solids content. Adjusting paint rheology for optimal in-can stability and application properties normally requires the use of thickeners, thixotropes and rheology modifiers of various types.

Liquid paints for brush, roller or spray application are typically pseudoplastic, or shear-thinning. Paint viscosity decreases as the shearing rate increases. Undisturbed in the can, high viscosity inhibits settling and flocculation of pigments and minerals. The shear of brushing or spraying reduces the viscosity so that a thin film is easily applied. Once on the substrate, the paint regains high viscosity so that sagging is avoided. Pseudoplasticity, however, implies a very rapid or instantaneous recovery of viscosity once shearing stops. This is not always desirable because rapid viscosity recovery will impede leveling and can retain brush marks, roller marks, or orange peel.

The proper balance of anti-sag and good leveling is often attained through the use of polymeric thickeners and mineral thixotropes which impart the desired degree of thixotropy to the paint. A thixotropic paint is shear thinning, for easy application and good coverage, but once shear is removed viscosity recovers over a period of time rather than immediately. When properly formulated, this provides just enough time for the paint film to remain sufficiently fluid for good leveling, while developing enough viscosity to prevent sagging.

Roller application, widely used on interior walls and ceilings, involves an additional rheological consideration – control of spattering. At the point of contact between roller and substrate, the paint is under pressure. As the roller moves, the pressure is released and small bubbles are formed between the substrate and the receeding roller edge. As the roller continues, the bubbles expand and their walls are stretched from roller to substrate until they resemble fibers. These paint "fibers" are eventually stretched long enough to break. Ideally, part of the fiber retracts into the substrate coating and the rest onto the roller. In reality, however, the fiber generally breaks in two places, releasing free paint droplets which fly off as spatter onto the painter and the surrounding area.

Spattering is a particular problem with latex paints containing conventional water-soluble, high molecular weight thickeners. Paint fibers being stretched between roller and substrate experience elongation flow rather than shear flow. The corresponding elongation viscosity arises from the stretching and aligning of the thickener polymer chains. The long chain, high viscosity polymers impart the high elongation viscosity which promotes spatter. Spattering is minimized with low molecular weight (short chain) grades of these thickeners, but at the expense of shear viscosity.

This is an issue concerning viscosity at both low and high shear. Adequate low shear viscosity is required to promote product uniformity. High shear viscosity is an important factor in spreading – paint coverage per given area – and in hiding. The most widely used conventional thickeners are very shear-thinning, such that very thin paint films can be applied. If the applied film is too thin, the dried film will give poor hiding and a second coat of paint will be required. One-coat paints require a relatively high high-shear viscosity so that an optimally thick wet paint film is applied.

In paints using conventional polymeric thickeners, a balance among viscosity, wet film build and spatter resistance is often achieved with intermediate viscosity polymers together with mineral thixotropes. An alternative which has gained increasing acceptance over the past two decades is the use of associative thickeners, as will be discussed.

Conventional thickeners – When conventional thickeners dissolve, their randomly coiled and entangled polymer chains occupy a large hydrodynamic volume immobilizing interstitial water. Although paint performance properties are a function of low and high shear viscosity, it is normally a moderate shear viscosity which guides processing quality control and the user's perception of paint consistency and quality. Moderate shear

⌘⌘⌘

viscosity represents the apparent viscosity when the paint is stirred and poured. When formulating to a given moderate shear viscosity with a conventional thickener of specific polymer type, molecular weight is proportional to low shear viscosity, but inversely proportional to high shear viscosity. Viewed another way, the higher the molecular weight, the lower the required use level to attain the desired consistency and physical stability. The larger polymer coils are more entangled and occupy a greater hydrodynamic volume so fewer are required for a given viscosity. Under high shear, however, the low numbers of the now uncoiled and disentangled polymer chains provide less resistance to shear flow than a greater number of shorter chains. The compromise, as noted above, has been to use medium viscosity polymer thickeners, often with a mineral thixotrope.

The most commonly used thickeners in water-based coatings are the nonionic cellulosics. Medium- and high viscosity hydroxyethyl cellulose (HEC) dominates this group. HEC is made by reacting ethylene oxide with approximately one-third the reactive hydroxyls along the cellulose backbone. Bioresistant grades are produced for the coatings industry by making the distribution of ethoxylated hydroxyls along the cellulose chain as uniform as possible. Hydroxypropyl methyl cellulose (HPMC) is the only other cellulosic with significant use in coatings. Ethyl hydroxyethyl cellulose and the anionic carboxymethyl cellulose are used in minor quantities. One derivative of guar gum, a natural polymer with a backbone similar to cellulose, finds some use in aqueous paints. Hydroxypropyl guar is more shear-thinning than the cellulosics for better suspension and sag resistance, but produces more roll spatter and is more susceptible to bacterial and enzymatic degradation.

The only conventional synthetic thickeners used in substantial amounts in water-based coatings are the alkali-soluble emulsions (ASE). These are supplied in liquid form as low viscosity water-insoluble latexes at low pH. In use they are neutralized with anmonia or a volatile aminoalcohol to provide thickening. The most commonly used ASEs are copolymers of methacrylic acid and ethyl acrylate. Some are lightly crosslinked with a small amount (<1%) of a polyfunctional monomer to enhance viscosity. The ASEs are similar to the cellulosics in rheology, but more bioresistant and somewhat more water sensitive.

Mineral thixotropes – The mineral thixotropes are not used as the primary thickener in water-based coatings as their efficiency is low compared to the polymers. They are,however, commonly used with conventional thickeners, particularly the cellulosics, to provide synergistic improvements in yield value (for suspension stabilization) and viscosity. Smectite and hormite clays provide thixotropic aqueous dispersions and can impart a controlled degree of thixotropy to paints thickened with the conventional polymers.

⌘⌘⌘

The most widely used mineral thixotrope in aqueous coatings is the hormite clay palygorskite, which is better known as attapulgite. The micronized attapulgite used in paint is composed of low aspect ratio needles which deagglomerate in water in proportion to the amount of applied shear. When shear is removed, these needles form a loosely cohesive random colloidal lattice. Undisturbed, this lattice provides yield value for suspension and emulsion stabilization and is shear thinning and thixotropic in response to shear cycling. Since attapulgite deagglomeration is mechanically rather than ionically driven, as for smectite clays, it is insensitive to most solutes and can be added at any convenient point in paint manufacture. Because optimum deagglomeration and dispersion of attapulgite is energy intensive, concentrated predispersed aqueous suspensions are sold.

Hydroclassified natural smectite clays and synthetic smectites are used in preference to attapulgite in those coatings where the finer tailoring of rheological properties they allow offsets their price premium and less convenient use requirements. The hydration and dispersion of a smectite clay into individual colloidal platelets is driven by shear and osmotic swelling. Once separated, the absence of shear allows these platelets to form a more uniform and cohesive structure than attapulgite's through ionic attraction of slightly positive platelet edges to negatively charged platelet faces. Solutes present during hydration can inhibit osmetic swelling and are generally proscribed. Solutes introduced after hydration can improve the cohesiveness of the colloidal structure (i.e., improve viscosity and yield value) but will have the opposite effect at greater concentration. Despite this relative lack of processing versatility compared to attapulgite, the smectites are used for their higher efficiency and the controllable synergism they provide with polymeric thickeners.

Precipitated silicas, from the reaction of sodium silicate and sulfuric acid under alkaline conditions, find some use as thixotropes in water-based coatings. The functional particles are clusters of strongly agglomerated primary amorphous silica particles. These clusters are chained together by weak hydrogen bonds into three-dimensional networks when dispersed in a liquid. In highly polar liquids, this network structure is ensured by bridging additives; in aqueous coatings, certain nonionic surfactants serve this function. Silica gels, from the reaction of sodium silicate and sulfuric acid under acid conditions, and fumed silicas are used in water-based coatings for flatting rather than rheology control.

Associative thickeners – Associative thickeners are relatively low molecular weight water soluble polymers with hydrophobic groups capping polymer and pendant chain ends. These thickeners provide little or no conventional (hydrodynamic) thickening. They instead form a network based on hydrophobe-hydrophobe affinity. Hydrophobe groups associate among the polymer chains, and they adsorb onto hydrophobic surfaces in the dispersed phases of the paint. These surfaces are primarily those of the latex binder, but can include hydrophobic pigments and minerals such as talc. Conventional and associative thickening in the presence of surfactant stabilized latex particles are compared in Figure 20.

Figure 20. Conventional vs. Associative Thickening of Latex Emulsion

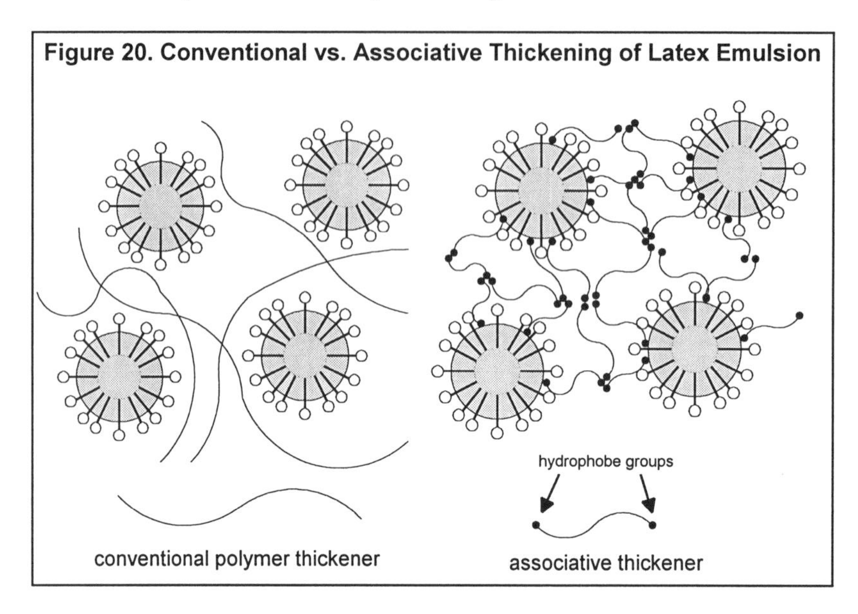

As a class, the associative thickeners have grown significantly in popularity during the past decade because they are generally less pseudoplastic than conventional thickeners and more Newtonian in their rheological effect. This, for instance, allows greater high-shear viscosity on paint application for optimal film build and hiding, and lower low-shear viscosity for good leveling. Associative thickeners also generally impart lower elongation viscosity, and are therefore preferred for spatter resistance.

Among the first associative thickeners were hydrophobically modified analogues of the alkali-soluble emulsions. These are lower in molecular weight than the conventional ASE thickeners, but on neutralization they do provide some hydrodynamic viscosity. These products are typically terpolymers of methacrylic or itaconic acid, ethyl acrylate and a macromonomer containing a polyethylene oxide chain terminated with an alky or alkylaryl hydrophobe group. The polymer's associative side chain is attached to its backbone in most products by either an ester or urethane linkage. More recently developed ASE-based thickeners have as associative side chains polyethylene oxide segments joined by urethane linkages, with terminal hydrophobe groups. These side chains mimic the structure of the

⌘⌘⌘

most popular class of associative thickeners, the hydrophobically modified ethoxylate urethane polymers (HEURs).

The HEURs have a backbone and side chains composed of polyethylene oxide segments joined through urethane linkages, terminated at all ends with hydrophobe groups. These products are currently the preferred associative thickeners because they impart the most Newtonian rheology and generally provide the best flow, leveling, film build and spatter resistance. This comes at the expense of sag resistance, suspending ability and thickening efficiency, however, when they are used as the sole rheological agent. The HEURs are sold as pourable aqueous solutions with an organic cosolvent added to suppress viscosity and ensure flowability.

The newest group of associative thickeners are schematically similar to the HEURs. The polyether segments may include ethylene oxide-propylene oxide block copolymer in place of polyethylene oxide, and the linkages are other than urethane (e.g. amide, ether).

The hydrophobically modified cellulosics consist of hydroxyethyl cellulose-based products, the first commercial associative thickeners, and more recently ethyl hydroxyethyl cellulose-based products. Relatively low molecular weight grades of the cellulosics are used, and these are modified via available hydroxyl groups on the polymer backbone. Hydrophobe substituents are one or two ethylene oxide units terminated with an alkyl or alkylaryl group. The hydrophobe modification makes these cellulosics more Newtonian for better film build and leveling.

Solvent system thickeners – Hydrogenated castor oil is used as a thixotropic thickener in paints based on nonpolar aliphatic solvents. Hydroxyl groups on the fatty acid chains of the castor oil triglyceride structure enable the formation of a colloidal network via hydrogen bonds. Polar attraction among functional groups accounts for the ability of polyamide-alkyd resins to function as thixotropic thickeners in nonpolar solvents.

Precipitated and fumed silicas are used to thicken solvent-based paints, with the best efficiency in nonpolar systems. Both can be used in polar solvents, although higher levels are generally required, with fumed silica providing thesomewhat better efficiency of the two. The organoclays are purified smectite clays cation exchanged with quaternary ammonium salts. They have been widely used in solvent-based coatings because the various grades available, depending upon the organic treatment, can thicken aromatic or polar solvents as well as nonpolar solvents. The conventional organoclays require a polar activator to facilitate deagglomeration of platelets and the formation of a thixotropic colloidal structure via hydrogen bonding. New organoclay products are suitably treated to eliminate the need for a polar activator.

⌘⌘⌘

Driers

The driers used in paint formulations are catalysts for the oxidation and polymerization reactions that promote binder film hardening after application. Most driers are soaps of transition metals, providing catalytic activity based on the ability of the metal cation to be oxidized from a stable lower valency to a less stable higher valency. The organic acids used to form these soaps provide compatibility in solvent-based paints and ensure emulsification into water-based systems. The earliest driers were linoleates, rosinates and tallates. These were largely replaced by naphthenates and particularly 2-ethylhexanoates. Currently, neodecanoates and so-called synthetic acid (similar to 2-ethylhexaonate) soaps are growing in use, in part because they can be produced with a higher metal concentration.

Paint driers are conventionally differentiated as primary or secondary. Primary driers are those that alone can promote oxidative film curing. They catalyze oxygen absorption into the film, and accelerate the decomposition of the formed peroxides into the free radicals which propagate the crosslinking mechanism. The most widely used primary driers are based on cobalt and manganese. Cobalt driers are by far the most reactive. They are surface driers in that they quickly dry the outermost layer of the paint film, preventing further drying – through drying – beneath. Cobalt used alone can thereby promote film wrinkling. Manganese driers are slower and provide some through drying, although they are, like cobalt, usually used with secondary driers to properly balance drying properties.

Lead driers have traditionally been the most common, although toxicity concerns now severely restrict their use. Lead is a secondary drier as its sole function is to catalyze oxygen uptake. This and its relatively low reactivity make it a good through drier. For this reason it was widely used in combinations with primary driers in the past. It is still used where toxicity and atmospheric sulfide staining do not rule it out. Calcium soaps have been used with lead driers to improve lead solubility and low temperature drying.

Zirconium driers have generally replaced lead, although their reactive chemistry is unrelated. Their funtionality in combination with primary driers is similar to lead's, but they are not discoloring and are not considered toxic. Aluminum driers are also used in place of lead, but the substitution is not straightforward. The paint must be formulated specifically to promote their activity.

Zinc soaps and zinc oxide, although not true driers, are used with cobalt driers to retard surface drying, accelerate through drying and provide harder cured films.

Iron driers are used in heat-cured paints where the poor color can be accomodated. Temperatures above 100°C are required for catalytic activity, but high film hardness is obtained.

⌘⌘⌘

Nonmetallic auxiliary driers are commonly used to improve the efficiency of primary driers and to maintain catalytic activity during in-can aging. Metallic soap driers typically lose some measure of activity between the time of manufacture and use, primarily through pigment adsorption. The heterocyclic 1,10-phenanthroline and 2,2'-dipyridyl, shown in Figure 21, complex cobalt and manganese through their nitrogens, acting in effect as chelating agents. The metal-organic complex protects the metal against deactivation through adsorption during storage, and modifies the metal's catalytic activity such that through dry is promoted.

Figure 21. Nonmetallic Auxiliary Driers

1,10-phenanthroline

2,2'-dipyridyl

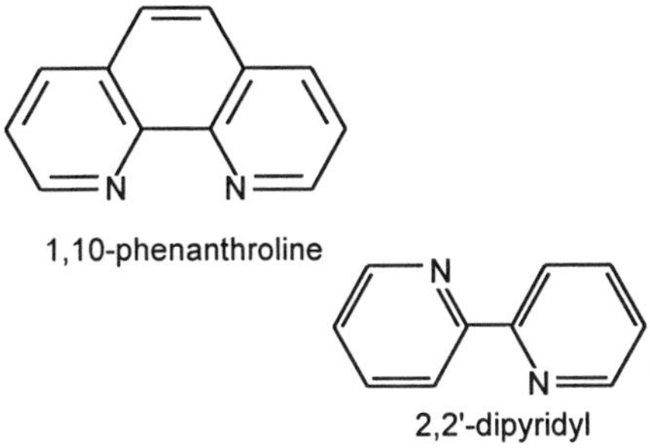

The organic driers have found general use in low VOC coatings based on lower molecular weight binders because they accelerate the catalytic acitvity of the primary driers for faster film cure. They serve as well to stabilize cobalt and manganese against loss of activity due to pigments, minerals, water and additives in water-based coatings.

Proprietary metallic driers are now available as another means of preventing loss of efficiency of cobalt and manganese soaps during paint storage. These "feeder" driers dissolve gradually into the paint vehicle to replace drier metals lost through adsorption.

Coalescing Agents

The dispersed polymer particles in a latex paint must coalesce after application to form a continuous film at room temperature or lower. Hard and high T_g polymers must be temporarily plasticized to enable film formation. Coalescing agents serve this function by softening the resin particles and lowering their minimum film formation temperature, but slowly volatilizing after application to allow the polymer film to attain final hardness. Slow evaporating ester-alcohol solvent is the currently most common coalescing agent because it efficently lowers the minimum film forming temperature with a wide variety of resins, has good hydrolytic stability, and is insoluble in water. The latter is important in preserving

⌘⌘⌘

coalescent efficiency by preventing loss of solvent from the polymer phase to the aqueous medium. Low volatility glycol ether ester and glycol ether solvents also find use as coalescing agents.

Plasticizers

Plasticizers are used in coatings primarily to improve the flexibility and toughness of the final polymer film. They are used more often in industrial, appliance and automotive coatings than in architectural paints. Plasticizers form strong polar associations with polymer molecules, but because there is no chemical bond, they are subject to loss through extraction or volatility. Plasticizers are therefore typically very low volatility water insoluble liquids which are selected for resin compatibility. The major coatings plasticizers can be characterized as the phthalic, trimellitic or adipic acid esters of 2-ethylhexanol (octyl) or isononanol. The single most commonly used plasticizer is dioctyl (i.e. ethylhexyl) phthalate.

The response of polymer film properties to plasticizer addition level is not necessarily linear, so optimal concentration must be determined in most cases. In general, increasing the concentration of plasticizer will increase film elongation and permeability, and decrease tensile strength. Film toughness and adhesion, however, will increase to a maximum and then decrease.

Biocides

Biocides are used in both solvent and aqueous paints as fungicides and algicides to keep the dried paint film free of discoloring mildew and/or algae. Biocides are used in water-based paints as preservatives as well. Here they prevent the bacterial propagation that can result in off-gassing in the can, and the enzymatic degradation of cellulosic thickeners that can result in viscosity loss. Ideally, paint biocides should be colorless, free of objectionable odors, nonyellowing in the paint film, of low toxicity, environmentally acceptable and cost effective. Few, if any, of those in common use today meet all these criteria.

Certain biocides are used exclusively as in-can preservatives, certain are used just as in-film fungicides, and others can serve both purposes. In the latter case, the spectrum of biocidal activity is typically related to concentration and solubility. Some biocides, for example, are effective against bacteria in one concentration and effective against fungi at a considerbly higher concentration. Preservatives are usually water soluble for best effect, while in-film fungicides need to have very low water solubility to resist leaching.

In the past, organomercury compounds were the most widely used paint biocides, functioning at very low concentrations as both preservatives and film fungicides. Because of their high toxicity, they were replaced in many cases by the less toxic but also less cost-effective organotin compounds, particularly tributyl tin oxide. Toxicity and environmental concerns have more recently restricted the use of these tin compounds as well.

As the continuing trend away from solvent-based paints in favor of aqueous products accelerates, the variety of available preservatives has kept pace. These can be generically described as nitrogen and/or sulfur-containing organic biocides. Many of these are actually formaldehyde releasers, which largely accounts for their effectiveness, but in some cases limits their use.

Broad spectrum biocides are generally preferred for water-based paints, but environmental, toxicity and cost considerations often necessitate tailoring of the biocide system for maximum effectiveness at minimum use levels. This involves identifying the particular types of microorganisms requiring control in both the manufacturing and end use environments. To avoid development of biocide resistance in the manufacturing environment, it is not uncommon for producers to periodically switch between preservative systems with different biocidal mechanisms.

Antifoams

The surfactants and thickeners used to stabilize water-based paints also stabilize air bubbles entrapped during manufacture or application. Even solvent-based coatings will develop bubbles under shear. Bubbles that remain in the applied coating result in cratering in the dried film. Vehicle-insoluble liquids – silicone oils, alkyl (C_6 - C_{10}) alcohols, and mineral oils – are used to destabilize the air bubbles so that they break. In aqueous systems, silicone/amorphous silica combinations are often used to good effect. Because antifoams can be emulsified and lose effectiveness on prolonged paint processing, they are sometimes added twice during manufacture – early in the process to prevent air entrapment, and near the end of the process to ensure performance during use.

Glycol

A slow evaporating glycol is usually added to latex paints to serve as both antifreeze and humectant. Ethylene glycol is most commonly used, followed by propylene glycol. As an antifreeze, the glycol depresses the freezing point of the aqueous medium. As water freezes, its volume expansion can push the latex particles together with sufficient force to cause coagulation by overcoming the stabilizing dispersant/surfactant layer. With a glycol added, even those paints which become cold enough to freeze tend to freeze to a slush that exerts less force on the latex particles.

As a humectant, the glycol controls evaporation from the paint film to facilitate wet lapping. When paint is applied by roller or brush, each brush-full or roller-full of paint overlaps the wet edge of the preceeding area of

⌘⌘⌘

coverage. Without the glycol, evaporation can quickly leave the previously applied film edge very viscous, although very weak due to only limited latex coalescence. The brush or roller shear from the subsequent lapped application can break up this film, leaving irregularities along the lapped edge.

Antiskinning Agents

The driers used in air-drying paints can also promote oxidative crosslinking at the surface of the paint in its container. This is prevented by the use of a small amount of volatile antioxidant, typically an oxime such as methylethyl ketoxime or butyraldoxime. Volatility and use level (<0.2%) are selected so as not to interfere with drying of the applied paint film. This however, can compromise antiskinning effectiveness in a container of paint which has been opened, partially used and reclosed for storage.

Corrosion Inhibitors

Moisture, ions and oxygen are required for corrosion of metallic surfaces to occur. Corrosion inhibitors are added to paints most often to prevent rusting of ferrous metals. Rusting is an oxidative process which occurs through an electrochemical mechanism. The overall corrosion reaction is:

$$4Fe + 3O_2 + 2H_2O \rightarrow 2Fe_2O_3 \cdot H_2O$$

A drop of water (actually conductive electrolyte solution from ambient salts) on steel, for example will create an electrolytic cell. Electron flow (current) will occur from the anode at the center of the drop, to the cathode at the edge. At the anode, iron is oxidized:

$$4Fe - 8e \rightarrow 4Fe^{2+}$$

while at the cathode oxygen is reduced:

$$4H_2O + 2O_2 + 8e \rightarrow 8OH^-$$

The cationic iron migrating toward the cathode, and the anionic hydroxyl migrating toward the anode form soluble ferrous hydroxide:

$$4Fe^{2+} + 8OH^- \rightarrow 4Fe(OH)_2$$

If sufficient oxygen is available, this is oxidized to the insoluble hydrated ferric oxide (rust):

$$4Fe(OH)_2 + O_2 \rightarrow 2Fe_2O_3 \cdot H_2O + 2H_2O$$

⌘⌘⌘

Coatings can inhibit rusting in relation to their impermeability to water, oxygen and electrolyte. Solvent-based coatings provide generally less permeable polymer films than water-based coatings, but none are completely inpermeable. Lamellar minerals and metals are used in coatings to reduce polymer film permeability by physically inhibiting diffusion through the film. Platy minerals, such as talc and mica, are used for this purpose, although micaceous iron oxide and aluminum flake are more widely employed.

Primers containing high levels of zinc dust are used to protect steel by a sacrificial cathodic mechanism. The less noble zinc acts as the electrochemical anode and corrodes instead of the steel substrate. The concentration of zinc powder in the dry primer film must be sufficiently high to allow the particles to make contact and allow an electrical current to be established between these particles and the steel.

Certain inorganic compounds, usually as the zinc salt, are used in paints as passivators to control rust formation. These inhibitors are slightly soluble in water, and are slowly dissolved from the paint film through moisture permeability. The ions formed are carried to the film-metal interface by moisture diffusion. There they cause the corrosion electrical potential to be elevated to the passive potential at which the corrosion rate is dramtically reduced by polarization at the anode.

Chromates have been the most widely used passivating inhibitors in coatings, although toxicity and carcinogenicity potential have prompted use of molybdates (zinc and calcium zinc) and zinc phosphate in their place in certain cases. Alkaline extenders, such as wollastonite, are used as synergists with passivators in some applications, apparently providing a beneficial pH buffering effect.

Flash Rust Inhibitors

When water-based paints are applied over exposed ferrous metal, the drying film may develop a scatter of light brown spots. This "flash" rusting is prevented with alkaline inhibitors, such as sodium nitrite, sodium benzoate, or 2-amino-2-methylpropan-1-ol (AMP). Alkaline minerals, wollastonite in particular, are also used to control flash rusting.

Photostabilizers

Many paints are prone to fading and film degradation from exposure to sunlight, particularly ultraviolet light. UV absorbers are therefore used to reduce ultraviolet absorption by the binder polymer and the resultant photoinitiated oxidative degradation. UV absorbers for exterior coatings ideally have high absorption at 280 nm through 380 nm, with none above 380 nm to avoid color effects. The UV absorber should also be nonvolatile and soluble in the coating film. Hydroxybezophenones, hydroxybenzo-

⌘⌘⌘

triazoles and hydroxyphenyltriazines are commonly used, with the latter generally providing the lowest volatility and best permanence.

Optimum photostability, for example in automotive clear topcoats, is obtained with synergistic combinations of UV absorbers and hindered amine light stabilizers (HALS) based on 2,2,6,6-tetramethylpiperidine. The UV absorber reduces the rate of generation of free radicals, while the light stabilizer reduces the rate of oxidative degradation by free radicals.

TECHNOLOGY

A critical factor in the design and development of a liquid coating is the dispersion of the colored and extender pigments in the vehicle. In order to provide a uniform film with the desired surface characteristics, the agglomerates of the primary and extender pigments must be broken down by mechanical shear into finely divided particles coated with the vehicle or binder.

Pigment Dispersion

Particle size reduction is accomplished in a variety of equipment from the traditional roller mills, ball mills, and pebble mills to the newer high-shear media mills and dispersers. Roller mills rotate in opposite directions at different speeds to create shearing action in the nip between the rolls; the narrower the nip, the higher the shearing energy and the finer the grind. Ball mills consist of steel cylindrical containers filled to slightly over one-third of their volume with steel balls, while a pebble mill is lined with porcelain and contains flint pebbles to approximately the same volume. During operation, another one-third of the mill's volume is filled with the coating formulation. Figure 22 is a simplified representation of a ball or pebble mill. Pigment deagglomeration and particle size reduction is accomplished primarily by shear forces as the balls or pebbles cascade down through the coating formulation. The fineness of grind achieved for any given formulation is a function of the length of time the mill is operated.

High-shear media mills, as in Figure 23, are presently the predominant type used in the production of paint and coatings. This equipment can produce high volumes of finished product at relatively low operating cost. These mills are highly versatile and can process a wide range of coatings formulations from very fluid compositions to high viscosity materials. Media mills consist of a hollow cylindrical shell with a rotating vertical or horizontal shaft and discs or pins located intermittently along the length of the shaft. These drive the grinding media through the coating formulation at very high velocity. The grinding media can be specially sized sand, or more commonly today, small steel or ceramic balls. Media mills can be incorporated into either batch or continuous production processes.

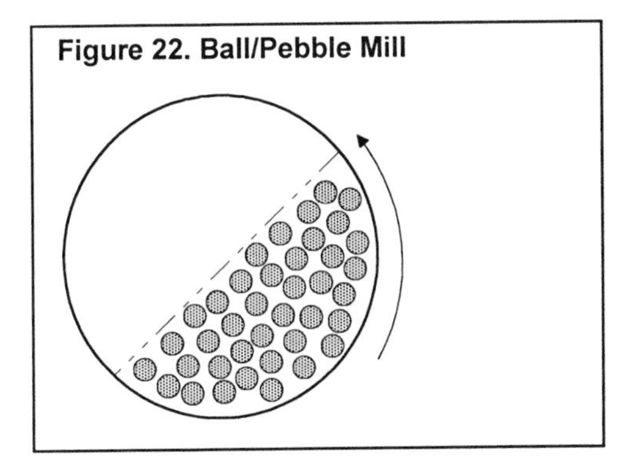
Figure 22. Ball/Pebble Mill

The high-speed disperser is utilized in batch operations due to its relative ease of operation and cleanup. The viscosity of the formulations run on this type of equipment must be in a well defined range. If the viscosity is too low, no shearing action will occur and if the viscosity is too high, the movement of the particles will not be rapid enough to break up the pigment agglomerates. Figure 24 depicts dispersion equipment with the preferred ratios of disperser disc size and placement to tank size.

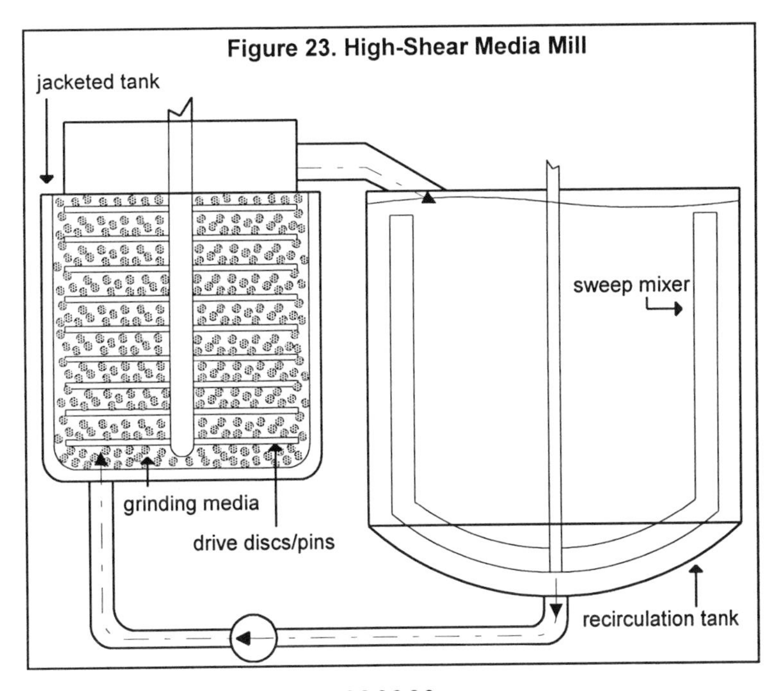

Figure 23. High-Shear Media Mill

⌘⌘⌘

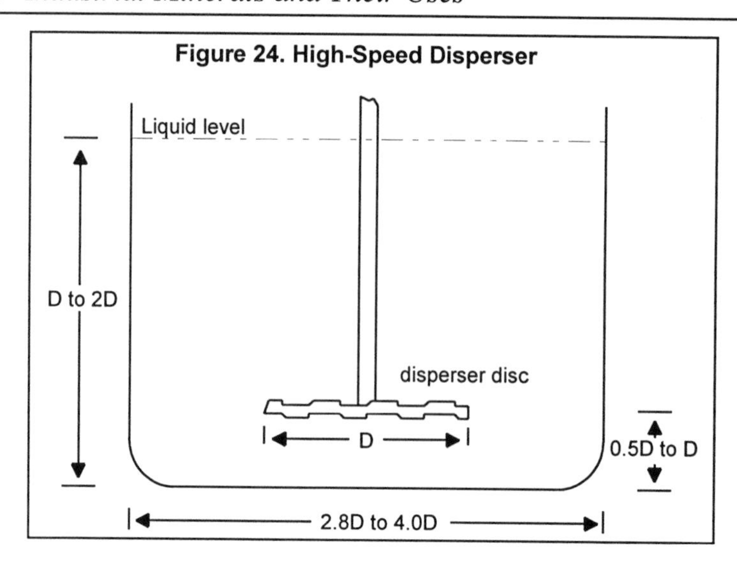

Figure 24. High-Speed Disperser

Application Forms

Paints and coatings are available in several different forms today based on cost, processing, application, end-use, performance and environmental considerations.

Conventional solvent systems – This includes conventional solvent-based paints as described above. This technology is mature and declining.

High solids solvent systems – This technology produces coatings containing less than 30% volatile organic solvents. Impractically high viscosity is avoided in high solids systems through the use of highly functional oligomeric binders carefully controlled in molecular weight and molecular weight distribution, and the use of reactive diluents (e.g. unsaturated monomers).

Two part catalyzed systems – These are coatings in which one component contains a reactive polymer such as an epoxy, while the other component contains the crosslinking agent. Once the two parts are mixed, the resulting mixture has a limited pot life during which it must be applied. Reaction occurs after the film is applied. With these systems, it is possible to apply relatively thick films at 100 percent solids and minimal hydrocarbon emissions.

Aqueous emulsion systems – Latex paints dominate the trade sales markets because of their ease of handling and applying, ease of cleanup, non-flammability, and lack of hydrocarbon solvent exposure.

⌘⌘⌘

Water-soluble binder systems – These coatings are based on binders that are soluble in water. Upon drying, the binder is crosslinked, usually through oxygen conversion. Convertible resins are modified through the addition of carboxylic, hydroxyl, or amino groups to render them water soluble.

Electrodeposition coatings – Electrodeposition (ED) coatings (E-Coat) are comprised of paint particles dispersed in an aqueous vehicle. Thermoset resins are generally used as the paint binder. The metal part to be coated can serve as either anode or cathode. In anodic systems, application of current causes the negatively charged paint particles to deposit on the anode. This occurs through precipitation by hydrogen ions generated there by the electrolysis of water. In cathodic systems, the positively charged paint particles deposit on the cathode through precipitation by generated hydroxide ions. The coated article is removed from the electrodeposition bath, rinsed and baked to cure.

The film thickness of ED coatings is essentially self-limiting. As film thickness increases, electrical resistance increases. Deposition rate decreases correspondingly until the film thickness is reached at which deposition stops. Because the entire surface of the metal object acts as an electrode, ED coatings are used to provide complete, void-free coverage. This is particularly important in the application of anticorrosion coatings. The corrosion protective primers used on automobiles are in most cases applied by electrodeposition. This system is also used to prime and topcoat many appliances.

Powder coating – The powders for powder coating are prepared by premixing all ingredients, (binders, pigments, extenders, additives) melt extruding, and then pulverizing. Powder coatings are generally thermoset, based on epoxy or polyester binders, although acrylic binders are used for particular applications, such as washing machines, where they provide superior detergent resistance. Thermoplastic powders – based on vinyl chloride copolymers, polyamides, fluoropolymers, and thermoplastic polyesters – find limited use because they are difficult to pulverize to fine particle sizes and are therefore applied in relatively thick films. They also provide relatively poor flow and leveling.

Most powder coatings are applied to bare metal as an electrostatic spray. The metal is grounded and the powder is given a negative electrostatic charge by the spray gun. The difference in potential causes the powder particles to adhere to the metal surface. Coverage tends to be uniform and self-limiting since the coating of powder acts as an insulator. The powder-covered article is baked, allowing the powder particles to fuse to a continuous film, flow, level, and crosslink. Over-coating previous coatings or heat-tolerant plastics generally requires preparation with a conductive primer.

A fluidized bed method is used to apply relatively thick thermoplastic powder coatings. The article to be coated is heated above the T_g of the binder resin and then suspended in a fluidized bed of coating powder. Powder particles fuse onto the object until the powder layer becomes sufficiently thick to act as a thermal insulator and prevent additional fusing. Since the surface may contain incompletely fused particles, the operation is completed in a separate oven. Electrostatic fluidized beds are also used, with charged powder and grounded substrates, which may or may not be preheated.

A recent development in thermoplastic powder coatings utilizes flame spray technology. A flame spray gun is used to heat the substrate and to instantaneously (a fraction of a second residence time) melt the powder as it is sprayed. The powder flows and fuses onto the article. Flame spray powder coating can be used on nonconductive, heat-tolerant surfaces such as concrete, wood, and certain plastics. It also has the advantage of portability over the other powder coating techniques, which rely on a large scale carefully controlled industrial setting. Flame spray apparatus can be used in the field to coat objects and to repair previously applied thermoplastic flame spray coatings.

Powder coatings are widely used as industrial finishes on appliances, machinery and outdoor fixtures and furniture because they provide a hard durable coating with essentially no VOC emissions and nearly 100% powder utilization – any overspray is easily collected and reused.

Radiation curing – Radiation-cure coatings crosslink rapidly at ambient temperature on exposure to radiation. This makes them particularly suitable for heat sensitive substrates such as paper, wood and certain plastics. The two types of radiation cure coatings are:

A) UV cure coatings – cure is initiated by excitation of UV/visible photoinitiators.

B) EB (electron beam) cure coatings – cure is initiated by ionization and excitation of the binder resin by high energy electrons.

Infrared and microwave cured coatings are on occasion also considered radiation cured, but in these systems the radiation is simply converted to heat, which initiates thermal curing.

Radiation cure coatings are typically solvent-free, high solids systems based on acrylated epoxy or urethane oligomers and diluent multifunctional acrylate monomers. Since radiation intensity varies significantly with

⌘⌘⌘

distance between the radiation source and the coated substrate, this technology is used almost exclusively on flat or cylindrical (i.e. rotatable) objects. Radiation curing does not lend itself well to decorative coatings and is used primarily with lithographic printing inks, clear abrasion-resistant coatings on vinyl flooring (using acrylated urethane oligomers) and clear coatings on plastics, such as glazing and eyeglass grade polycarbonates (using alkoxysilane-acrylate monomers and colloid silica).

TEST METHODS

The test methods employed in the evaluation of paints and coatings fall into two basic catagories:

1) Those tests used to evaluate the properties and characteristics of a coatings formulation for a particular application.

2) A selected subset of these tests utilized for the quality control of production material.

The most widely used quality control tests for paints and coating are as follows.

Solids content –The "solids" of a paint include those non-volatile components which remain part of the paint film. Testing the solids content of a paint is important to ensuring good control of the manufacturing process. Paints which deviate from the specified solids content may have variations in color and film durability. The solids content is also important in calculating the amount of coverage a paint will impart.

Specific gravity – Specific gravity is an important, easy-to-measure property which indicates the weight of the paint for a given volume.

Specific gravity is used as an in-process quality control test and as an end-product test. Running a specific gravity is an easy way to determine if all the major components were added in the correct amounts.

Specific gravity can be measured very easily by filling a tared container of known volume and then weighing. For exact measurements to a high level of precision, pycnometers are used. Hydrometers are also used for quick measurements.

Fineness of grind – During the dispersion process it is important to be able to measure the degree of breakdown of agglomerates and aggregate particles. If the grind is not fine enough, poor color uniformity, gloss or hiding power can result. One fast method to measure the "fineness of grind" is to use the Hegman Gauge, illustrated in Figure 25. This gauge is a steel plate having a groove of declining depth from zero to 100 microns. Samples are drawn over the gauge with a straight edge. The depth of the groove at which particles are evident is taken as the Hegman reading for particle size.

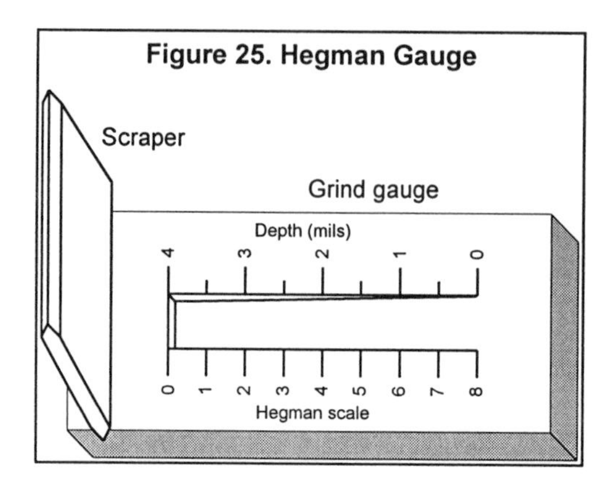

Figure 25. Hegman Gauge

Viscosity – Viscosity measurements are performed as both process tests and end product tests. Many plant viscosity tests, such as the Ford Cup and Zahn Cup test, simply measure the time required for a given volume of sample to pass through a standard size orifice. The Brookfield viscometer is used to measure viscosity from shear resistance on a standard spindle rotating at a constant speed. A Stormer viscometer measures the time required for a standard paddle to revolve 100 revolutions from a constant force (weight and pulley). The Stormer is sometimes preferred when measuring the viscosity of a heterogeneous mixture. For homogeneous varnishes, the Gardner-Holdt bubble viscometer is used. It quickly indicates varnish viscosity by measuring the time required for a bubble to rise to the top of a standard tube after being inverted under standard conditions.

The problem with most of these viscometers is that they measure viscosity at relatively low shear. Although they can be used as good production control tests and may relate to degree of dispersion, paints are usually applied at high shear by brush, roller, or spray. Since most paints are pseudoplastic, the viscosity tests described may not relate to the paint's application rheology.

Wet hiding power – To measure the ability of the wet coating to produce an opaque film, it is uniformly applied to an impervious black and white striped surface using a brush until complete hiding is achieved. The weight of the coating required to accomplish this is determined and the hiding power in square feet is calculated as follows:

⌘⌘⌘

$$\text{Hiding Power} = \frac{W \times 454 \times A}{G}$$

where W = weight per gallon, A = area of surface covered, G = grams of paint applied.

Wet hiding power is inversely proportional to coverage. The more paint needed, the less area that can be coated per gallon.

Dry hiding power – The dry hiding procedure uses the same striped panel and coating procedure as above. However, five brushouts are made applying a different amount of paint each time. The panels are allowed to dry for 24 hours and the contrast ratio is determined for each brushout. Contrast ratio is equal to the reflectance over the black area divided by the reflectance over the white area. The weight of the paint is plotted against the contrast ratio and the amount of the paint required to give a contrast ratio of 98 is read from the resulting curve. The dry hiding power is then calculated from the same formulas as for wet hiding power. Dry hiding is promoted by increased light scattering, increased light absorption and uniformity of film thickness.

Determination of specular gloss – A 60° gloss reading on a standardized test panel is used to determine film smoothness and thus visual gloss. Increase in surface roughness or imperfections promotes diffuse reflection and lower gloss.

Evaluation of Color – The color of a finished coating may be qualitatively evaluated by visual comparison to an established standard using appropriate test panels. Most color comparisons, however, are being made quantatively by measurement of the test panels on color measurement instruments which calculate color coordinates and determine the color difference.

ASTM Procedures – The following is a list of recognized ASTM Procedures used primarily for the evaluation of coatings formulations and determination of coating characteristics.

D 281-84: Test Method for Oil Absorption of Pigments by Spatula Rub-Out.
D 332-87: Test Method for Relative Tinting Strength of White Pigments by Visual Observation.
D 344-89: Test Method for Relative Hiding Power of Paints by the Visual Evaluation of Brushouts.
D 387-86: Test Method for Color and Strength of Color Pigments with a Mechanical Miller.
D 523-89: Test Method for Specular Gloss.
D 609-90: Method for Preparation of Steel Panels for Testing Paint, Varnish, Lacquer and Related Products.

⌘⌘⌘

D 610-85: Test Method for Evaluating Degree of Rusting on Painted Steel Surfaces.
D 658-91: Test Method for Abrasion Resistance of Organic Coatings by Air Blast Abrasives.
D 660-87: Test Method for Evaluating Degree of Checking of Exterior Paints.
D 661-86: Test Method for Evaluating Degree of Cracking of Exterior Paints.
D 661-86: Test Method for Evaluating Degree of Erosion of Exterior Paints.
D 714-87: Test Method for Evaluating Degree of Blistering of Paints.
D 772-86: Test Method for Evaluating Degree of Flaking of Exterior Paints.
D 869-85: Test Method for Evaluating Degree of Settling of Paints.
D 870-87: Practice for Testing Water Resistance of Coatings using Water Immersion.
D 1006-73: Practice for Conducting Exterior Exposure Tests of Paints on Wood.
D 1212-91: Methods for Measurement of Wet Film Thickness of Organic Coatings.
D 1308-87: Test Method for Effect of Household Chemicals on Clear and Pigmented Organic Finishes.
D 1360-90: Test Method for Fire Retardancy of Paints.
D 1474-85: Test Method for Indentation Hardness of Organic Coatings.
D 1475-90: Test Method for Density of Paint, Varnish, Lacquer, and Related Products.
D 1543-86: Test Method for Color Permanence of White Architectural Enamels.
D 1735-87: Practice for Testing Water Resistance of Coatings Using Water Fog Apparatus.
D 2197-86: Test Method for Adhesion of Organic Coatings by Scrape Adhesion.
D 2369-90: Test Method for Volatile Content of Coatings.
D 2486-89: Test Method for Scrub Resistance of Interior Latex Flat Wall Paint.
D 2793-69: Test Method for Block Resistance of Organic Coatings and Wood Substrates.
D 2805-88: Standard Test Method for Hiding Power of Paints by Reflectometry.
D 4062-88: Test Method for Leveling Characteristics of Paints by Draw Down Method.
D 2931-84: Guide for Testing Latex Flat Wall Paints.
D 2932-80: Guide for Testing Exterior Solvent Reducible House and Trim Coatings.
D 3170-87: Test Method for Chipping Resistance of Coatings.
D 3258-80: Test Method for Porosity of Paint Films.
D3450-90: Test Method for Washability Properties of Interior Architectural Coatings.
D 3794-79: Practice for Testing Coil Coatings.
D 4400-89: Test Method for Sag Resistance of Paints Using a Multinotch Applicator.
D 4414-84: Practice for Measurement of Wet Film Thickness by Notch Gauges.

⌘⌘⌘

D 4938-89: Test Method for Erosion Testing of Anti-Fouling Paints Using High Velocity Water.
D 4939-89: Test Method for Subjecting Marine Anti-Fouling Coating to Biofouling and Fluid Shear Forces in Natural Seawater.
D 4946-89: Test Method for Blocking Resistance of Architectural Paints.
D 4958-91: Test Method for the Comparison of Brush Drag of Latex Paints.
D 5150-91: Test Method for Hiding Power of Architectural Paints Applied by Roller.
D 5178-91: Test Method for Mar Resistance of Organic Coatings.

BIBLIOGRAPHY

Burns, R.M., Bradley, M.W., 1955, *Protective Coatings for Metals*, Reinhold, New York
Dick, J.S., 1987, *Compounding Materials for the Polymer Industries*, Noyes Publications, Park Ridge, NJ
Eckert, C.H., Berson, J.P., Harris, T.S., 1983, *Extender and Filler Pigments*, Kline, Fairfield, NJ
Glaser, M.A., Brownstead, E.J., 1959, *Modern Organic Finishes*, Federated Society of Paint Technology, New York
Grim, R.E., 1953, *Clay Mineralogy*, McGraw-Hill, New York
Koleske, J.V.(ed.), 1995, *Paint and Coating Testing Manual*, ASTM, Philadelphia
Lambourne, R.(ed.), 1987, *Paint and Surface Coatings: Theory and Practice*, Halsted Press, New York
Lewis, P.A.(ed.), 1988, *Pigment Handbook*, Wiley, New York
Solomon, D.H., Hawthorne, D.G., 1983, *Chemistry of Pigments and Fillers*, Wiley, New York
Turner, G.P.A., 1988, *Introduction to Paint Chemistry and Principles of Paint Technology*, 3rd ed., Chapman and Hall, New York
Von Fisher, W., Bobalek, E.G., 1953, *Organic Protective Coatings*, Reinhold, New York
Wicks, Z.W., Jones, F.N., Pappas, S.P., 1992, *Organic Coatings: Science and Technology*, Wiley, New York

⌘⌘⌘

NOTES

⌘⌘⌘

FIVE

PAPERMAKING

Janis Anderson
R.T. Vanderbilt Company, Inc.
Norwalk, CT

Ever since words were put into written form, mankind has been searching for suitable surfaces to write upon. Although early cave paintings, cruciform tablets and heiroglyphics on stone walls in Egypt still exist after thousands of years, these surfaces all lack portability. A more convenient medium was obviously needed. Among the earliest paper-like materials were papyrus, parchment and vellum. Papyrus was a woven mat of reeds beaten into a flat surface that could be written upon. Parchment and vellum were made from goat and sheep skins. Leaves, the bark of certain trees, thin metal plates, wax tablets and wood all served as paper-like materials in earlier eras.

Legend has it that paper as we know it was first produced in China by Ts' ai Lun during the first century AD. A screen was dipped into a vat containing a mixture of beaten hemp and flax fibers in water. The screen was lifted out of the vat allowing the water to drain, and the wet paper mat was removed from the screen and air dried. This basic process was used to make paper until the early 19th century.

The Chinese exported paper to Europe and the Middle East, guarding the secret of papermaking until the 600s AD. After capturing some Chinese papermakers and forcing them to reveal their secrets, the Arabs dominated the paper industry for more than 400 years. The flow of the paper trade between the Middle East and Europe came to a halt with the Crusades. Early papermills sprang up in Europe around 1100 AD as Crusaders brought the technology to Europe.

Hand-made paper production continued in Europe at a rather slow pace until the mid-15th century with the advent of Johann Gutenberg's printing press. Although Gutenberg's most famous work, the Bible, was printed on parchment, the printing press soon created an unprecedented need for paper. Papermills were started all over Europe, but the demand still outstripped supply. Rags, the source of paper fiber, became scarce, and riots ocurred in some cities when availability was limited. New and better sources of pulp were required.

Around 1720, Reaumer in England observed that wasp's nests were similar in nature to paper. He didn't pursue the prospect of making pulp from wood, but the idea was planted in the minds of other experimenters. It wasn't until 1853 that Watt and Burgess patented a process for making wood pulp. Their process is still used to produce stone groundwood. As its name implies, the pulp is produced by grinding wood with a high speed grindstone. This produces a rather coarse, poor color, weak pulp that is used for newsprint, directory paper and coated publication papers. The first U.S. groundwood mill was built in 1867.

Pulps have since been made by various chemical processes, covering the full pH range from acid to neutral to alkaline. Each pulping method produces characteristic pulp properties, depending on the type of wood used, and each pulp type lends itself to a different end use. The most widely used chemical cooking process in North America today is the alkaline "kraft process". The kraft process was invented in 1884, and the first U.S. kraft mill started production in 1909. As will be detailed later, semi-chemical, chemi-mechanical, and non-sulfur cooking processes are also used. These methods are often considered more environmentally friendly than some of the older processes.

In 1798, Louis Robert of France invented a machine, the forerunner of the modern Fourdrinier, which could make paper continuously rather than one sheet at a time. The story goes that Robert was a frustrated personel manager tired of dealing with papermakers' constant fighting over who was the best and who could make the most sheets per day. The French Revolution and the Napoleonic era which followed prompted Robert to move work on his machine to England. In 1804, Henry and Sealy Fourdrinier bought the patent rights to Robert's machine, improved it, and made it more practically useful. Their Fourdrinier papermachine was the first of the several continuous forming processes used today. The first U.S. Fourdrinier machine was built in Saugerties, New York in 1826.

Modern high speed paper machines produce grades ranging from unbleached kraft for grocery sacks and corrugated cardboard, to tissue, newsprint, coated magazine papers, and uncoated printing and writing papers. The 1994 production of paper and paperboard in United States was approximately 86 million tons. A breakdown by grades is as follows:

Grade	**10^6 short tons**
Newsprint	7.1
Coated Papers	8.1
Uncoated Groundwood	1.8
Other Printing and Writing	1.6

⌘⌘⌘

Uncoated Free sheet	12.5
Other Packaging Papers	4.7
Tissue	6.0
Unbleached Kraft Paperboard	22.1
Semi-Chemical Medium	5.5
Bleached Paperboard	5.0
Recycled Paperboard	12.0
Total	86.4 x 10^6 tons

PULPING

"Pulp" is the common name for the fibrous portion of paper. Many different materials have been used over the centuries as sources of pulp. These include cotton linters, old rags, cellulose from woody and non-woody plants and more recently, polyester. Prior the 1850's, pulp was made from rags, cotton or flax fibers that were too short for thread making, or rice stalks. The individual fibers were liberated by beating the mass in water. Flax and cotton fibers are long and strong and produce high quality papers noted for their longevity. The vast majority of modern papermaking pulp is made from wood. The common papermaking softwoods are the pines, spruces, firs, hemlocks and balsams. Softwood fibers are significantly longer and stronger than hardwood fibers. The common papermaking hardwoods are the maples, oaks, birches, beeches and aspens. Hardwood fibers are short and narrow. They fill in the spaces between the longer softwood fibers in the paper web.

Wood pulp is made by several processes ranging from mechanical to semi-chemical to fully chemical methods. Each pulping process produces a pulp with properties unique to itself.

Stone Groundwood

The oldest pulping method is the stone groundwood process. Most stone groundwood is made from softwood, ideally spruce and fir. Groundwood pulping yields a coarse, weak pulp that produces paper with excellent printability. The large bulky fibers are easily compressible during the printing process and have good ink receptivity. Stone groundwood is used extensively for newsprint, directory papers and base paper for coated publication grades. Stone groundwood pulp has the tendency to discolor rapidly upon exposure to light. This is primarily due to the presence of lignin, the complex phenolic compounds that hold the cellulose fibers together in the wood. Mechanical pulping methods do not remove lignin from the fibers. The lignin is dark brown to black in color and darkens on exposure to air and light. The well-known phenomenon of newspaper yellowing and embrittlement with age is due to the discoloration and degradation of the lignin in the pulp. Newsprint paper can have as much as 85% of its pulp as groundwood, with just enough chemically pulped softwood to keep the sheet together on the press.

The stone groundwood process begins with debarking of the logs in drum barkers. These are drums 12 to 15 feet in diameter and 50 to 100 feet long. The logs tumble against each other as the barker slowly rotates. The action of the logs against each other and the sides of the drum, along with the washing action of water sprays, removes the majority of the bark. The logs are inspected and if necessary returned to the barker to reprocess until the bark has been removed sufficiently. The debarked logs are stored for use at some later time or transported by conveyor or water to the groundwood mill. The waste bark is transported to waste piles for eventual burning to generate electricity and process steam.

Figure 1. A Grinder Configuration Used in the Stone Groundwood Process

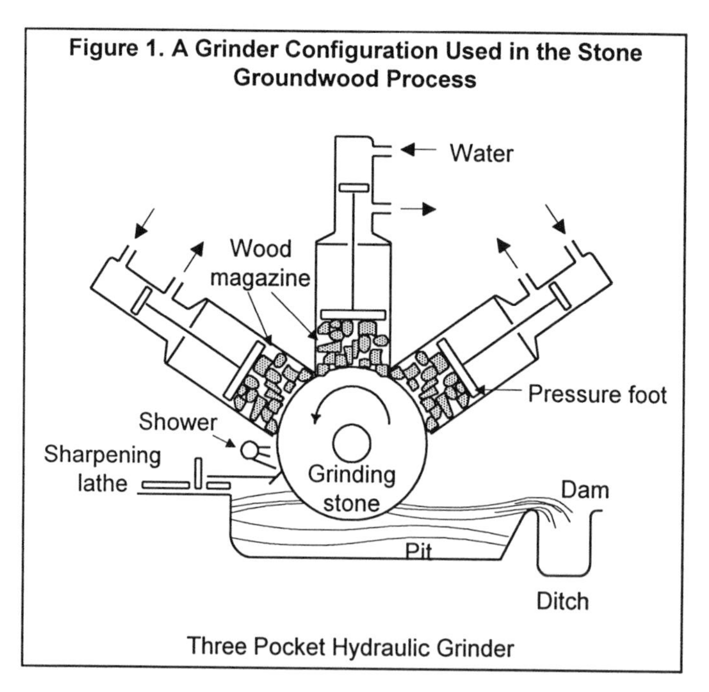

Three Pocket Hydraulic Grinder

The wood grinders come in various configurations, one of which is shown in Figure 1. All have in common wood holding pockets, a stone grinding wheel, a pulp outlet, and water. Large volumes of water are used to move the wood, and dilute and clean the pulp. The logs are forced against the grind stone, reducing them to the pulp. This process is highly energy intensive, with many mills still operating on hydroelectric power. The

⌘⌘⌘

groundwood pulp is screened, washed and centrifugally cleaned. Reductive bleaching is employed for some pulps using sodium hydrosulfite. This improves the brightness slightly, but since there is no removal of the lignin, the long-term effect is limited, and yellowing still occurs.

Refiner Pulping

Refiner mechanical and thermomechanical pulping processes are modifications on the stone groundwood method. Refiner mechanical pulping begins with wood chips rather than debarked logs. The chips are reduced to a pulp by passing through a series of attrition mills. The attrition mills are comprised of closely gapped high speed disks surfaced with cutting knives. The chips are fed into the refiner and are disintegrated as they pass between the rotating disks. Like stone groundwood pulping, refiner mechanical pulping is extremely energy intensive. Thermomechanical pulping employs heat in conjunction with refiners. The chips are steamed under pressure prior to refining. Certain strength properties can be improved by adjusting steam pressure and temperature and the solids (consistency) of the pulp. Refiner mechanical and thermomechanical pulps are stronger than stone groundwood and are preferred for some types of paper.

Semi-Chemical Pulping

Chemical pulping processes are used to extract all or part of the non-cellulose part of the wood from the cellulose. Various methods are employed in all pH ranges from highly acid through neutral to highly alkaline. Each process yields a pulp suited to a particular type of paper. Semi-chemical processes combine chemical treatment with mechanical refining. The most frequently used semi-chemical process is NSSC or Neutral Sulfite Semi-Chemical. Other methods are acid sulfite, bisulfate, alkaline sulfite, sulfate, cold soda and non-sulfur processes. In all cases, wood chips are pretreated with one or more chemicals, followed by refining to disintegrate the fibers.

The benefit of partial chemical treatment over refiner groundwood, is removal of some of the lignin, producing a pulp that will bleach to a higher brightness. Fiber strength is also higher because semi-chemical refining requires less severe mechanical disintegration. With improved processing, however, comes the loss of pulp yield. Stone groundwood has a yield of 95-98%. The semi-chemical pulping processes produce yields from 60 to 80%.

NSSC pulp was first made in a commercial pulp mill in 1925. It is used primarily for making corrugating medium. By the 1960's about 2.5 million tons of NSSC pulp were produced per year in the United States. NSSC is used for hardwood pulps. Softwood does not lend itself as well to to the process because of high chemical demand and higher refiner energy requirements. Softwoods pulped by NSSC also tend to be lower in pulp strength than other chemical pulping methods. The best hardwood species are low-lignin hardwoods such as aspen, beech, birch and maple. High-lignin hardwoods like oak produce low strength pulps.

NSSC can be either a batch or continuous process. In either case, the chips are first pre-steamed at atmospheric pressure. The pulping liquor, a solution of sodium sulfite and sodium carbonate in water, is introduced to the chips in the digester, and the chips are digested at about 180°C for 1 to 6 hours. The digested pulp is refined, washed, and then bleached or pumped to storage.

Chemical Pulping

Fully cooked chemical pulps are produced in all pH ranges from acid to alkaline. The major processes used in the United States and Canada are the sulfite and sulfate (or kraft) processes.

Sulfite process – Sulfite pulping was invented in the mid 1800s by B. Tilghman using a mixture of sulfurous acid and calcium bisulfate. Wood pulped by this method produced a strong easily bleached product. Unfortunately, the chemical conditions in the digester were extremely corrosive and Tilghman did not commercialize his process. The problems of corrosivity were eventually solved using lead-lined or ceramic-lined digesters. The first commercial sulfite mill was built in Sweden in 1874. The advantages of the sulfite process are high yield for a cooked pulp, low chemical cost, and easy bleachability of the pulp. Disadvantages include the highly corrosive nature of the chemicals, lower pulp strength than kraft, and limited useable wood species, with balsam, fir, spruce and Western hemlock preferred. Hardwoods do not lend themselves well to sulfite pulping.

Sulfite processes vary by the alkali used. These include calcium, magnesium, sodium and ammonium. The alkali affects the process in various ways. Solubility varies from highly soluble sodium and ammonium throughout the full pH range, to calcium that is soluble only at very low pH. Ammonium-based sulfite has the fastest pulping rate, sodium the slowest. Chemical recovery is easiest with magnesium, most difficult with sodium.

For sulfite pulping, the cooking liquor is prepared by burning sulfur to produce sulfur dioxide, and then dissolving the cooled gas in a solution of the alkali. The chips are loaded into the digester and pre-steamed to remove air. The cooking liquor is then fed to the digester and the digester heated to cooking temperature, which depends on the alkali used. After the cook is complete, in 3 to 6 hours, the digester is discharged to the blow pit. The pulp is washed to remove the pulping liquors and then cleaned and bleached.

Sulfite pulping has become less common in recent years due environmental concerns. The production of sulfur dioxide and large amounts of difficult to dispose spent acid and black liquor have given the sulfite

⌘⌘⌘

pulping methods an environmentally unfriendly reputation. Nevertheless, some interesting by-products are produced from the waste black liquor. Vanillin, lignosulfonates, alcohol, sugars, yeast, dispersants, leather tanning chemicals, and ore flotation agents are among the many black liquor derivatives.

Kraft process – There are two types of alkaline pulping processes. The soda process is based on sodium hydroxide and is the older of the two. It is little used today. The sulfate or "kraft" process is based on sodium hydroxide and sodium sulfate. It is the largest single pulping method employed for fully cooked chemical pulps.

Kraft pulping was developed in the 1870s in Germany. The German word "kraft" means strong, as the strength of the pulp is an important property of kraft pulp. The kraft process has grown significantly, at the expense of soda and sulfite, since the 1930s. Kraft has many advantages over the other pulping methods. A wide range of hardwood and softwoods can be pulped, cooking times are shorter, less pitch is generated, the pulp has good strength properties, spent liquors are more easily recovered, and turpentine and tall oil are obtained as softwood byproducts. The Kraft process is outlined in Figure 2.

The primary cooking chemicals in the kraft process are sodium hydroxide and sodium sulfide. The sodium sulfate which is added to the spent liquor is reduced to sodium sulfide in the recovery furnace. Sodium carbonate is also present but does not participate in the pulping process; it is converted to sodium hydroxide during the recausticizing process.

In the kraft mill the wood is debarked and chipped, and the chips are pre-heated in a steaming vessel to remove air and assure good liquor permeation. The steamed chips are loaded into the top of the digester along with white liquor, which consists of sodium hydroxide plus sodium sulfide in water. After heating the chips to 170°C for 1 to 2 hours, the pulp is discharged into the blow tank. The pulp is washed free of black liquor and cleaned, then unbleached pulp is either sent to storage or to the bleach plant.

The recovery of black liquor, a solution of lignin and spent pulping chemicals, is a major portion of the kraft process. The black liquor is first concentrated from 15% to 65% solids by various evaporators. This is kept heated so that its tar-like viscosity is reduced to a flowable state. The make-up sodium sulfate is added to the liquor which is then burned in a specially designed recovery furnace/power boiler. The high BTU black liquor burns in the upper section of the furnace, generating heat for process steam and electrical generation. The ash consists of soda ash (Na_2CO_3) and the salt cake (Na_2SO_4) that is reduced to sodium sulfide in the char bed at the bottom of the boiler. This reduced ash is dissolved in water to form green liquor.

⌘⌘⌘

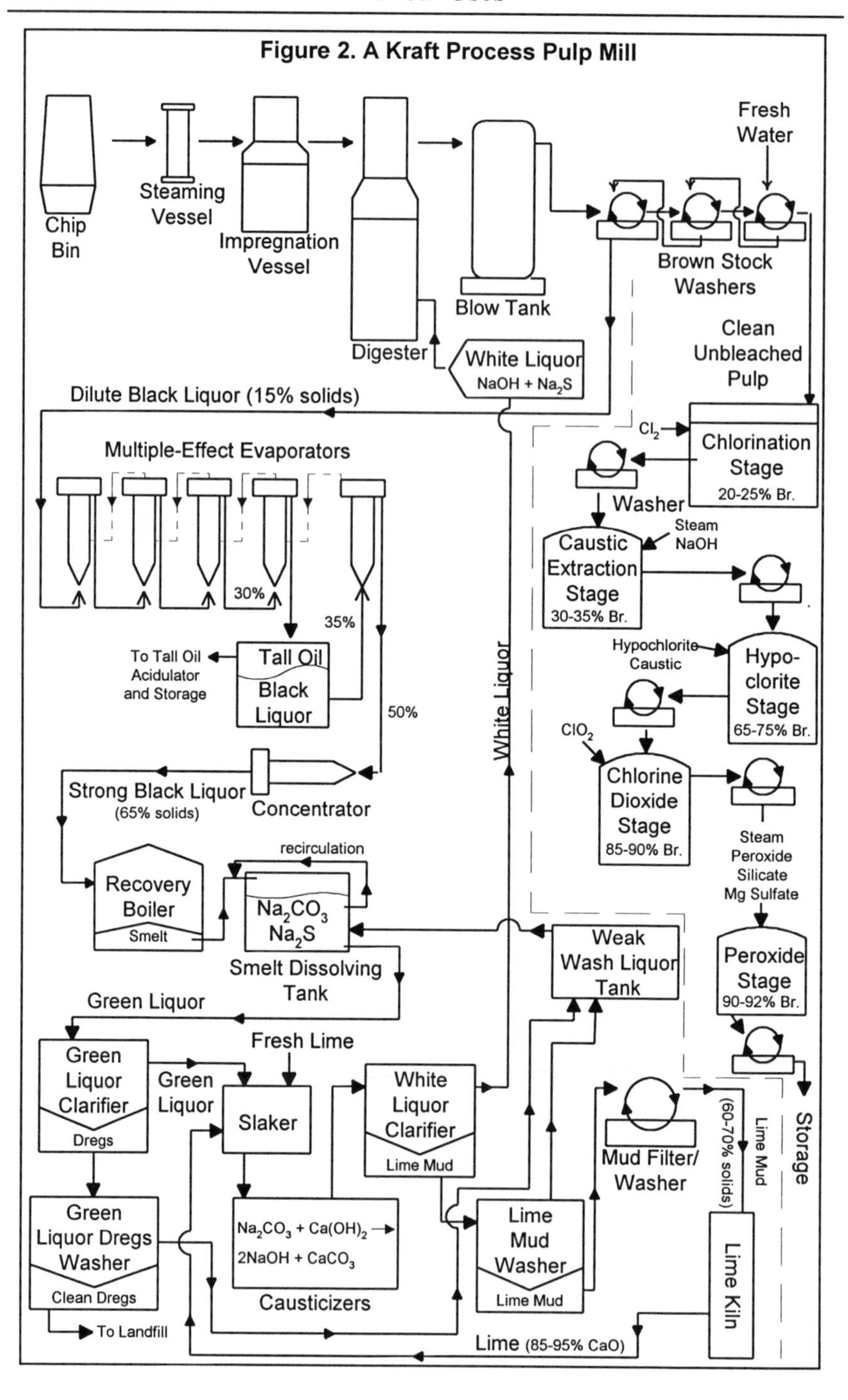

Figure 2. A Kraft Process Pulp Mill
Chip Bin
Steaming Vessel
Impregnation Vessel
Digester
Blow Tank
Fresh Water
Brown Stock Washers
Clean Unbleached Pulp
White Liquor
NaOH + Na2S
Dilute Black Liquor (15% solids)
Multiple-Effect Evaporators
30%
35%
50%
To Tall Oil Acidulator and Storage
Tall Oil
Black Liquor
Strong Black Liquor
(65% solids)
Concentrator
Recovery Boiler
Smelt
recirculation
Na2CO3
Na2S
Smelt Dissolving Tank
Green Liquor
Green Liquor Clarifier
Dregs
Green Liquor
Fresh Lime
Slaker
Green Liquor Dregs Washer
Clean Dregs
To Landfill
Na2CO3 + Ca(OH)2 →
2NaOH + CaCO3
Causticizers
White Liquor Clarifier
Lime Mud
Lime Mud Washer
Lime Mud
White Liquor
Weak Wash Liquor Tank
Mud Filter/ Washer
(60-70% solids)
Lime Mud
Lime Kiln
Lime (85-95% CaO)
Cl2
Chlorination Stage
20-25% Br.
Washer
Steam
NaOH
Caustic Extraction Stage
30-35% Br.
Hypochlorite
Caustic
Hypo-clorite Stage
65-75% Br.
ClO2
Chlorine Dioxide Stage
85-90% Br.
Steam
Peroxide
Silicate
Mg Sulfate
Peroxide Stage
90-92% Br.
Storage

⌘⌘⌘

The green liquor, $Na_2CO_3 + Na_2S$, must be converted back to white liquor, $NaOH + Na_2S$, prior to reuse in the digester. This process of conversion of green liquor to white is known as recausticizing. Recausticizing begins with crude calcium carbonate which is reduced to calcium oxide in a rotary lime kiln. The calcium oxide is dissolved in water to produce slaked lime, $Ca(OH)_2$. The green liquor is mixed with slaked lime in the slaker. The sodium carbonate reacts with the calcium hydroxide to produce sodium hydroxide and calcium carbonate, which precipitates out as lime mud. The white liquor is clarified to remove any residual lime mud, and is pumped back to the pulp mill. The lime mud is returned to the lime kiln to be converted back to lime.

The method appears to be quite efficient but there are losses along the way. The make-up in sulfur and sodium losses is made in the recovery furnace by the addition of salt cake. The make-up in the recausticizing area is made at the lime kiln with fresh crude calcium carbonate.

The pulp produced by the kraft process is widely used throughout the paper and board industries. Unbleached kraft is used for grocery sacks, wrapping papers, raw material packaging, and linerboard for corrugated boxes. Bleached kraft pulps are used for printing and writing papers, sanitary tissue, and bleached paperboard.

Bleaching of kraft pulp involves the use of various chemicals to remove residual colored components. The yield of kraft pulp is low compared to other methods, because of the high degree of delignification. A yield of 45-50% is the norm. Bleaching further reduces the yield.

Bleaching – Traditionally, kraft pulps have been bleached by chlorine and chlorine dioxide, with sodium hydroxide extractions to fully remove the colored materials. The advent of the controversy over the dangers of dioxin have produced a spate of new bleaching agents. Oxygen, ozone, peroxide and chlorine dioxide have replaced all or part of the chlorine in the first bleaching stage. These other pulping agents have been shown to reduce the amount of dioxin and other organic chlorine materials produced by the bleaching process.

In a chlorine bleach operation, the unbleached pulp is mixed with gaseous chlorine at 3% to 5% air dry pulp. After about 1 hour, sodium hydroxide is added to extract the dark colored material the chlorine has liberated. The extracted pulp is washed and dewatered, and then bleached with either chlorine dioxide or sodium hypochlorite. Extractions with sodium hydroxide alternate with the chlorine dioxide or hypochlorite until the pulp achieves the desired brightness without sacrificing too much strength. The pulp is washed, cleaned, screened and sent to storage or the paper mill.

⌘⌘⌘

PAPERMAKING

Modern papermaking is more science than art. Except for specialty hand made papers, all paper is made by machines in modern, sophisticated, computer controlled factories. Modern papermakers are highly trained technicians with an intimate knowledge of the chemistry of the wet end, the dynamics of water removal, the intracacies of high speed manufacturing equipment, the economics of efficient steam generation and useage, and the use of computers. These are just a few of the areas of expertise required to keep the machines, which may be as long as three football fields, operating continuously.

Making paper by a continuous process dates back to 1798, with Robert's invention of the forerunner of the modern Fourdrinier former. Many different types of formers are in use today. They are of three basic types, the Fourdrinier former, the twin wire former, and the cylinder former. Each has many variations on the basic design, and is especially suited to particular papers. Fourdrinier and twin wire formers are used to make papers that range from extremely lightweight tissue grades to linerboard. Cylinder machines are used to make heavyweight papers and paperboards that are formed in several layers.

There are three main parts to all paper machines: the forming section, the press section, and the dryer section. Other parts, such as coaters, size presses, calenders, and reels vary from machine to machine depending on the grade of paper being made.

The primary purpose of the main sections of the paper machine is the removal of water from the paper web. The forming section reduces the water content from 90% to 75-80%. The press section reduces the water content further, to 50-60%, and the driers reduce the moisture content to the final 5-8%. These figures vary depending on the grade of paper produced.

Stock Preparation

Prior to the forming section of the Fourdrinier or twin wire machine is the stock preparation area. In this area between the pulp mill and paper machine, the various pulps are refined to reduce the drainage rate and improve formation by increasing the fiber-to-fiber bonding area, and then blended to produce what is known as "furnish" or in older terminology, "stuff". The furnish also includes pH controlling additives such as alum or soda ash, sizing agents, dyes, filler pigments wet end starch, pitch control agents and biocides. This mixture is usually about 3 to 4% consistency. Consistency, to a papermaker, is percent oven dry solids of a pulp mixture. The furnish is diluted with white water which has been recycled from the forming section of the machine. The diluted stock is passed through a series of centrifugal cleaners to remove dirt, fiber bundles and other undesirable particles which would otherwise cause holes, spots or breaks of the paper web on the machine. The cleaned stock is deaerated then pumped through a final screen immediately prior to the headbox. The final additive prior to the headbox is the retention chemicals. These are added as late as possible so that the flocs they create do not pass through any devices that could break them by shear.

The Fourdrinier Former

The Fourdrinier former is likely the most widely used forming method employed in modern papermaking. It has proven uniquely versatile in the variety of paper grades it can produce. Wire speeds up to 5000 feet per minute are possible with some lightweight grades. The following description follows the pulp flow from the headbox to the couch roll just prior to the press section. A Fourdrinier wet end is shown in Figure 3.

Figure 3. A Fourdrinier Wet End

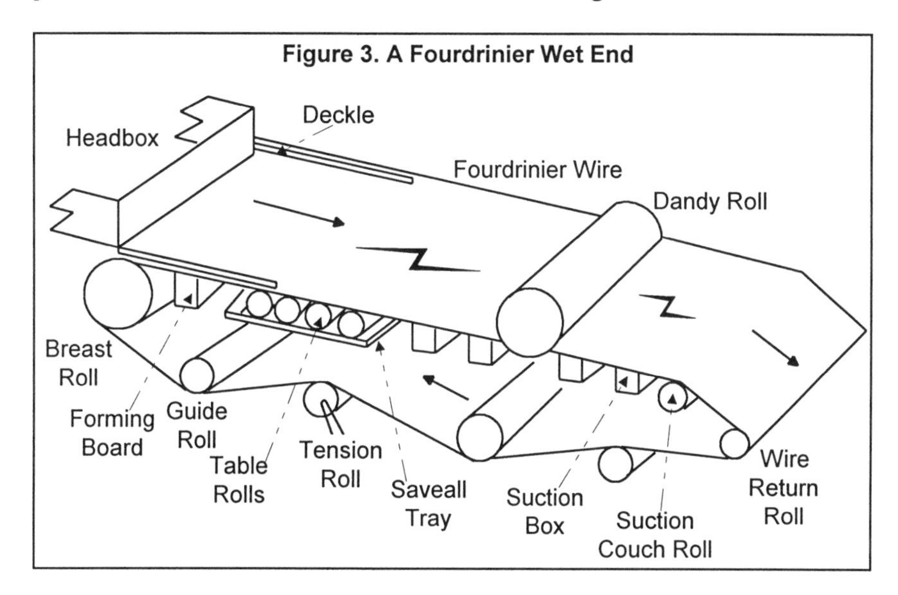

The headbox is a pressurized stainless steel-lined box that controls the flow of pulp and chemicals onto the forming wire. Pressurizing the headbox alllows the machine to run at higher speeds than would be otherwise possible. In order to equalize the speed at which the stock flows from the headbox with the speed of the wire, an artificial hydrostatic head must be created. Without pressurizing, the theoretical headbox height for a machine making paper at 1800 fpm would be 14.1 feet; for a 3000 fpm machine, the headbox height would have to be 39.1 feet. Pressurizing provides the required head in boxes that are 3 to 4 feet high. Certain of the newer designs are even smaller.

⌘⌘⌘

The headbox distributes the stock across the full width of the wire through a slit called the slice. The slice opening can be adjusted to give a consistent cross machine profile. The "wire" is a continuous belt made of woven polyester or similar synthetic fiber, or of phosphor-bronze for certain paper grades. Many different weave patterns are used, from the basic square weave to complex multilayer weaves designed to optimize water drainage, paper formation, and retention of fines and fillers.

Directly below the slice are the breast roll and forming board. The breast roll is the largest roll in the forming area. It turns the wire and supports it firmly as the dilute stock first contacts it. The forming board is a foil with a small surface angle that creates a slight vacuum to draw water through the wire as the paper web begins to drain. As the web moves down the wire, consistency increases rapidly making drainage more difficult. A series of angled foils, smaller versons of the forming board, or table rolls create suction that helps pull water through the paper web. Unfortunately, this also pulls fine fibers and filler particles through the web. After the foils or table rolls, the wire passes over suction boxes that draw out additional water. On top of the wire on some machines, over the suction boxes, rides the dandy roll. This has several functions, including smoothing the top side of the sheet and reducing sheet two-sidedness by pulling some of the fines and fillers from the bottom back toward the center of the sheet. The dandy roll can also be used to put a watermark on the sheet.

The final water removal device of the forming section is the suction couch. The couch roll sets the sheet by vacuum so that it has enough wet strength to travel unsupported through the press section without breaking. This is critical, as the paper web is still quite wet and weak at this stage. The couch roll on many machines is the only mechanically driven roll in the forming section. It is the one that moves the continuous wire belt. The many other rolls are free turning and move along with the wire as it travels over them.

Water Recovery

The largest volume raw material in the papermaking process is water. Significant amounts are recovered and reused throughout the mill. The first area to capture water from the paper web is directly under the forming board and foils or table rolls, and are called the trays. The first tray, under the forming board, catches the water containing the greatest concentration of fines and fillers. The first pass retention of the machine is determined from the difference in the consistency at the headbox and that at the trays. The water from the trays flows into a silo under the wire pit. The silo water is used to dilute the fresh furnish in the stock preparation areas. The water removed by the suction boxes and couch roll is blended with the silo water. Any run-off from the wire cleaning showers is caught in the wire pit and flows back to the silo. Under the couch is the couch pit. This captures the edge trim and the full paper web off the wire during start-up and web breaks farther down the machine. Since this material is at 20% consistency, it is dewatered by a disk filter called a save-all. The recovered stock is added to the broke for reuse. Broke is, in addition to the save-all stock, paper from farther down the machine such as winder cuttings or sheet cutter trimmings that can be repulped.

The Twin Wire Former

The twin wire formers are a variaton of the Fourdrinier. The Fourdrinier tends to make a sheet that is two-sided, because drainage occurs in only one direction. The sheet is uneven in distribution of fine fibers, dyes, and filler pigments. This can be especially undesirable in colored papers, since dyes have a greater affinity for the finer fibers and pigment particles. Because water drainage on the Fourdrinier concentrates these fines on the wire side of the sheet, the color of darkly dyed papers tends to be uneven with the bottom and top sides distinctly different.

The twin wire former has two continuous wire belts running very close together with the stock in between. This configuration allows for drainage in both directions, avoiding two sidedness. Twin wire configurations can also be used to make a multilayer paper wherein the two layers may be of the same or different stocks. This is used to make heavier weight paper such as linerboard.

The Cylinder Former

Cylinder machines are often used to make multilayer paperboard. The cylinder machine consists of a series of vats for stocks that may be of the same or different composition depending on the paper grade. A wire covered cylinder rotates in the vat of stock. The stock is picked up onto the cylinder by applying vacuum at a point where the cylinder surface exits the stock. Vacuum is drawn on the stock to drain it until a point at the top of the rotation where a continuous felt contacts the cylinder. The paper web is released from the cylinder and is carried along on the underside of the felt. The surface then reenters the stock and the process begins again. On a multilayer machine, several cylinders are lined up such that the felt passes over all of them in sequence building up a thicker paper as it moves along. The paper web is then transferred to the press section for further water removal.

The Press Section

The press section of a paper machine is the second major dewatering process after drainage. The paper web travels from the end of the forming fabric through a series of two to four presses. These presses are composed of two

⌘⌘⌘

rolls and a felt. The felt is a continuous loop. The press rolls are loaded to exert maximum pressure to drive out as much water as possible without crushing the sheet. The function of the felt is to both support the sheet and carry away the water that the presses have removed. Older, slower machines had the first open draw, where the paper was unsupported, between the couch roll and the first press. This area of unsupported paper was a site of frequent web breaks. This area also created a speed increase. The stretching of the paper between the couch roll and the first press can be the equivalent of 75 to 100 fpm or more in speed difference between the wire and the reel. As machines increased in speed and basis weights got lighter, it became apparent that the sheeet would not remain intact between the couch and the first press. This problem has been corrected by the use of a suction roll pick-up felt. The felt contacts the wire just below the couch roll. The suction of the roll on the inside of the felt picks the sheet up off the wire and the felt supports the sheet through the first press. The paper is then transferred to the second press to pass through the second felt and so forth through the third and, if necessary, fourth press. This pressing operation reduces the water content from 80% to about 60%.

The Dryer Section

From the press section, the paper goes to the dryer sections. Steam heated or infrared dryers are the most expensive dewatering methods on the paper machine. The dryers willl reduce water in the sheet from 60% to 5-8%. Steam dryers are a series of cylinders over which the paper web passes, while being held to the surface by felts. Drying also occurs between the cylinders by heated air blown into the gaps. This is known as pocket ventilation.

A widely used dryer system for tissue and ultralight paper grades is the Yankee Dryer. This is a single large diameter (10 to 20 feet) dryer, steam heated like the smaller cylinder dryers. The sheet leaves the press section and is carried over the Yankee, being doctored off on the opposite side. The Yankee is not a felted dryer, so the sheet is often held onto the dryer surface by air impingement. This improves the drying rate by improving sheet-to-dryer suface contact.

Infrared drying is used in certain specialty applications, or when space is limited since an IR dryer takes up significantly less space than cylinder dryers of equivalent capacity. Because of the high cost of electricity or gas to fire these dryers, they are not widely used. The IR dryer also represents a potential fire hazard because of operating temperatures in excess of 800°F.

Other parts of the paper machine vary according to the end use of the paper. These include the size press, calender, reel, and supercalender.

⌘⌘⌘

The Size Press

The size press is a type of coater that applies a light coating of a particular additive to one or both sides of the sheet. Materials run on a size press include starch and surface size. Starch is used to prevent picking-out of fibers when the sheet is typed on. Surface sizing with materials such as styrene maleic anhydride or cellulose gum serve to increase the resistance to penetration by aqueous fluids such as ink. Surface sizes can also be used to achieve erasability of bond papers by keeping ink from penetrating into the paper.

The size press is usually composed of two rolls between which the paper web passes. The sizing solution is sprayed onto the sheet, the rolls press the size into the sheet, and the excess is captured for reuse. Several size press configurations are shown in Figure 4.

Figure 4. Size Press Configurations

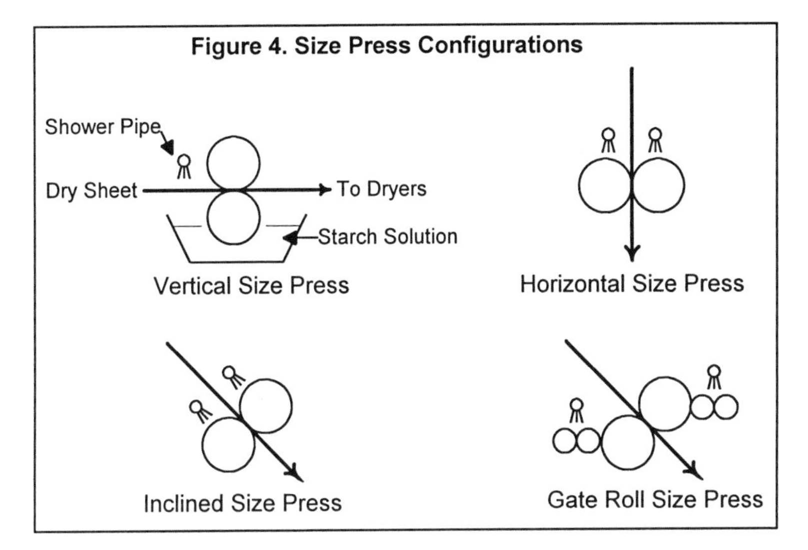

The Calender Stack

After the paper has been dried, it is often calendered to even the cross-machine profile. Calender stacks are made of two or more rolls which are set at close tolerances. The paper web pases between the rolls and is smoothed and levelled. On some paperboard machines, calender sizing is done. A light coating of starch or CMC is put onto the underside of the board for cure control.

The Reel

The final item on the paper machine is the reel. The reel is the roll that holds the finished paper coming off the machine. Finished paper must be removed from the reel to be useful to the final customer. Depending on the end use,

⌘⌘⌘

paper is rewound onto cores and cut to size for web fed printing presses, cut to sheet size for sheet fed presses or typing use. Tissue is cut to size and formed into many items for sanitary uses. Unbleached kraft papers are formed into sacks or used as linerboard for corrugated boxes.

The Supercalender

An important auxilliary piece of equipment used primarily in making coated papers, but also used for some highly filled uncoated papers, is the supercalender. The supercalender is made of alternate polished metal and compressed paper or cotton rolls. The stacks are variable in size with from 5 to 20 rolls. Supercalendering imparts gloss to coated and filled uncoated papers. A supercalender stack is pictured in Figure 5.

Figure 5. A Supercalender

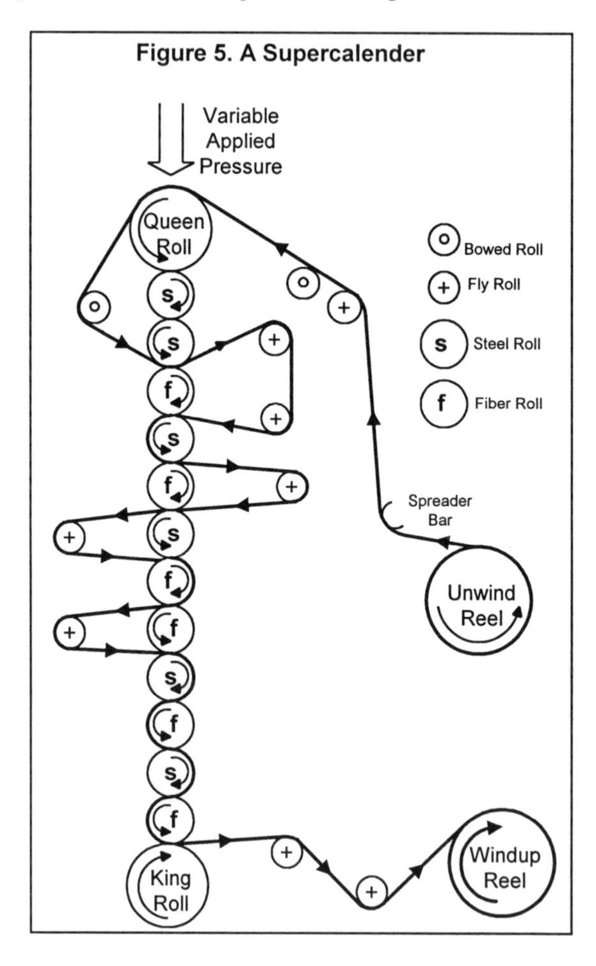

⌘⌘⌘

A combination of nip pressure, steam pressure, speed of the paper web through the calender and the coating composition determines the finished sheet quality. Gloss and smoothness are increased by supercalendering, but brightness, opacity and caliper are decreased. These properties must be balanced to produce the best overall quality sheet.

PAPER PIGMENTS

In the pulp and paper industry, mineral raw materials are traditionally called "pigments", whether or not they are used to affect color. The use of mineral pigments in the pulp and paper industry fall into five major catagories: filling, coating, pitch control, deinking of recycled papers, and microparticle retention. The minerals comonly used for each job are:

Paper Filling: kaolins, titanium dioxide, talc, calcium carbonate, calcined clays, synthetic silicas and aluminas.

Paper Coating: kaolins, titanium dioxide, talc, calcium carbonate, plastic pigment, blanc fixe, satin white, zinc oxide, structured kaolin pigments.

Pitch Control: talc, fine particle sized clays, bentonite.

Deinking of Recycled Papers: talc, kaolin.

Micro Particle Retention: bentonite.

PAPER FILLING

Paper filling pigments are added primarily to reduce the cost of the paper. Mineral fillers in general, except for titanium dioxide (TiO_2) are less expensive than fiber and therefore reduce the cost of the paper. Of course, many other benefits are obtained from these minerals, including improvements in brightness, opacity, smoothness, print quality and economy.

Opacity

Opacity is an optical reflectance phenomenon. It is the resistance of an object to the transmission of light. For paper lower opacity means greater "show-through". Sheet opacity is a function of the refractive index differences among air, cellulose fiber, and filler. As light enters the paper, it is either reflected, bent (refracted), absorbed, or passed through. Mineral fillers contribute to opacity by aiding in light reflection, refraction, and absorption. The higher the filler refractive index, the more opaque the sheet will be.

Brightness/Whiteness

A printed image appears better if it contrasts well with the background. A brighter – higher reflectance at all wavelengths of light – printed paper looks better than a lower brightness sheet. Mineral fillers are generally higher in brightness than bleached pulp, so their use in paper can improve sheet brightness. Paper fillers range in GE brightness from 80% to 99+%. Whiteness is a function of shade and color. Two minerals can have the same GE brightness, but one will look brighter because of shade differences. Clays tend to be yellow-white, while talc, titanium dioxide and some of the synthetics are more blue-white. The human eye perceives blue as bright, so a blue-white filler will look brighter than a yellow-white one. The shade of the mineral filler will affect paper color.

Gloss

Ther phenomenon of gloss is both optical and physical. The surface luster of a sheet provides a pleasing effect. A paper with good gloss has high reflectance, and good continuity of that reflectance. Too much gloss is known as glare and is undesirable. Gloss relates to the smoothness or evenness of the paper surface. Mineral filllers can enhance or decrease gloss as desired.

Print Quality

Mineral fillers affect aspects of print quality such as ink receptivity and holdout. Certain fillers accept ink better than others, while others absorb it too well and tend to draw the ink into the sheet rather than keeping it on the surface.

Formation

The finer particle size of filler compared to fiber aids in sheet formation. The minerals are difficult to retain in the paper web, but once the mat of fibers begins to form the filler particles are caught in the fiber interstices. They fill in the spaces creating a more tightly formed sheet. Polymeric retention aids are often used to flocculate the fibers and filler particles so that the filler agglomerates are large enough to stay betweeen the fibers without sacrificing the opacifying affect of the fine particles. The fillers can also be chosen to increase or decrease sheet bulk. Book and magazine publishers often have set requirements for the caliper of paper they use. That is, no more than a specified number of sheets may occupy a certain thickness.

⌘⌘⌘

Economy

Mineral fillers are generally less costly than virgin fiber, so it is less expensive to make paper at something less than 100% fiber. Fillers can also substantially reduce drying costs. Mineral fillers release associated water in drainage and pressing operations more readily than do cellulose fibers. Less steam is therefore required for drying.

Mineral fillers can benefit paper quality, but they also have some limitations. Sheet strength is reduced in proportion to substitution of filler for fiber. Strength comes from fiber-to-fiber bonds, the number of which are reduced when fiber is replaced by filler. Depending on the type of paper being made, anywhere from 5% to 50% filler is used. The optimum level in each case is determined by paper forming efficiency and the intended end use of the paper.

The Perfect Filler

TAPPI (Technical Association of the Pulp and Paper Industry) defined the "perfect filler" many years ago as having the following properties:

1) It should have a reflectance of 100% in all wave lengths of light to provide maximum brightness and whiteness.

2) It should have a very high refractive index to provide maximum opacifying and brightening power.

3) It should be completely free from grit or extraneous matter and have a particle size distribution close to 0.3 micron, approximately half the wavelength of light.

4) It should have a low specific gravity and be soft and non-abrasive.

5) It should be capable of imparting to paper a surface capable of taking any finish from the lowest matte to the highest gloss.

6) It should be completely chemically inert and insoluble.

7) It should be completely retained in the paper web so that there is no loss.

8) It should be reasonable in cost.

No such perfect filler exists; each mineral filler has its own good and bad characteristics. The papermaker chooses a filler or blend of fillers to give the overall best performance for the intended end use of the paper.

⌘⌘⌘

Papermaking and pH

Fine papers, i.e. printing and writing grades, are made by either at acid pH (4.0 to 5.5) or neutral-alkaline pH (7.0 to 8.5). The pH range is determined by the type of internal size used. Internal sizing gives the sheet strength while in the wet state, especially when passing through the size press. Sizing also gives finished paper resistance to penetration by inks and other aqueous liquids.

Since the 1800's, sizing has been achieved by the wet end addition of rosin. Rosin size is emulsified in water prior to addition to the wet end of the paper machine. The introduction of preemulsified sizes in the 1970's greatly simplified the sizing process. The rosin size must be anchored to the fibers in order to be effective as a sizing agent. This is accomplished by reaction with aluminum in the form of papermaker's alum, $Al_2(SO_4)_3$. The pH of the wet end is important to this process as the chemical nature of the aluminum ion is a function of pH. At a pH of 4.0 to 5.5, aluminum is Al^{+3} which can react with the rosin in the size. This aluminum rosinate is precipitated onto the fibers creating the water resistant barrier. The precise mechanism of this process has been debated for many years.

Unfortunately, the low pH papermaking process has been associated with early deterioration of paper. Papers made in the past two centuries have been found to be very short-lived. This is especially troublesome to librarians and others concerned with archival records.

In the 1950s, experimenters began studying alternate sizing agents that would allow for papermaking at neutral to alkaline pH. What have become known as synthetic or cellulose reactive sizes were developed. These sizes are of two types, alkylketene dimer (AKA), and alkenylsuccinic anhydride (ASA). They are believed to anchor to the paper fibers by direct reaction rather than precipitation, as with rosin alum sizes. These synthetics achieve higher degrees of sizing at lower dosages than rosin sizes. The chemistry of the wet end is much more exacting for neutral alkaline sizing than for acid sizing and must be monitored very carefully to achieve sizing. Very strict attention must be paid to the various additives accompanying the size, their type, quantity and location of addition. The likely driving force behind the switch from acid to alkaline papermaking is the prospect of using calcium carbonate as an inexpensive functional filler.

Minerals in Acid Papermaking

The major fillers in acid papermaking are kaolin, calcined clay, talc and titanium dioxide. Minor fillers are the synthetic silicas and aluminas.

Kaolin – Kaolin, or clay as it is commonly known among papermakers, is the largest single filler pigment in the North American paper industry. With the growth of alkaline papermaking, calcium carbonate use is increasing

⌘⌘⌘

rapidly and may one day overtake clay, but at present clay still predominates. The North American paper industry uses approximately 4.5 million tons of clay each year, which is about 60% of all the kaolin produced. Clay accounts for approximately 75% of all paper fillers.

Clay is used in almost all filler applications because of its high quality and reasonable price. Filler clays usually fall into the category of "soft clay", having a particle size of 60% minus 2 microns, and a GE brightness of 80 to 90 depending on grade. They may be air floated or water processed and are delivered to the papermill either as powders with dispersant added, or 70% solids aqueous slurries. This greatly facilitates preparation of the clay for the paper machine.

Air floated clays are beneficiated by dry grinding and air classification. The quality of the raw material, specifically in the areas of brightness and the presence of colored impurities, is very important to an air floated clay. Dry processing can only make minor improvements in brightness. Impurities such as quartz, titanium dioxide, mica or iron oxide cannot be removed by air flotation except as oversized particles. In general, air floated clays are lower quality than water washed clays and find only limited use in paper making.

Wet processing, also known as water washing, is the more widely used method of beneficiation of paper making clays. Water washing begins in the clay pit where the clay is removed from the pit surface using wet drilling methods. This wetted clay is dispersed in water at 30% solids. The heavier grit and other impurities are removed by settling and screening. The degritted clay is fractionated into various particle size ranges by centrifuging. The coarser fractions are used for filler grades. The finer, brighter fractions are used for coating grades. After the fractionation, the clays are further purified by ultra-high intensity magnetic separation which removes iron, titanium dioxide, mica, etc. from the clay slurry. Bleaching with either zinc or sodium hydrosulfite can be done to further improve brightness. The clay, at 30% solids, is then flocculated and dewatered to 60% solids by vacuum filtration. The clay concentrate is either spray dried or increased to 70% solids by addition of dispersing agents and dry clay.

The introduction of 70% solids slurry clays in the 1970s was a boon to both the papermakers and the clay processors. The papermakers welcomed the ease of handling of the slurry and the reduced manpower requirements in the additives preparation area. The clay producers were able to realize large cost savings by not having to spray dry most of the clay prior to shipment.

Filler clays are the mainstay of acid papermaking filler requirements. Clays are significantly cheaper than fiber. Addition rates of 8 to 10% based on total sheet weight are typical for fine papers. The use of clay offers several benefits, including increased brightness, opacity, smoothness, print quality, drainage on the wire, and drying rates, plus reduced bulk and improved formation. Disadvantages include reduced retention, strength loss, and adsorption of certain chemical additives onto the surface of the clay. This necessitates higher additive levels to compensate for the amount lost through adsorption.

Titanium Dioxide – The second most important mineral filler for acid papermaking is synthetic anatase titanium dioxide, TiO_2. Anatase has become the opacifying filler of choice for acid papermakers. It is preferred over rutile despite its lower refractive index (2.5 vs. 2.7), because it has the blue-white color preferred to rutile's yellow-white.

Anatase TiO_2 is available to the papermaker either as a powder which has had dispersing agents added to make the slurrying process easier, or as a high solids slurry. Slurry is often preferred, especially by high volume users, because it saves the time and effort of the slurry makedown step. The optimum amount of dispersant is assured, the deagglomeration has been achieved by the manufacturer and the need for dry product storage is eliminated.

Opacification of white papers is a necessity for most uses. Fiber alone has a refractive index very close to that of air and would therefore produce a translucent sheet. Lighter basis weights make this an even greater problem. TiO_2, with its high refractive index, effectively opacifies the sheet. TiO_2 has a very fine particle size of approximately 0.25 microns, about one half the wavelength of visible light. It has a very high brightness of 97 to 100 GE, and high reflectance through most of the visible spectrum (400 to 700 nm). It fulfills most of the requirements of the ideal paper filler, except for price. Its high price is nevertheless well worth the benefits to the papermaker. Typical usage levels of 2 to 3% TiO_2 based on paper weight yield opacity increases of 10 to 15%.

Among the advantages of TiO_2 for filling printing and writing papers are, first and foremost, improved opacity, plus improved brightness and color, reduced show-through and improved printability. These combine to give a better appearance to the sheet. For bond, ledger and writing papers, TiO_2 gives the improvements in optical properties listed above without sacrificing sizing efficiency or sheet strength.

Calcined Clay – Calcined kaolin is employed as a TiO_2 extender. Calcined clay can replace TiO_2 up to a 1:1 ratio without sacrificing optical properties. Calcined clay is produced from water-washed filler grades that are spray dried then heated in rotary kilns at either 650 to 700°C for partially calcined grades or 1000 to 1100°C for fully calcined grades. The partially calcined grades are more popular for paper filling. They are bulkier, less abrasive, have lower specific gravity, but also lower brightness than the fully calcined

⌘⌘⌘

grades. The fully calcined products are used in paper coatings to extend TiO_2 and impart gloss.

Calcined clays act as TiO_2 extenders by spacing the ultrafine TiO_2 particles. The clay particles are significantly larger than TiO_2 and act as a core that is then surrounded by TiO_2 particles. The clay keeps the TiO_2 particles apart, thereby improving the overall light scattering of the sheet. Calcined clay has a slightly higher refractive index than regular clay and as a TiO_2 extender, can be added at up to 50% of the TiO_2 content without adversely affecting opacity or other optical properties. With calcined clay at about one quarter the price of anatase TiO_2, this can provide significant cost savings.

Talc – High brightness, ultrafine platy talc can also be used as a TiO_2 extender for fine papers. Its brightness and refractive index are lower than calcined clay, but its large thin plates space TiO_2 particles more effectively than calcined clay. Both talcs and calcined clays have similar hydrophobic natures. They both naturally resist being wet by water. This hydrophobicity makes preparation of slurries at levels greater than 50 to 55% challenging. Talc is usually supplied in powder form for either dry addition to the paper stock or preparation of dilute slurries prior to addition. Calcined clays are available from the producers as predispersed 50 to 55% aqueous slurries.

Talc's advantages as a paper filler include its smoothness and very low abrasion. Low abrasivity helps to prolong wire life and other parts of the paper making machinery especially slitters. The paper is smoother when made with talc rather than calcined clay.

Synthetics – Synthetic fillers, other than TiO_2, are the silicas and aluminas. The silicas are either amorphous or crystalline, and are formed by precipitating various forms of silicon dioxide from acidified soluble silicates. The synthetics can be hydrated silica, simple alkaline earth metal silicates, or aluminum silicates.

The silica products tend to have refractive indicies similar to the clay and talc but lower than calcined clay. The bulkiness of the particles can aid in TiO_2 extension, but they are most noted for their high brightness.

Alumina trihydrate (ATH) is made from waste generated by the aluminum metal purification process. It has high brightness but low refractive index. At low levels of substitution for TiO_2, 25% or less, optical properties can be maintained. Like the synthetic silicas, ATH can be used as a paper brightning pigment when opacity is not as important. ATH is used in NCR (no carbon required) papers where it favorably affects the special dyes used in the paper. ATH is also known for its flame retardant properties. ATH retards burning by the release of water at rather low temperatures. It is added to impart flame retardance to certain specialty papers.

⌘⌘⌘

Minerals in Alkaline Papermaking

One of the many advantages proposed for the conversion from acid to alkaline papermaking is the ability to use calcium carbonate as a filler. Its solubility in weakly acidic systems make it unusable except at pH 7.0 or above.

Calcium carbonate – Natural ground calcium carbonate (GCC) is produced in a wide range of particle sizes and is available either as powder or slurry. High brightness and low cost make GCC very popular with alkaline papermakers. There are, however, problems associated with its use, especially in the area of deposit formation on the papermachine. Most pitch deposits found on an alkaline papermachine are composed mainly of calcium carbonate that has deposited along with other sticky components in the furnish.

The other form of calcium carbonate available to the alkaline papermaker is synthetically produced precipitated calcium carbonate (PCC). PCC is made from limestone that has reduced to lime in a rotary kiln, slaked to calcium hydroxide then converted to calcium carbonate by reaction with CO_2. PCC can be precipitated in many different crystal structures. The particle size and shape can be tailored to the customer's specific needs.

Satellite PCC plants are being built by certain manufacturers at or near to the papermill site. Lime from the recausticizing area of the kraft pulp mill is recycled to the PCC plant where it is converted to PCC. One of the many advantages of this process is the savings in shipping costs. Since it is difficult to produce high solids slurries of PCC, the on site mills can provide PCC slurry at low solids to the papermill without incurring the costs associated with shipping large volumes of water. A disadvantage of on-site plants is that the papermill is committed to long term use of PCC. Some mills have found that PCC alone does not fulfill the filler requirements and some GCC must be used along with the PCC. As a condition of PCC plant construction, the papermill must purchase some contracted amount of PCC, and it must be paid for even if not used. PCC has found more use in coatings for alkaline papers than filler and will be discussed further in the section on coatings.

Calcium carbonate as a filler for alkaline papermaking is preferred because of its high brightness. Its refractive index of 1.6 puts it between regular filler clay and calcined clay, giving it some opacifying efficiency. It improves ink receptivity, thus improving print quality. It can be added at higher levels than fillers used for acid papermaking, because the fiber stength is greater under alkaline conditions and less fiber can be used. Filling levels of 25% or more can be achieved with calcium carbonate. These higher filler levels aid in increased drainage on the wire and faster drying. This can be employed to either reduce production costs or increase machine speed giving increased production.

Other fillers – Kaolin and TiO_2 are used in alkaline papermaking, but to a much lesser extent that in acid papermaking. Kaolins are cheaper in some cases than calcium carbonate, and satisfy some filling requirements for alkaline. Calcium carbonate does not opacify well so TiO_2 must be used for some papers. The extender pigments, especially calcined clay and talc, are used and perform better than calcium carbonate when opacifying is needed.

PAPER COATING

A major sector of fine papers manufacturing is that of pigment coated grades. Pigment coatings improve printing properties of the sheet. Approximately one fifth of the fine papers produced are pigment coated. This is the fastest growing segment of the paper industry.

The coating of paper followed by supercalendering, does many things to the sheet to improve its quality. Brightness can be improved by applying a higher brightness coating over a lower brightness base sheet, while the pigments in the coating contribute to light scattering. The coating must be uniformly applied to maximize the overall effect.

The opacity of coated paper is becoming more important as basis weights continue to go down and postal rates continue to go up. Printers are required to use lighter basis weight sheets without sacrificing print quality. By choosing the right combination of coating pigments and a good quality base sheet, the papermaker can produce a good light-weight sheet without compromising on opacity or other properties.

Gloss of the sheet is affected by the pigments used in the coating color. Clays, especially delaminated clays, contribute to development of gloss. Depending on the end use, sheet gloss can range from matte for art papers to high gloss enamels for cover papers. Gloss is an optical property that affects other sheet properties as well. It is an indication of how uniformly the sheet is covered by the coating, and how smooth the surface is. The smoothness of the surface will determine how well the paper will contact the printing press plates and how evenly the ink will be transferred.

The requirements of coated papers can be achieved by the proper selection of coating pigments, binders and other additives, combined at the right proportions, having the right rheology for the type of coater used, all applied to the right kind of base sheet. No coating, no matter how prepared and applied, can make up for a poor quality base sheet.

⌘⌘⌘

Coating application processes have changed significantly over the years. Three major coating processes are used today: roll coaters, air knife coaters, and blade coaters.

Roll Coaters

Roll coaters were first used on papermachines in the late 1920's. Roll coaters, as their name implies, use a series of rolls to apply coating to the paper. The coating color is added to a reservoir between the first two rolls that are set to meter the amount of coating to be applied to the sheet. The next several rolls oscillate as they turn distributing the coating evenly across the whole width of the rolls. The paper web passes between the final two rolls, which transfer the coating onto the paper surface. Depending upon the specific process, both sides of the paper may be coated simultaneously or each side can be coated sequentially. A double roll coater is illustrated in Figure 6.

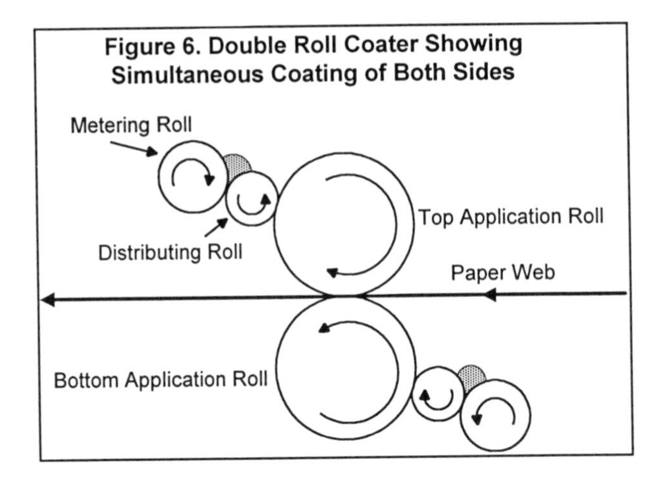

Figure 6. Double Roll Coater Showing Simultaneous Coating of Both Sides

Coating colors for roll coaters can run from 50 to 65% solids, with coating weights dependent on solids. Ten to fifteen pounds of coating per 3000 ft^2 is possible with this method. Roll coaters are speed limited to about 1500 fpm.

Air-knife Coaters

The air-knife, or air doctor coaters were first introduced by S.D. Warren Paper Company. The air-knife coater has an applicator roll which is submerged in a trough of coating color. The paper web passes over the applicator picking up coating off the roll. The excess coating is removed by the air-knife. Excess coating is returned to the trough under the applicator roll. Coating solids for this process tend to be low, in the range of 30 to

50%. Low running viscosity is necessary for good runability of the air-knife coater, and is controlled by the coating solids. Low coating weights are a result of the low coating color solids. Most air-knife coaters are off-machine rather than on machine as for roll and blade coaters. An air doctor coater is pictured in Figure 7.

Figure 7. The Air Doctor Coating Process

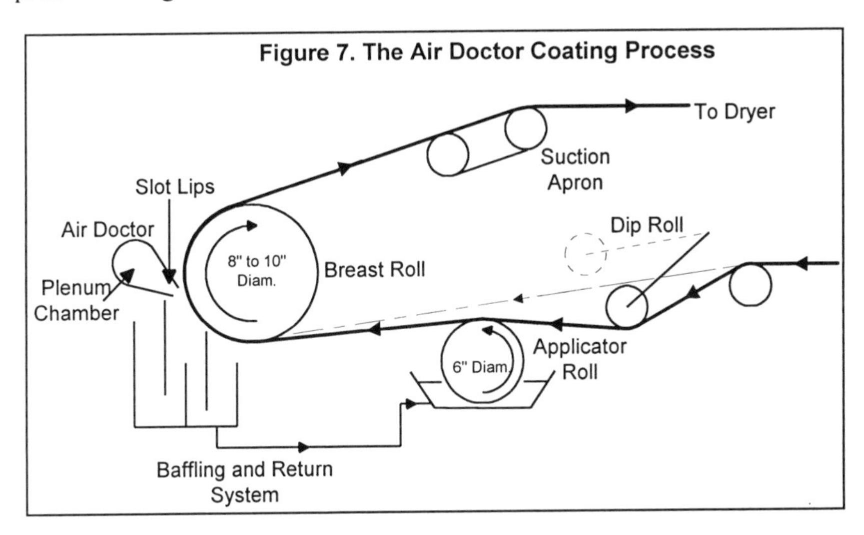

Blade Coaters

Blade coaters were first introduced in the 1940s. Blade coaters have several configurations, but most are of the trailing blade type. In this process, the coating color is applied to the paper web and the excess removed by a doctor blade.

The simplest trailing blade coater is the puddle type. The puddle coater gets its name from the fact that the coating color is applied to the paper from a trough that is created between the applicator roll and the doctor blade. The paper passes over the roll, through the puddle of coating, and the excess coating is doctored off by the blade. The coating thickness is determined by the coating color solids and the gap between the blade and the paper web.

A variation of the trailing blade coater which has found wide acceptance is the inverted blade, flooded nip coater, as in Figure 8. This coater is comprised of an applicator roll that turns in a pan of coating, a backing roll and an inverted trailing blade. The applicator roll is under the backing roll. The paper passes between the applicator and backing rolls where it is coated. Excess coating is removed by the inverted doctor blade. Coating weights are controlled by the coating solids and viscosity, the backing roll to blade gap, the applicator roll speed, the web speed and the blade angle. High

⌘⌘⌘

solids (60 to 65%) coatings with stable high shear viscosity are used on these types of coaters.

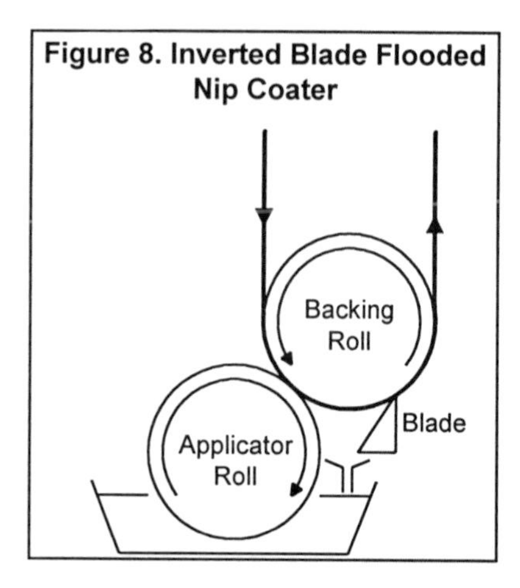

Figure 8. Inverted Blade Flooded Nip Coater

Many variations on the blade coater have been introduced. Among the more popular is the Bill blade coater. This trailing blade coater can coat both sides of the paper simultaneously. The web passes through a coating head that has an orifice on each side that allows coating to contact both sides of the sheet. Excess coating is doctored off by blades on either side of the sheet. Two different coatings can be applied to the sheet at the same time with this type of coater.

Blade coaters have the advantage of running at higher speeds than roll or air-knife coaters. They tend to give more even coating distribution of coating over the paper surface, and more variations in coating weights can be achieved. Their disadvantages include the rather complex sheet runs which must be used in order to achieve coating of both sides of the sheet in other than the Bill-blade configuration. An extremely uniform coating color is a necessity. Fine lumps of coating can get caught under the blade leaving scratches and uncoated areas. This problem can be addressed by careful screening of the coating prior to use along with careful coating preparation that assures good mixing to fully break up pigment agglomerates. The coatings must have good high shear stability, neither thinning or thickening excessively. This can be controlled through the right choice of pigments, binders and flow control additives.

Minerals in Paper Coatings

Titanim dioxide, kaolin clays, and calcium carbonate are the most widely used minerals in paper coatings.

⌘⌘⌘

Titanium dioxide – Rutile TiO_2, rather than anatase, is used in paper coatings, because of its higher refractive index, 2.7 vs. 2.5. For paper coating, synthetic rather than natural rutile is used, because it is white and has better uniformity of crystal structure. Natural rutile is black.

Synthetic rutile is made from titanium-rich sands, which are primarily of Australian origin. Synthetic TiO_2 is produced by either the sulfate or the chloride process. The sulfate process is older and is becoming less common. In the United States, it has been deemed environmentally unfriendly and has been almost completely phased out. The sulfate process is primarily used to produce synthetic anatase. The chloride process is used for making both anatase and rutile.

The chloride process begins with the high temperature reaction of titanium ore with gaseous chlorine producing titanium tetrachloride, iron chlorides, and chlorides of other metals present in the ore. The titanium tetrachloride is recovered from the other metallic chlorides by fractional distillation. The purified titanium chloride is reacted with oxygen to form titanium dioxide plus gaseous chlorine, which is recycled back to the beginning of the process. The TiO_2 precipitates preferentially in the rutile crystal habit. Anatase can be made by seeding the liquor with anatase crystals.

Rutile TiO_2 is provided to the papermill in 50 pound bags, dry bulk, or as a predispersed high solids slurry. The slurry form is easier to handle, requiring no makedown equipment and much less storage space.

Titanium dioxide's primary purpose in a paper coating is opacification. It also contributes to brightness, but if brightness is required in a coating there are several high brightness pigments available at much lower cost. Of the commonly used coating pigments, TiO_2 has by far the highest refractive index, and therefore contributes the greatest amount of opacification per pound. Overuse of TiO_2 must be avoided because too many particles of TiO_2 lead to an actual loss of opacity through agglomeration and crowding. The preferred approach is to use TiO_2 at the lowest level possible and supplement with extenders to get the required optical properties. Other benefits of rutile use are its decrease in paper show-through, better print-to-background contrast, better half-tone print quality, and the ability to produce lighter basis weights without loss of opacity.

Kaolins – By far, the largest volume coating pigment is clay in its various forms. The primary coating clays are fine particle, high brightness water-washed clays. These are rated by color and particle size into grades from No. 3, the lowest brightness, coarsest grade, to Fine No. 1, the highest brightness, finest particle size grade. Coating clays are the workhorses of the industry. They are used in high solids formulations for coated glossy publication paper, and high brightness high quality coatings for enamel cover stock. These clays are produced to papermakers' expectations with consistently high quality and little or no variation in properties year to year. Coating clays provide coating brightness, viscosity control, gloss, opacity, smoothness, and print quality improvement. Most coatings are at least 75 parts clay, with some as high as 100 parts, depending on the end use of the paper.

Delaminated clays are produced from water-washed clays that have been specially treated to separate the booklets of clay platelets into individual platelets. As a paper coating component, the thin delaminated platelets impart high gloss to supercalendered paper. Brightness and TiO_2 extension of delaminated clay are also good. One disadvantage is that the platy structure contributes to coating dilatency. At high solids, under high shear, coatings containing extremely platy pigments like delaminated clay thicken, restricting flow. The amount of delaminated clay that can be used in a coating formula is consequently limited.

Calcined clay has a higher refractive index than regular coating clay, and contributes more to opacity. It is also a TiO_2 extender. The method used to produce calcined clay results in a particle that adds bulk to the coating and opacity at lower coating weights. Calcined clay contributes to print quality as well. Unfortunately, the bulkiness and anhydrous crystal structure prevent the preparation of calcined clay slurries at greater than about 50% solids. At higher solids, rheological stability at high shear is difficult to maintain. The amount of calcined clay that can be used in a high solids coating is therefore limited to far less than is ideal based on the benefits to the coated paper quality.

Calcium carbonate – The use of calcium carbonate in paper coatings had been restricted to specialty applications prior to the advent of alkaline papermaking. A large amount of the furnish for the coated paper's base stock is composed of coated broke. The calcium carbonate in the coating was incompatable with the acid pH in the wet end.

In the alkaline papermaking era, calcium carbonate has made great strides in use for paper coatings. The demand is for the precipitated grades rather than natural ground material. PCC comes in a wide range of particle sizes and shapes and can be tailored in the satellite plants to the papermakers' needs. The PCC plants are a cooperative effort between the pigment supplier and the papermaker that may be unique in the history of supplier/customer relationships. In the ideal case, waste products from the mill's recausticizing plant are reacted and precipitated as pigment grade calcium carbonate, slurried and returned to the papermill. The PCC manufacturer has a captive market, which has worked to the benefit of both parties.

⌘⌘⌘

PCC has high brightness, ultrafine particle size, and produces good gloss and printability. A major factor in the use of calcium carbonate is its cost. The product is relatively inexpensive yet of very high quality.

Talc – Ultrafine platy talcs have been used sparingly as coating pigments. Talcs give good surface smoothness, TiO_2 extension and gloss when calendered. Talc also gives a matte surface to uncaldendered papers.

Talcs are inherently difficult to disperse at high solids because of their platy and hydrophobic nature. This limits their usefulness in high solids systems. Slurries of greater than 50% are difficult to prepare and require uneconomical amounts of wetting and dispersing chemicals plus slurry antisettling agents. This generally limits talc to use in certain specialty coated grades, such as rotogravure papers, where it can improve printability. Talc has been more commonly used in Europe than North America because it is more abundant in supply than kaolin there. European papermakers have learned to overcome most formulating difficulties.

Other Coating Pigments

Several less common coating pigments are used in specialty applications. These include blanc fixe, satin white, zinc oxide, plastic pigment, aluminum trihydrate and the newly introduced structured kaolin pigments.

Blanc fixe – Blanc fixe, synthetic barium sulfate, is a specialty coating pigment that finds its primary use in coatings for photographic print paper. It has a high refractive index, high brightness and high density. It is highly purified chemically and inert in the paper coating. Blanc fixe is supplied to the papermaker as a 50 to 70% solids paste. It is mixed with photographic gelatin as a binder and applied to one side of the paper.

Satin white – Satin white, calcium sulfoaluminate, is a synthetic precipitated product made from a mixture of lime and alum. It is an extremely bright pigment of variable composition depending on the method of precipitation and drying. The crystal habit is acicular, with needles that continue to grow with time. The needles form a bulky, loosely packed coating structure. Satin white suffers from low solids slurries with very high viscosity, limited binder compatibility, and chemical instability. Satin white was used in paper coatings until the 1960's. Today, it is used almost exclusively as a coating for playing cards and a few specialty art papers.

Zinc oxide – Zinc oxide is a specialty coating pigment that is used for electroconductive papers, although many of the electroconductive copy methods have been replaced by xerography and other similar processes. Zinc oxide accepts and retains an electrostatic charge when exposed to high voltage corona. The charged surface is selectively sanitized when exposed to an image under proper illumination. Toner adheres to the non-electrostatically charged areas producing an image.

Zinc oxide for this use is prepared by the "French" process; metallic zinc that is burned in air to produce fine particle size zinc oxide. This product is sensitive to UV and blue light and must be surface modified to respond to all colors during the copying process. Silver halide dyes and solvent based binder systems are used in conjunction with the zinc oxide pigment. High levels of pigment must be applied to the sheet to insure sufficient coverage for image reproduction.

Plastic pigment – Plastic pigments, also known as opaque polymers, are not mineral pigments but microspheres of polystyrene and similar plastic materials. These microspheres act to opacify by their internal structure. They are either hollow spheres or water filled spheres. Light contacting the spheres pases through several layers and changes of medium. Each time this occurs, the light is refracted creating opacity. These products have high brightness, provide good gloss, act as TiO_2 extenders, improve print gloss and reduce print mottle. Because of their cost, they are most often used in high quality enamel papers.

Alumina trihydrate – Alumina trihydrate (ATH) as a coating pigment offers high brightness and good gloss. It finds a nitch as a substitute for plastic pigment in some specialty applications.

ATH is made from bauxite. The aluminum is solubilized in sodium hydroxide and the insoluble impurties removed by sedimentation and filtration. The ATH is precipitated out of the sodium aluminate solution by cooling and seeding with ATH crystals. The sodium aluminate decomposes into alumina trihydrate and sodium hydroxide. The filter cake is washed and slurried or dried. This procedure produces a highly purified product that is 95 GE brightness, and an extremely fine particle size.

ATH has another somewhat unique property in that it can be used to produce flame proof coatings. At elevated temperatures, the ATH releases its water of crystallization in such a fashion that it cools the paper to below the ignition temperature, effectively retarding the burning.

Structured kaolin pigments – New variations of kaolin are available called structured pigments. These are based on calcined clay and are clusters of clay particles plus polymer. Special functional groups are added to the clay surface that optimize light scattering and TiO_2 extension. These high bulk pigments improve brightness and opacity. They are more effective opacifiers in light weight rather than heavy coatings, and are finding use in rotogravure papers, lightweight coated and coated board.

⌘⌘⌘

PITCH CONTROL

An important use of certain minerals in the pulp and papermaking system is that of pitch control. "Pitch" as defined by the papermaker is any organic materials that are unwanted in the system. These materials are found throughout the pulp mill and papermill and often are not of wood origin. In the pulp mill, pitch refers to unsaponifiable matter from the pulping process, usually consisting of resin acids. These resin acids are insoluble in water and sometimes are present as oil-in-water emulsions. The emulsified resin acids are very unstable and easily coagulate into large masses. Such factors as pH, water temperature, and excessive shear can upset the fragile balance of the system causing the emulsion to break down. The resin acids then agglomerate into large masses that can stick to the walls of tanks and chests, and plug screens, pipes and refiners. The pitch can also become attached to the pulp fibers and be carried over into the paper mill.

Various methods can be used to reduce the pitch problems in the pulp mill area. Good pulp washing is the primary means of removing the resin acids. Keeping excessive amounts of processing additives, especially defoamers, out of the system will eliminate problems that come from these pitch-like materials. When insufficient washing capacity is available, other pitch control methods must be employed. One is the use of certain minerals, especially platy talc and ultrafine particle size clay, which adsorb pitch onto their oleophillic surfaces. The unwanted organic substances in the pulp mill process water become attached to the mineral surfaces and are thus deactivated. This type of pitch control is widely used in North American pulp mills. Several of the talc suppliers also provide designs for equipment that can be used to slurry the talc. The dilute talc slurry can be easily pumped to various points in the pulp mill.

The amount of pitch present varies depending on the species of wood being pulped, the age of the wood, the time of year and the pulping process being employed. Hardwood pulps have more pitch than softwood. Sulfite pulps have higher pitch contents that sulfate pulps. For kraft pulps 20 to 30 pounds of talc per air dry ton of pulp usually is sufficient to control the pitch. For sulfite pulps, 100 to 150 pounds per air dry ton is usually required. The mechanical pulps require more than kraft pulps but less than the sulfite pulps.

Talc is not added at only one location in the pulp mill, as much more effective pitch control is achieved if the total talc dose is distributed to several points. Key target areas are the screening areas, points of major pH change such as between the bleach towers and caustic extraction stages, and pumps and refiners which create large shear forces on the fibers which can liberate pitch from their internal surfaces. The best pitch control strategy is to add the talc just prior to the trouble spot. The talc is there where it is

⌘⌘⌘

needed most. The talc-pitch particles are then removed during the subsequent washing and dewatering operations. A small amount of pitch will pass into the paper mill but will become incorporated into the finished paper.

The other method of pitch control is the use of surfactants that keep the pitch dispersed so that it does not agglomerate and deposit.

Talc is used to control pitch in the papermill, as well. In the papermill, the pitch is usually more of the unwanted chemical nature. Papermill pitch consists of various organic materials. Often these are additives that have been added for a certain purpose, have done their job, and then build up in the system if they have not been retained in the paper web. These materials can plug forming fabrics and press felts, and deposit on various machine surfaces. These unwanted organic materials include, in addition to organics carried over from the pulp mill, defoamers, sizing materials, latex from coated broke, adhesives from recycled materials and other "stickies". Because these materials are similar in chemical nature to pulp mill pitch they can be controlled in a similar manner. Addition of 10 to 20 pounds of talc per ton of paper will control most papermill pitch problems. Since many of these substances build up from the constant recycling of white water, treatment of the white water itself with talc is usually effective. One real advantage of using talc as a papermill pitch control agent is that the talc also acts as a filler, so the pitch is carried out of the mill in the finished paper.

Ultrafine airfloated clay is used to a limited extent as a pulp mill pitch control agent. The clay has a very high surface area that adsorbs pitch in a manner similar to talc.

PAPER RECYCLING

The concern over disposal of solid waste has grown to major proportions in the 1990s. In an attempt to reduce the volume of material in our landfills, certain types of waste have been targeted for recycling. Paper, especially newsprint, has been deemed the biggest culprit. This waste disposal problem, along with limited fiber resources in some areas, has fueled the widespread adoption of paper recycling. Recycled paper currently supplies nearly 55% of the fiber used in North American paper mills.

The Europeans have led the way in recycling fiber for many years. Prior to the 1980s only a very limited amount of paper was recycled in the United States. The concept of an unlimited supply of virgin fiber and nearly unlimited space to dump waste, made the prospect of recycling in America uninteresting to many papermakers.

Recycling paper is a rather complex chemical process that varies depending on the type of paper being recycled. Newsprint must be deinked.

⌘⌘⌘

Magazine paper must be both deinked and the coating pigments and binders removed. Copier papers present a real challenge in removing the toner from the fiber. Each of these papers, plus many others, has its own unique deinking scheme and chemical requirements. Because most deinking processes include the removal of organic substances besides ink, problems similar to those encountered in pitch control must be countered.

Ultrafine particle size platy talcs have been used extensively as recycling additives. The highly adsorbant talcs attract inks, adhesives and other organics that may be part of the recycled stock. Talcs also function in flotation deinking processes as foam stabilizers, and since talc tends to float on water, it helps to carry the ink particles to the surface. The talc-waste particles are large enough to be easily removed by filtration or other cleaning methods.

Kaolin clays have also found use in recycling. Most newsprint recycling plants use a small portion of coated magazine stock mixed with the newsprint. The recyclers have found that the clays from the magazine coatings act as nuclei for agglomeration of ink particles, making the agglomerates large enough to be easily removed from the pulp.

MICROPARTICLE RETENTION

A relatively new innovation in the retention and drainage processes used in the wet end of paper machines is known as microparticle retention. Microparticle retention uses a combination of highly charged water soluble polymers plus a water swellable bentonite clay. The water soluble polymers are cationic polyacrylamide or cationic potato starch. The cationic polymer is added to the paper stock early in the machine approach system forming large flocs of fiber and filler. Screens just prior to the machine break these flocs into microflocs. Immediately prior to the headbox, a dilute slurry of the bentonite is added which reacts with the mircoflocs to form a bridged structure. This large flocced network improves retention and drainage on the wire.

Advantages reported for microparticle retention include increased water removal, increased first pass retention, decreased steam usage from higher consistency sheet entering the driers, and in some cases, increased machine speed.

BIBLIOGRAPHY

Anonymous, 1994, "Month in Statistics", *Pulp and Paper*, July, p 11.

Bierman, C.J., 1993, *Essentials of Pulping and Papermaking*, Academic Press, New York.

⌘⌘⌘

Berube, R.R, 1991, "Chemically Structured Coating Pigments Improve Paper Appearance and Performance", *Paper Age*, May, pp 16-18, 40.

Booth, G.L. et al, 1976, *Pigmented Coating Processes for Paper and Board*, Technical Associationof the Pulp and Paper Industry, New York.

Casey, J.P. , 1981, *Pulp and Paper Chemistry and Chemical Technology*, Vol. 1-4, John Wiley & Sons, New York.

Ferguson, K.(ed.), 1994, *New Trends and Dvelopments in Papermaking*, Miller Freeman, San Francisco.

Hagermeyer, R.W.(ed.), 1976, *Paper Coating Pigments,* TAPPI Monograph Series No. 38, Technical Association of the Pulp and Paper Industry, Atlanta.

Hagermeyer, R.W. (ed), 1984, *Pigments for Paper*, TAPPI Press, Atlanta.

Harriman, E.J., 1988, "The Less Well Known Mineral Pigments in Papermaking", *Industrial Minerals, 8th Industrial Minerals Congress*, pp 104-110.

Kellogg, T.L., Hardy, R.E., 1993, "Structured Pigments Offer Improved Opacity, Titanium Dioxide Extension", *Pulp and Paper*, May, pp 71-74.

Lowry, P.M., 1988, "Wet End Balance Through the Use of Multi-Component Retention Systems", *TAPPI Papermakers Conference Procedings 1988*, pp 231-234.

McCourt, A.M., et al, 1993, "A Practical View of a Microparticle System in Supercalendered Paper", *TAPPI Journal*, Vol 76 No. 10, pp 165-168.

Mc Vey, H. and Harben, P., 1989, "Industrial Minerals For Paper", *Industrial Minerals*, Dec., pp 41-47.

O'Brien, J. 1994, "Second Generation ATH Pigment Improves Coating Formulations", *Paper Age*, Oct., pp 20,21.

Patrick, K. L., 1994, "Satellite PCC Plant, Union Camp Mill Shift Alkaline Machines to High Gear", *Pulp and Paper*, Apr., pp 81-87.

Patrick, K.L.(ed.), 1991, *Paper Coating Trends in the Worldwide Paper Industry*, Miller Freeman, San Francisco.

Thompson, R.R. (ed), 1947, *The Progress of Paper*, The Lockwood Trade Journal Co., New York.

Willets, W.R., et al, 1958, *Paper Loading Materials*, TAPPI Monograph Series No. 19, Technical Association of the Pulp and Paper Industry, New York.

⌘⌘⌘

SIX

RUBBER

Peter A. Ciullo
R.T. Vanderbilt Company, Inc.
Norwalk, CT

Norman Hewitt
PPG Industries, Inc.
Pittsburgh, PA

From Columbus onward, European explorers of Central and South America found the natives exploiting the elastic and water resistant properties of the dried latex from certain trees. The indigenous peoples already knew how to crudely waterproof fabrics and boots by coating them with latex and then drying. They also rolled dried latex into the bouncing balls used for sport. This dried latex became quite a curiosity in Europe, especially among the natural scientists. It received the name "rubber" in 1770 when John Priestly discovered that it could rub out pencil marks.

By the early nineteenth century, rubber was recognized as a flexible, tough, waterproof, and air-impermeable material. Commercial exploitation, however, was stymied by the fact that its toughness and elasticity made it difficult to process. More importantly, articles made from it became stiff and hard in cold weather, and soft and sticky in hot weather. The quest to make useful goods from rubber led Thomas Hancock of Great Britain to invent the rubber band and, in 1820, a machine to facilitate rubber processing. His "masticator" subjected the rubber to intensive shearing that softened it sufficiently to allow mixing and shaping. This development was followed, in 1839, by the discovery of vulcanization, which is generally credited to both Hancock and Charles Goodyear of the United States. Vulcanization – heating an intimate mixture of rubber and sulfur to crosslink the rubber polymer network – greatly improved rubber strength and elasticity and eliminated its deficiencies at temperature extremes. Upon this mechanical and chemical foundation, the rubber industry was born.

The source of rubber latex at that time was the *Hevea brasiliensis* tree, which is native to the Amazon valley. Brazil became the primary source of rubber, but as rubber use grew questions arose as to this country's ability to insure adequate supply from its wild rubber trees. In 1876, Henry Wickham collected 70,000 Hevea seeds in Brazil and sent them to Kew Gardens in London for germination. Few seedlings resulted, but those that did allowed the British to establish a plantation system throughout the Far East.

Dunlop's patenting of the pneumatic tire in England in 1888 ushered in the age of the bicycle as a prelude to the era of automobiles. Tires need rubber, and the demand grew sufficiently great that in the early years of the twentieth century all sources of wild rubber in tropical America and Africa were being tapped. This demand, and the higher prices it caused, turned the plantations in Ceylon (Sri Lanka), Malaya (Malaysia), Singapore, and the East Indies (Indonesia) into prosperous enterprises. By 1914, plantation rubber had overtaken the production of wild rubber, and by 1920 it accounted for 90% of the world's supply.

The ready availability of high quality plantation rubber facilitated the advances in production methods and product quality which catalyzed the development of better automobiles and their expanding reliance on rubber products. The demand for rubber also prompted research into the synthesis of practical substitutes. As early as the 1880s, organic chemists had identified isoprene as the main structural unit of rubber. By 1890, several researchers had made synthetic rubber-like polyisoprenes. With wild and plantation rubber readily available, however, synthetic alternatives remained mostly of academic interest. The Allied blockade of Germany during World War I changed this by demonstrating to Germany, and the world, the strategic importance of rubber in war. The Germans produced more than 2000 tons of methyl rubber by polymerizing 2,3-dimethyl-1,3-butadiene, but its properties were poor and its production was abandoned at the war's end.

The lesson was learned, nevertheless. The governments of Germany and Russia, the military nations most susceptible to loss of natural rubber through naval blockade, instituted programs to develop synthetic alternatives. This was given additional impetus in the mid-1920s by a forced rise in rubber prices due to British restrictions on plantation production. Work in Germany and Russia concentrated on polymers of 1,3-butadiene, since it was less expensive and easier to manufacture than isoprene (2-methyl-1,3-butadiene). This led to the production in the early 1930s of sodium catalyzed polybutadiene as SK rubber in Russia and Buna (from butadiene and Na) in Germany. These polymers were hard, tough, and difficult to process. Copolymerization of 1,3-butadiene with other monomers was pursued to obtain more easily processed and rubber-like products. In Germany this led within a few years to Buna S, a copolymer of butadiene and styrene, and Buna N, a copolymer of butadiene and acrylonitrile.

By the time war broke out in 1939, both Germany and Russia could satisfy their rubber needs with reasonably satisfactory synthetic products. As the war spread to the Far East, the U.S. government realized its supply of natural rubber was at risk. In 1940 it established the Rubber Reserve Company as a government corporation. This organization was charged with stockpiling natural rubber and instituting a synthetic rubber research and development program. Based on the technology developed in Germany and Russia prior to the war, styrene-butadiene polymers soon became the focus of development efforts as the best general-purpose alternative to natural rubber.

The Japanese also identified the strategic significance of natural rubber to both the United States and Great Britain. Following Pearl Harbor, the Japanese promptly cut off most of the Far East supply. It was only the rapid development of synthetic rubber production in the U.S. that negated this. Production of styrene-butadiene rubber (SBR), then called GR-S, began in a government plant in 1942. Over the next three years, government financed construction of 15 SBR plants brought annual production to more than 700,000 tons.

Although SBR was the workhorse substitute for natural rubber during the war years, it was supplemented by neoprene, butyl rubber, and nitrile rubber. Neoprene, poly(2-chloro-1,3-butadiene) had been introduced by DuPont in 1931 as DuPrene and was actually the first commercially successful synthetic elastomer. Neoprene provided high tensile strength like natural rubber and significantly better weather and oil resistance, but it was too expensive for consideration as a general purpose elastomer. Its production rose from 9,000 tons to 45,000 tons between 1942 and 1945. Butyl rubber had been invented at Standard Oil when it was discovered that the saturated chemical structure of rubber-like high molecular weight isobutylene polymers could be vulcanized by incorporating a small amount of isoprene into the polymer. The resulting elastomer showed unique and useful properties, particularly low gas permeability. Construction of the first butyl rubber plant had already begun in 1941, but the facility was nationalized and put into full production by 1943. Most of the wartime output went into inner tubes for military and essential civilian vehicles. From a base of 1400 tons in 1943, butyl rubber production rose rapidly to over 47,000 tons in 1945. Nitrile rubber, the current common name for acrylonitrile-butadiene rubber, showed such great potential as a special-purpose oil- and solvent-resistant rubber that its commercialization in Germany was followed by production at four new U.S. plants by the time of America's entry into the war. The expanding industrial market for this elastomer was immediately pre-empted by strict allocation for military use. By the end of the war, production had risen to 15,000 tons per year.

World War II represents the accelerated adolescence of the rubber industry, the preparation for its great postwar maturation and growth. Developments in organic and polymer chemistry provided insight into the

⌘⌘⌘

properties of natural rubber, which led to neoprene, SBR, nitrile and butyl rubber. These were made practically and commercially significant, however, only by the complementary developments in compounding. Amine vulcanization accelerators were discovered at Diamond Rubber in 1906. The value of carbon black to the improvement of rubber physical properties was discovered at Diamond Rubber in 1912. Internal mixers for compounding were first offered by Fernley Banbury in 1916. Guanidines and thiazoles began use as accelerators in 1921. By the end of the Second World War, compounding art and science had fully bridged the gap between elastomer and rubber product.

COMPOUNDING MATERIALS

Today there is a wide variety of rubber polymers, each with its own set of characteristic attributes, and each offered with modifications designed to enhance one or more of those attributes. In most cases, nevertheless, the elastomer by itself lacks one or more property necessary to produce a sale-able product. A number of materials must be added to make it commercially useful. Compounding is the means by which elastomer and additives are combined to ensure efficient manufacture of the best possible product. The design of a compound formula is the basic function of rubber technology.

The compounding of rubber products starts with the choice of elastomer, filler (reinforcing or extending), crosslinking chemicals, and various additives which, when mixed together, will provide a compound with the desired properties and performance. Mixing is followed by forming operations such as milling, extrusion, and calendering. These lead to the final processing step of vulcanization or curing in which the compound changes from a thermoplastic to a thermoset or crosslinked state.

ELASTOMERS

The first step in effective compounding is the selection of the proper elastomer. The inherent properties of each elastomer determine its suitability for any given application. These properties are a function of elastomer chemistry, which in turn is determined by manipulation of synthesis variables. The synthesis of the natural rubber polymer is, of course, dictated by nature and genetics, although adjustments to the process of converting the latex to solid rubber enables some latitude in modifying properties.

Most synthetic elastomers are made by either solution or emulsion polymerization. In some cases, both methods are used to provide respectively characteristic grades of the given polymer. In general, emulsion polymerization provides broader molecular weight distribution and better milling qualities. Solution polymerization generally results in narrower

⌘⌘⌘

molecular weight distribution, poorer milling qualities, but better physical properties in the cured state.

Figures 1a and 1b gives the most commonly used elastomers along with their industry accepted abbreviation and structural formula.

Figure 1a. Common Elastomers

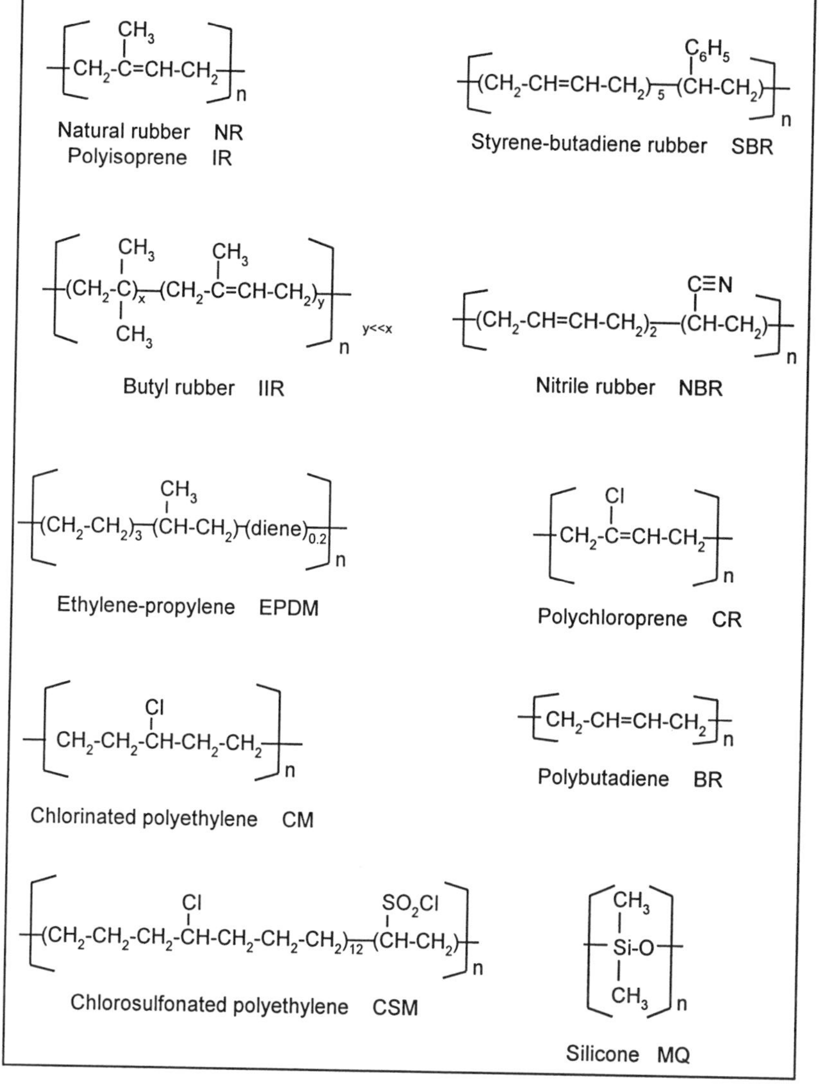

$-[CH_2-C(CH_3)=CH-CH_2]_n-$

Natural rubber NR
Polyisoprene IR

$-[(CH_2-CH=CH-CH_2)_5-(CH(C_6H_5)-CH_2)]_n-$

Styrene-butadiene rubber SBR

$-[(CH_2-C(CH_3)_2)_x-(CH_2-C(CH_3)=CH-CH_2)_y]_n-$ y<<x

Butyl rubber IIR

$-[(CH_2-CH=CH-CH_2)_2-(CH(C{\equiv}N)-CH_2)]_n-$

Nitrile rubber NBR

$-[(CH_2-CH_2)_3-(CH(CH_3)-CH_2)(diene)_{0.2}]_n-$

Ethylene-propylene EPDM

$-[CH_2-C(Cl)=CH-CH_2]_n-$

Polychloroprene CR

$-[CH_2-CH_2-CH(Cl)-CH_2-CH_2]_n-$

Chlorinated polyethylene CM

$-[CH_2-CH=CH-CH_2]_n-$

Polybutadiene BR

$-[(CH_2-CH_2-CH_2-CH(Cl)-CH_2-CH_2-CH_2)_{12}-(CH(SO_2Cl)-CH_2)]_n-$

Chlorosulfonated polyethylene CSM

$-[Si(CH_3)_2-O]_n-$

Silicone MQ

⌘⌘⌘

Natural Rubber (NR)

Natural rubber is the prototype of all elastomers. It is extracted in the form of latex from the bark of the Hevea tree. The rubber is collected from the latex in a series of steps involving preservation, concentration, coagulation, dewatering, drying, cleaning, and blending. Because of its natural derivation, it is sold in a variety of grades based on purity (color and presence of extraneous matter), viscosity, viscosity stability, oxidation resistance, and rate of cure. Modified natural rubbers are also sold, with treatment usually performed at the latex stage. These include epoxidized natural rubber (ENR); deproteinized natural rubber (DNR); oil extended natural rubber (OENR), into which 10-40% of process oils have been incorporated; Heveaplus MG rubber – natural rubber with grafted poly(methyl methacrylate) side chains; and thermoplastic natural rubber (TNR) – blends of natural rubber and polypropylene.

Figure 1b. Common Specialty Elastomers

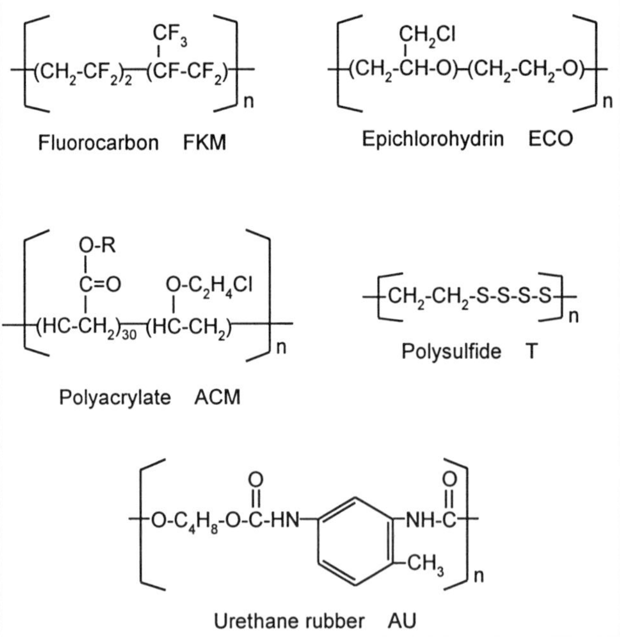

$$\left[(CH_2\text{-}CF_2)_2 \text{—} (CF(CF_3)\text{-}CF_2) \right]_n$$

Fluorocarbon FKM

$$\left[(CH_2\text{-}CH(CH_2Cl)\text{-}O)(CH_2\text{-}CH_2\text{-}O) \right]_n$$

Epichlorohydrin ECO

$$\left[(HC(C(=O)\text{-}O\text{-}R)\text{-}CH_2)_{30} (HC(O\text{-}C_2H_4Cl)\text{-}CH_2) \right]_n$$

Polyacrylate ACM

$$\left[CH_2\text{-}CH_2\text{-}S\text{-}S\text{-}S\text{-}S \right]_n$$

Polysulfide T

$$\left[O\text{-}C_4H_8\text{-}O\text{-}C(=O)\text{-}HN\text{-}C_6H_3(CH_3)\text{-}NH\text{-}C(=O) \right]_n$$

Urethane rubber AU

The natural rubber polymer is nearly 100% *cis*-1,4 polyisoprene with Mw ranging from 1 to 2.5 x 10^6. Due to its high structural regularity, natural rubber tends to crystallize spontaneously at low temperatures or when it is stretched. Low temperature crystallization causes stiffening, but is easily reversed by warming. The strain-induced crystallization gives natural rubber products high tensile strength and resistance to cutting, tearing, and abrasion.

Like other high molecular weight polymers, natural rubber can be pictured as a tangle of randomly oriented sinuous polymer chains. The "length" of these chains is a function of their thermodynamic behavior and is determined as the statistically most probable distance between each end. This chain length reflects the preferred configuration of the individual polymer molecule. The application of force to a rubber sample effectively changes the chain length. When the force is removed, the chain tries to regain its preferred configuration. In simple terms, this can be compared to the compression or extension of a spring. This effect is the basis of rubber's elasticity.

Elasticity is one of the fundamentally important properties of natural rubber. Rubber is unique in the extent to which it can be distorted, and the rapidity and degree to which it recovers to its original shape and dimensions. It is, however, not perfectly elastic. The rapid recovery is not complete. Part of the distortion is recovered more slowly and part is retained. The extent of this permanent distortion, called permanent set, depends upon the rate and duration of the applied force. The slower the force, and the longer it is maintained, the greater is the permanent set. Because of rubber's elasticity, however, the permanent set may not be complete even after long periods of applied force. This quality is of obvious value in gaskets and seals.

The rubber's polymer network allows elasticity and flexibility to be combined with crystallization-induced strength and toughness when stretched. The elastic nature of this network also accounts for the exceptional resilience of cured rubber products. This resilience means less kinetic energy is lost as heat during repeated stress deformation. Products made from natural rubber are less likely than most other elastomers to fail from excessive heat buildup or fatigue when exposed to severe dynamic conditions. This has secured the place of natural rubber as the preferred sidewall elastomer in radial tires.

As already noted, this rubber polymer network was originally an impediment to rubber processing. Mixing additives with a tough, elastic piece of raw rubber was a substantial challenge. The solution came with the discovery of its thermoplastic behavior. High shear and heat turn the rubber soft and plastic through a combination of extension, disentanglement, and oxidative cleavage of polymer chains. In this state it is considerably more receptive to the incorporation of additives so that the rubber's natural attributes can be modified and optimized as desired. The commercial utility of natural rubber has in fact grown from the ease with which its useful properties can be changed or improved by compounding techniques.

Another important and almost unique quality of uncured natural rubber compounds is building tack. When two fresh surfaces of milled rubber are

⌘⌘⌘

pressed together they bond into a single piece. This facilitates the building of composite articles from separate components. In tire manufacture, for example, the separate pieces of green (uncured) tire are held together solely by building tack. During cure they fuse into a single unit.

Following World War II, natural rubber's virtual 100% market share dropped under the competitive pressure of synthetic elastomers. It nevertheless still accounts for about 1/3 of all elastomers used because of its unique balance of properties. This level of demand is supported mainly by tires and tire products, which account for over 70% of natural rubber consumption. Natural rubber is used in the carcass of passenger car cross-ply tires for its building tack, ply adhesion, and good tear resistance. It is also used in the sidewalls of radial ply tires for its fatigue resistance and low heat buildup. In tires for commercial and industrial vehicles, natural rubber content increases with tire size. Almost 100% natural rubber is used in the large truck and earthmover tires which require low heat buildup and maximum cut resistance. Natural rubber is also used in industrial goods, such as hoses, conveyor belts, and rubberized fabrics; engineering products, for resilient load bearing and shock or vibration absorption components; and latex products such as gloves, condoms, and adhesives.

Styrene-Butadiene Rubber (SBR)

SBR polymers are widely produced by both emulsion and solution polymerization. Emulsion polymerization is carried out either hot, at about 50°C, or cold, at about 5°C, depending upon the initiating system used. SBR made in emulsion usually contains about 23% styrene randomly dispersed with butadiene in the polymer chains. SBR made in solution contains about the same amount of styrene, but both random and block copolymers can be made. Block styrene is thermoplastic and at processing temperatures helps to soften and smooth out the elastomer. Both cold emulsion SBR and solution SBR are offered in oil-extended versions. These have up to 50% petroleum base oil on polymer weight incorporated within the polymer network. Oil extension of SBR improves processing characteristics, primarily allowing easier mixing, without sacrificing physical properties.

SBR was originally produced by the hot emulsion method, and was characterized as more difficult to mill, mix, or calender than natural rubber, deficient in building tack, and having relatively poor inherent physical properties. Processability and physical properties were found to be greatly improved by the addition of process oil and reinforcing pigments. "Cold" SBR generally has a higher average molecular weight and narrower molecular weight distribution. It thereby offers better abrasion and wear resistance plus greater tensile and modulus than "hot" SBR. Since higher molecular weight can make cold SBR more difficult to process, it is commonly offered in oil-extended form. Solution SBRs can be tailored in polymer structure and properties to a much greater degree than their emulsion counterparts. The random copolymers offer narrower molecular weight distribution, low chain branching, and lighter color than emulsion SBR. They are comparable in tensile, modulus, and elongation, but offer lower heat buildup, better flex, and higher resilience. Certain grades of solution SBR even address the polymer's characteristic lack of building tack, although it is still inferior to that of natural rubber.

The processing of SBR compounds in general is similar to that of natural rubber in the procedures and additives used. SBR is typically compounded with better abrasion, crack initiation, and heat resistance than natural rubber. SBR extrusions are smoother and maintain their shape better than those of natural rubber.

SBR was originally developed as a general purpose elastomer and it still retains this distinction. It is the largest volume and most widely used elastomer worldwide. Its single largest application is in passenger car tires, particularly in tread compounds for superior traction and treadwear. Substantial quantities are also used in footwear, foamed products, wire and cable jacketing, belting, hoses, and mechanical goods.

Polybutadiene Rubber (BR)

This elastomer was originally made by emulsion polymerization, generally with poor results. It was difficult to process and did not extrude well. Polybutadiene became commercially successful only after it was made by solution polymerization using stereospecific Ziegler-Natta catalysts. This provided a polymer with greater than 90% *cis*-1,4-polybutadiene configuration. This structure hardens at much lower temperatures (with T_g of -100°C) than natural rubber and most other commercial elastomers. This gives better low temperature flexibility and higher resilience at ambient temperatures than most elastomers. Greater resilience means less heat buildup under continuous dynamic deformation as well. This high-*cis* BR was also found to possess superior abrasion resistance and a great tolerance for high levels of extender oil and carbon black. High-*cis* BR was originally blended with natural rubber simply to improve the latter's processing properties, but it was found that the BR conferred many of its desirable properties to the blend. The same was found to be true in blends with SBR.

The 1,3-butadiene monomer can polymerize in three isomeric forms: by *cis* 1,4 addition, *trans* 1,4 addition, and 1,2 addition leaving a pendant vinyl group. By selection of catalyst and control of processing conditions, polybutadienes are now sold with various distributions of each isomer within the polymer chain, and with varying levels of chain linearity, molecular weight and molecular weight distribution. Each combination of chemical properties is designed to enhance one or more of BR's primary attributes.

⌘⌘⌘

The largest volume use of polybutadiene is in passenger car tires, primarily in blends with SBR or natural rubber to improve hysteresis (resistance to heat buildup), abrasion resistance, and cut growth resistance of tire treads. The type of BR used depends on which properties are most important to the particular compound. High-*cis* and medium-*cis* BR have excellent abrasion resistance, low rolling resistance, but poor wet traction. High-vinyl BRs offer good wet traction and low rolling resistance, but poor abrasion resistance. Medium-vinyl BRs balance reasonable wet traction with good abrasion resistance and low rolling resistance. Polybutadiene is also used for improved durability and abrasion and flex crack resistance in tire chaffer, sidewalls and carcasses, as well as in elastomer blends for belting. High- and medium-*cis* BRs are also used in the manufacture of high impact polystyrene. Three to twelve percent BR is grafted onto the styrene chain as it polymerizes, confering high impact strength to the resultant polymer.

Butyl Rubber (IIR)

Butyl rubber is the common name for the copolymer of isobutylene with 1 to 3% isoprene produced by cold (-100°C) cationic solution polymerization. The isoprene provides the unsaturation required for vulcanization. Most of butyl rubber's distinguishing characteristics are a result of its low level of chemical unsaturation. The essentially saturated hydrocarbon backbone of the IIR polymer will effectively repel water and polar liquids but show an affinity for aliphatic and some cyclic hydrocarbons. Products of butyl rubber will therefore be swollen by hydrocarbon solvents and oils, but show resistance to moisture, mineral acids, polar oxygenated solvents, synthetic hydraulic fluids, vegetable oils, and ester-type plasticizers. It is likewise highly resistant to the diffusion or solution of gas molecules. Air impermeability is the primary property of commercial utility. The low level of chemical unsaturation also imparts high resistance to ozone. Sulfur-cured butyl rubber has relatively poor thermal stability, softening with prolonged exposure at temperatures above 150°C because the low unsaturation prevents oxidative crosslinking. Curing with phenol-formaldehyde resins instead of sulfur, however, provides products with very high heat resistance, the property responsible for a large market in tire-curing bladders.

The molecular structure of the polyisobutylene chain provides less flexibility and greater delayed elastic response to deformation than most elastomers. This imparts vibration damping and shock-absorption properties to butyl rubber products.

The unique properties of butyl rubber are used to advantage in tire inner tubes and air cushions (air impermeability), sheet roofing and cable insulation (ozone and weather resistance), tire-curing bladders, hoses for high temperature service, and conveyor belts for hot materials (thermal stability with resin cure).

⌘⌘⌘

Halobutyl Rubber (CIIR, BIIR)

Halobutyl rubbers are produced by the controlled chlorination (chlorobutyl; CIIR) or bromination (bromobutyl; BIIR) of butyl rubber. Halogenation produces an allylic chlorine or bromine adjacent to the double bond of the isoprene unit. The halogenated butyl rubbers share many of the attributes of their butyl rubber parent: superior air impermeability, resistance to chemicals, moisture, and ozone, and vibration damping. The presence of halogen provides new crosslinking chemistry and the ability for adhesion to and vulcanization with general purpose, highly unsaturated elastomers. Bromobutyl is generally faster curing than chlorobutyl and somewhat more versatile in the curing systems that can be used to tailor product properties. Halobutyl rubbers typically use a zinc oxide-based cure, but bromobutyl can be vulcanized with peroxide or magnesium oxide as well.

The primary application for halobutyl rubber is in tires. The combination of low gas and moisture permeability, high heat and flex resistance, and ability to covulcanize with highly chemically unsaturated rubber has secured the use of these rubbers in the innerliners of tubeless tires. Passenger tires use chlorobutyl alone or in a blend with 20 to 40% natural rubber. High-service steel-belted truck tires use 100% bromobutyl innerliner compounds. Chlorobutyl is also used for truck inner tubes for its superior heat resistance compared to butyl rubber. Halobutyl rubbers are added to sidewall compounds for improved ozone and flex resistance, and to certain tread compounds for improved wet skid resistance and traction.

Halobutyl rubbers are also used in vibration damping mounts and pads, steam and automatic dishwasher hoses, chemical-resistant tank linings, and heat-resistant conveyor belts. They are also widely used in pharmaceutical closures because of their low gas and moisture permeability, resistance to aging and weathering, chemical and biological inertness, low extractables (with metal oxide cures), and good self-sealing and low fragmentation during needle penetration.

Neoprene (CR)

Neoprene is the common name for the polymers of chloroprene (2-chloro-1,3-butadiene). These are produced by emulsion polymerization. The chloroprene monomer can polymerize in four isomeric forms: *trans* 1,4 addition; *cis* 1,4 addition; 1,2 addition, leaving a pendant vinyl group and allylic chlorine; and 3,4 addition. Neoprene is typically 88-92% *trans*, with degree of polymer crystallinity proportional to the *trans* content. *Cis* addition accounts for 7-12% of the structure and 3,4 addition makes up about 1%. The approximately 1.5% of 1,2 addition is believed to provide the principal sites of vulcanization.

The high structural regularity (high *trans* content) of neoprene allows the strain-induced crystallization that results, as for natural rubber, in high

⌘⌘⌘

tensile strength. The 2-chloro substituent, instead of natural rubber's 2-methyl, results in a higher freezing point (poorer low temperature resistance) and alters vulcanization requirements. Neoprenes are generally cured with zinc oxide and magnesium oxide, or lead oxide for enhanced water resistance. The presence of chlorine in the polymer structure improves resistance to oil, weathering, ozone and heat. The improved oxidation resistance is due to the reduced activity of the double bonds caused by the chlorine. Except for low temperature resistance and price, neoprene would be considered nearly as versatile as natural rubber.

There are three types of general purpose neoprenes – G, W, and T types – with selected features modified to offer a range of processing, curing and performace properties. Products are made from neoprene because it offers good building tack, good oil, abrasion, chemical, heat, weather, and flex resistance, and physical toughness. Neoprene is widely used in hoses of all types (water, oil, air, automotive, industrial), wire and cable jacketing, power transmission and conveyor belting, bridge and building bearings, pipe gaskets, footwear, roof coatings, and coated fabrics.

Nitrile Rubber (NBR)

Nitrile rubber is the generic name given to emulsion polymerized copolymers of acrylonitrile and butadiene. Its single most important property is exceptional resistance to attack by most oils and solvents. It also offers better gas impermeability, abrasion resistance, and thermal stability than the general purpose elastomers like natural rubber and SBR. These attributes arise from the highly polar character of acrylonitrite, the content of which determines the polymer's particular balance of properties. Commercial nitrile rubbers are available with acrylonitrile/butadiene ratios ranging from 18:82 to 45:55. As acrylonitrile content increases, oil resistance, solvent resistance, tensile strength, hardness, abrasion resistance, heat resistance, and gas impermeability improve, but compression set resistance, resilience and low temperature flex deteriorate. Selection of the particular grade of NBR needed is generally based on oil resistance vs. low temperature performance. Blends of different grades are common to achieve the desired balance of properties.

Nitrile elastomers do not crystallize when stretched and so require reinforcing fillers to develop optimum tensile strength, abrasion resistance, and tear resistance. They also possess poor building tack. Although nitrile rubbers are broadly oil- and solvent-resistant, they are susceptible to attack by certain strongly polar liquids, to which the nonpolar rubbers, such as SBR or natural rubber, are resistant. Nitrile rubber is poorly compatible with natural rubber, but can be blended in all proportions with SBR. This decreases overall oil resistance, but increases resistance to polar liquids in proportion to the SBR content. Nitrile polymers are increasingly used as additives to plastics to provide elastomeric properties. Blends with polyvinyl chloride are popular for confering improved abrasion, tensile, tear, and flex properties.

Nitrile rubber is used primarily in soling, plus hoses, tubing, linings, and seals used for the conveyance or retention of oils and solvents.

Ethylene Propylene Rubbers (EPM, EPDM)

The first commercial ethylene propylene rubbers were made by the random copolymerization of ethylene and propylene in solution using Ziegler-Natta catalysts. Since these compounds (EPM) were fully saturated, they were highly resistant to oxidation, ozone, heat, weathering, and polar liquids. They could be cured, however, only by peroxide. The greater versatility of a sulfur curable elastomer was sought, and found by the incorporation of limited amounts of a third monomer into the polymer. Dienes were found which were compatible with the EPM polymerization process, and which had one double bond that would preferentially polymerize to leave a pendant double bond available for vulcanization. The latter criteria left the polymer backbone saturated and capable of offering the same high level of stability as EPM. The dienes most commonly used today to make the ethylene propylene diene polymers (EPDM) are 1,4 hexadiene, ethylidene, norbornene and dicyclopentadiene, each conferring a different rate and state of cure to the polymer.

An extensive range of EPDM polymers are produced by varying the molecular weight, molecular weight distribution, ethylene/propylene ratio and level and type of diene termonomer. Elastomers are available containing from 50% to more than 75% ethylene by weight. Polymers with lower ethylene content are amorphous and easy to process. Higher ethylene content gives crystalline polymers with better physical properties, but more difficulty in processing. The amount of terpolymer is typically 1.5 to 4%, but can be as high as 11% in ultra-fast curing grades. Most EPDMs are incompatible with diene rubbers (eg. natural, SBR, NBR) because of their relatively slow cure rate. The ultra-fast cure EPDMs overcome this.

In general, the ethylene propylene rubbers are compounded to provide good low-temperature flexibility, high tensile strength, high tear and abrasion resistance, excellent weatherability (ozone, water, oxidation resistance), good electrical properties, high compression set resistance, and high heat resistance. The high molecular weight crystalline EPDMs can incorporate high levels of fillers. EPMs and EPDMs have low resistance to hydrocarbon oils and their lack of building tack must be compensated by the use of resin.

The ethylene propylene rubbers are probably the most versatile of the general purpose elastomers, and can be compounded in nearly the full spectrum of applications not requiring resistance to hydrocarbon oils. The high volume use is in tire sidewalls as an additive to improve ozone resistance. Other major uses which exploit their desirable properties and versatility are single-ply roofing and ditch liners, automotive seals, gaskets, weatherstripping and boots, appliance parts, and hosing. Outside the rubber industry, they are used as viscosity modifiers in lubricating oils; they improve low temperature impact strength when added to polyolefins at low levels; and they form thermoplastic elastomers when blended in higher ratios with polyolefins or other thermoplastic resins.

Polyisoprene (IR)

Polyisoprene is made by solution polymerization of isoprene (2-methyl-1,3-butadiene). The isoprene monomer, the structural unit of the natural rubber polymer, can polymerize in four isomeric forms: *trans* 1,4 addition, *cis* 1,4 addition, 1,2 addition, leaving a pendant vinyl group, and 3,4 addition. The production of a synthetic analogue to natural rubber was stymied for over 100 years because polymerization of isoprene resulted in mixtures of isomeric forms. In the 1950s, rubber-like elastomers with >90% *cis* 1,4 isoprene configuration were finally produced using stereospecific catalyts.

Polyisoprene compounds, like those of natural rubber, exhibit good building tack, high tensile strength, good hysteresis, and good hot tensile and hot tear strength. The characteristics which differentiate polyisoprene from natural rubber arise from the former's closely controlled synthesis. Polyisoprene is chemically purer – it does not contain the proteins and fatty acids of its natural counterpart. Molecular weight is lower than natural rubber's, and lot-to-lot uniformity is better. Polyisoprene is therefore easier to process, gives a less variable (although generally slower) cure, is more compatible in blends with EPDM and solution SBR, and provides less green strength (pre-cure) than natural rubber. Polyisoprene is added to SBR compounds to improve tear strength, tensile strength, and resilience while decreasing heat buildup. Blends of polyisoprene and fast curing EPDM combine high ozone resistance with the good tack and cured adhesion uncharacteristic of EPDM alone.

Polyisoprene is typically used in favor of natural rubber in applications requiring consistent cure rates, tight process control, or improved extrusion, molding, and calendering. Tires are the leading consumer. The synthetic elastomer can be produced with the very low level of branching, high molecular weight, and relatively narrower molecular weight distribution that contributes to lower heat buildup compared to natural rubber. For this reason, certain grades of polyisoprene are used as a alternative to natural rubber in the tread of high service tires (truck, aircraft, off-road) without sacrificing abrasion resistance, groove cracking, rib tearing, cold flex properties, or weathering resistance. Footwear and mechanical goods are also major uses. Because of polyisoprene's high purity and the high gum (unfilled) tensile strength of its compounds, it is widely used in medical goods and food-contact items. These include baby bottle nipples, milk tubing, and hospital sheeting.

Chlorinated and Chlorosulfonated Polyethylene (CM, CSM)

Chlorinated polyethylene (CM) is produced by the chlorination of high density polyethylene either in solvent solution or aqueous suspension. The substituent chlorine on the saturated olefin backbone enhances heat and oil resistance. The chlorine also provides flame resistance. The polymer is thermoplastic when processed on conventional elastomer equipment, and compounds can be molded, calendered or extruded. Chlorinated polyethylene is most often steam cured using a peroxide curing system. The major end use is wire and cable applications, particularly flexible cords for up to 600 volts. Other major uses are in automotive hose, sheet goods and as an impact modifier in plastics.

Simultaneous chlorination and chlorosulfonation of high density polyethylene in an inert solvent yields chlorosulfonated polyethylene (CSM). Various grades are available with chlorine contents ranging from 24 to 43% and sulfur from 1.0 to 1.4%. CSM is widely used in the rubber industry because its compounds are odorless, light stable, and easily colored, highly resistant to oxygen, ozone, weathering, and corrosive chemicals, and resistant to abrasion, heat, low temperatures, oil, and grease. They also have excellent electrical properties and provide high tensile properties without highly reinforcing fillers. The particular degree and balance of these attributes is governed primarily by chlorine content. Increasing chlorine gives increasing flame and oil resistance, decreasing heat and electrical resistance and low temperature flexibility, and slightly decreasing ozone resistance.

Uncured chlorosulfonated polyethylene is more thermoplastic than other commonly used elastomers. It is generally tougher at room temperature, but softens more rapidly on heating. Vulcanization can be obtained with peroxides, as with other chemically saturated polymers, or with sulfur crosslinking at the sulfonyl chloride groups. The former promotes better resistance to heat and compression set, the latter can result in high tensile strength and excellent mechanical toughness. Compounds can also be cured at ambient temperatures by the combination of water (atmospheric moisture) and a divalent metal oxide (magnesia), which form sulfonate salt bridges. These ionic cures are slow but develop high modulus.

Ambient temperature ionic curing of CSM has led to its major use in single-ply roofing membranes, and geomembranes for reservoir liners and waste dump liners. Roofing membranes are installed in an uncured thermoplastic state, with seaming by heat or solvents. Weathering slowly cures the membrane, progressively increasing durability and toughness.

⌘⌘⌘

Because these compounds are easily colored, they can be made with heat reflective or absorptive colors as required. An alternative approach to weather resistant roofing is to coat a neoprene or EPDM base with CM paint.

The combination of colorability, toughness, enviromental durability and resistance to flame, oil, radiation and corrosive chemicals has also secured CSM's widespread use in automotive hoses, tubes, gasketing and wiring, electrical wire insulation for up to 600 volts, wire insulation in nuclear power stations, industrial hoses and tank linings, coated fabrics, hot conveyor belting, and construction coatings and gaskets.

Silicone Rubber

Because of its unique properties and somewhat higher price compared to the other common elastomers, silicone rubber is usually classed as a specialty elastomer, although it is increasingly used as a cost-effective alternative in a variety of applications. Two types of silicone elastomers are available, each providing the same fundamental properties. These are the thermosetting rubbers that are vulcanized with heat, and RTV (room temperature vulcanizing) rubbers.

The basic silicone polymer is dimethylpolysiloxane with a backbone of silicon-oxygen linkages and two methyl groups on each silicon. The silicon-oxygen backbone provides a high degree of inertness to ozone, oxygen, heat (up to 315°C), UV light, moisture, and general weathering effects, while the methyl substituents confer a high degree of flexibility. The basic polymer properties are modified by replacing minor amounts of the methyl substituents with phenyls and/or vinyls. Phenyl groups improve low temperature flexibility (to as low as -100°C) without sacrificing high temperature properties. Vinyl groups improve compression set resistance and facilitate vulcanization. Of the available silicone elastomers – methyl silicone (MQ), methyl-vinyl silicone (VMQ), methyl-phenylsilicone (PMQ), methyl-phenyl-vinyl silicone (PVMQ), and fluoro-vinyl-methyl silicone (FVMQ) – the methyl-vinyl types are most widely used.

Thermal vulcanization typically uses peroxides to crosslink at the vinyl groups of the high molecular weight solid silicone rubbers. Compounded products offer the attributes noted above plus superior resistance to compression set, excellent biocompatability, vibration damping over a wide temperature range, and thermal ablative properties. The latter enables the silicone rubber to form a thermally insulating surface char on exposure to temperatures up to 5,000°C. The rubber remains elastomeric beneath the char. Silicone elastomers generally offer poorer tensile, tear, and abrasion properties than the more common organic rubbers, but this is routinely improved by reinforcement with fumed silica, which also improves electrical insulation properties.

⌘⌘⌘

Room temperature vulcanizing (RTV) silicones are low molecular weight dimethylpolysiloxane liquids with reactive end groups. As with the heat cured polymers, there can be minor substitution of methyl groups with phenyls – for improved low temperature flexibility – or with fluoroalkyl groups – for improved oil and solvent resistance and even broader temperature service. Vulcanization of the RTV silicones is obtained from either a condensation or an addition reaction. Condensation cures can be either moisture independent or moisture dependent. For moisture independent compounds, the reactive polymer end group is usually silanol. The crosslinking agent may be a silicone with silanol end groups, using an organic base as the condensation catalyst, an alkoxysilicate (e.g., ethyl silicate), using a metallic salt catalyst, or a polyfunctional aminoxy silicone, often requiring no catalyst. These compounds are known as "two-package" RTVs since the curing agent and/or catalyst is kept separate and added to the compound just prior to use.

Moisture-curing compounds, also known as "one-package" RTVs, are compounded from silanol-terminated polymer with a polyfunctional silane curing agent (e.g., methyltriacetoxysilane) and condensation catalyst, or from a polymer end-stopped with the curing agent. Crosslinking occurs on exposure to atmospheric moisture, starting at the surface and progressing inward with diffusion of moisture into the compound.

Addition-cured RTVs are typically compounded from a dimethylvinylsiloxy-terminated polymer, a polyfunctional silicon hydride crosslinker, and a metal ion catalyst. Vulcanization is independent of moisture and air and forms no volatile byproducts. These products are usually sold as two-package RTVs, but are also available as one-package compounds containing an inhibitor which is volatilized or deactivated by heat to trigger the cure.

Most fabricators of silicone rubber products do not do their own compounding, but purchase premixed compounds requiring only catalyst and/or curing. Solid (thermally cured) rubbers are used in automotive underhood applications, primarily for their heat resistance. Products include ignition cables, coolant and heater hoses, O-rings, and seals. Similar applications are found in aircraft seals, connectors, cushions, and hoses, and in home appliance O-rings, seals, and gaskets. Long service life plus circuit integrity (from ablative charing) and no toxic gas generation have secured the place of silicone rubber in wire and cable insulation for electric power generation and transmission, for naval shipboard cable of all types, and for appliance wiring. The inherent inertness and biocompatibility of silicone rubbers have enabled their use in food contact and medical products. These include baby bottle nipples, belts and hoses for conveying foods and food ingredients, surgical tubing, subdermal implants, and prosthetic devices.

⌘⌘⌘

RTV silicones are used by the automotive, appliance, and aerospace industries for electronic potting compounds and formed-in-place gaskets, to form molds for the manufacture of plastic parts, and widely in construction adhesives, sealants, roof coatings, and glazing.

Special Purpose Elastomers

The special purpose rubbers are typically premium priced but cost effective in supplying one or more unique property required for specific demanding applications.

Fluoroelastomers – These are chemically saturated co- and terpolymers of vinylidene fluoride, hexafluoropropylene, tetrafluoroethylene, perfluoro (methyl vinyl) ether, and propylene in various combinations. They are designed to provide extraordinary levels of resistance to oil, chemicals, and heat. They are generally classified into four groups: A, B, F, and Specialty. The lettered groups have increasing fluid resistance, reflecting their increasing flourine levels of 66%, 68%, and 70% respectively. The Specialty group contains further enhanced properties, such as improved low temperature flexibility. Some fluoroelastomers incorporate a bromine-containing curesite monomer and can be vulcanized with peroxides. The rest are most often cured with bisphenol. Because of their exceptional resistance to heat aging and a broad range of chemicals, fuels and solvents, fluoroelastomers are used in a wide variety of demanding automotive, aerospace, and industrial applications. These include seals, gaskets, liners, hoses, protective fabric coatings, diaphragms, roll covers, and cable jacketing.

Polysulfides – The polysulfide rubbers (T) are made by addition of organic halides to a hot aqueous solution of sodium polysulfide. With agitation, the polymer precipitates in the form of small particles which are then washed, coagulated and dried. Linear polysulfides are made by copolymerizing ethylene dichloride and di(chloroethyl) formal with the sulfur linkage. Branched polysulfides are made from di(chloroethyl) formal and 2% trichloropropane. Vulcanized polysulfides are used for their excellent combination of low temperature flexibility, gas and water impermeability, and resistance to ozone, sunlight, heat, weathering, and most oils, solvents and fuels. Linear polymers are cured by zinc oxide, with crosslinking at terminal hydroxy groups. Branched polymers are cured by peroxide (usually zinc peroxide) with crosslinking at terminal thiols. Liquid polysulfide polymers with thiol terminals are also produced. These can be cured at room temperature using peroxides or other oxygen donating curing agents to provide properties similar to the other polysulfide rubbers. The major applications for polysulfides are in products contacting fuels, solvents, or oils, and/or requiring low temperature flexibility and high impermeability to moisture and gases. These include sealants and putties, diaphragms and gaskets, roller coatings, hoses, potting compounds, anticorrosion coatings, and solid rocket fuel binders.

Epichlorohydrin Rubbers – These rubbers are the homopolymer of epichlorohydrin (CO) and the copolymer of epichlorohydrin and ethylene oxide (ECO). The chemically saturated polyether backbone plus the polar chloromethyl substituents provide a combination of excellent low temperature flexibility, good oil, solvent, fuel, ozone, and fire resistance, and high air impermeability. The homopolymer, with 38% chlorine, has better fluid resistance and flame retardancy. The copolymer, with 26% chlorine and the unsubstituted ethylene oxide groups in the backbone, has better low temperature flexibility. Vulcanization of these rubbers is by nucleophilic displacement of chlorine by a thiourea, triazine, or thiadiazole. The copolymer is also available with allyl glycidyl ether included as a curesite monomer, allowing peroxide or sulfur cures. Because of their characteristic attributes and moderate price (for a specialty elastomer), the epichlorohydrin rubbers are most widely used in automotive applications. These include seals, gaskets, hoses and tubing. Other applications are coated fabrics and roll covers.

Urethane Rubbers – Polyurethane eleastomers are produced by reacting a diisocyanate typically toluenediisocyanate or methylene di(4-phenyl)-isocyanate, with either a polyether polyol (e.g., 1,4-butanediol) or a polyester polyol (e.g., from adipic acid and ethylene glycol). The low molecular weight polyols produce hard polymers, while the high molecular weight polyols give soft polymers. A slight excess of polyol is used to provide hydroxy termination for peroxide cures. For sulfur-curing elastomers, some of the polyol is replaced with an unsaturated diol. The polyurethanes in general are characterized by high tensile strength and abrasion resistance combined with good oil and tear resistance. Urethane rubber is used in solid tires, seals, potting compounds, engineering mechanical goods, elastic thread, footwear, and sheets.

Polyacrylate Rubber – The acrylic elastomers (ACM) are poly(alkyl)acrylates. The alkyl groups can range from C_2 to C_8, but the most common acrylic monomers used are the ethyl, n-butyl, 2-methoxyethyl, and 2-ethoxyethyl acrylates. The polyacrylate rubbers generally contain a small amount of curesite monomer for versatility in crosslinking. This is typically a monomer with a reactive halogen (e.g., allylic chlorine), or in some cases an epoxide. With a chemically saturated backbone, the acrylic rubbers offer superior resistance to ozone, heat, UV light, and oxidation. The polar acrylate group provides resistance to oils, aliphatic solvents, and the sulfur-bearing lubricants which can crosslink unsaturated elastomers. These rubbers also offer superior flex life, impermeability to many gases, and good colorability. Applications include hoses, gaskets, seals and tank linings resistant to oils and high pressure (sulfur-containing) lubricants, as well as belting and fabric coatings.

SULFUR-BASED CURE SYSTEMS

Vulcanization, the introduction of crosslinks into the elastomer's polymer matrix, is a fundamental determinant of rubber properties. In short, it converts a substance that is plastic and moldable into one that is flexible and elastic. Vulcanization increases tensile strength, modulus (stiffness), hardness, abrasion resistance and rebound, and decreases elongation, hysteresis (heat buildup), compression set, and solubility. Except for tensile and tear strength, which usually show a specific optimum crosslink density, vulcanization-induced changes are proportional to the number of crosslinks and their length. Excessive crosslinking can convert the elastomer to a hard, brittle solid. Longer (polysulfide) crosslinks promote better tensile and tear strength and better fatigue properties. Shorter crosslinks (mono- and disulfide) provide better oxidative and thermal stability and lower compression set.

Sulfur is, by far, the most commonly used curing agent for elastomers with chemically unsaturated polymer backbones, particularly the more common diene rubbers: natural, SBR, polybutadiene, nitrile, polychloroprene, and polyisoprene. Other curing agents can be used with unsaturated elastomers, but sulfur dominates because it is low in cost and toxicity, broadly compatible with other compounding additives, and able to predictably provide the desired vulcanization properties. A typical sulfur vulcanization system is composed of sulfur, a metal oxide (usually zinc oxide), a fatty acid (to solubilize the oxide's metal) and one or more organic accelerators. Sulfur, nevertheless, will perform its crosslinking function even without promoters, as long as there is sufficient heat and time provided. Lacking the promoters, however, vulcanization could take many hours, or even days, while today's systems take only minutes.

Sulfur is available to the compounder in two forms: amorphous and rhombic. The amorphous form, also known as insoluble sulfur is a metastable high polymer that is insoluble in rubber and most solvents. Rhombic sulfur, a ring of eight sulfur atoms, is soluble in rubber and the form normally used for vulcanization. About 1 to 3 phr (parts per 100 parts of rubber elastomer) of sulfur are used for most rubber products. Vulcanization by sulfur alone is believed to be primarily via a thermally induced free-radical reaction. The sulfur will slowly react allylic to the site of unsaturation to form polysulfide substituents. These then are subject to

⌘⌘⌘

heat-induced homolytic cleavage to thiyl (R-S·) and polysulfenyl (R-Sx·) free radicals, which in turn react with adjacent polymer chains to form monosulfide and (mostly) polysulfide links.

Activators

The rate at which sulfur will react with the unsaturated polymer chains can be increased by the addition of activators: a metal oxide plus fatty acid. The most common combination is zinc oxide and stearic acid, with the primary function of the fatty acid being to solubilize the zinc in the elastomer. In the presence of the metal, it is believed that the sulfur reacts as a cation at the double bond which results in charged and uncharged polysulfides, the latter of which could in turn form free radicals. Metal activated sulfur vulcanization will proceed more rapidly than crosslinking by sulfur alone, but still too slow for most production purposes. The metal oxide/fatty acid is, in practice, used not to activate the sulfur itself, but to activate the organic compounds used as vulcanization accelerators.

Accelerators

Vulcanization accelerators, in simplest terms, hasten the cleavage of the sulfur ring and formation of thiyl and polysulfenyl radicals. The accelerators react in the form of their more active zinc salts, due to the nearly ubiquitous presence of zinc oxide in sulfur vulcanized compounds. Structures of the more commonly used accelerators – the guanidines, thiazoles, sulfenamides, thiurams, and dithiocarbamates – are shown in Figures 2a and 2b. The choice of accelerator will affect the scorch (premature vulcanization) safety, the cure rate, and the length and number of crosslinks which form. These properties, which are generally related to the speed with which the accelerator is converted to its very active salt form, are compared in Table 1.

Table 1. Accelerator Comparison

Accelerator Type	Scorch Safety	Cure Rate	Crosslink Length
None		very slow	very long
Guanidines	moderate	moderate	medium-long
Mercaptobenzothiazoles	moderate	moderate	medium
Sulfenamides	long	fast	short-medium
Thiurams	short	very fast	short
Dithiocarbamates	least	very fast	short

The concept of acceleration was discovered in 1906 when it was found that addition of aniline to a sulfur/zinc oxide system significantly decreased vulcanization times. Dialkyl amines were also found to provide acceleration, and so aniline and the dialkyl amines were derivatized to the range of

⌘⌘⌘

accelerators presented in Figure 2. Aniline begat the guanidines, thiazoles, and sulfenamides (essentially amine "blocked" forms of mercaptobenzothiazole). The amines begat the thiurams and dithiocarbamates. The thiazoles (MBT and MBTS) and the sulfenamides (CBTS, BBTS, MBS) account for most of current accelerator production, with the sulfenamides the more widely used. The superior scorch safety of the sulfenamides is provided by the time required to thermally decompose into MBT and amine.

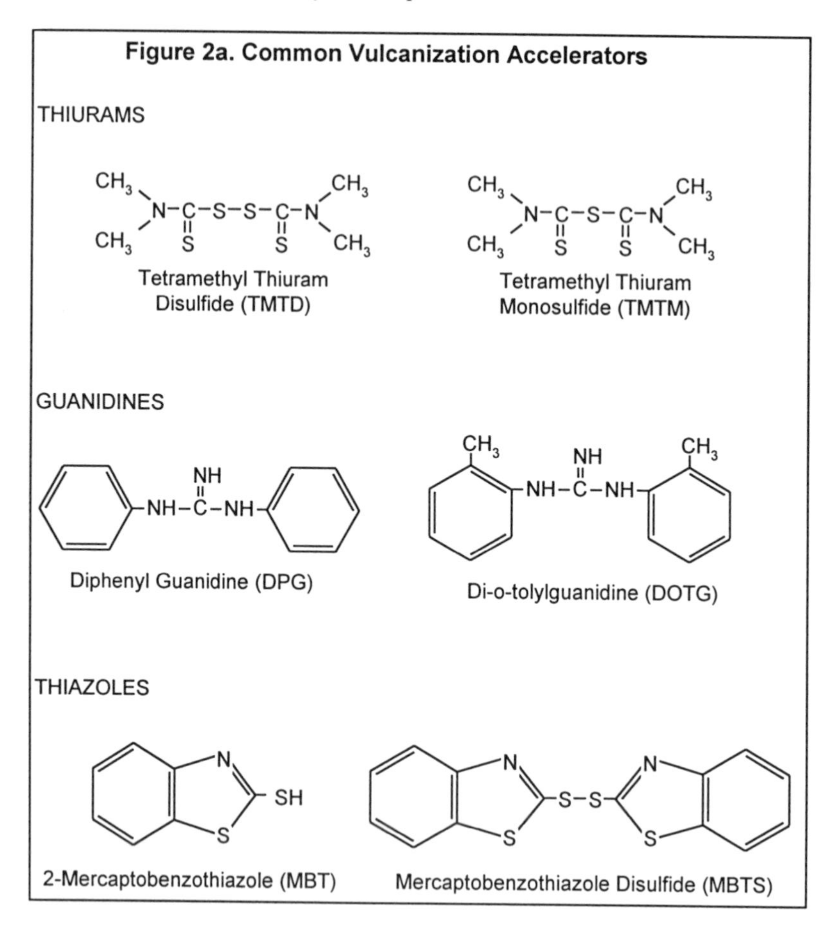

Figure 2a. Common Vulcanization Accelerators

While the sulfenamides are the most common primary accelerators, they are frequently used with secondary accelerators (a.k.a. "kickers") which act as activators for the cure system. The thiurams and dithiocarbamates, which are short on scorch safety when used as primary accelerators, are used at relatively low levels to help activate the cure.

⌘⌘⌘

Retarders

The desired level of scorch safety can be obtained through use of scorch retarders in addition to the accelerators chosen. The traditional retarders are acids – benzoic acid, salicylic acid, phthalic anhydride – which interfere with the activity of the accelerators. However, these acids do not only retard the scorch time, they also slow the cure rate and reduce the ultimate state of cure. With the widespread use of sulfenamide accelerators has come a new type of retarder which further improves upon these accelerators' superior scorch safety without significantly reducing cure rate or ultimate state of cure. For example, the retarder N-(cylcohexylthio)phthalimide will react with the MBT formed by the dissociation of the sulfenamide to form cyclohexyldithiobenzothiazole (CDB). The CDB is a good delayed action accelerator in its own right. It can be pictured as a sulfenamide analogue, with the cyclohexylthio group in place of an amine group. CDB, in turn, dissociates to accelerate the cure. The extended scorch safety derives from the total time required for the sulfenamide to dissociate, its liberated MBT to react with the retarder to form CDB, and the CDB to itself dissociate.

Figure 2b. Common Vulcanization Accelerators

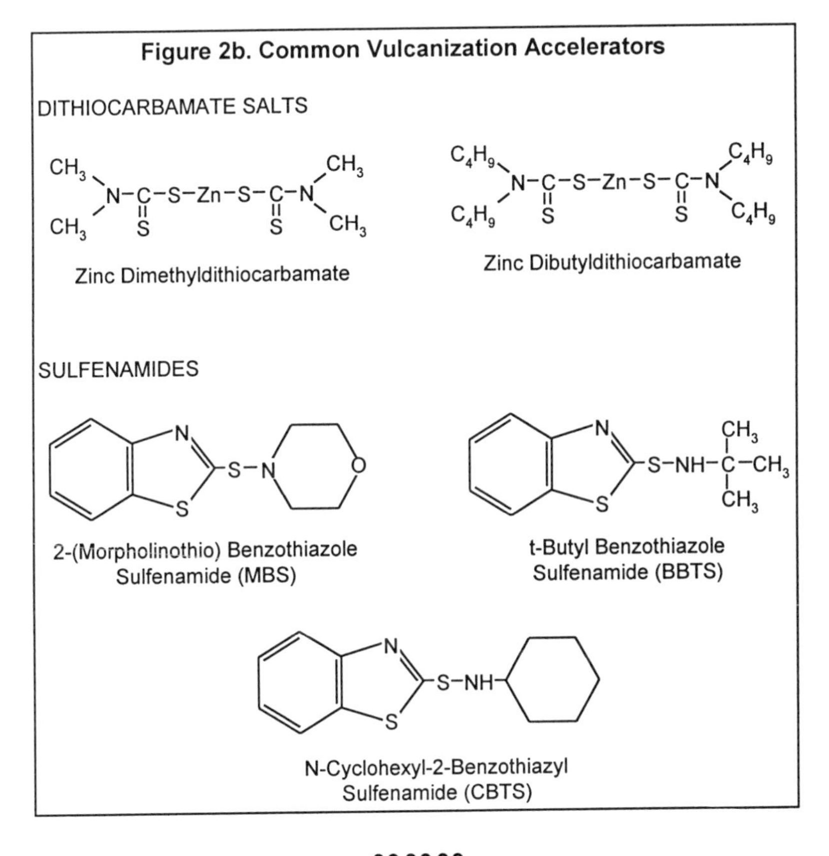

⌘⌘⌘

Crosslink Length

The length of the sulfide crosslinks formed during vulcanization will affect rubber properties. Mono- and disulfide crosslinks are more stable (less prone to scission) than polysulfide links and so promote better thermal and aging characteristics. Polysulfide links, on the other hand, provide somewhat better molecular flexibility. This can result in better dynamic fatigue resistance. Physical properties aside, the more stable crosslinks are often preferred to provide reversion resistance. Reversion is the cleavage of sulfide crosslinks during vulcanization which results from extending the cure beyond the time required to obtain the desired optimized balance of vulcanizate properties. When overcured in this way, certain elastomers, particularly natural rubber, will revert to the soft, more plastic, less elastic condition characteristic of the uncured compound.

The choice of accelerator will influence the length of crosslinks, but this can be further controlled by adjusting the accelerator:sulfur ratio. Increasing this ratio progressively favors shorter crosslinks. This can be alternatively accomplished by using sulfur donors in place of most or all of the elemental sulfur for vulcanization. The thiuram accelerators, particularly the disulfides, and dithiodimorpholine are commonly used for this purpose. The use of a cure system designed to form predominately short crosslinks is referred to as efficient vulcanization (EV).

NON-SULFUR CURE SYSTEMS

Elastomers with chemically saturated polymer backbones cannot be crosslinked with sulfur and so require alternate curing agents. The most widely used of these are the peroxides, which can also be used with unsaturated elastomers. Metal oxides or difunctional compounds are used, as well, in special cases.

Peroxides

Peroxides cure by decomposing on heating into oxy radicals which abstract a hydrogen from the elastomer to generate a polymer radical. The polymer radicals then react to form carbon-carbon crosslinks. With unsaturated elastomers, this occurs preferentially at the site of allylic hydrogens. The rate of crosslinking is directly proportional to the rate of decomposition of the peroxide. Cure rates and curing temperatures therefore depend on the stability of the peroxide, which decreases in the order dialkyl > perketal > perester or diaryl. The most commonly used of these crosslinkers is dicumyl peroxide.

Although carbon-carbon crosslinks are more thermally stable than sulfur crosslinks, they provide generally poorer tensile and tear strength. Peroxides are also incompatible with many of the antioxidants used in rubber, since the antioxidants are designed to scavenge and destroy oxy and peroxy free radicals (see Antioxidants, below). Peroxides cannot be used with butyl rubber because they cause chain scission and depolymerization.

Difunctional Compounds

Certain difunctional compounds are used to crosslink elastomers by reacting to bridge polymer chains. For example, diamines (e.g., hexamethylenediamine carbamate) are used as crosslinks for fluoroelastomers; p-quinone dioxime is oxidized to p-dinitrosobenzene as the active crosslink for bridging at the polymer double bonds of butyl rubber; and methylol terminated phenol-formaldehyde resins will likewise bridge butyl rubber chains (with $SnCl_2$ activation) as well as other unsaturated elastomers.

Metal Oxides

Metal oxides, usually zinc oxide but on occasion lead oxide for improved water resistance, are used as crosslinking agents for halogenated elastomers such as neoprene, halobutyl rubber, and chlorosulfonated polyethylene. The metal oxide abstracts the allylic halogen of adjacent polymer chains to form an oxygen crosslink plus the metal chloride salt.

FILLERS

A rubber compound contains, on average, less than 5 lbs. of chemical additives per 100 lbs. of elastomer. Filler loading is typically 10-15 times higher. Of the ingredients used to modify the properties of rubber products, the filler plays a dominant role. The term "filler" is misleading, implying, as it does, a material intended primarily to occupy space and act as a cheap diluent of the more costly elastomer. Most of the rubber fillers used today offer some functional benefit that contributes to the processability or utility of the rubber product Styrene-butadiene rubber, for example, currently the highest volume elastomer, has virtually no commercial use as an unfilled compound.

Filler Properties

The characteristics which determine the properties a filler will impart to a rubber compound are particle size, surface area, structure, and surface activity.

Particle Size – If the size of the filler particle greatly exceeds the polymer interchain distance, it introduces an area of localized stress. This can contribute to elastomer chain rupture on flexing or stretching. Fillers with particle size greater than 10,000 nm (10 μm) are therefore generally avoided because they can reduce performance rather than reinforce or extend. Fillers with particle sizes between 1,000 and 10,000 nm (1 to 10 μm) are used primarily as diluents and usually have no significant affect, positive or

⌘⌘⌘

negative, on rubber properties. Semi-reinforcing fillers, which range from 100 to 1000 nm (0.1 to 1μm) improve strength and modulus properties. The truly reinforcing fillers, which range from 10 nm to 100 nm (0.01 to 01 μm) significantly improve rubber properties.

Of the approximately 2.1 million tons of fillers used in rubber each year, 70% is carbon black, 15% is kaolin clay, 8% is calcium carbonate, 4% is the precipitated silicas and silicates, and the balance is a variety of miscellaneous minerals. Figure 3 classifies the various fillers by particle size and consequent reinforcement potential.

Figure 3. Filler Classification Chart

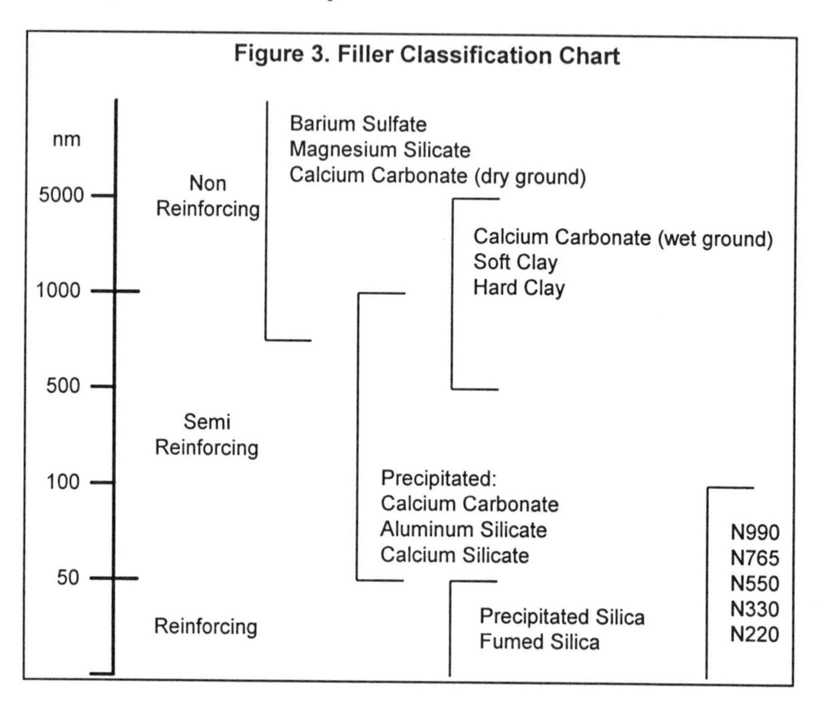

Most talcs and dry-ground calcium carbonates are degrading fillers because of their large particles size, although the planar shape of the talc particles contributes some improvement in reinforcement potential. The soft clays would fall into a class of diluent fillers that do not contribute reinforcement, yet are not so large that they degrade properties.

The hard clays contribute some reinforcement to rubber compounds, primarily because of their smaller particle size, and are normally classified as semi-reinforcing fillers. Ultrafine precipitated calcium carbonates would also fit into the semi-reinforcing class. The carbon blacks and precipitated silicates and silicas are available in various particle sizes that range from semi-reinforcing to highly reinforcing. They generally exist as structural agglomerates or aggregates rather than individual spherical particles.

⌘⌘⌘

Surface Area – Particle size is generally the inverse of surface area. A filler must make intimate contact with the elastomer chains if it is going to contribute to reinforcement of the rubber-filler composite. Fillers that have a high surface area have more contact area available, and therefore have a higher potential to reinforce the rubber chains. The shape of the particle is also important. Particles with a planar shape have more surface available for contacting the rubber than spherical particles with an equivalent average particle diameter. Clays have planar-shaped particles that align with the rubber chains during mixing and processing, and thus contribute more reinforcement than a spherical-shaped calcium carbonate particle of similar average particle size. Particles of carbon black or precipitated silica are generally spherical, but their aggregates are anisometric and are considerably smaller than the particles of clay. They thus have more surface area per unit weight available to make contact with the polymer. Rubber grade carbon blacks vary from 6 to 250 m^2/g. Most reinforcing precipitated silicas range from 125 to 200 m^2/g; a typical hard clay ranges from 20 to 25 m^2/g.

Structure – The shape of an individual particle of reinforcing filler (e.g. carbon black or precipitated silica) is of less importance than the filler's effective shape once dispersed in elastomer. The blacks and precipitated inorganics used for reinforcement have generally round primary particles but function as anisometric acicular aggregates. The round particles clump together into chains or bundles that can be very dense or open and lattice-like. These aggregate properties – shape, density, size – define their structure. A high structure filler would have aggregates favoring high particle count, and for those particles to be joined in chain-like clusters, from which random branching of additional particle chains may occur. In simplest terms, the more an aggregate deviates from a solid spherical shape and the larger its size, the higher is its structure. The higher its structure, in turn, the greater its reinforcing potential

For reinforcing fillers which exist as aggregates rather than discreet particles, carbon black and silica in particular, a certain amount of structure that existed at manufacture is lost after compounding. The high shear forces encountered in rubber milling will break down the weaker aggregates and agglomerates of aggregates. The structure that exists in the rubber compound, the persistent structure, is what affects processability and properties.

Surface Activity – A filler can offer high surface area and high structure, but still provide relatively poor reinforcement if it has low specific surface activity. The specific activity of the filler surface per cm^2 of filler-elastomer interface is determined by the physical and chemical nature of the filler surface in relation to that of the elastomer. Nonpolar fillers are best suited to nonplar elastomers; polar fillers work best in polar elastomers. Beyond this

⌘⌘⌘

general chemical compatibility is the potential for reaction between the elastomer and active sites on the filler surfaces. Carbon black particles, for example, have carboxyl, lactone quinone, and other organic functional groups which promote a high affinity of rubber to filler. This, together with the high surface area of the black, means that there will be intimate elastomer-black contact. The black also has a limited number of chemically active sites (less than 5% of total surface) which arise from broken carbon-carbon bonds as a consequence of the methods used to manufacture the black. The close contact of elastomer and carbon black will allow these active sites to chemically react with elastomer chains. The carbon black particle effectively becomes a crosslink. The non-black fillers generally offer less affinity and less surface activity toward the common elastomers. This can be compensated to a greater or lesser extent by certain surface treatments.

Figure 4. Silanes Used With Sulfur Cure

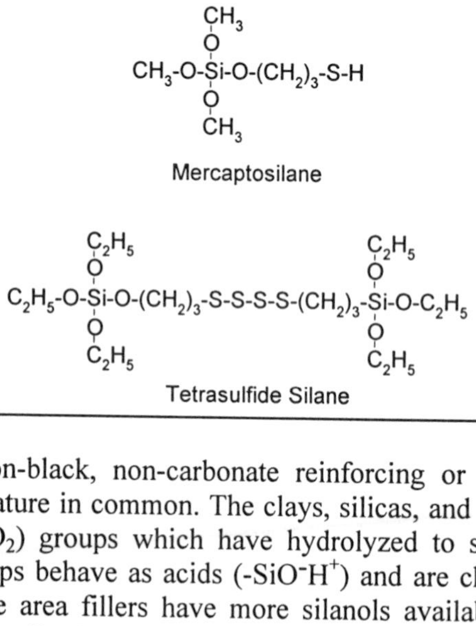

$$\begin{array}{c} CH_3 \\ | \\ O \\ | \\ CH_3\text{-}O\text{-}Si\text{-}O\text{-}(CH_2)_3\text{-}S\text{-}H \\ | \\ O \\ | \\ CH_3 \end{array}$$

Mercaptosilane

$$\begin{array}{ccc} C_2H_5 & & C_2H_5 \\ | & & | \\ O & & O \\ | & & | \\ C_2H_5\text{-}O\text{-}Si & \text{-}O\text{-}(CH_2)_3\text{-}S\text{-}S\text{-}S\text{-}S\text{-}(CH_2)_3\text{-} & Si\text{-}O\text{-}C_2H_5 \\ | & & | \\ O & & O \\ | & & | \\ C_2H_5 & & C_2H_5 \end{array}$$

Tetrasulfide Silane

All of the non-black, non-carbonate reinforcing or semi-reinforcing fillers have one feature in common. The clays, silicas, and silicates all have surface silica (SiO_2) groups which have hydrolyzed to silanols (-SiOH). These silanol groups behave as acids ($-SiO^-H^+$) and are chemically active. The higher surface area fillers have more silanols available and are thus more reactive. One effective means to modify the surface chemistry of non-black fillers is by the addition of silane coupling agents. These react with the silanols on the filler surface to give a strong bond, and also contain a functional group that will bond to the rubber during vulcanization. The result is filler-polymer bonding (crosslinking) that increases modulus and tensile strength and improves abrasion resistance. Modification of the filler surface also improves wetting and dispersion. The silane coupling agents are

⌘⌘⌘

expensive, and as a result they are used primarily in higher priced rubber compounds, primarily to promote maximum abrasion resistance.

There are two silane coupling agents commonly used in sulfur-cured compounds filled with non-black fillers. These are mercaptosilane and tetrasulfide silane, shown in Figure 4. The methoxy or ethoxy groups react during mixing with the silanol groups on the surface of silica, silicate, or clay particles to give a strong bond. The sulfur-containing group of each structure reacts during vulcanization to give bonding to the polymer.

The mercaptosilane is more active on a phr weight basis. About 1.5 to 2.0 times as much tetrasulfide silane is required to achieve similar modulus and abrasion resistance levels. The mercaptosilane has a strong mercapto odor that makes working with the neat liquid product unpleasant. It is normally used as a concentrate carried on an inert carrier, or in the form of pretreated filler.

Figure 5. Silane Coupling

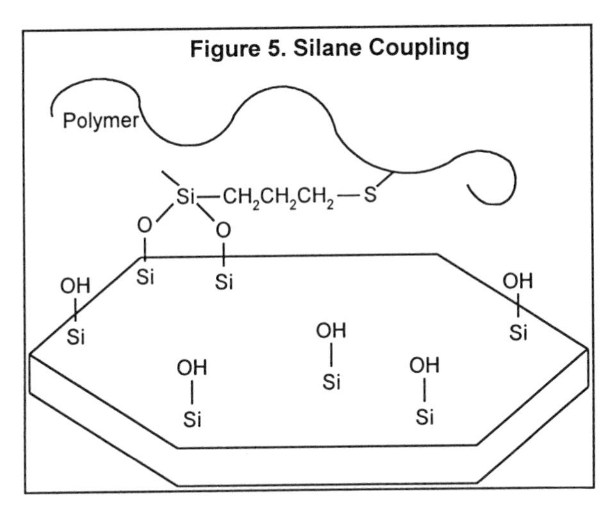

The silane coupling bond is depicted in Figure 5. The chemistry of the bond would be similar whether the filler was a clay pretreated with mercaptosilane or a high surface area silica that had the silane added during mixing. The addition sequence is very important when adding silane coupling agents during mixing of the compound, especially when using the tetrasulfide. To make the most efficient use of these high-cost chemicals, it is important to mix the polymer, filler, and silane coupling agent for 1 to 2 minutes before adding any other ingredients that will interfere with the reaction between the filler and the silane. If the mixing temperature gets too high (above about 150 to 160°C) the tetrasulfide group can be broken prematurely, reducing its ability to bond to the polymer during vulcanization.

⌘⌘⌘

Silanes with other functional groups such as amino, vinyl, methacryl, or epoxy groups are also used, primarily with non-sulfur cure systems. The functional group is tailored to the type of cure system to maximize bonding to the polymer during vulcanization.

Silanols show similarities to carboxylic acid groups in their reactions with amines, alcohols, and metal ions. While water adsorbed on the surface of filler particles will reduce silanol reactivity, hot compounding volatilizes some of this water, leaving a very reactive surface. Some of the reactions with silanols can have a profound effect on the properties of the rubber compound, especially where the chemical involved is an important part of the cure system. Most of the accelerators used in sulfur cure systems contain an amine group. Strong adsorption or reaction with filler particles can decrease the amount of accelerator available for vulcanization reactions. This can give slower cure rates and a reduced state of cure. Similar effects can result from the reaction of zinc ions with filler particles. For example, the active silanols on the surface of a precipitated silica particle will react with zinc stearate to give two known reaction products: one is a zinc stearate complex, and the other is a bridging of zinc across two silanol groups, as shown in Figure 6.

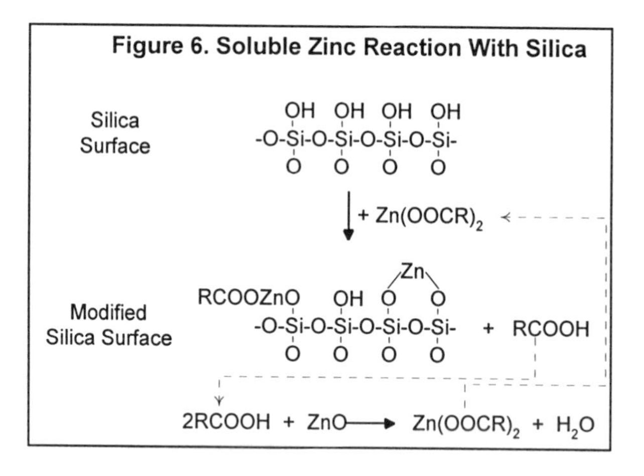

Excess stearic acid is released to combine with more zinc oxide. If allowed to proceed to completion, this mechanism would eventually rob all the zinc, leaving none to activate the vulcanization reaction. Lower surface area non-black fillers also react with the zinc ion, but show less effect because they have fewer silanol groups available.

These negative effects on the cure system can be reduced or completely avoided by adding other chemicals that will tie up the silanol groups and reduce their activity. Such additives commonly used in non-black compounds include diethylene glycol (DEG) and polyethylene glycol

⌘⌘⌘

(PEG), hexamethylene tetramine (Hexa), hexamethoxy methyl melamine (HMMM), and triethanolamine (TEA). These are mixed into the compound prior to the addition of the zinc oxide and accelerators. Magnesium oxide in nitrile and neoprene compounds also reduces the tendency for the fillers to rob zinc from the cure system.

Many of these additives also reduce the polarity of the filler surface and thus improve wetting and dispersion in non-polar polymers. Polar oils or aromatic resins also generally improve filler dispersion and the properties of compounds containing non-black fillers. Other techniques used to adjust the cure of compounds containing non-black fillers include increasing the level of amine accelerators and adding zinc oxide in a second mixing stage.

Filler Effects

The principal characteristics of rubber fillers – particle size, surface area, structure, and surface activity – are interdependent in improving rubber properties. In considering fillers of adequately small particle size, reinforcement potential can be qualitatively considered as the product of surface area, surface activity, and persistent structure or anisometry (planar or acicular nature).

The general influence of each of the three filler characteristics above on rubber properties can be summarized as follows:

1) Increasing surface area (decreasing particle size) gives lower resilience and higher Mooney viscosity, tensile strength, abrasion resistance, tear resistance, and hysteresis.

2) Increasing surface activity (including surface treatment) gives higher abrasion resistance, chemical adsorption or reaction, modulus (at elongation >300%), and hysteresis.

3) Increasing persistent structure/anisometry gives higher Mooney viscosity, modulus (at elongation <300%), and hysteresis, lower extrusion shrinkage, tear resistance, and resilience, and longer incorporation time.

In general terms, the effect of a filler on rubber physical properties can be related mainly to how many polymer chains are attached to the filler surface and how strongly they are attached. Filler surface area and activity are the main determinants, supplemented by structure. Since the filler particles can be considered crosslinks for the elastomer chains, the presence or absence of a coupling agent on the surface of non-black fillers is also important.

⌘⌘⌘

Modulus/Tensile Strength – Modulus is a measure of the force required to stretch a defined specimen of rubber to a given percent elongation. Most often, modulus is reported at 300% elongation (four times original length). This can be alternatively viewed as the resistance to a given elongating force. For an uncompounded elastomer, elongation is primarily a function of stretching and disentangling the randomly oriented polymer chains and breaking the weak chain-chain attractions. Vulcanized, but unfilled, elastomers more strongly resist elongation because the sulfur crosslinks must be stretched and broken to allow chain extension and separation. The introduction of a filler into the vulcanizate provides further resistance to elongation. A filler with low surface activity will increase resistance to elongation by the viscous drag its surface provides to the polymer trying to stretch and slide around it. Higher surface area, greater anisometry or structure, and higher loading (the latter two effectively increasing the surface area exposed to the elastomer) will all increase the modulus. Fillers with strong chain attachments, via active sites or coupling agents, provide the most resistance to the chain extension and separation required for elongation.

It is helpful to visualize the filler particles acting as giant crosslinks. Figure 7 is a schematic representation of such a system with the filler particles simplified to spheres for convenience. Before stretching (Step 1), the polymer chains are in random configuration. Chains A, B and C have multiple points of attachment to the filler particles, corresponding to the latter's active sites. On elongation, resistance is supplied as the energy required to detach the chain segments from these active sites (Step 2 and 3). The amount of energy required to attain maximum elongation, and then required to overcome the stress distribution implied in Step 3 to cleave chain-chain and chain-filler attachments, likewise explains the higher tensile strength (force required for elongation to sample rupture) of a system of this type.

After the elongating force has been removed, the elastomer chains return to their preferred random orientation (Step 4), except that now they have the minimum number of points of attachment to the filler as a consequence of having been extended, as in Step 3. Less force would now be required to return these chains to ultimate extension, because the intermediate points of attachment that existed in Steps 1 and 2 have been eliminated. This accounts for the phenomenon known to rubber technologists as stress softening. With repeated stress-relaxation cycling, a decrease in modulus from the initial maximum is obtained. Stress softening is a temporary effect. After a period without strain, the rubber will recover to near its original modulus, as the active filler sites again attach to polymer segments. A percentage of the original modulus is permanently lost, however, due to irrecoverable chain and bond cleavage.

⌘⌘⌘

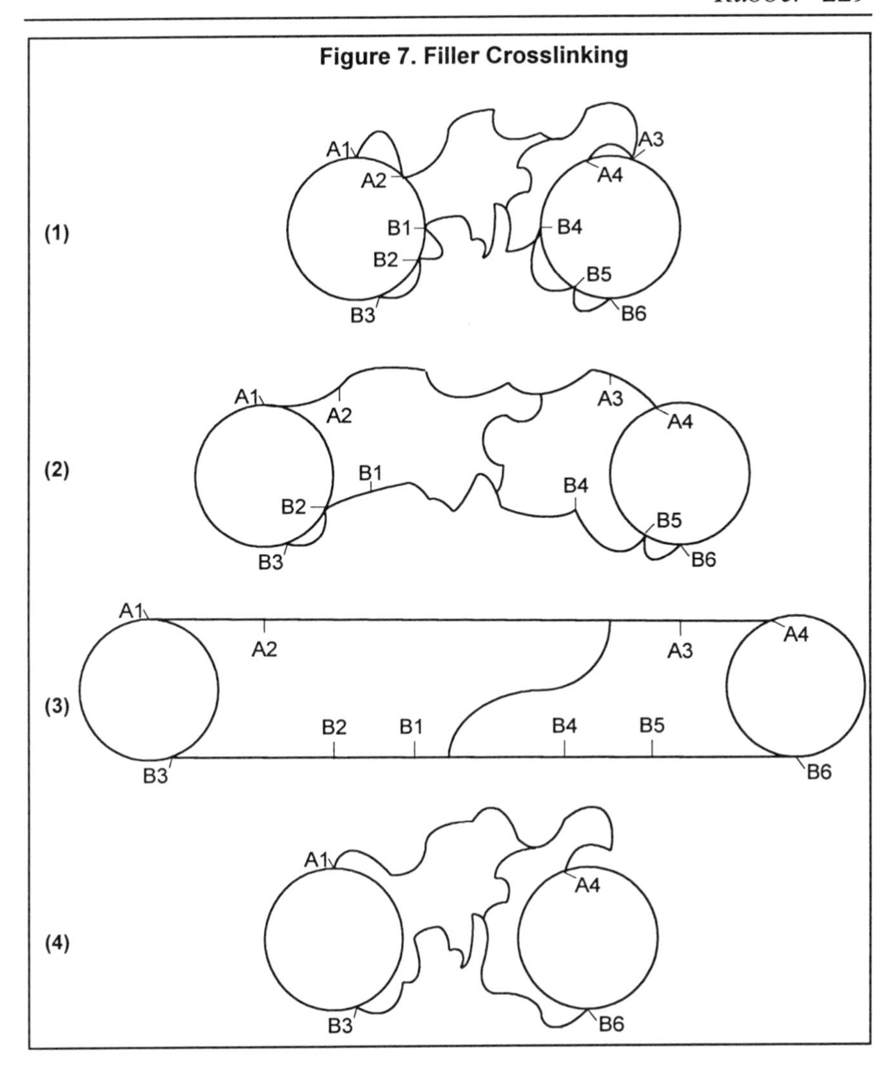

Hardness/Flexural Strength – The analogous resistance to elastomer chain deformation accounts for the increase in rubber hardness obtained in proportion to filler loading, surface area, structure activity/surface treatment, and structural anisometry. Hardness is typically measured as resistance to surface indentation under specific conditions, so the force is supplied as a point of compression rather than an area of tension. Similar filler particle dynamics apply, nevertheless. Flexural strength is likewise related as a measure of a sample's resistance to deformation (flex) under a compressive force. With a force supplied to an end-supported test sample, the upper surface is under compression and the lower surface is in tensicn

⌘⌘⌘

(elongation). Both compression and elongation are effectively resisted in proportion to the durability of the filler-elastomer bond. High structure or anisometric fillers provide greater resistance to compression because this force is generally perpendicular to their more rigid planar or elongate dimension.

Impact Strength – Impact strength or impact resistance is the measure of a hard rubber's resistance to fracture under sudden impact or force. The impact force is one of compression on one side of the rubber and tension on the other, so that fracture is a result of failure similar to that occuring when a tensile sample is elongated to break. The same filler-elastomer consideration as in tensile and flexural testing therefore apply.

Tear Strength/Flex Resistance – Tear strength and flex resistance are essentially a measure of resistance to crack or slit propagation. Both typically measure the growth of a crack under tension, although flex testing, with multiple flexing cycles, introduces the additional factor of heat generation and fatigue. Large or poorly bound fillers will act as flaws and initiate or propagate cracks under test conditions. Small particles size, high surface area, and high surface activity allow the filler particles to act as barriers to the propagation of microcracks, and provide the higher tensile strength required to resist failure.

Resilience/Hysteresis – Resilience is essentially a measure of rubber elasticity – the characteristic ability to quickly return to original shape following deformation. Unfilled elastomers are at their peak resilience because there is no obstacle to elastomer chain extension and contraction (return to randomness). The introduction of a filler creates such an obstacle in proportion to the strength of the particle-polymer interaction. Resilience is therefore generally in inverse proportion to filler loading and reinforcement.

If resilience is considered the ratio of energy release on recovery to the energy impressed on deformation, hysteresis can be viewed as a measure of the amount of impressed energy that is absorbed. An unfilled (elastic) elastomer will convert most of the energy of deformation into the mechanical energy used in returning to the original shape. The degree to which input energy is converted to heat, due to friction from polymer chains sliding past each other, is a function of polymer chain morphology. All fillers increase the hysteresis of rubbers, the most reinforcing fillers generally producing the largest increases. This is due to the energy consumed in polymer-filler friction and in dislodging polymer segments from active filler surfaces. This is heat energy which would otherwise have a mechanical manifestation as elastic rebound. In short, hysteresis is inversely related to resilience.

⌘⌘⌘

Abrasion Resistance – Abrasion resistance is a function of filler structure, particle size, and surface activity (filler-elastomer adhesion). At equivalent loadings, large particles will be more easily dislodged from the rubber surface than finer particles. Finer particle sizes also provide more uniform distribution and more particles per unit surface area. Loss of a larger particle will expose more of the relatively soft surrounding elastomer matrix to wear. The effect is acute on the edge of the depression left by the dislodged filler particle. This is the area most susceptable to elongation, crack initiation and ultimate loss. For a fine filler, the better the rubber-particle bond – a function of surface area and surface activity or coupling agent treatment – the more difficult it will be to dislodge. Abrasion resistance under dynamic conditions is also affected by compound hysteresis which, in turn, is a function of filler selection.

The classic illustration of the role of hysteresis vs. resilience in abrasion resistance is wear resistance of tire treads. The tread is deformed at the area of contact with the road. Wear comes from the sliding of the rubber across the road just to the rear of the area of contact. This sliding occurs during elastic recovery of the deformation. Greater resilience would give quicker recovery and this would expose greater tread area to sliding and consequent wear. A tread compound with greater hysteresis will inhibit elastic recovery and thereby decrease the area of sliding and the amount of wear. This must be balanced against the ability of the tread to recover sufficiently to again deform and dissipate heat with the conclusion of one revolution, otherwise heat buildup will be excessive.

Filler Types

The use of fillers in rubber products is nearly as old as the use of rubber itself. Whether the initial motivation was function or economics is not known. Prior to the discovery of vulcanization, the tackiness of rubber goods was subject to correction by the traditional remedy for wet or sticky surfaces – sprinkling with talcum powder. The technological leap from talc on the outside to incorporation of talc on the inside of the rubber to serve the same purpose is credible. Hancock's masticator, and the rubber mixing machinery it spawned, facilitated the incorporation of all types of fine particulates, including ground limestone, clays, barite, zinc oxide, zinc sulfide, iron oxides, asbestos, mica, litharge (lead oxide) and kieselguhr (diatomite). Whether the original reason for adding these powders was reduction in tack, adjustment of color, or simply reducing the cost of the rubber compound, two conclusions were inevitably drawn. Some fillers could be added in large amounts without detracting from desirable rubber properties, and others would actually improve rubber properties, usually by increasing hardness and durability.

⌘⌘⌘

In time, the fillers which were likely first used for their color, zinc oxide and carbon black, became recognized for their reinforcement potential. In the early years of the 20th century, prior to World War I, zinc oxide was the most widely used reinforcing filler in rubber; the abrasion resistance it provided made it the preferred filler in tire treads. It was also during this period that zinc oxide was discovered to be the activator for the newly emerging organic accelerators. The reinforcing effect of carbon black was quantified during the first decade of this century, but remained unexploited for about ten years. The zinc oxide-filled tire tread invariably outlasted the canvas-based carcass, so any improvements offered by black-filled tread were less than compelling. It was only after B.F. Goodrich Co. purchased the patent rights to tire carcass cord and acquired carbon black compounding technology in 1912 that reinforcement requirements changed. The more durable carcass construction allowed room for improvement in tread wear. By the end of World War I, carbon black had widely displaced zinc oxide in pneumatic tire treads, although it took more than five years thereafter for the tire manufacturers to convince the general public that black tires were better than white.

Zinc oxide filler use was reduced to white and colored compounds. This started eroding in the 1930s with the introduction of precipitated calcium carbonate, and was all but eliminated in the 1940s with the development of precipitated calcium silicate and precipitated silica. Carbon black, meanwhile, became firmly established as the premier reinforcing filler.

Carbon Black – Carbon black is essentially elemental carbon in the form of fine amorphous particles. Each particle is composed of randomly oriented microcrystalline layered arrays of condensed carbon rings. Because of their random orientation, many arrays expose open layer edges with unsatisfied carbon bonds at the particle surface. This in turn provides the sites for chemical activity. Individual round carbon black particles do not exist as discrete entities but form aggregates, which may be clumps or chains of various sizes and configurations. The functional carbon black "particle", therefore, is actually the aggregate. Average particle size and aggregate configuration (structure) are the major determinants of the utility of a given carbon black in a specific rubber compound. The major differences between commercial grades result from control of these averages.

Prior to World War II, the predominant reinforcing black was made from small flames of natural gas impinging on iron channels. The deposit scraped from these channels was known as channel black. The finest particle size used for rubber, about 24 nm average, was Hard Processing Channel (HPC) because it produced the stiffest stocks. Larger particles sizes were available as Medium Processing Channel (MPC) at 26 nm average, and Easy Processing Channel (EPC) at 29 nm average. Channel blacks are no

⌘⌘⌘

longer used in rubber because of their high cost and the availability of suitable alternatives.

Furnace blacks began displacing channel blacks for rubber reinforcing in the early 1940s. Furnace blacks were at first, as early as the 1920s, made by burning natural gas in large horizontal furnaces, yielding a relatively coarse (60 to 80 nm average) semi-reinforcing product. Synthetic rubber development during World War II, particularly the reinforcemnt requirements of the otherwise unusable SBR, promoted the production of finer grades. Ultimately, natural gas was abandoned as a feedstock in favor of residual aromatic petroleum oils. This liberated carbon black manufacture from proximity to natural gas sources and allowed location of plants more convenient to the consuming industries. Oil furnace blacks today account for about 95% of the carbon black used in rubber.

The thermal process, introduced in 1922, makes the largest particle size and lowest structure blacks. Thermal blacks are made in a large cylindrical furnace by the thermal decomposition of natural gas in the absence of flame or air. Thermal blacks range from 100 to 500 nm average and are generally used as low cost functional extender fillers. Because of their relatively large size and low structure they can be used at higher loadings and provide better resilience and lower hysteresis than the more reinforcing furnace blacks.

Table 2. - Carbon Blacks

ASTM Series Classification	Old Classification	Size Avg, (nm)
N100	SAF (Super Abrasion Furnace)	11 to 19
N200	ISAF (Intermediate Abrasion Furnace)	20 to 25
N300	HAF (High Abrasion Furnace)	26 to 30
N400	FF (Fine Furnace)	31 to 39
N500	FEF (Fast Extruding Furnace)	40 to 48
N600	GPF (General Purpose Furnace)	49 to 60
N700	SRF (Semi-Reinforcing Furnace)	61 to 100
N800	FT (Fine Thermal)	101 to 200
N900	MT (Medium Thermal)	201 to 500

The carbon blacks that provide the highest ratio of reinforcement to surface area are those produced using the shortest reaction time. Short reaction time promotes randomness of carbon ring layer orientation within the particle and the consequent occurence of layer edges, with unsatified carbon bonds, at the surface. Furnace black production accomodates the manipulation of reaction time and conditions that results in some degree of tailoring of product surface activity. Heat-treating (graphitizing) high activity blacks has shown a marked reduction in surface activity. This is believed due to the realigning of carbon layers such as that the maximum number of carbon-carbon bonds are formed and the layer planar surfaces rather than the edges are exposed at the particle surface. The production of thermal blacks requires a sufficiently long reaction time that they are largely graphitized in this way. Thermal blacks consequently provide characteristicaly low surface activity and reinforcement potential.

Table 3. - Comparison of Carbon Blacks in SBR

Compound	
SBR 1500	100
Carbon Black	50
Naphthenic Oil	10
Zinc Oxide	3
Stearic Acid	2
Agerite® Stalite® S	1
Vanax® NS	1
Sulfur	2

Carbon Black	N110	N220	N351	N550	N660	N762	N990
Mooney @ 100°C							
Viscosity, (ML1+4)	65	64	59	60	52	48	46
Die Swell, 0.25" diam die							
45 rpm, 70°C							
% Swell	9.6	8.0	7.2	4.0	4.4	6.0	5.6
Physical Properties							
300% Modulus, MPa	13.2	12.9	14.3	12.3	9.9	7.9	3.7
(psi)	1920	1870	2070	1790	1440	1140	530
Tensile Str, MPa	27.4	27.5	25.2	20.2	21.4	22.5	12.6
(psi)	3960	3990	3670	2920	3110	3260	1820
Elongation, %	500	530	470	480	540	630	730
Hardness, Shore A	60	61	59	57	53	50	44
Tear, Die C, kN/m	36	37	33	35	33	34	26
(pli)	205	210	190	200	190	195	150
Dynamic Testing							
Rebound, Zwick, %	44	46	53	58	60	61	63
Heat Buildup, °C	137	134	117	112	107	108	104
Firestone, °F	278	273	245	233	224	226	220

Vanderbilt Rubber Handbook

Until 1968, carbon black nomenclature was informal and based on a variety of characteristics, including level of abrasion resistance, level of reinforcement, vulcanizate modulus, processing properties, general usefulness, particles size, and electrical conductivity. In 1968, the ASTM Committee on Carbon Black established a common nomenclature system consisting of a prefix followed by a three digit number. The prefix is either N, for normal curing, or S for slow curing. At the time, rubber grade channel

⌘⌘⌘

blacks, which are slow curing, were still used. They have since been discontinued; all current rubber grade blacks carry the N prefix.

The first of the three digits indicates a range of average particle size in nanometers. The second and third digits are assigned by the ASTM Committee to new products as they are developed. In general, lower structure blacks are assigned lower numbers and higher structure blacks, higher numbers, although there are some exceptions. Table 2 shows the series of carbon blacks used today by ASTM classification, the old letter classification and size ranges.

Table 3 compares the physical properties of a typical SBR-compound as a function of carbon black size and type. The test methods used are described below under Physical Testing of Rubber.

Kaolin Clay – Kaolin clay is typically used to reduce rubber compound cost while improving physical or processing properties.

Rubber filler clays are classified as either "hard" or "soft" in relation to their particle size and stiffening affect in rubber. A hard clay will have a median particle size of approximately 250 to 500 nm, and will impart high modulus, high tensile strength, stiffness, and good abrasion resistance to rubber compounds. Soft clay has a median particle size of approximately 1000 to 2000 nm and is used where high loadings (for economy) and faster extrusion rates are more important than strength.

The anisometry (platelet shape) and particle size of the clays account for their affect on modulus and hardness. More hard clay than soft is used in rubber because of its semi-reinforcing effect and its utility as a low cost complemenent to other fillers. It is used to improve the tensile and modulus of ground calcium carbonate compounds and will substitute for a portion of the more expensive carbon black or precipitated silica in certain compounds without sacrificing physical properties.

Table 4 compares calcium carbonate, hard clay and soft clay in an SBR compound. The effects of particle size and shape are evident. The calcium carbonate represents large (approximately 2500 nm average) isometric particles providing little improvement in compound properties compared to the clays. The hard clay provides greater tensile and tear strength than the soft clay.

Table 5 compares the ability of a hard clay and a dry-ground calcium carbonate to reduce the cost of a 65 Durometer shoe sole compound reinforced with precipitated silica. All three compounds show equivalent modulus and tensile strength. The calcium carbonate, however, has significantly reduces tear strength and abrasion resistance while the clay maintains these properties.

⌘⌘⌘

Table 4. Comparison of Clay and Calcium Carbonate in SBR

Compound	
SBR 1006	100
Vanfre® AP2	2
Zinc Oxide	5
Stearic Acid	2
Agerite® Stalite® S	1.5
Sulfur	2
Methyl Cumate®	0.1
Altax®	1.5

Filler/ phr	300% Mod. MPa (psi)	Tensile MPa (psi)	Elong %	Hardness Shore A	Tear[1] kN/m (pli)	Comp. Set[2], %	Mooney t_5/M_L[3]
Hard Clay[A]							
50	1.7 (250)	9.0 (1310)	770	50	13.2 (75)	46	27/25
75	2.3 (330)	11.8 (1710)	830	57	17.6 (100	48	27/29
100	3.2 (460)	12.8 (1860)	810	61	24.6 (140)	50	25/32
150	4.8 (700)	10.1 (1470)	700	67	37.0 (210)	51	19/47
Soft Clay[B]							
50	1.8 (260)	3.3 (480)	540	50	11.4 (65)	43	27/23
75	2.1 (310)	7.0 (970)	720	55	13.2 (75)	49	26/27
100	2.8 (410)	8.8 (1270)	740	62	20.2 (115)	50	24/32
150	3.3 (480)	7.6 (1100)	730	67	22.9 (130)	54	23/40
Calcium Carbonate[C]							
50	1.4 (200)	1.7 (250)	390	49	6.2 (35)	40	24/21
75	1.2 (170)	1.8 (260)	420	52	5.3 (30)	43	24/22
100	1.2 (170)	1.6 (230)	410	53	5.3 (30)	45	24/23
150	1.1 (160)	1.4 (200)	420	59	7.0 (40)	50	24/25

[1]Die A [2]22 hours, 70°C [3]132°C
[A]Airfloat, Dixie Clay [B]Airfloat, McNamee Clay [C]Atomite

Vanderbilt Rubber Handbook

Most of the clay used in rubber is airfloat. Some water-washed clay is used because its lower level of impurities provides better color and less die wear with rubber extrusions. Aminosilane and mercaptosilane treated hard clays provide better reinforcement than untreated clay, and in some applications can rival furnace blacks. The combination of platey morphology and chemical reactivity enable the silane-treated kaolins to impart a unique blend of properties to elastomers. These include high modulus, low hysteresis, good abrasion resistance, low viscosity, low set, and resistance to heat and oxidative aging. The combination of high modulus and low hysteresis, in particular, have earned them a place alone or as a partial carbon black replacement in products requiring good dynamic properties. These include transmission and V-belts, as well as non-tread tire components.

⌘⌘⌘

Table 5. Silica plus Lower Cost Fillers in Shoe Sole Compound

Compound			
SBR 1502		100	
Diethylene Glycol		3	
Stearic Acid		1	
Zinc Carbonate		1	
Sulfur		2	
MBTS		0.7	
MBT		0.4	
TMTM		0.3	
DPG		0.1	
HI-SIL®233 (ppt silica), phr	45	30	30
Hard Clay, phr	--	23	--
Calcium Carbonate (dry-ground), phr	--	--	23
t90% Cure, 150°C	12	7	3
Cure Time, 150°C	15	11	10
Hardness	66	65	65
300% Modulus, MPa	3.6	3.1	3.1
Tensile Strength, MPa	20.5	20.9	20.4
Elongation, %	735	770	720
Trouser Tear, kN/m	27	25	8
Pico Abrasion Index	68	67	48
Rebound, 23°C	48	43	49
Rebound, 100°C	54	50	60

PPG

Table 6 compares an aminosilane and mercaptosilane treated clay to an untreated hard clay in a synthetic polyisoprene compound. The improvements in modulus, tear strength, compression set and reduction in elongation with the treated clays are all indicative of the improved elastomer-clay bonding and consequent reduced polymer mobility. The ultimate tensile strength is not significantly affected since the rupture strength of the polymer-polymer linkages remains unchanged.

The high modulus with low hysteresis of the treated clays vs. carbon black, as seen in a natural rubber compound in Table 7, suggest that the silane mediated polymer-clay bond is stronger than the polymer-black bonds at the carbon black active sites. Deformation results in loss of fewer polymer-filler bonds with the treated clays, so that energy loss as heat is less compared to the carbon black. This stronger polymer-clay bond is also consistent with the higher modulus and abrasion resistance.

⌘⌘⌘

Table 6. - Comparison of Treated to Untreated Hard Clay

Compound	
Natsyn® 400 (Polyisoprene)	100
Hard Clay	75
Zinc Oxide	5
Stearic Acid	2
AgeRite® White	1
Sulfur	2.75
Amax®	1.25
Methyl Tuads®	0.2

Clay Treatment	**Aminosilane**[A]	**Mercaptosilane**[B]	**None**[C]
Hardness, Shore A	62	65	60
300% Modulus, psi	1630	1770	770
Tensile Strength, psi	2960	3240	3080
Elongation, %	480	480	580
Die C Tear, ppi	245	235	155
Compression Set B (22 hours, 70°C)	12.4	14.0	29.6
Mooney Scorch, min. (MS-3/121°C)	19	23	18
Mooney Viscosity (ML-4/100°C)	58	42	33

[A]Nulok® 321 [B]Nucap® 200 [C]Suprex®

Huber

Calcined clay and delaminated clay are also used in rubber. The former is used mostly in wire and cable coverings where it provides excellent dielectric and water resistance properties, although poor reinforcement. Delaminated clay is individual clay platelets instead of the platelet stacks characteristic of other clay products. As such it represents the most planar or anisometric form of clay available and imparts particularly high stiffness and low die swell. This is demonstrated in comparison to a hard clay in Table 8.

Calcium Carbonate – Calcium carbonates for rubber, often referred to as "whiting", fall into two general classifications. The first is wet or dry ground natural limestone, spanning average particle sizes of 5000 nm down to about 700 nm. The second is precipitated calcium carbonate (PCC) with fine and ultrafine products extending the average size range down to 40 nm.

Ground natural products show low anisometry (specific shape depends on grinding process), low surface area and low surface activity. They are widely used in rubber, nevertheless, because of their low cost, and because they can be used at very high loadings with little loss of compound softness,

⌘⌘⌘

elongation or resilience. This all follows from the relatively poor polymer-filler adhesion potential, as does poor abrasion and tear resistance. Dry-ground limestone is probably the least expensive compounding material available and more can be loaded into rubber than any other filler. Water-ground limestone is somewhat more expensive, but offers better uniformity and finer particles size.

Table 7. Comparison of Treated Hard Clay to Carbon Black

Compound	
Natural Rubber (No. 1 smoked sheets)	100
Zinc Oxide	5
Stearic Acid	4
Cumar® MH 2½	7.5
Sulfur	3
MBT	1

Clay Treatment	**Aminosilane**[A]	**Mercaptosilane**[B]	
Carbon Black	--	--	**N330**
Level, phr	75	75	45
Hardness, Shore A	56	55	57
300% Modulus, psi	1630	1360	1400
Tensile Strength, psi	3320	3320	3880
Elongation, %	520	560	590
Die C Tear, ppi	235	215	660
NBS Abrasion Index, %	83	79	204
Heat Buildup, °F (Firestone)	183	142	219
Mooney Scorch, min. (MS-3/121°C)	12	9	10
Mooney Viscosity (ML-4/100°C)	46	45	61

[A]Nulok® 321 [B]Nucap® 200

Huber

The much smaller size of precipitated calcium carbonates provide a corresponding increase in surface area. The ultrafine PCC products (<100 nm) can provide surface areas equivalent to the hard clays. Manipulation of manufacturing conditions allows the production of precipitated calcium carbonates of two distinct particle shapes. Precipitated calcites are essentially isometric particles, while precipitated aragonites are acicular. The natural aragonite form is found in oyster shells and pearls. Precipitated calcite is the form generally used in rubber compounding. Table 9 compares three typical PCC products to a fine ground natural calcium carbonate.

⌘⌘⌘

Tensile strength, tear strength, and modulus are all a function of particle size, while hardness is nearly unaffected.

Table 8. Comparison of Delaminated Clay to Hard Clay

Compound	
Natural Rubber (smoked sheets)	100
Clay	156
Zinc Oxide	5
Stearic Acid	4
Cumar® MH 2½	7.5
Sulfur	3
MBT	1

	Delaminated Clay[A]	Hard Clay[B]
Hardness, Shore A	74	70
300% Modulus, psi	2400	2200
Tensile Strength, psi	3080	3000
Elongation, %	360	390
NBS Abrasion Index, %	53	58
Olsen Stiffness (2" span, in-lbs)	27	17
Die Swell (% comparison)	100	150

[A]Polyfil® DL [B]Suprex®

Huber

Stearate-treated versions of both the ground and precipitated calcium carbonates are available. The surface coating controls moisture absorption, improves dispersion and promotes better elastomer-particle contact, and protects the calcium carbonate from decomposition by acidic ingredients or compounding reaction products. Table 10 compares a calcium stearate-coated ultra fine PCC with a hard clay in natural rubber. The PCC provides superior tensile strength, tear resistance, resilience, abrasion resistance, and flex-crack resistance. Since calcium carbonate offers no reactivity toward silanes, analogues to the silane treated clays and synthetic silicas do not exist. There is, however, an ultrafine PCC with a reactive surface coating of chemically bonded carboxylated polybutadiene polymer. With either sulfur or peroxide vulcanization, crosslinking occurs between the elastomer and the carboxylated polybutadiene coating. This provides the highest level of reinforcement among the PCC products, comparable to thermal blacks. Table 11 compares a coated PCC of this type to an MT black at equal volume loading in neoprene.

⌘⌘⌘

Table 9. Comparison of Natural and Precipitated Calcium Carbonates

Compound	
SBR 1502	100
Zinc Oxide	5
Stearic Acid	1
Cumar® MH 1½	15
Sulfur	2.75
Methyl Tuads®	0.35
Altax®	1.5

	Ultrafine PCC[A]	PCC[B]	Natural[C]
Avg Particle Size, um	0.07	0.7	3.0
Surface Area, m^2/g	28	8	4
Level, phr	150[1]	175[1]	175
Hardness	70	75	70
300% Modulus, psi	550	425	240
Tensile Strength, psi	1880	1050	650
Elongation, %	600	550	650
Crescent Tear, ppi	200	125	45

[1]Optimal loading
[A]Multifex®MM [B]Albaglos® [C]Vicron® 15-15

Specialty Minerals

Precipitated Silica – Precipitated silica is an amorphous form of silicon dioxide produced by reacting sodium silicate solution with either sulfuric acid or a mixture of carbon dioxide and hydrochloric acid. The discreet silica particles which initially form – the primary particles – fuse into aggregates, which in turn form loose agglomerates. The precipitate is filtered, washed of residual sodium sulfate or sodium chloride, dried, and milled. Like the carbon blacks, the precipitated silicas used in rubber are bought as agglomerates, which after milling exist in the elastomer as aggregates. The particles size reported for precipitated silica products is typically that of the primary particles (10 to 30 nm) rather than the aggregates (30 to 150 nm) which are the functional particles. Surface area (BET) is therefore frequently used in place of particle size to classify the various grades.

Precipitated silicas are also classified according to structure, which is a function of the manipulation of primary particle size and aggregate water pore volume during manufacture. For convenience, structure has been related to oil absorption as given in Table 12 although oil absorption is generally proportional to surface area. Fully reinforcing silicas for elastomers would be characterized as high structure or very high structure with surface areas at or above about 150 m^2/g and oil absorption at or above about 200ml/100g. A comparison of rubber properties as a function of

⌘⌘⌘

precipitated silica surface area in an SBR shoe sole compound is shown in Table 13. The 60 m^2/g product does not provide the reinforcement (tensile, tear, abrasion) obtained with the higher surface area silicas.

Table 10. Comparison of Treated and Untreated Ultrafine PCC

Compound	
Natural Rubber (smoked sheet)	100
PCC	56
Zinc Oxide	5
Stearic Acid	1
Sulfur	2.5
MBTS	0.55
DPG	0.20
Permanax®WSL	1

		Calcium Stearate PCC[A]	Hard Clay
Cure @ 141°C, min.		12.5	17.5
Hardness, IRHD		53	53
400% Modulus, MPa		8.1	9.3
Tensile Strength, MPa		29.4	21.6
Elongation, %		700	605
DuPont Abrasion Index		80	70
Mooney Scorch/120°C, min.		28	56
Resilience, %	78	72	
Crescent Tear Cold, N		242	44
Cescent Tear 100°C, N		123	39
Flex Crack Resist., kc (De Mattia)		63	17

[A]Winofil® S

Zeneca

Despite similarities in size and structure between precipitated silicas and carbon blacks, fundamental differences in surface activity exist. The silica surface is highly polar and hydrophilic and contains adsorbed water. The surface hydroxyl groups are acidic and tend to retard cure rate. The adsorbed moisture volatilizes at compounding temperatures. The tendency of silica surfaces to react with the zinc oxide acceleration activator was discussed previously. High molecular weight polyethylene glycols are the most common additives used with precipitated silicas to reduce their reactivity toward zinc oxide and organic accelerators, and to reduce their polarity.

The use of additives to make the surface of precipitated silica less hydrophilic and more "rubberphilic" facilitates incorporation, dispersion, and more intimate filler-elastomer contact during compounding. This provides an improvement in rubber physical properties, as would be expected from a high surface area, high oil absorption filler. However,

⌘⌘⌘

reinforcement comparable to that obtained with carbon black requires a polymer-filler bonding mechanism comparable to that provided at the carbon black active sites. With precipitated silicas, this comes by way of a reactive silane.

Table 11. Comparison of Treated Ultrafine PCC and Thermal Black

Compound	
Natural Rubber (smoked sheet)	100
PCC	56
Zinc Oxide	5
Stearic Acid	1
Sulfur	2.5
MBTS	0.55
DPG	0.20
Permanax® WSL	1

	Carboxylated Polybutadiene PCC[A]	Medium Thermal Black
Loading[1], phr	104	72
Hardness, IRHD	73	64
300% Modulus, MPa	5.3	7.5
Tensile Strength, MPa	10.1	9.4
Elongation, %	584	452
Tear Strength 23°C, N/mm	53.2	45.6
Tear Strength 100°C, N/mm	16.1	11.8
Compression Set, % (25% strain, 70°C, 22hr)	12	13

[1]Equal volume loading (40 volumes)
[A]Winofil® FX

Zeneca

The effectiveness of silane coupling of precipitated silica to elastomer is demonstrated in Table 14. This is a peroxide cured nitrile automotive hose compound for use in high temperature, oil resistant service. A semi-reinforcing silica (55 m^2/g surface area) untreated and with two different silanes, is compared to a reinforcing carbon black. The silica with no silane shows properties that are deficient in several areas. But with 1 phr of methacrylsilane added, the properties are better than the N550 control compound for abrasion resistance, heat aging resistance, and volume change in ASTM #3 Oil. The silica with 1 phr of mercaptosilane did not perform quite as well in this peroxide cured compound illustrating that the silane functional group should be matched to the cure system. A sulfur-cured nitrile would be a more appropriate application of the mercaptosilane treatment.

⌘⌘⌘

Table 12. Definition of Precipitated Silica Structure

Silica Structure Level	Oil Absorption (ml/100g)
VHS	> 200
HS	175-200
MS	125-175
LS	75-125
VLS	<75

Common practice has been to use the functional silanes as separate compounding ingredients. This necessitates careful control of addition sequence and procedures because the silane must react with the silica surface in preference to all other components in order to be optimally effective. Aggresive competitors for the silica surface are amines, glycols, resins, metal oxides, and other silica particles (agglomeration).

Table 13. Comparison of Silica Surface Area in Shoe Sole Compound

Compound	
SBR 1502	100
Precipitated Silica	45
Diethylene Glycol	3
Stearic Acid	1
Zinc Carbonate	1
Sulfur	2
MBTS	0.7
MBT	0.4
TMTM	0.3
DPG	0.1

Silica Surface Area (m^2/g)	**60**	**150**	**185**	**220**
Mooney Viscosity (ML4, 100°C)	52	105	120	129
t90% Cure, 150°C	2	12	13	14
Hardness	58	66	67	68
300% Modulus, MPa	4.0	3.6	3.0	3.1
Tensile Strength, MPa	6.9	20.5	19.9	18.6
Elongation, %	450	735	750	750
Trouser Tear, kN/m	3	27	36	49
Pico Abrasion Index	25	68	70	78
Rebound, 23°C	60.2	48.4	49.4	50.6
Rebound, 100°C	74.4	54.4	54.6	55.2

PPG

Precipitated silicas pretreated with silane are now available to simplify compounding. The pretreatment also makes the silica less hydrophilic, for easier dispersion and lower viscosity. Comparison of a commercial

⌘⌘⌘

mercaptosilane pretreated precipitated silica vs. an untreated control in a SBR compound is shown in Table 15.

Table 14. Effect of Silane Treatment On Semi-Reinforcing Precipitated Silica

Compound	
Nitrile Rubber (40 ACN, 65 ML4)	100
Filler	50
Magnesium Oxide	5
DOP Plasticizer	10
Stearic Acid	2
Polyethylene Glycol	1
TMQ Antioxidant	1.5
OPDA Antioxidant	1.5
Zinc Oxide	5
Sulfur	0.1
Sartomer® SR-350	5
DiCup® 40C	4

Filler	N550	Silica[A]	Silica[A]	Silica[A]
Methacrylsilane[B]	--	--	1	--
Mercaptosilane[C]	--	--	--	1
t90% Cure, 165°C	11.8	10.5	10.5	11.9
Hardness	70	69	72	68
100% Modulus, MPa	5.8	3.4	5.0	4.5
Tensile Strength, MPa	21.2	14.3	19.5	14.2
Elongation, %	270	385	267	282
Die C Tear, kN/m	46.5	37.1	39.9	39.2
Pico Abrasion Index	145	68	167	79
Compression Set (72hr, 100°C)	13.0	14.5	12.5	12.5
Aged 14 days @ 135°C				
Hardness	89	86	85	85
Tensile	16.3	17.7	18.0	13.8
Elongation	42	78	70	59
Volume Change				
#3 Oil, 72 hrs, 150°C	6.2	3.9	3.8	3.6
Fuel B, 48 hrs, 23°C	28.6	30.6	29.3	29.8

[A]HiSil® 532EP (55 m^2/g surface area) [B]A174 [C]A189

PPG

Miscellaneous Fillers – Although kaolin clay, calcium carbonate, and precipitated silica account for most of the non-black fillers used in rubber today, there are a number of other fillers routinely used for their low cost or unique funtionality.

⌘⌘⌘

Table 15. Comparison of Pretreated to Untreated Precipitated Silica

Compound	
SBR 1502	100
HPPD	2
Stearic Acid	2
HA Oil	3
Aromatic Resin	10
Zinc Oxide	3
Polyethylene Glycol	1.5
Sulfur	2
MOR	1.5
TMTM	0.6

Silica	Mercaptosilane Pretreated[A]	Untreated[B]
Loading, phr	51.5[1]	50
Mooney Viscosity, ML100	65	95
t90% Cure, 150°C	9	15
Hardness	70	73
300% Modulus, psi	1200	450
Tensile Strength, psi	3400	2700
Elongation, %	510	650
Pico Abrasion Index	85	60
Flexometer Heat Buildup, °C	25	45
Rebound %, 23°C	53	49
Rebound %, 100°C	75	61

[1]Equals 50 phr silica
[A]Ciptane® [B]HiSil® 210

PPG

Talc – Although widely used as a reinforcing filler in plastics, relatively little talc is used for this purpose in rubber. Platy talcs are white, hydrophobic, and alkaline, with greater anisometry than kaolin clay. They are readily treated with silanes and other coupling agents. However, they tend to be too large in particle size for effective elastomer reinforcement. Micronized talcs with median particle size of 1 to 2 microns and essentially all particles <10 microns are available and are used, although they compete with the generally less expensive clays.

Barite – Barite, ground natural barium sulfate, is used in acid resistant compounds because of its inertness, and as a high gravity filler where weight is desired. It has little effect on cure, hardness, stiffness, or aging. Precipitated barium sulfate, also known as blanc fixe, is available in fine enough particle size to be semi-reinforcing. It provides the same softness and resilience as barite but better tensile strength and tear resistance.

⌘⌘⌘

Diatomite – As silica, diatomite is chemically inert, but its high adsorptive capacity for accelerators can affect cure. It imparts stiffness, hardness and low die swell. Diatomite is used as a filler in silicone rubber, and because of its adsorptive capacity, as a process aid in high oil rubber compounds.

Mica – Because of its platy nature and high aspect ratio, mica is occasionally used as a filler or semi-reinforcer, depending upon particle size.

Fumed Silica – Fumed silica is generally finer in primary particle size and higher in surace area than precipitated silica. As a reinforcing agent, it provides an aggregate structure similar to that of carbon black and precipitated silica. Due to its pyrogenic manufacture, it has lower moisture content and fewer surface hydroxyls than precipitated silica. Its high price confines it to premium applications, particulary silicone rubbers. Surface treated grades are available.

Precipitated Silicates – These products are coarser and less structured than the precipitated silicas and, as such, are only semi-reinforcing, but can be used at high loadings. Precipitated calcium silicate and precipitated sodium aluminum silicate are the most common alternatives.

ANTIDEGRADANTS

The proper choice of elastomer, cure system and filler will ensure that the desired properties are obtained from a rubber product. Antidegradants are used to prevent these properties from changing during service. The primary degradant effects are oxidation and ozone attack.

There are certain general criteria, beyond cost, guiding the selection of antioxidants and antiozonants. The volatility of the antidegradant determines its potential for loss during compounding and service, which in turn affects the rubber product's maximum service life. The solubility of the antidegradant is important, since it will ideally be soluble in the rubber compound, but insoluble (unextractable) in any liquids the product will routinely contact. Chemical stability is essential to maximum product service life. This includes stability to heat, light, oxygen and other compounding additives. Volatility, solubility, and stability are all involved in determining whether or not the antidegradant is considered discoloring or staining. Discoloration refers to the color imparted to the rubber itself, and is an issue only for non-black stocks. Staining is discoloration imparted to another surface in contact with or near the rubber article.

Antioxidants

Most elastomers are subject to oxidation, although unsaturated polymers oxidize more readily than saturated polymers. Oxidation is a cyclic free radical chain process that proceeds by two mechanisms. Chain scission of the polymer backbone causes softening and weakening. Radical-induced crosslinking causes hardening and embrittlement. In most cases, both mechanisms occur, and the one which predominates determines compound properties. Chain scission, for example, is the primary mechanism in natural rubber while crosslinking is predominant with SBR. Heat, UV light, humidity, and pro-oxidant metal ions (eg. copper, iron, manganese, cobalt) will each accelerate oxidation.

Oxidation is initiated by the formation of carbon radicals as a result of chain cleavage. The carbon radicals add oxygen to form peroxy radicals. These abstract hydrogens from the rubber to form hydroperoxides and generate the additional carbon radicals which continue the cycle. The hydroperoxides decompose into two oxy radicals (R-O· + ·OH) which just accelerate the oxidative degradation.

Primary antioxidants scavenge radicals before they can react with the elastomer. They can be generally classed as secondary amines and substituted phenols with very reactive hydrogens. More specifically, the primary antioxidants can be categorized and differentiated as in Figure 8. The aromatic amines are the most effective primary antioxidants, but are also discoloring. Of these, the diphenylamines and polymeric dihydroquinolines are the least discoloring while the p-phenylene diamines (PPDs) are highly discoloring. The PPDs, nevertheless, are the most commonly used antidegradants because they are also antiflex agents, metal ion deactivators, and antiozonants. The phenolic antioxidants are usually used in non-black compounds where the amine antioxidants cannot be used. The phenolics can produce colored reaction products, but the discoloration is significantly less than produced by the amines. The hindered bisphenols are the most effective, and the most expensive. Certain mercaptoimidazoles and dithiopropionates are used as synergists with the primary antioxidants.

Secondary antioxidants, shown in Figure 9, destroy the hydroperoxides before they can decompose into radicals. The alkyl and alkylaryl phosphites are the preferred secondary antioxidants today. They are used only to protect the raw elastomer in storage and during manufacture, since they are destroyed during vulcanization. Combinations of phenols and phosphites are widely used synergistically to protect synthetic rubbers. The dithiocarbamates are no longer generally used as secondary antioxidants because they are very active accelerators. They are used in specific cases, however, such as zinc dibutyldithiocarbamate in butyl rubber production, and nickel dibutyldithiocarbamate for oxidation and static ozone protection in nitrile rubber.

Protection against oxidation is not a linear function of antioxidant concentration. Laboratory aging studies will normally indicate an optimum use level, beyond which there is little or no benefit. Amine antioxidants are generally used at 1 to 2 phr and the phenolics at 0.5 to 2 phr.

⌘⌘⌘

Figure 8. Primary Antioxidants

STAINING / DISCOLORING

SUBSTITUTED PARAPHENYLENEDIAMINES

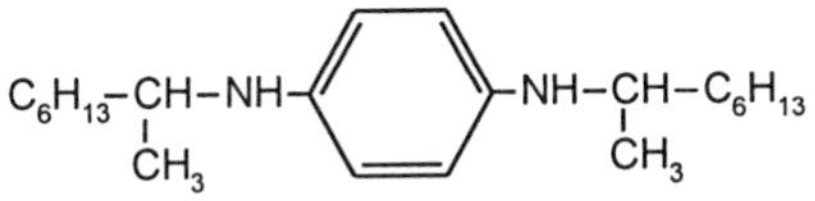

N,N'-Bis(1-methylheptyl)-p-phenylenediamine

POLYMERIZED DIHYDROQUINOLINES

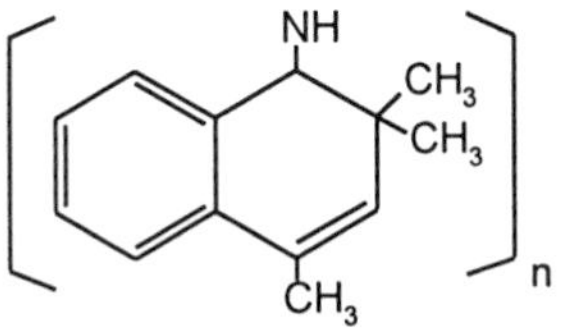

Poly-2,2,4-trimethyl-1,2-dihydroquinoline

DIPHENYLAMINE DERIVATIVES

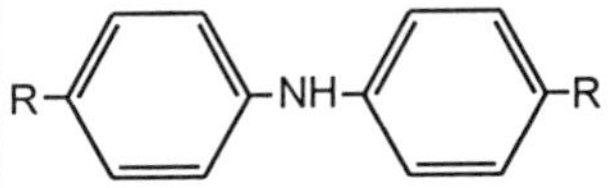

Alkylated diphenylamine

NON-STAINING / NON-DISCOLORING

HINDERED BISPHENOLS

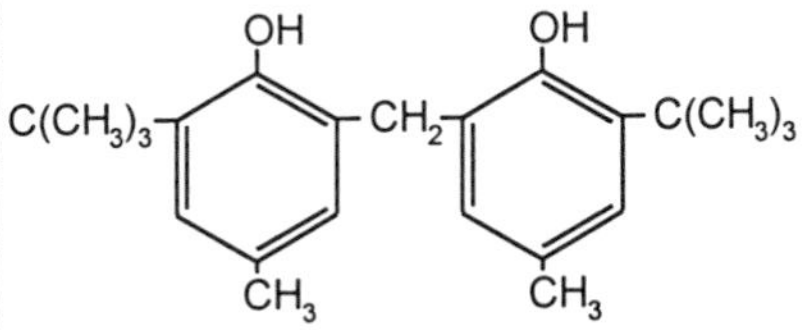

2,2'-Methylenebis(4-methyl-6-*t*-butylphenol)

SUBSTITUTED HYDROQUINONES

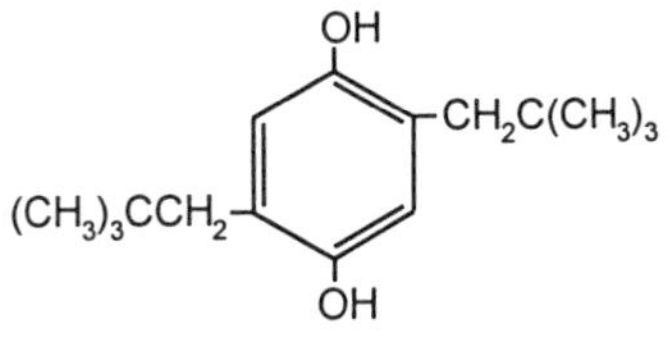

2,5-Di-t-amylhydroquinone

HINDERED PHENOLS

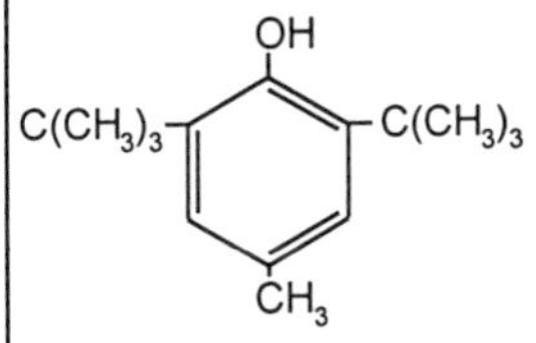

2,4-Di-*t*-butyl-4-methylphenol (BHT)

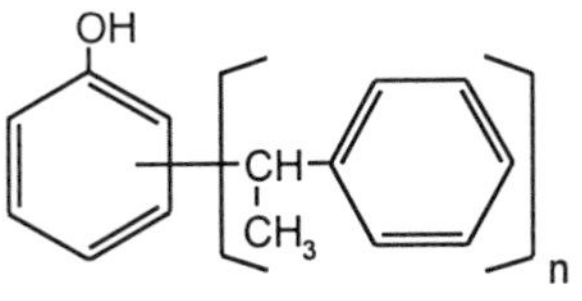

Styrenated phenols

⌘⌘⌘

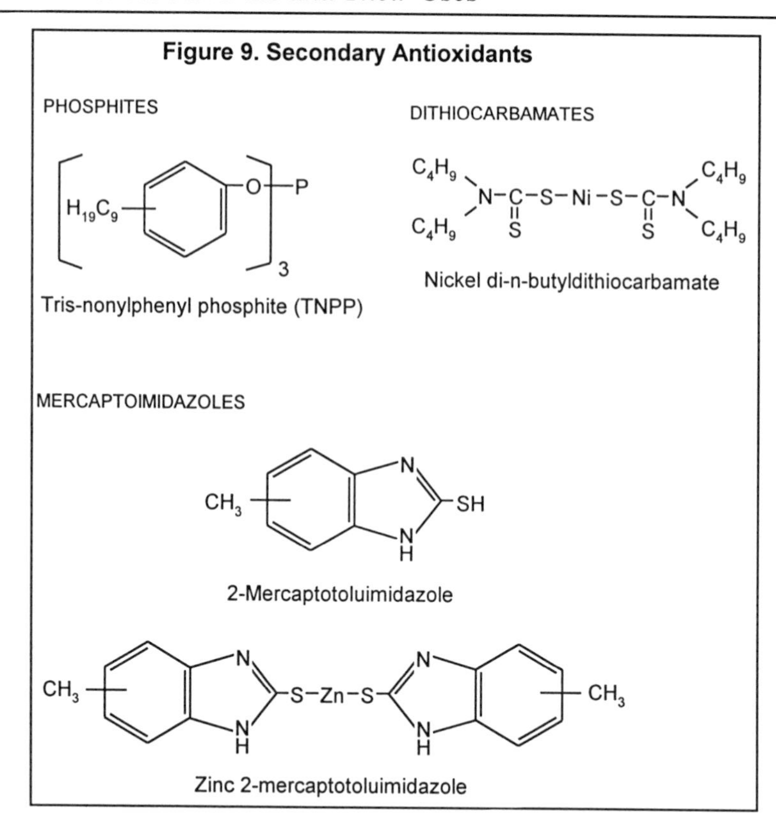

Antiozonants

Ozone reacts with and cleaves the double bonds of elastomers. Under strain, this causes cracks to form and propagate, leading ultimately to product failure. Elastomers are subject to ozone attack in proportion to their level of unsaturation. Highly unsaturated rubbers (natural, SBR) are readily cracked. Rubbers with low unsaturation (butyl) or a deactivated double bond (neoprene) are moderately ozone resistant. Saturated elastomers (EPDM) are totally resistant.

The requirements for static ozone resistance vs. dynamic ozone resistance are very different, so that choice of antiozonant depends greatly on the expected service of the rubber product. Static protection is provided by petroleum waxes, usually paraffin and/or microcrystalline waxes. The waxes work by blooming to the rubber surface to form a physical barrier to ozone attack. The choice of wax or wax blend is based upon migration temperature where mobility and solubility of the wax in the rubber are balanced so that sufficient bloom occurs for optimum protection. Because the wax film is inextensible, it will rupture under deformation and expose the elastomer. Waxes protect only under static conditions.

⌘⌘⌘

The substituted PPDs, described above, are used as chemical antiozonants when service conditions involve continuous dynamic deformation. For a particular rubber, the optimum PPD is one which is soluble in the elastomer, and which will diffuse to the product surface as needed. At the surface, a protective film is formed by the reaction between ozone and the PPD. The ability of the PPD to migrate to the surface as this film is degraded or otherwise broken, and thus react with ozone to repair it, is essential to long term protection. The alkylaryl PPDs are the most widely used for their protection against both oxidation and ozone.

Waxes and PPDs are commonly used together for applications, such as tires, where both static and dynamic protection is required. Waxes are typically used at 1 to 2 phr and PPDs at 1.5 to 3 phr.

When the staining and discoloration of PPDs cannot be tolerated (e.g., white sidewalls) ozone protection is obtained by adding ozone-resistant polymers. These work by relieving stress and inhibitting crack propagation; 30 phr or more of polymer is usually required.

OTHER ADDITIVES

While the foregoing components of a rubber compound determine its primary service characteristics and service life, there are a number of other ingredients used under specific circumstances to modify the processing, appearance, or performance of the rubber product.

Processing Aids – Oils are added during compounding and in some cases are incorporated during elastomer manufacture. They primarily serve a plasticizing function, although at high levels they can also reduce compound cost. Certain elastomers, most notably EPDM and SBR to a lesser extent, can incorporate a relatively large amount of oil into their polymer matrix without significantly affecting the physical properties obtained in the finished rubber product. A very high molecular weight EPDM, for example, that would be nearly impossible to process in a rubber mill will become workable by the incorporation of extender oil during manufacture. Petroleum oils, vegetable oils, and synthetics, such as esters and high molecular weight sulfonic acids, are often added during compounding to reduce elastomer viscosity, promote mold flow and release, improve milling and extrusion, facilitate incorporation of fillers, improve low temperature flexibility, and provide softer vulcanizates.

The petroleum oils are used primarily with nonpolar elastomers and are either naphthenic, paraffinic, or aromatic, depending upon elastomer compatibility, volatility, and expected product service conditions. Vegetable oils are typically from castor or tall oils. Vulcanized vegetable oil products have the unique ability to promote flow under shear but to resist high temperature flow. The synthetic ester plasticizers, such as dioctyl phthalate and dibutyl phthalate are used with polar elastomers.

Although processing aids are used principally to aid processing, some will affect rubber properties to some extent and are utilized for this secondary function as well. For example, a mineral oil plasticizer may also be used to lower modulus, increase resilience, improve elongation, or reduce hysteresis.

Tackifiers – Elastomers with little or no natural tack can be improved by the addition of resins or resin oils (natural abietic acid derivatives), polyindene (coumarone-indene) resins, polyterpene resins and phenol-formaldehyde resins.

Flame Retardants – Most hydrocarbon rubbers will burn, so flame retardants are used if expected service conditions include the possibility of exposure to fire. Fillers such as alumina trihydrate, magnesium hydroxide, and zinc borate are used because they emit inert vapor at high temperatures. Most often, primary flame resistance is suppled by chlorinated paraffins or brominated aromatic resins in combination with antimony trioxide.

Odorants – Small amounts of essential oils are added to rubber compounds to mask characteristic or objectional odors. The most common applications are wearing apparel and drug sundries.

Blowing Agents – For sponge and microporous rubbers, cellular structures are obtained in highly plasticized compounds in either of two ways. Sodium bicarbonate and oleic acid are used to react and liberate carbon dioxide at curing temperatures. Complex azo compounds, which decompose to release nitrogen at curing temperatures, are used to provide a more uniform, closed cell structure than the bicarbonate.

Colors – Non-black products are colored by mineral pigments and organic dyes. White articles usually contain titanium dioxide, alone or supplemented in whole or part by zinc oxide, zinc sulfide, lithopone (precipitated zinc sulfide/barium sulfate), or talc. A trace of ultramarine blue is sometimes used to correct off-white tints. Iron oxides are used for colors other than white.

Abrasives – Rubber products requiring abrasive properties, such as ink erasers and polishing wheels, use abrasive fillers like ground quartz, pumice, or carborundum.

⌘⌘⌘

RUBBER PROCESSING

The production of rubber goods involves two basic operations – compounding and curing. Rubber polymers behave as viscoelastic fluids when sheared at elevated temperatures. This enables the incorporation of the various fillers and chemical additives in the process of compounding.

Mastication

The first step in rubber compounding is mastication or polymer "breakdown". This is essentially development of the polymer's viscoelasticity to make it receptive to the additives. Most synthetic elastomers are produced with the uniformity in chemistry, viscosity and stability that minimizes or precludes the need for mastication. This step is, nevertheless, usually necessary for compounds containing a blend of polymers to provide a uniform mixture prior to further compounding.

Natural rubber is characteristically widely variable in viscosity from lot to lot, so that several lots are usually blended and masticated to control viscosity and physical properties of the compound. The breakdown of natural rubber promotes viscoelasticity by extending or disentangling the polymer chains or by severing the chains. Breakdown cycles for both natural and synthetic rubber are often facilitated by the inclusion of chemical plasticizers.

Polymer breakdown and subsequent compounding was historically accomplished on open roll mills, although today these operations are almost always performed in an internal mixer. The open mill consists of two metal rolls which are jacketed for temperature control. These rolls turn toward each other at fixed separations and often at different speeds. This provides high shear mixing forces. Banbury® Mixers are the most commonly used internal mixers today. They consist of two rotor blades turning toward each other in an enclosed metal cavity. The cavity is fed from a loading chute through which the rubber and in later steps the fillers and compounding chemicals are added. Mixing shear, and consequent mixing time, is determined by rotor shape, size, and speed. A large Banbury can produce 500 kg of finished compound in a matter of minutes.

Masterbatching

As a compounding process, masterbatching means incorporation of the other compounding ingredients into the rubber, except for the curing system. By omitting the cure system, the masterbatch may be intensively mixed without fear that the high temperatures generated will cause premature vulcanization. The objectives of masterbatching are homogeneous blending of polymer and chemical additives, the best possible deagglomeration and dispersion of fillers, and the development of the proper final viscosity. This viscosity, in turn, is a function of the homogeneity achieved in the mixing of the polymer and additives. Because considerable heat is generated during milling, the masterbatch operation must often be terminated before its objectives are attained. This is to avoid heat degradation. In such cases, the compound is remilled before addition of the curing system. Masterbatches are typically prepared in a Banbury mixer, after which they are ejected and cooled.

Separate from the compounding process, masterbatches are often prepared of individual additives that are difficult to disperse during compounding, zinc oxide for example. A concentrated dispersion of the additives in polymer is preformed. This can then be added to the rubber compound for faster incorporation of the additive because it is essentially predispersed.

Remilling

If the masterbatch shows inadequate dispersion or is too high in viscosity for finish mixing, it is remilled to prepare it for the next step. Remilling is most commonly required with high carbon black, low oil natural rubber tire compounds.

Finish Mixing

The finish mixing step completes the rubber compound by the incorporation of the curing system. This is usually done in an internal mixer where mixing rate and time are balanced to ensure homogeneity without generating temperatures which will cause premature curing. While internal batch mixers like the Banbury are still the standard of rubber compounding, continuous mixers are gaining popularity for their potential to significantly reduce operating costs and to yield exceptionally uniform product quality. Because continuous mixers require continuous accurate weighing and feeding of all components, preparation of ingredients to accomodate these handling requirements is very important. The rubber or masterbatch, for instance, is used in pellet or granular form.

Extruding

Screw-type extruders are used both in the compounding operations and to form articles for vulcanization. In masterbatching, remilling, and finish mixing, the hot batch of rubber generally goes directly to an extruder. The extruder screw will force the rubber through a die to form pellets, or through a pair of rolls to form a slab or sheet. In either case, the extrusion operation allows the compound to cool down while it is turned into a more easily handled form.

Heated mixing extruders can be fitted to batch mixers to improve compound uniformity and reduce viscosity, and they can warm up material for calender feed. Rubber compound in strip or pellet form is fed to the extruder where it is heated and extruded in slug or rope form into the nip of a rubber calender. This type of extruder can also be used to take cold finished compound in strip, pellet, or slab form, heat it, blend it, and extrude it in the desired shape. This can be a finished shape, such as a rubber hose, which can be continuously vulcanized right after the extruder. This can also be a component for fabrication with other parts into a more complex item such as a tire. Where the shape must be very precise, the compound may be first warmed on a mill or warm-up extruder and then fed to a shaping extruder.

Calendering

Rubber calenders consist of at least three rolls which can be adjusted for gap, speed and temperature. They are used to form rubber sheeting to required lengths and thickness for subsequent building operations. They are also used for frictioning or skim coating fabrics. Cord and wire are coated in this way for making plies used in the construction of tires and conveyor belts.

Vulcanization

After the rubber compound has been processed and formed, it is vulcanized. This process involves three stages: induction, curing, and reversion or overcure. The induction period is the time at vulcanization temperature during which no measurable crosslinking occurs. It determines the safety margin of the compound against "scorch" during the processing steps preceeding crosslinking. Scorch is premature vulcanization that can occur due to the effects of heat and time. Because these effects are cumulative, the time until scorch will slowly decrease with each processing step. Scorching, or precuring, results in a tough and unworkable batch, which must be scrapped. The time to scorch is dictated by the processing and the additives used, so that a well designed compound will have a scorch time slightly longer than the equivalent of its maximum anticipated cumulative heat history.

The amount of time the compound must be cured, the cure time, is determined in part by the "rate of cure". This is the rate at which crosslinking and the consequent development of stiffness (modulus) occurs. Cure time is also a function of the desired "state of cure". As vulcanization (crosslinking) proceeds, the various properties developed by vulcanization are not optimized simultaneously. At any given time during vulcanization, the state of cure is a measure of the development of these properties. Cure time is the time required for the compound to reach a state of cure where the desired balance of properties has been attained.

⌘⌘⌘

When a compound is cured beyond the point where its balance of properties has been optimized it becomes overcured. For most elastomers, overcure means the compound becomes harder, weaker and less elastic. With other elastomers, particularly most natural rubber compounds, overcure results in reversion. The compound softens, becoming less elastic and more plastic.

Preparation – Rubber stocks about to be assembled for vulcanization may be in a number of forms, such as calendered strips or sheets, plied up slabs, extruded tubes or partially combined with fabrics by frictioning or skimming operations. Tires, tubes, hose, belts, footwear, rolls and many other articles are built to proximate finished form before vulcanization. The various stocks that are combined to form such articles must be dimensionally stable, free from bloom, and possess the required building tack, all of which depend upon proper compounding and processing. The building tack in SBR and some heavily loaded rubber compounds can be improved to some extent by the use of pine tar, rosin esters and various coal tar or petroleum derivatives. The use of cements or solvents is also practiced in some cases to assure against ply separation or splice failure during cure.

Vulcanization or curing may be done by a number of methods, depending on the compound in process, and the size, shape and overall structure of the finished article.

Press Curing – Press curing includes the molding of articles by compression, transfer, or injection methods. Blocked-in articles which are cured directly between press platens also come under this classification, as well as unblocked slabs. The heat source is generally saturated steam. Electrically heated platens are also used in some installations. Radio frequency waves have been suggested as a means of curing or for pre-heating blanks to reduce the press cure time.

Compounds for press curing must be carefully designed to flow properly without scorching before the desired shape is reached, and to cure rapidly so that a maximum turnover is obtained. They must be easily removed from the molds after cure and exhibit attractive surfaces in the finished form. The flow characteristics depend on the plasticity of the uncured stock as well as selection of the proper acceleration to permit complete filing of the cavity before the cure begins.

Flat Belting and Slab Cures – In articles of this type, the press platens and side strips control the dimensions of the finished article. Belting structures of both transmission and conveyor types are built to their approximate finished shapes before curing. The fabrics they contain have been dried and stretched

⌘⌘⌘

before frictioning and skim coating, and an additional stretch is applied to each section in the press prior to and during cure.

Slabs for soling, tile, or matting must possess suitable flow characteristics and they must be properly accelerated to permit "cycling" in multistage presses without developing porosity as the pressure is released and reapplied during cure. Slabs for such articles are built to their approximate finished thickness on the calender or sheeting mill, and heel blanks or similar items may be cut by dies from strips taken directly from a warm-up mill.

Open Steam Curing – Open steam curing is used in the production of hose, wire and cable, and other extruded articles such as tubing or channel stripping. With this method, the article may be in direct contact with the steam, wrapped with fabric tape, encased in an extruded covering of lead, or supported by soapstone in a shielded pan. Equipment may consist of a jacketed autoclave or a closed chamber in which the articles are placed and steam is introduced.

Another method of open steam curing is used in the production of insulated wire and cable. This method, known as the CV (continuous vulcanization) process, utilizes jacketed tubes which may be 60 meters (200 feet) in length and operate under internal steam pressure upward of 1.4 MPa (200 psi) as compared to the 200 to 550 kPa (30 to 80 psi) range of the conventional steam vulcanizer.

In operating a closed-chamber steam vulcanizer, the curing cycle consists of a rise to a predetermined pressure, a holding period at the required curing pressure, and a blowdown to atmospheric pressure. In the CV process, extruded wire or cable covering goes directly from a tuber into steam at curing pressure and, with the latest equipment design, passes through a water or condensation seal prior to emerging from the tube at atmospheric pressure.

Dry Heat Curing – Articles such as coated fabrics and footwear may be cured in heated dry air. Coated fabrics are generally festooned in heaters and cured at atmospheric pressure, although by the use of certain dithiocarbamate accelerators, curing in a roll as it comes from the calender can be accomplished at room or slightly elevated temperatures. Footwear heaters are generally operated at 200 kPa (30 psi) air pressure, and ammonia gas is sometimes used to produce a glossy surface on the finished article.

Continuous Curing of Extrusions – The direct, continuous curing of extrusions as the rubber exits the extruder is growing in popularity. In addition to minimizing handling and labor costs, very fast line speeds can be obtained. These factors make for economical extruded articles when a particular item can be run for long periods without changing compounds or profiles (dies).

Hot air can be used to provide the heat for vulcanization. This method gives a clean, attractive surface and has a wide processing latitude to easily make a variety of shapes. The negative aspects are the poor heat transfer of air and the possibility of oxidation during cure. Shearing head extruders can be used to heat the compound to vulcanization temperature before it exits the die, thus overcoming some of the problems with poor heat transfer of air. Fludized beds improve heat transfer over air alone. High velocity air is pumped upward in the curing trough to fluidize small glass beads (Ballotini). Beads adhering to the extrudate are normally removed by spray or brushed upon exit from the fluidized bed. Complex profiles can make complete removal difficult.

Liquid Cure Medium (LCM) uses a eutectic mixture of salts to heat the extrudate. The molten salt bath usually has a metal conveyor to keep the rubber compound submersed. This method has good heat transfer to the rubber. However, cleaning salt off the extrudate can be difficult and satisfactory disposal of dilute salt rinse is becoming increasingly difficult. This method is advantageous for articles with the low compression set obtained by using peroxides. Sulfur can be used as a coagent to minimize surface tack. Deformation problems versus other atmospheric methods are sometimes minimized.

Microwave with Hot Air Soak is another current method. Particularly effective for thick section extruders, it also retains the clean extrudate surface of the straight hot air method, but with significantly faster heating. Nevertheless, microwave methods put additional constraints on compounding. In a microwave field, heating occurs simultaneously throughout the rubber. Polar rubbers such as neoprene and nitrile will usually heat very rapidly in the microwave unit. Hydrocarbon rubbers heat slower unless they contain 20 to 40 volume percent carbon black. Localized hot spots can develop if the black loading is too high or if dispersion is poor. High zinc oxide levels, to 20 phr, can also improve microwave heating.

PHYSICAL TESTING OF RUBBER

Most testing of rubber and rubber compounds is conducted to determine processing characteristics or to measure physical properties after vulcanization. Processability of a rubber compound is dependent on the compound's viscosity and elasticity. Generally, the physical properties of vulcanized rubber compounds are measured by static and dynamic mechanical tests designed to simulate the mechanical conditions of finished

⌘⌘⌘

rubber products. Following are the test procedures and instruments most widely used to measure and evaluate processability and finished properties.

Processability

The fabrication of rubber products generally involve the mixing and processing of unvulcanized compounds through complex equipment. Tests to measure the processability of unvulcanized rubber are chiefly concerned with rheological properties. That is, the response of the rubber compound to the forces and temperatures imposed on it during the operations of mixing, extrusion, calendering, and curing. Generally, these tests are used for control purposes in factory operations to ensure that subsequent processing and curing steps are carried out uniformly.

The earliest and still one of the most popular rheological instruments is the shearing disk viscometer. This is generally referred to as the Mooney Viscometer named after Melvin Mooney (1893–1968) of U.S. Rubber.

Mooney Viscosity, Mooney Scorch, Cure Index – The Mooney Viscometer is widely used throughout the rubber industry as a standard instrument for determining the relative viscosity of rubber and rubber-like materials in the raw or compounded (but uncured) state. It is also used for determining the curing characteristics of vulcanizable compounds. The viscometer measures the force, or torque, required to rotate a metal disk, or rotor, within a shallow cylindrical cavity filled with rubber compound (ASTM D-1646).

A typical curve from a Mooney Viscometer test is shown in Figure 10. The terms which are used throughout the rubber industry to describe stock flow and cure characteristics are defined on it, and discussed further below. Mooney Viscosity is reported in Mooney units, which are related to the torque required to rotate the disk (rotor). The value reported is not a true viscosity, but it does indicate relative viscosity, or resistance to flow, of compounds measured under the same conditions. The test temperature and the running time when the viscosity measurement is taken must be defined when comparing Mooney Viscosity values of different compounds. A common test temperature is 100°C (212°F). The running time should be long enough so that the reported viscosity is the minimum viscosity (MV) on the curve. Four minutes is a common running time. Some rubber processors use different times and temperatures.

Mooney Scorch Time is the total running time to reach a viscosity of five or ten units above the minimum (t_5 or t_{10}). It is used as a rough measure of the processable life of a compound under normal processing conditions – the greater the scorch time, the longer the processable life. A common test temperature for Mooney Scorch measurements is 121°C (250°F), but some processors may use different temperatures because of particular factory experience.

⌘⌘⌘

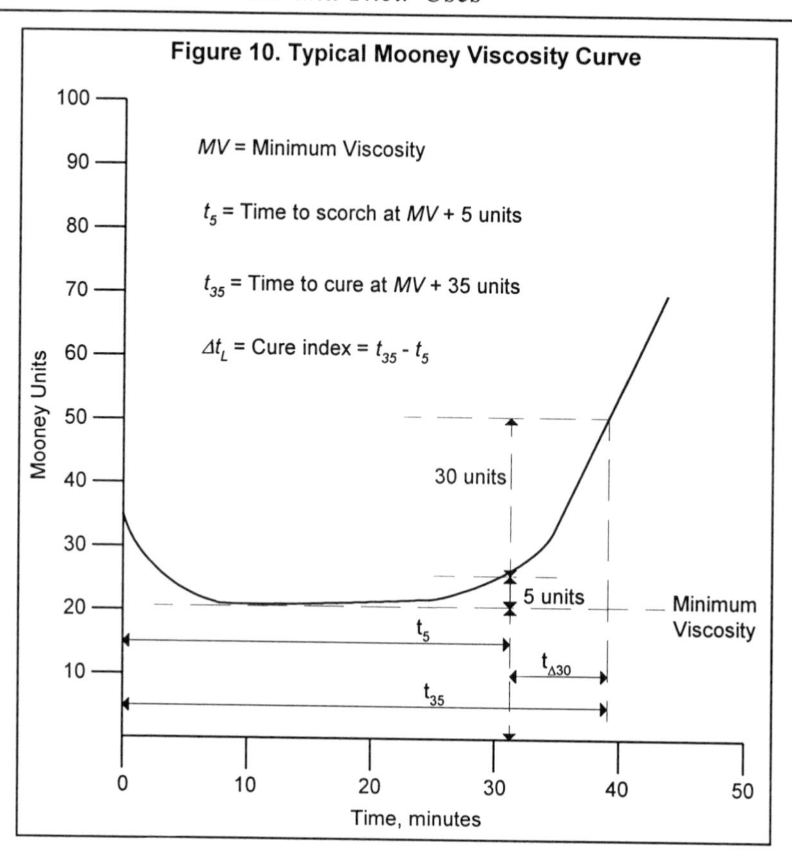

Mooney Cure Time is defined as the total running time to reach a viscosity of 35 units above the minimum (t_{35}). It is used to estimate curing time at the test temperature.

Cure Index, t_L, is the difference between the Cure Time and the Scorch Time (t_{35}-t_5). It is used as a measure of cure rate of the compound – the smaller the Cure Index, the faster the cure rate.

The Mooney Viscometer test is one of the oldest processability tests in the rubber industry. Many processors now prefer the oscillating disk rheometer (ODR) for measuring cure time and cure rate, but the Mooney Viscosmeter is usually preferred for viscosity and scorch time measurements.

Oscillating Disk Rheometer – The oscillating disk rheometer (ODR) measures the complete curing characteristics of an elastomer compound, from a green (uncured) stock to a fully-cured vulcanizate, at a specified temperature (ASTM D-2084). Data from the ODR are often used in conjunction with Mooney viscosity and scorch to characterize the

⌘⌘⌘

processing and curing behavior of elastomer compounds. Mooney Viscometer measurements are usually made at processing temperatures, while ODR measurements are usually made at curing temperatures.

The ODR measures the force (torque) required to oscillate a biconical disk (rotor) back and forth within a shallow cavity filled with rubber compound. As the compound cures, its viscosity increases and more torque is needed to move the rotor. The instrument plots a continuous curve of torque versus time. This curve is often called an ODR trace. From an ODR trace, the rubber compounder can estimate stock viscosity, scorch time, cure rate and cure state. Important test variables are the arc through which the disc oscillates, the speed of oscillation, and the test temperature. The usual arc is 1° (actually, this means 1° on either side of the starting point, or a total swing of 2°); some laboratories use 3° or 5°. Normal oscillation rate is 100 cycles per minute. Test temperature should be close to the expected cure temperature.

A typical ODR trace is shown in Figure 11. The measurements which can be made from this trace and the terms used to describe them are:

Area A – This gives an indication of compound viscosity.

Area B – This indicates the rate of cure of the compound.

Area C – This indicates the state of cure of the vulcanizate.

M_L , *Minimum torque* – A measure of the viscosity of the uncured compound.

M_H, *Maximum torque* – A measure of cure state. With some compounds, maximum torque can be related to vulcanizate modulus and hardness.

t_{s1} – Time for torque to increase 1 dn.m (0.1N.m) or 1 lbf-in above M_L – a measure of scorch time or processing safety. Some laboratories use t_{s2} (i.e., time for torque to increase 2 dN.m or 2 1lbf-in above ML) instead of t_{s1}.

t_{50}, t_{90} – Time to reach 50% or 90% of maximum torque development, calculated as time to 0.5 M_H or 0.9 M_H – a measure of cure rate or an estimate of cure time at the test temperature. The shorter the time, the faster the cure rate.

Some rubber compounders use other measures of cure rate and cure time which can be derived from the ODR trace. For example:

⌘⌘⌘

t'_{50}, t'_{90} – Time for torque to reach $M_L + 0.5\ (M_H - M_L)$ or $M_L + 0.9\ (M_H - M_L)$.

Cure Rate Index – The slope of the linear portion of the rising curve between t_{s1} and t_{90}.

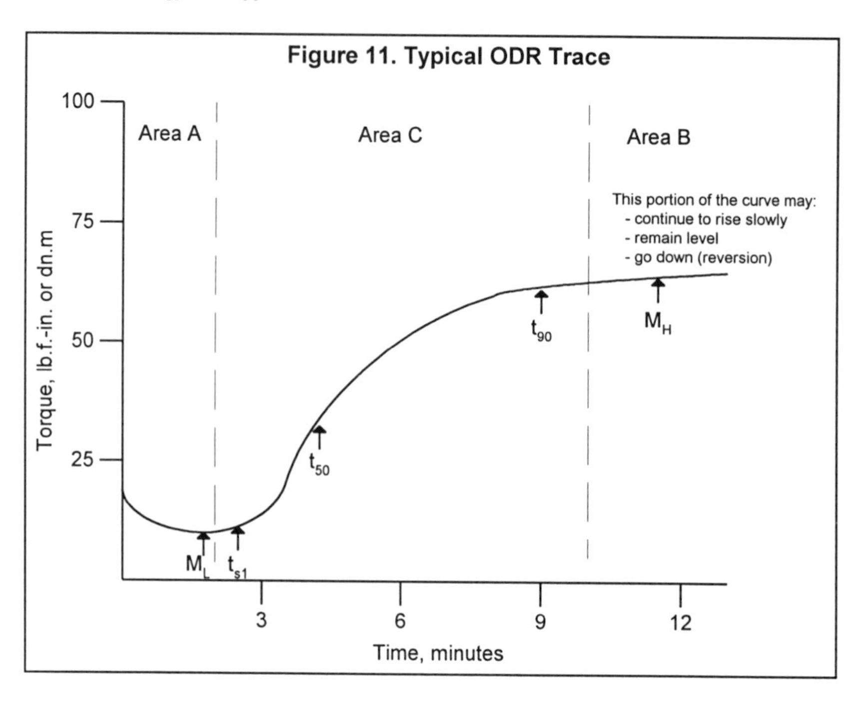

Vulcanizate Tests

By far the most frequently measured rubber properties are hardness, tensile strength, and elongation. These are often referred to as the physical properties of a vulcanizate. These are supplemented by additional tests designed to be more or less predictive of actual service performance and service life.

Hardness – Hardness, as applied to rubber products, is the relative resistance of the surface to indentation under specified conditions. Hardness of rubber is usually measured with a small spring-loaded hardness gauge known as a durometer (ASTM D-2240). The durometer may be handheld or mounted on a stand. The measurement is made by pressing the indentor against the sample and reading the scale, which is calibrated in arbitrary units ranging from 0 (soft) to 100 (hard). A Type A durometer is used for most soft rubber products; there is also a Type D durometer for hard rubber and plastic-like materials. On the A scale, a gum rubber band would measure around 40, a tire tread 60, and a shoe sole 80.

⌘⌘⌘

Hardness is probably the most often-measured property of elastomer vulcanizates. It appears in almost every specification and is widely used by rubber goods manufacturers for quality control. However, its practical significance is questionable, because surface indentation rarely bears any relationship to end-use performance. Also, hardness is a very imprecise measurement; it is not uncommon to find 5 or more points difference in readings by different people on the same piece of rubber. Nevertheless, hardness will probably remain as a standard test in the rubber industry because it is simple and quick to measure and has been correlated with product quality over many years of experience.

Stress-Strain/Tensile Properties – Somewhat more meaningful and useful to the industry are stress-strain data or tensile properties. Stress-strain curves are generated through the pulling (or tensile straining) of a "dumbbell" (ASTM D-412). This dumbbell is died out of a flat cured sheet about 2 mm thick. The stress required to break the dumbbell is the tensile strength. The strain or % stretch registered at the instant of rupture is called ultimate elongation, or simply elongation. It is common practice to record and tabulate the stress values at 100%, 200%, etc. instead of plotting curves.

Tensile properties are normally measured at room temperature (about 23°C). Equipment is available for doing this at lower or higher temperatures, which have a pronounced effect on the hardness and stiffness of a vulcanizate.

Tensile Strength is defined as the force per unit of the original cross-sectional area which is applied at the time of rupture of a test specimen. It is expressed in pounds force per square inch, kilograms force per square centimeter, or megaPascals. Tensile strength is an important characteristic as a tool for compound development, manufacturing control and determination of susceptibility to deterioration by oil, heat, weather, and other environmental factors. It is not necessarily an indication of quality, but is often used as such. It is true that the low range of tensile strength values is indicative of low quality compounds and the high range indicative of high quality. However, in the mid-range (about 1500 to 3000 psi), it is difficult to correlate tensile strength with other mechanical properties.

Ultimate Elongation or Elongation at Break is another property measured in conjunction with tensile strength. Elongation, or strain, is defined as the extension between benchmarks produced by a tensile force applied to a specimen. It is expressed as a percentage of the original distance between the benchmarks. Ultimate elongation is the elongation at the moment of rupture. Like tensile strength, elongation is used as a tool for quality control and determination of resistance to environmental effects. With the exception of rubber bands, rubber products are seldom stretched anywhere near their ultimate elongation in service.

⌘⌘⌘

Tensile Stress, more commonly called Modulus is the stress required to produce a certain strain, or elongation. Modulus values can be taken at any elongation off the stress/strain curve. In rubber testing, modulus is usually reported as 100% Modulus, 300% Modulus, or 600% Modulus, meaning the stress at 100%, 300%, or 600% elongation.

Stress-strain properties are normally measured on vulcanizates. The tensile properties of uncured stocks are sometimes used as a measure of green strength – i. e., the strength of the stock for handling before cure.

The most commonly used tensile test shape is the dumbbell. However, a straight strip, a molded o-ring, or a ring die-cut out of a slab may also be used. Since tensile test results are somewhat variable, it is common to test two or three specimens out of each cured slab and average the results.

Tensile test machines vary from simple devices, which simply record the force required to stretch the specimen while the operator measures elongation manually, to complex instruments which plot out the entire stress-strain curve or automatically calculate and print out the key tensile properties. Basically, the specimen is clamped between two grips, then one grip moves away from the other at a controlled rate of speed, stretching the specimen until it breaks. The standard strain rate (rate at which the grips are separated) for rubber is 8.5 millimeters per second, or 20 inches per minute.

Heat Resistance – The phrase "heat resistance" as used in the rubber industry really means resistance to irreversible changes in properties as a result of prolonged exposure to high temperature. A rubber vulcanizate heated to high temperature can undergo two different kinds of temporary or reversible change: expansion, and softening (plastic flow). A rubber vulcanizate aged at high temperature can undergo irreversible chemical changes: further crosslinking, cleavage of polymer chains by oxidation, or loss of some ingredients by evaporation or migration. Comparisons of heat resistance generally deal with the chemical stability of the vulcanized polymer in terms of crosslinking and oxidation. The current commercial elastomers differ greatly in inherent heat resistance.

The normal means of measuring heat resistance is to determine stress-strain properties before and after heat aging. Interpretation of results is done in various ways, depending on what the chemist or end user feels is most relevant. Heat resistance is expressed in several ways:

- Percent change in elongation caused by heat aging.
- Percent change in tensile strength caused by heat aging.
- Points change in hardness caused by heat aging.
- Percent of original elongation retained through heat aging.
- Time of aging in hours until breaking elongation drops to 100%.

In an elastomer which hardens during heat aging, loss of rubber-like flexibility is equated with reduction in elongation to 100%.

⌘⌘⌘

Compression Set – Compression Set is defined as the residual deformation of a material after removal of an applied compressive stress. Resistance to compression set is the ability of an elastomeric material to recover to its original thickness after having been compressed for an extended period.

For optimum performance in service, compression set values should be as low as possible. Low set values mean that the material has recovered nearly to its original height, and there is very little residual deformation. This is particularly important in applications where a rubber part is expected to provide a seal under a compressive force, and the sealing force is removed and reapplied repeatedly.

Compression set tests may be run either by applying a known constant force to the test specimen (Method A of ASTM D-395) or by compressing the specimen to a known constant deflection (Method B of ASTM D-395). The specimen is held in the compressed position for a specified period of time at a specified temperature, after which the compressive force is removed and the specimen is allowed to recover at room temperature for 30 minutes. In Method A, compression set is the difference between the original and final thickness of specimen, expressed as a percentage of the original thickness. In Method B, compression set is calculated by expressing the difference between the original and final thickness of the specimen as a percentage of the deflection to which the specimen was subjected. The standard specimen for a compression set test is a round pellet. O-rings of specified dimensions can also be tested.

Compression set tests are usually run at elevated temperatures, to simulate conditions or aging effects. Common test conditions are 70 hours at 70°C (158°F) or 100°C (212°F), although heat-resistant materials such as fluoroelastomers may be tested for longer periods of time at temperatures up to 200°C (392°F) or more. If the end product is expected to perform at low temperatures, e.g. below 0°C (32°F), compression set is measured at the expected service temperature.

Although there are differences of opinion about its validity, the compression set test is the easiest to run and therefore the most popular method of measuring the ability of a compound to return to its original shape and dimensions after being deformed for a long time. Low compression set data do not necessarily correlate with high resilience or low creep.

Abrasion Resistance – Abrasion Resistance is defined as the resistance of a rubber composition to wearing away by contact with a moving abrasive surface. It is usually reported as an Abrasion Resistance Index, which is a ratio of the abrasion resistance of the test compound compared to that of a reference standard measured under the same test conditions.

Abrasion resistance is measured under definite conditions of load, speed and type of abrasive surface. The standard laboratory tests generally cannot

⌘⌘⌘

be used to predict service life, because factors affecting abrasion are complex and vary greatly from application to application. Nevertheless, abrasion resistance tests are useful in making quality control checks on rubber products intended for rough service.

The most frequently used abrasion tester is the National Bureau of Standards (NBS) Abrader. Test samples are pressed under a specified load against a rotating drum covered with abrasive paper. The number of revolutions of the drum required to wear away a specified thickeness of the test specimen is recorded and compared to the number of revolutions required to wear away the same thickness of standard reference material (ASTM D-1630).

The Pico Abrader is used for measuring abrasion resistance of soft vulcanized rubber compounds and other elastomeric materials. A pair of tungsten carbide knives of specified geometry and controlled sharpness are rubbed over the surface of a pellet-shaped sample in a rotary fashion under controlled conditions of load, speed and time. The volume loss of the test material is measured and compared to a reference compound (ASTM D-2228).

The Taber Abrader measures wear by weight and/or thickness loss. Specified abrasive wheels are held under load of 500 or 1000 grams against a test specimen mounted on a rotating turntable. The equipment is described in ASTM D-1044, but the complete method is not followed for elastomer evaluations.

Tear Strength or Tear Resistance – Tear Strength or Tear Resistance of rubber is defined as the maximum force required to tear a test specimen in a direction normal to (perpendicular to) the direction of the stress. Tear strength is expressed as force per unit of specimen thickness – pounds force per inch (lbf/in), kilograms force per centimeter (kgf/cm), or kiloNewtons per meter (kN/m).

There are several types of tear specimens, with no apparent correlation between types. Some are cut (nicked) to provide a starting point for tearing, while others are not. The most frequently used sample is an unnicked 90 degree angle specimen listed in ASTM D-624 as Die C and sometimes referred to as Graves Tear. Other specimens used are a razor-nicked crescent specimen – Die A of ASTM D-624, also referred to as Winkleman Tear – and a trouser tear specimen, so called because it resembles a pair of men's trousers, of ASTM D-470.

Tear test specimens are pulled on a tensile tester at a cross-head rate of 8.5 millimeters per second (20 inches per minute). The maximum force required to initiate or propagate tear is recorded as force per unit thickness. Tear strength is frequently used to indicate relative toughness of different compounds. However, it is difficult to correlate with end-use performance.

⌘⌘⌘

Resilience – Resilience is the energy returned by a vulcanized elastomer when it is suddenly released from a state of strain or deformation. High resilience is what causes the bounce or snap often associated with products of natural rubber. In more specific terms, resilience is defined as the ratio of energy given up on recovery from deformation to the energy required to produce the deformation, and is usually expressed as percent. Elastomers are used in many applications because of their resilience properties. High or low resilience may be required, depending on the application. The lower the resilience, the less vibration transmitted. Low resilience is required in dampeners, while high resilience is needed in tires for fuel economy (low rolling resistance).

One instrument used to measure resilience is a vertical rebound tester. In this test, a metal plunger is allowed to fall on the test specimen and the height to which the plunger rebounds is recorded (ASTM D-2632). Another instrument used to measure resilience is the Yerzley Oscillograph. Besides resilience, this instrument can be used to measure other mechanical properties such as static and dynamic modulus, kinetic energy and creep (ASTM D-945). Specimens are loaded by an unbalanced lever, and the resultant cyclic reflections are recorded on a chart. Testing can be done over a temperature range of -40°C to +121°C (-40°F to +250°F).

Hardness Change – Durometer hardness is sometimes used to follow changes in stiffness at low temperatures. This is a simple technique, and may be useful as a check on factory operations, but it is not very precise.

Impact Strength or Impact Resistance – Impact strength or impact resistance is the resistance of a material to fracture under a sudden impact or shock force. It is a measure of the toughness of the material. Impact strength is usually only determined for plastics and hard elastomer compounds (i.e., those in the D durometer hardness range), where the impact resistance of the material in actual service is important. Soft elastomer compounds will bend, not break, on impact unless they have been cooled below their brittleness temperature. Several types of impact testers and test methods are available. Basically, they all measure the energy required to break a test specimen in a single sharp blow, and can be performed over a temperature range from -70 to +120°C (-94 to +248°F). The most widely used tests are Izod Impact and Charpy Impact; both procedures are given in ASTM D-256.

In the Izod Impact Test, a rectangular specimen is clamped securely at one end, then struck by a weight at the end of a pendulum. As the pendulum continues its swing, it moves a pointer which records the amount of force used up in breaking the specimen. Impact strength is reported as force per unit of specimen width (e.g., N/m or lbf/in). The specimen is usually notched in the center, but unnotched specimens can also be tested.

⌘⌘⌘

The Charpy Impact Test is similar to the Izod test, except that the specimen is supported on both ends like a simple beam. The striking force swings on a pendulum. Impact strength is reported as force per unit width.

Flex Resistance – The term flex resistance does not mean resistance to flexing or resistance to bending. It means the ability to withstand numerous flexing cycles without damage or deterioration. Flex-crack resistance means the ability to sustain numerous flexing cycles without the occurrence of cracks in the surface resulting from stress and ozone attack. Flex-cut-growth resistance means the ability to withstand numerous flexing cycles with a cut in the stressed surface with little or no growth of that cut.

The most common type of failure or damage resulting from repeated flexing is the formation of surface cracks, known as flex cracking. Shoe soles and tires are examples of rubber products that may show flex cracking. Another mode of failure due to flexing is the separation of rubber from fabric in a fabric-supported product such as a conveyor belt.

The two most commonly used flex resistance tests are the DeMattia Flex Test (ASTM D-813) and the Ross Flex Test (ASTM D-1052). The Demattia flexer alternately pushes the test specimen together, bending it in the middle, then pulls it back to straighten it. The Ross flexer repeatedly bends the specimen over a metal rod, then straightens it. In both tests, a small cut or nick of prescribed size and shape may be made in the center of the test specimen. Data generated from these tests include: flex life, the number of test cycles a specimen can withstand before it reaches a specified state of failure; and crack growth rate, the rate at which the cut propagates itself as the sample is flexed. These tests are particularly useful in evaluating compositions that are intended for use in products which will undergo repeated flexing or bending in service.

Another, more elaborate, test is the Monsanto Fatigue to Failure Tester. This instrument measures the ultimate fatigue life as cycles to failure. Tensile-like samples are flexed or stretched at 100 cycles per minute at a preselected extension ratio. Samples can be extended over a range of 10% to 120%. The fatigue performance of compounds can be measured and compared either at constant extension ratio (strain) or at constant strain energies (work input). This instrument is more likely to be used as a research and development tool than as a factory control tester.

Heat Buildup and Compression Fatigue – Heat buildup in elastomers is the accumulation of thermal energy and resultant temperature increase in a rubber product due to internal friction when the product is subjected to repeated, rapid cyclic deformation. The loss in properties of an elastomer due to heat buildup and other molecular effects of dynamic deformation is known as fatigue. Elastomeric compositions which show the least heat

⌘⌘⌘

buildup and the least change in properties due to fatigue are usually considered the best candidates for use in dynamic applications, such as tires.

The Goodrich Flexometer is used for heat buildup testing (ASTM D-623). A cylindrical test specimen is alternately compressed and then released in rapid cycles for a specified period of time. The temperature rise within the test specimen is recorded. Additional data obtained are permanent set and static compression. The Goodrich Flexometer Test is sometimes called the Compression Fatigue Test, because the test specimen is fatigued, or weakened, by the rapidly cycling compression stress.

Weather and Ozone Resistance Tests

Weathering is a complex combination of erosion and UV light-induced oxidation. No elastomer is immune to this, especially if it is highly filled and light in color. Ozone attack, on the other hand, happens only to stressed vulcanizates of unsaturated polymers when ozone is present in the atmosphere. Saturated elastomers, like EPDM, are virtually immune to ozone attack.

Tests for weather and ozone resistance depend on visual observation of specimens (e.g., for the appearance or growth of checks or cracks), and are therefore somewhat inaccurate and qualitative. The differences are typically so gross, nevertheless, that they can be demonstrated dramatically. A given ozone containing atmosphere can cause visible attack on one vulcanizate in a few minutes and no attack on another in a week.

Outdoor Exposure – Elastomer compositions progressively degrade in physical properties upon prolonged exposure to the elements (air, sunlight, rain, etc.) because of chemical changes in the elastomer molecule. These changes are accelerated by oxidation from ozone and oxygen in the atmosphere, by ultraviolet rays, by temperature variations, and other environmental factors, but would occur to some extent simply with the passage of time. Typical effects are surface hardening, crazing and cracking, plus gradual changes in tensile strength, elongation, and other physical properties. Some elastomers become soft and gummy, others become hard and inflexible.

In order to determine the actual effects of normal weathering, specimens are exposed outdoors in different locations with different climates. Changes in color, cracking, crazing, chalking, mildew growth, and stress-strain properties are recorded at various time intervals. Typical check periods are 1, 2, 5, 10, and 20 years exposure. Test specimens are usually stretched at low strains and mounted on wooden supporting racks for outdoor exposure.

Three procedures are specified by ASTM D-518:

Procedure A – Straight strips elongated 20%.

⌘⌘⌘

Procedure B – Bent loops, where the elongation varies from 0 to 25% at different points on the loop.
Procedure C – Tapered samples elongated to strains of 10 to 20%.

The specimens in their supporting racks are usually exposed facing South for maximum sunlight at an angle of 45°. Samples may also be buried in the ground to determine resistance to moisture, soil chemicals, insects, bacteria and oxidation effects in the absence of direct sunlight.

Weather-Ometer and Fade-Ometer – The Weather-Ometer is a special machine in which test specimens are continuously or intermittently exposed to water spray and artificial light. The light, produced by a carbon arc, is essentially the same wavelenghts as natural sunlight, but with increased intensity in the ultraviolet range. In the Fade-Ometer, specimens are exposed to constant controlled humidity (instead of water spray), along with artificial light. Specimens may be exposed unstrained or under a slight elongation. Both tests are described in ASTM D-570.

As in the other accelerated aging tests, tensile properties are measured after aging and compared with original properties. In addition, visual observations are made for cracking, crazing or color changes, as is done with specimens aged outdoors. The light intensity in the Weather-Ometer and Fade-Ometer can be related to sunlight intensity at various locations throughout the world.

Deterioration by Ozone – Ozone is a highly reactive gas which can cause rubber products to crack and fail prematurely unless they are protected by antiozonants or made of an ozone-resistant elastomer. Ozone is present in the atmosphere, and is produced constantly by arcing in high voltage electrical devices.

To test for resistance to ozone attack, samples are stretched to 20 or 40% strain on a test rack, or bent in a loop to produce a surface strain. The specimens are then placed in the ozone chamber, a special oven equipped with an ozone generator. Ozone concentration in the chamber can be controlled at any desired level – usual test concentrations are 0.5, 1, 3, or 100 ppm (parts ozone per million parts of air, by volume). Test temperature is usually 40°C (104°F). The test specimens are inspected at various time intervals until initial cracking occurs. Testing may continue until the specimen actually breaks, or may be stopped at some pre-determined degree of attack.

Dynamic testing provides a more rigorous test for ozone attack. Here, a special fabric-backed specimen is continuously flexed over a roller within the ozone chamber. Any protective chemical films which might build up on the surface of the specimen in static testing are quickly broken by the

⌘⌘⌘

continuous flexing in dynamic testing. Interpretation of results is the same as for static testing. ASTM D-1149 covers static testing and ASTM D-3395 covers dynamic testing in a controlled ozone atmosphere.

Accelerated Aging

Rubber technologists usually can't, or don't want to, wait five, ten or twenty years to find out how well their compounds will withstand the ravages of weather and time. Therefore, a variety of laboratory accelerated aging tests have been developed to predict the service life and aging characteristics of elastomers and rubber-like materials. These accelerated tests generally do not give exact correlations with service performance, but they do enable screening and comparison of compounds within a practical time scale. And, some useful correlations have been made with long term outdoor aging studies. Some of the more common accelerated aging tests are:

Air Oven – Specimens are exposed to the deteriorating influence of air in an oven at specified elevated temperatures for known periods of time, after which their physical properties are determined (ASTM D573). These are compared with the properties determined on original (unaged) specimens. Changes in hardness and percentage changes in tensile strength and elongation at break are reported.

Heating in Air in a Test Tube – In this method (ASTM D-865), the test specimens are heated in air, but they are confined within individual test tubes. The isolation provided by the test tube prevents cross-contamination of compounds due to loss of volatile materials (e.g., antioxidants or curatives) and their subsequent migration into other rubber specimens being tested. With this test, characteristics may be determined in a way that is free of some of the complications inherent in community-type aging devices where numerous compounds are aged in the same enclosure. Here again, hardness and tensile properties measured after aging are compared with original properties and the percent change is calculated.

Deterioration by Heat and Oxygen – In this test (ASTM D-572), pure oxygen under pressure is used to accelerate oxidative deterioration of the rubber specimen. Test specimens are placed in a metal vessel designed to retain an internal atmosphere of pure oxygen gas under pressure up to 2070 kPa (300 psi) at temperatures of 70° or 80°C (158° or 176°F). After the prescribed aging period, hardness and tensile properties are measured and compared with original properties. Usually, hardness change and percent change in tensile strength and elongation are reported.

⌘⌘⌘

Fluid Resistance

No elastomer can withstand and contain every fluid known to industry. Stated in another way, any elastomer can be severely attacked and rendered unserviceable by a selected fluid or liquid. Measurement of fluid resistance requires definition of terms. One elastomer is better than another only with reference to one class of chemically similar fluids, like aromatic oils, ketones, oxidizing acids, or brake fluid.

Fluid Resistance (Chemical Resistance) – Fluid resistance is a general term describing the extent to which a rubber product retains its original physical characteristics and ability to function when it is exposed to oil, chemicals, water, organic fluids or any liquid which it is likely to encounter in actual service. Fluid resistance tests may not give a direct correlation with service performance, because service conditions cannot easily be defined or controlled. However, they yield comparative data on which to base judgements about expected performance, and they serve as useful tools for screening compounds for particular service conditions.

Determining fluid resistance involves measuring a specimen's weight, volume and physical properties before and after exposure to the selected fluids for a specified time at a specified temperature. The effect of the fluid on the specimen is judged on the basis of the percent of the original properties retained or lost. Properties most often checked are volume change, weight change, and changes in hardness, tensile strength and elongation at break. Changes in tear strength, compression set, and low temperature properties are also frequently evaluated.

ASTM D-471 describes the testing of rubber in contact with fluids. It also lists the composition of various fluds which have been given special ASTM designations and are used as reference standards throughout the rubber industry.

ASTM Oils are specific petroleum-base oils which have different aniline points covering the range of commercial lubricating oils. The aniline point of a petroleum oil appears to correlate with the swelling action of that oil on an elastomer. In general, the lower the aniline point, the more severe the swelling action.

ASTM Reference Fuels have been selected to provide the maximum and minimum swelling effects produced by commercial gasolines. Reference Fuel A has a mild effect on elastomeric vulcanizates and produces results of the same order as highly paraffinic, straight run gasolines. Reference Fuel B causes more severe swelling, exceeding the swelling action of commercial gasoline. Reference Fuel C causes even more severe swelling, and is typical of the highy aromatic premium grades of automotive gasoline.

ASTM Service Fluids are special fluids which are either non-petroleum or are compounded from special petroleum hydrocarbon fractions. They

⌘⌘⌘

have swelling characteristics outside the range of ASTM Oils or Reference Fluels.

Oil Resistance – Oil resistance is a special case of fluid resistance. Broadly defined, oil resistance is the ability of a rubber product to perform its intended function while in contact with oil. This recognizes that the elastomer may be swollen or weakened to some extent by the oil.

The tire rubbers (natural rubber, SBR), when placed in oil, absorb the fluid slowly until either the oil is all gone or the rubber has disintegrated. They never reach equilibrium. The so-called oil-resistant elastomers absorb some oil, especially at elevated temperature, but only a limited amount. With some it may be negligible. Most end uses requiring oil-resistant elastomers can tolerate appreciable swelling or volume increase. Hence, volume increase is not a very significant way to measure the ability of a rubber article to perform its intended function in oil.

Low-Temperature Properties

As elastomer compositions are cooled below room temperature, they stiffen and become more difficult to bend, twist, or stretch. This stiffening is gradual until the stiffening temperature (also called the second order transition temperature) is reached. Then, further decrease in temperature causes a very sharp increase in stiffness. At very low temperatures, the elastomer composition becomes brittle and will crack or break if subjected to sudden impact or bending; the temperature at which this occurs is known as the brittle point or brittleness temperature.

The brittle point has no relationship to stiffness as shown by stiffness vs. temperature curves. For example, stiffness measurements may indicate a high degree of flexibility at a certain temperature, but impact tests may show the composition to be brittle at the same or higher temperature. This lack of relationship is not surprising, however, because stiffness measurements involve loading at low speed and low deflection, while brittleness tests involve loading at high speed and high deflection.

Following are some of the tests used to determine flexibility behavior of elastomer compositions at low temperatures. Results can be used to predict actual field performance.

Brittleness Temperature – This test (ASTM D-746) is used to determine the temperature at which the composition exhibits brittle failure. A small rectangular strip is clamped in a holder at one end and cooled to a predetermined temperature in a chilled liquid bath. The specimen is then given a sudden sharp blow by a solenoid-actuated striking arm. If the specimen does not fracture, the temperature of the liquid bath is lowered and the process is repeated until the specimen breaks when struck.

⌘⌘⌘

Torsional Stiffness – Torsional stiffness tests measure the modulus of rigidity of an elastomer composition over a broad temperature range. Two widely-used tests are ASTM D-1043, which uses the Clash-Berg Tester, and the ASTM D-1053, which uses the Gehman Torsional Stiffness Tester. In both tests, the sample is chilled to a preset temperature, then twisted with a known force. The amount of twist is measured and related to the stiffness (modulus of rigidity) of the sample. The test temperature is then changed and the test is repeated, until a complete curve of stiffness vs. temperature is plotted. The temperature which produces a stiffness of 69 MPa (10000 psi) is sometimes taken as the Stiffness Temperature.

Temperature Retraction – This test (ASTM D-1329) can be used to measure viscoelastic properties of elastomers and rubber-like materials at low temperatures, and also to evaluate crystallization effects. The test specimen is stretched to a specified elongation and frozen at -70°C (-94°F). Then, it is gradually warmed until it begins to retract toward its unstretched dimensions. The temperature at which the specimen retracts 10% has been found to correlate with brittleness temperature.

BIBLIOGRAPHY

Ohm, R.F., 1990, *The Vanderbilt Rubber Handbook*, R.T. Vanderbilt Co., Norwalk, CT

Morton, M., 1987, *Rubber Technology*, Van Nostrand-Reinhold, New York

Morton, M., 1959, *Introduction to Rubber Technology*, Reinhold, New York

Bateman, L., 1963, *The Chemistry and Physics of Rubber-Like Substances*, Wiley, New York

Dick, J.S., 1987, *Compounding Materials for the Polymer Industries*, Noyes, Park Ridge, NJ

Stephens, H.L., 1985, *Textbook for Intermediate Correspondence Course*, American Chemical Society, Rubber Division, Washington, DC

Katz, H.S., Milewski, J.V., 1987, *Handbook of Fillers for Plastics*, Van Nostrand-Reinhold, New York

Hagemeyer, R.W., 1984. *Pigments for Paper*, TAPPI Press, Atlanta

Eirich, F.R., 1978, *Science and Technology of Rubber*, Academic Press, New York

Mark, H.F., et al, 1971, *Encyclopedia of Polymer Science and Technology*, Interscience, New York

⌘⌘⌘

SEVEN

ADHESIVES AND SEALANTS

Hector Cochrane, Ph.D.
Oliver Products Company
Grand Rapids, MI

The use of adhesives and sealants dates back thousands of years. The ancient Egyptians used gum arabic and animal glues for bonding papyrus reeds and veneering furniture (Dick, 1987). For several hundred years the crevices between the wooden planks of boat hulls have been caulked with rope and pitch to prevent water leaks.

The first commercial adhesive factory was built in England in the early 1700s. Very little changed in the development of adhesives and sealants until the development of synthetic polymers during the past 80 years (Dick, 1987).

1912	Phenolic resin adhesives used in plywood
1920s	Cellulose ester adhesives
1927	Goodrich rubber adhesives
1930	Urea formaldehyde adhesives
1939	Polyvinyl acetate polymers
1940	Chlorinated rubber
1940s	Melamine formaldehyde resins
1945 to 50	Polyesters, polyurethane, epoxy, and silicone polymers
1950 to 65	Polyethylene, ethylene vinyl acetate, polyvinyl ether, and cyanoacrylate

Definition and Purpose of Adhesives and Sealants

Adhesives are materials that are capable of binding other substances together by surface attachments. The surfaces bonded together can be metal, fabric, film, plastic, wood, stone, glass etc., and combinations of these materials. Normally the thickness of the adhesive layer is less than 1/8 inch.

Caulks and sealants are used to fill joints, gaps, and cavities between materials. They hold in or hold out air or other gases, liquids (most commonly water), or solids. Sealants and caulks normally fill gaps of 1/8 inch to about 1inch. Caulks are materials used for sealing gaps where there is very little joint movement (<$\pm$ 3%) and therefore require no or only minor elastomeric properties, as with oil based window putties. Sealants are elastomeric materials with adhesive properties to bond surfaces. Depending on the polymer base, they are used for joints which have movements from $\pm$5% to $\pm$50%. Sealant adhesives are elastomeric materials which both fill and bond joints and gaps. They provide structural strength when joining substrates, as in the use of polyurethane sealants for automobile windshields. Mastic sealants are elastomeric materials which never cure.

General Properties

For adhesives and adhesive sealants to be effective they must form strong bonds with the surfaces they are joining. Even though the primary function of sealants, caulks and mastics is to occupy space between substrates, they must also have good adhesion to the surfaces of the cavity so that during joint expansion and contraction the forces of stress are not concentrated at the sealant/substrate interface, but are transmitted and absorbed by the elastomeric component of the sealant. Adhesives and sealants must therefore form good adhesive bonds with adjacent surfaces to be effective. Most adhesives and sealants are liquids or pastes when applied to ensure that they wet the adherend surface. This allows the adhesive or sealant molecules to be very close to the surface atoms of the adherend so that the chemical and/or van der Waal's bonds can be formed.

The ability of an adhesive to wet a surface can be tested very simply by placing a droplet of the adhesive onto the surface and observing the contact angle of the droplet, as shown in Figure 1.

If the adhesive wets the surface easily, the droplet will flow across the surface and have a low contact angle, $\theta_A < 90^\circ$, as shown with Adhesive A. Conversely, a liquid which cannot wet the surface well will have a very high contact angle, $\theta_B > 90^\circ$, as shown with Adhesive B. Indeed if the liquid completely wets the surface the droplet angle will be 0. Conversely, if it does not wet the surface at all, the angle will be 180°, as with water on Teflon film.

The ability of an adhesive to wet a surface can be calculated from the properties of the adhesive and the surface as follows.

(I) $$\gamma_{SV} = \gamma_{SL} + \gamma_{LV}\cos\theta$$

where

γ_{SV} = free energy of the surface in presence of adhesive A vapor.

γ_{SL} = free energy of the surface in presence of liquid adhesive A.

γ_{LV} = free energy (surface tension) of adhesive A.

⌘⌘⌘

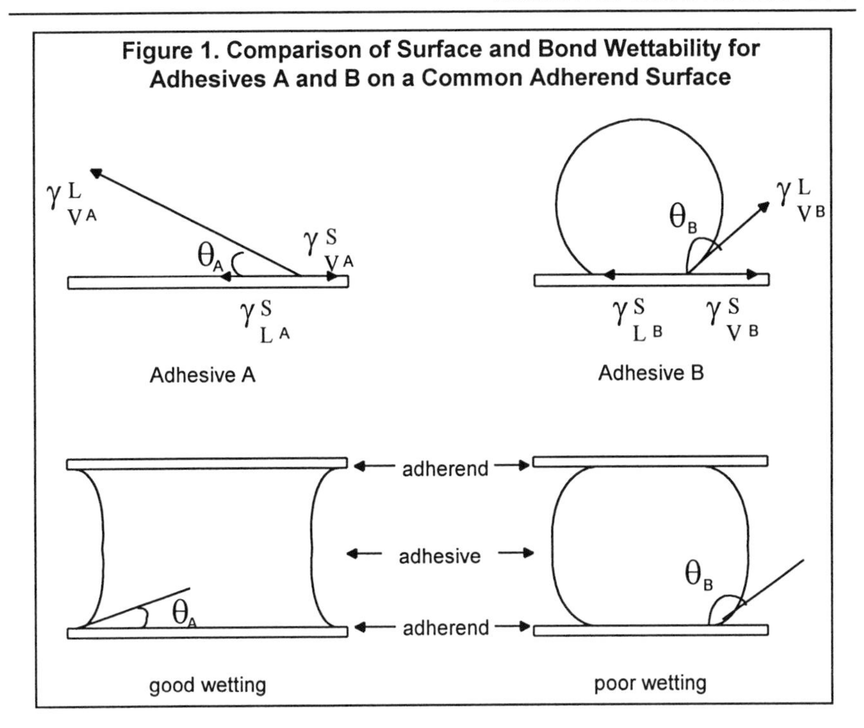

The reversible work of adhesion, W_A, between a liquid adhesive and a solid surface and their surface free energy is:

(II) $W_A = \gamma_{SV} + \gamma_{LV} - \gamma_{SL}$

Combining equations (I) and (II):

$$W_A = \gamma_{LV}(1+\cos\theta)$$

If the adhesive completely wets the solid surface, i.e. $\theta = 0°$, then :

$$W_A = 2\gamma_{LV} = \text{twice the surface tension of the adhesive.}$$

If the adhesive doesn't wet the solid surface at all, $\theta = 180°$, then:

$$W_A = \gamma_{LV}(1+\cos 180°) = \gamma_{LV}(1-1) = 0$$

The ability of adhesive to wet the surface then depends on the relative free energies of the liquid adhesive and the surface on which it is placed.

Table 1 shows the surface free energies of some simple liquids and substrate surfaces. For good surface wetting by a liquid, the free energy of the

⌘⌘⌘

surface should be at least 10 dynes/cm greater than that of the liquid (Gilbertson, 1991). For example, epoxy resin adhesives wet polar metal surfaces giving strong adhesive bonds. They do not wet low surface free energy nonpolar surfaces such as polyethylene, resulting in very weak bonds if any.

Table 1. Surface Tensions Values of Liquids and Solids

Liquids[1]	Temperature °C	Surface tension dyne/cm
n – Octane	20	21.8
Ethanol	20	22.8
Benzene	20	28.9
Water	20	72.8
Solids[2]		
Polytetrafluoroethylene (Teflon)		19 – 20
Silicone rubber		24
Polypropylene		29 – 31
Polyethylene		30 – 31
Polyvinyl chloride		39
Polyester		41 – 44
Styrene butadiene rubber		48

[1] (Adamson, 1967)
[2] (Gilbertson, 1991)

Two ways to improve the wettability of an adhesive are:

1) Heat the adhesive to lower both its viscosity and surface tension. This is inappropriate with a heat-cure adhesive.

2) Add surfactants to reduce the surface tension of the liquid adhesive. Care must be taken not to add too much surfactant, however, or it may weaken the adhesive bond.

The ability of the adhesive to wet the surface can also be increased by increasing the energy of the surface by:

1) Removing grease and dirt from the surface.

⌘⌘⌘

2) Chemically modifying the surface.
   a) Wire brushing the surface (metal) to remove the oxide layer.
   b) Chemically etching metal surfaces.
   c) Oxidizing the surface of plastic film and sheet by corona treatment, which can increase the surface energy by 7 to 10 dynes/cm.
   d) Applying a primer coat to a surface to make it more compatible with the adhesive.

If the substrate is very smooth, the adhesive bond strength can be increased by roughening its surface. This not only increases the area of the adhesive/substrate joint, but can improve the bond through mechanical interlocking.

Adhesives and sealants must have suitable rheology (flow) properties for their specific applications. Rheology control involves a number of considerations, including:

1) Method of application – spray, brush, extrusion from tube, hot melt adhesive guns.
2) Application temperatures, -30° to +300°F (-34° to +150°C).
3) Cure temperatures, -30° to +300°F (-34° to +150°C).
4) The ability to wet the substrate and flow into crevices to displace water and prevent air bubbles in the finished joint.
5) The ability to remain where applied.
6) Controlled penetration of porous surfaces, to give good mechanical bonds.
7) The ability to remain on the surface of porous surfaces to prevent bond line starvation.
8) Self-levelling on horizontal surfaces (e.g., potting compounds, epoxy encapsulation of computer chips).
9) Non-sagging on vertical and horizontal surfaces.

The flow properties of adhesives and sealants depend on their composition. Critical flow properties are listed in Table 2.

Adhesives can have a wide range of viscosity from, 1,000 to 50,000 mPa.s, while sealants may range from 50,000 to 750,000 mPa.s. Hot melt adhesives are solids at room temperature and have high viscosities, 20,000 to 70,000 mPas, at their melting temperatures of 200 to 300°F (93° to 150°C).

The viscosities of adhesives and sealants can be measured with various types of equipment such as the spindle (e.g. Brookfield), cone and plate, bob and couette viscometers, and torque rheometers. The viscosity value of an adhesive or sealant is dependent on the shear rate at which the measurement is made, as shown in Figure 2.

⌘⌘⌘

Table 2. Critical Flow Properties for Adhesives and Sealants

	Adhesive	Sealant
Extrudability	–	MIL 7502 (ASTM C 731)
Sag Properties	Leneta (ASTM D 4400)	Boeing (ASTM D 2202)
Recovery Rate	–	–
Viscosity (Brookfield)	ASTM D 2196 ASTM D 2556	ASTM D – 2196

Ideal liquids, such as water, are Newtonian in that their measured viscosity is independent of shear. A few systems, such as water in sand, show increasing viscosity with increasing shear rate. These systems are called dilatant. Most paints, adhesives, and sealants show a decrease in viscosity as the shear rate is increased. These systems are called pseudoplastic.

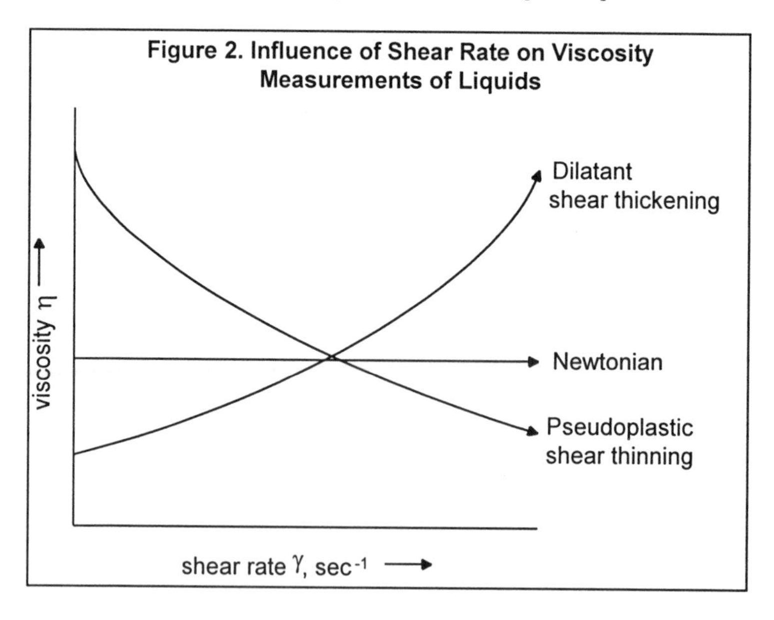

Frequently, the flow properties of adhesives and sealants are measured with a Brookfield viscometer at two different spindle speeds at a rotation ratio

⌘⌘⌘

of 10:1. The ratio of the low spindle speed viscosity to the high spindle speed viscosity is termed the Shear Thinning or Thixotropic Index and is a measure of shear sensitivity (pseudoplasticity); see Figure 3.

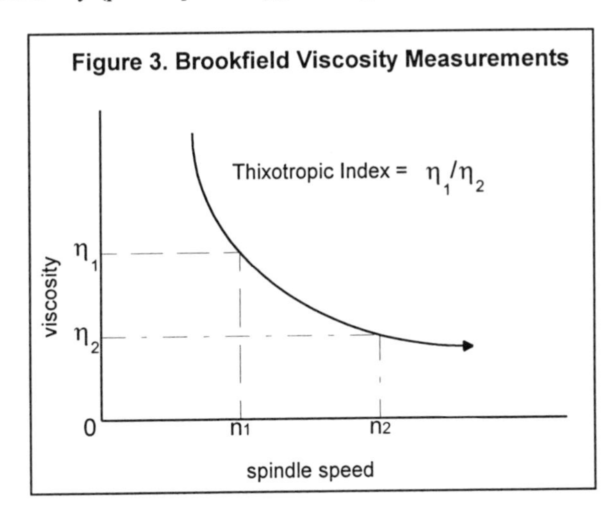

Figure 3. Brookfield Viscosity Measurements

These Brookfield viscosity measurements provide no direct information on the extrusion, sagging or application properties of the adhesive or sealant, as these processes all occur at different shear rates as shown in Table 3 (Barnes, 1989).

Table 3. Typical Shear Rates for Various Coating Phenomena

	Shear Rates, sec^{-1}
Sedimentation	$10^{-6} - 10^{-4}$
Leveling	$10^{-2} - 10^{-1}$
Draining (sagging)	$10^{-1} - 10^{+1}$
Extruding	$10^{0} - 10^{2}$
Mixing, Stirring	$10^{1} - 10^{3}$
Spraying, Brushing	$10^{3} - 10^{4}$

An ideal adhesive or sealant should have the rheological properties shown in Figure 4 (Cabot, 1994). When a force (stress) is applied to an adhesive or sealant, no flow of the material occurs until a certain stress level is reached. This value is called the yield point. This parameter can be correlated with the sag resistance of the adhesive or sealant. The higher the yield value, the better the sag , and vice versa.

⌘⌘⌘

Figure 4. Rheological Properties of Adhesives and Sealants

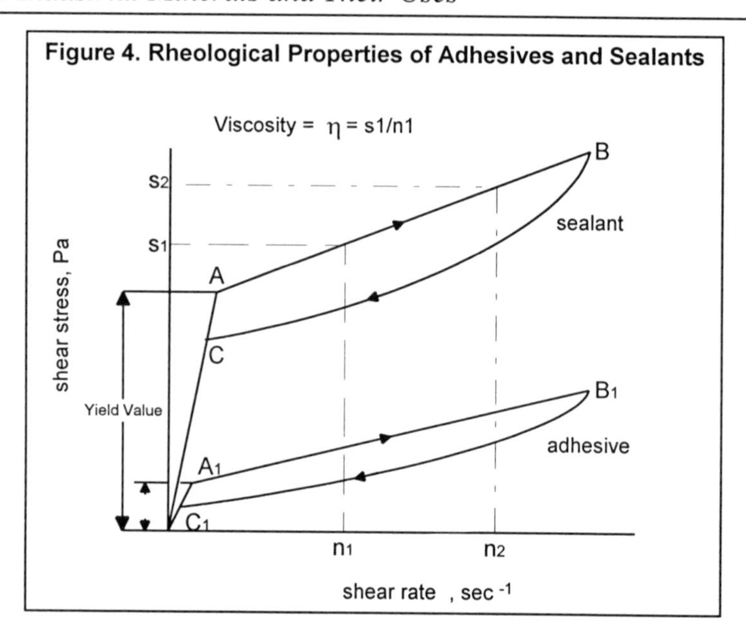

As the stress is increased further, the system begins to flow and will exhibit thixotropic behavior as shown by the flow curves. The greater the applied stress, the greater the shear strain or movement of the system and the lower the measured viscosity (line AB). As the shear stress is reduced, the viscosity of the system increases (line BC). There is a time lag in the recovery of the system which is called the recovery rate. In fact, the sag resistance of the system is related to the yield value and the rate of recovery after the adhesive or sealant has been sheared during application, whether by extrusion, brushing, or spraying. The ideal adhesive or sealant should have a high yield point, a high shear thinning index, and a fast recovery rate.

Examples of the rheological properties of silicone sealants reinforced with precipitated and fumed silicas are shown in Table 4 and Figure 5 (Cabot, 1994).

The more open, branched chain three dimensional aggregate structure of the fumed silica makes the material a much more effective thixotrope and reinforcing filler than the precipitated silica which has tight grape cluster-like aggregates.

The cure rate of an adhesive or sealant is very important and affects how the material is used commercially. The adhesive or sealant user may be looking for a product that can cure under very cold conditions (-30°F/-34°C) or can cure in seconds or minutes on an automobile or appliance production line. Cure rate and conditions depend on product chemistry. There are four basic types of adhesive and sealants: evaporative, chemically reactive, hot melt, and pressure sensitive.

⌘⌘⌘

Table 4. Precipitated vs. Fumed Silica in a Silicone Rubber Sealant

Formulation	1	2	3
Type of Silicas	Precipitated	Precipitated	Fumed
Surface Area, m^2/g	136	136	108
Bulk Density, lbs/cu. ft.	11.2	11.2	3.2
Loading, Parts	14	28	14
Processing and Rheological Properties			
Incorporation Time, min.	4	7	11
Dispersion Level	Poor	Poor	Excellent
Skin Time, min.	20	20	26
Yield Value, Pa	0	0	410
Viscosity, 1rpm, Pa.s	36.5	49.4	45
STI	2	1.5	1.5
Slump Value, cm	>>5	>>5	0
Extrusion Rate @ 40 psi, g/min.	Too high to measure		250
Cured Physical Properties			
Shore A Durometer	14	22	20
Die B Tear, kN/m	3.0 (17)	6.1 (35)	4.4 (25)
Peel Adhesion, kN/m	0.3(1.5)	0.7 (4.0)	1.1 (6.0)
Modulus @ 50mpa	0.21 (30)	0.33 (45)	0.29 (40)
Tensile, Mpa	0.58 (85)	1.52 (220)	1.63 (240)
Lap Shear, Mpa	0.6(95)	0.6 (95)	1.9 (275)
% Elongation	240	290	450
Clarity, %T at 500 nm	9	5	47

Evaporative – Evaporative adhesives and sealants are composed of polymers dissolved in an organic solvent or dispersed as an aqueous colloid (i.e. latex). After application the solvent or water must be lost by evaporation or diffusion into a porous substrate, such as paper or wood, so that the polymer forms a good glue line with good cohesive strength. The maximum polymeric content of a solvent based system is normally 30%. At higher levels the viscosity of the system may be too high resulting in poor wetting of the substrate surface. Latex systems have lower viscosities and can tolerate a polymer concentration up to about 50%. Above this level the latex particles may agglomerate and settle. The cure time for evaporative systems is hours to several days and can be decreased by raising the temperature of the environment in which the joint is being held.

⌘⌘⌘

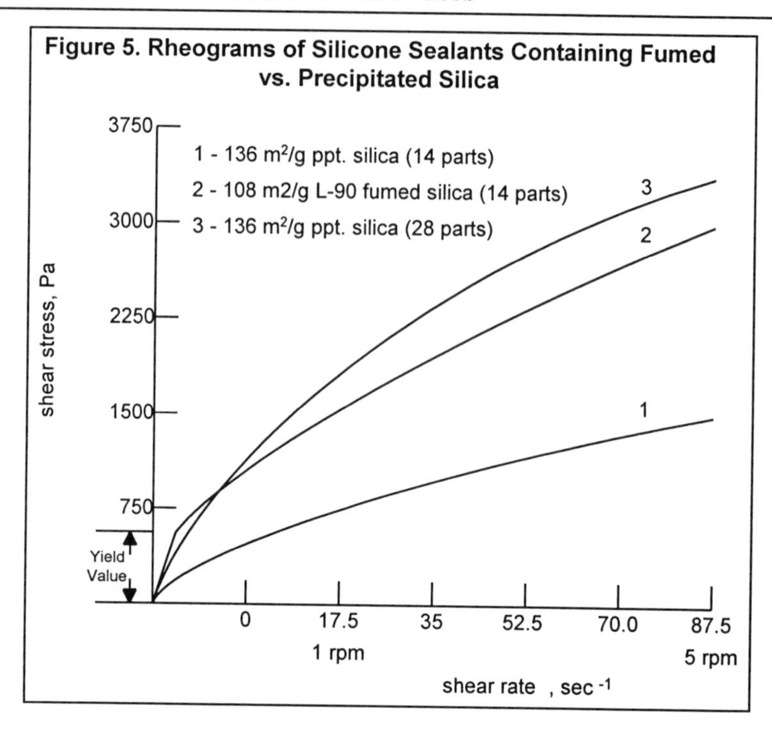

Figure 5. Rheograms of Silicone Sealants Containing Fumed vs. Precipitated Silica

Chemically Reactive – Chemically reactive adhesives and sealants usually low molecular weight liquid components which react on application to form a polymerized glue line with good cohesive strength. There are two general types of chemically reactive products. One reacts by polymer condensation and gives off low molecular weight by-products, usually water or alcohol, such as for phenol-formaldehyde resin and two part silicone rubber potting sealants. The other type of reactive product polymerizes without the formation of low molecular weight by-products, often through addition polymerization. Examples of this type are epoxies, cyanoacrylates, urethanes, and polyesters. Their cure times can be adjusted from minutes to hours. Some one component systems are moisture-cure and these depend on atmospheric moisture to initiate the cure mechanism. The rate of cure then depends on the relative humidity and temperature of the cure environment as well as the rate of diffusion of moisture from the surface to the center of the sealant/adhesive layer. As expected thin sections cure faster than thick ones. One component moisture-cure sealants include silicones and polyurethanes. Cure times can range from hours to days.

Important cure related considerations for evaporative and chemically reactive adhesive and sealant types include:

⌘⌘⌘

a) Pot-life, the working time that the adhesive or sealant can be tooled be fore it starts to cure and becomes unworkable.
b) Skin-time, the time to form a dry surface skin at the adhesive/sealant air interface.
c) Tack-free time, the time for an adhesive/sealant to cure so that it's sur face is no longer tacky to the touch.
d) Time to cure, so that it can support a mechanical load.
e) Cure conditions, temperature, relative humidity.
f) Evolution of by-products during cure, such as water or alcohol.
g) Shelf life, how long the uncured adhesive or sealant can be stored in its package before its properties start to change; crosslinking of one compo nent silicone sealants, for example.

Hot Melt – Hot melt adhesives are based on thermoplastic polymers that are solids at room temperature. Hot melts are applied at high temperatures, under which they melt and wet the substrate surface. On cooling, the polymer returns to the solid state providing good cohesive strength. Two major advantages of the hot melt systems are that they set quickly after being applied (seconds to minutes) and require no cure and hence have no by-products.

Pressure Sensitive – Pressure sensitive adhesives and sealants are composed of a viscoelastic material which is fluid enough to wet a surface under a slight applied pressure but cohesive enough to form a moderate strength bond to hold the wetted surfaces together after the applied pressure is removed. These systems can be separated and reformed at will especially if the cohesive strength of the bond is stronger than the adhesive strength.

Although the bond strength of the pressure sensitive adhesives and sealants are much weaker than those of the three other systems, they form instantaneous bonds, require no curing, and have no by-products to be removed from the joint area.

ADHESIVE AND SEALANT APPLICATIONS

The primary markets for adhesives and sealants may be divided into packaging/converting/disposables, construction, and transportation.

Packaging, Converting and Disposables (Asselin, 1990)
Corrugated cases and cartons – In this market, part of the adhesive is applied by the case and carton manufacturer and the remainder of the adhesive is applied by the packager to close the box. The major products used are water-borne and hot melt adhesives. The most frequently used water-borne adhesives are polyvinyl acetate homopolymers modified with polyvinyl

⌘⌘⌘

alcohol and plasticizers. Water based starch, dextrin and ethylene vinyl acetate adhesives are also used. The hot melt adhesives are primarily based on ethylene vinyl acetate polymers. These are modified with tackifiers, antioxidants and waxes to control set time, tack and adhesion.

The continued emphasis is on faster packaging line speeds and new adhesive application methods such as custom designed spray and gravure coating in place of conventional extrusion equipment. Water-borne adhesives with higher solids contents for faster drying are being developed to complement their lower application equipment cost, lower energy requirements, and lower prices compared to hot melt adhesives. However, hot melt adhesives are expected to continue gaining market share in the future due to their faster set times allowing faster packaging line speeds and hence greater productivity.

Can and bottle labels – Can and bottle labeling adhesives are used for packaging food and household products such as beverages, detergents, and personal care products. The major adhesives used are liquid and hot melt products. Some pressure sensitive adhesives are used as well.

The beverage industry – soft drink producers, breweries and wineries – are the largest market for liquid and hot melt adhesives. Other significant markets include detergents, cooking oils, and motor oils. The liquid adhesives used include starch, dextrin, casein and resins. Starch and dextrin adhesives are being replaced because of their lower adhesion levels. Hot melts currently account for 10 to 15% of the market, but are expected to have significant growth in the future.

The type of adhesive used depends on the bottle material and labeling conditions. Casein adhesives are normally used for labeling typically cold and wet wine bottles. They give good adhesion and some water resistance. Resin and hot melt adhesives are used for their good adhesion to plastic bottle surfaces in labeling PVC and PET bottles. The easiest labeling conditions are on dry glass at room temperature. This market is dominated by the lower cost starch and starch-resin adhesives. Many labelers, nevertheless, prefer hot melt adhesives for quick set and a clean wrinkle-free label.

Major future drivers in the bottle labeling market include waste reduction and recycling of plastic bottles, the prime market for resin labeling adhesives, and the ever demanding quest for faster bottling speeds. Future growth in the latter market is expected for both resin and hot melt adhesives.

Envelopes – The U.S. envelope market is expected to continue growing at 3%/year over the next few years. In 1989, 180 billion envelopes were produced using 41,000 metric tons of adhesives, termed “gums” within the industry. There are four different types of adhesives used in envelopes, these are: seam adhesives, window adhesives, seal adhesives, and stamp adhesives.

⌘⌘⌘

Seam adhesives are used in manufacturing the side, center, bottom and diagonal seams of the envelopes. Several different types are used including dextrin, resin, and dextrin/resin and starch/resin blends. The specific type used depends on the required adhesion level, application method and machine speed. As the envelope producing machine speed increases, dextrin and dextrin/resin are being replaced by starch/resin blends which are applied by roller, stencil or extrusion.

Window adhesives attach the clear window film to the rest of the envelope. The window film is usually polyester, orientated polypropylene or polystyrene. Formulated homopolymers or copolymers are used as adhesives.

Envelope sealing adhesives are usually remoistenable, and typically blends of dextrin and compatible polyvinyl acetate homopolymers as well as all dextrin or all resin types. Some water-borne and hot melt pressure sensitive adhesives based on acrylic or rubber resin liquids and rubber resin hot melts are also used. Natural rubber latex is the basis for self sealing envelopes.

Stamp adhesives are usually remoistenable and based on dextrin, resin, or dextrin/resin blends.

New trends in the envelope industry include the manufacture of high strength lightweight envelopes based on duPont's Tyvek® spunbonded olefin. These require special seam and seal adhesives. Laser printing of envelopes, which can generate local temperatures of up to 230°C, requires temperature-stable adhesives. Finally, there is the continued push for higher speed envelope manufacturing machines, which will require adhesives with accordingly fast application and bonding..

Bookbinding – This is a mature market for adhesives and new adhesive penetration is related to superior performance. The bookbinding industry produces a wide range of products including catalogues, directories, magazines, soft cover books, and hard cover books. The product value varies from low value throw away catalogues to high value archive quality books. The ultimate properties required from the adhesive depend on the type of publication. The operations in book binding include: binding, case making, casing in, cover attachment, and reinforcement.

Binding and cover attachments use hot melts, emulsion and animal glues. Ethylene vinyl acetate (EVA) and block copolymers are used for these operations. Case making normally uses animal glues to bond the cover materials to the cover boards. Casing in, the attachment of the book block end papers to the cover, involves the use of emulsions or starch based adhesives. Reinforcement involves the attachment of paper or cheesecloth to the spine of the book. Hot melt adhesives are used for this operation.

The animal glues are applied at elevated temperatures and build tack through heat loss but set through water loss. Water-borne starch pastes and resin emulsions are applied at room temperature and set up by water loss.

⌘⌘⌘

Major requirements for book binding adhesives beyond adequate adhesion to the various substrates in the book include flexibility and the ability to perform at cold and hot temperatures. The process of binding books is highly automated except in the case of special limited volume editions. New adhesives must therefor be adaptable to use on existing equipment, or new equipment must be developed for their use. New products such as thermoset polyurethane hot melts provide excellent adhesion and flexibility but new application technology is required for their use in high volume applications with minimum equipment down time.

Disposables – This fast growing market is divided into three distinct segments: disposable diapers, feminine care products, and medical and hospital products.

The disposable diaper market can be subdivided into baby diapers and adult incontinence garments. The baby diaper industry grew at a rate of 12%/year in the 1980s and is expected to grow at half this rate in the 1990s. This slowdown will be partly compensated by the faster projected growth in the adult incontinence market as the U.S. population ages. Adhesives are used in several parts of the baby diaper including construction, leg and waist elastic attachment, and leakage shield attachment. Hot melt adhesives are used for their fast setting, moisture resistance, flexibility and non irritating properties. The primary adhesives are based on synthetic rubber but some ethylene vinyl acetate and polyolefin based hot melts are also used. The biodegradability of the 18 billion diapers/year produced in the USA has received much attention due to the shortage of landfill space. Significant progress has been made in this area and market forces may necessitate the development of a biodegradable hot melt adhesive for diaper use.

In the feminine care market hot melt adhesives are preferred products for the same reason listed for baby diapers. The hot melts used include synthetic rubber, modified amorphous polypropylene, EVA copolymers, and polyethylene. The hot melts are applied by spray and slot coating extrusion. Some pressure sensitive adhesives are also used.

The medical and hospital product market includes surgical caps, gowns, dressing tapes, bandages, and patient drapes. A wide variety of adhesives are used including hot melts, water-borne and some natural and synthetic rubber. The polymers used are acrylics, polyvinyl ether, rubbers, and vinyl acrylic esters. The adhesives must show good adhesion on various materials and rough surfaces, allow moisture vapor transmission, have low odor, have clean removal, and must not irritate the skin.

Tapes and labels – The tape and label industry is one of the most dynamic applications for adhesives. The market was estimated to be 5 billion square

⌘⌘⌘

meters in 1990. Tapes are used in many markets including packaging, masking, automotive, industrial, and medical.

The three main forms of adhesives for tapes and labels are:

1) Solvent-borne, based on acrylic copolymers or rubber/resins dissolved in hydrocarbon solvents.
2) Water-borne pressure sensitive adhesives, based on acrylic copolymers or rubber/resin polymers.
3) Hot melt, based on block copolymers, rubbers and acrylic polymers.

Polybutadiene, silicone and vinyl ether adhesives are also used occasionally in tape and label applications. Solvent-borne adhesives usually give better performance than water-borne or hot melt adhesives, but environmental laws limiting the volatile organic content (VOC) of products and the desire to reduce worker exposure to hazardous materials (i.e solvents) are hurting these products. Attempts are being made to reduce VOC by developing new formulation with solids contents of 60+% compared to the normal 30 to 40%.

Water-borne styrene butadiene rubber (SBR) products with high tack and adhesion have gained a considerable share of the paper label market. New tackified acrylic emulsion adhesives matching the tack and adhesion of SBR adhesives are competing with them in the electronic data processing (EDP) label market. Also, new ethylene vinyl acetate/acrylic copolymer emulsions are being used on PVC face stocks to provide improved resistance to plasticizer migration. In the hot melt area, acrylic adhesives are starting to be used in medical and clear film applications.

Future product needs continue to change depending on the market segment; for example, industrial tapes want higher cohesive strength, packaging tapes which are recyclable and less expensive; and medical tapes with low skin irritation and clean release.

Two fast growing markets for PSA's are in EDP labels (bar code labels used to track inventory or change pricing) and primary product identification (food, personal care, and household items). One of the major changes in the tape and label market is the introduction of new face stocks based on polystyrene, polypropylene, and polyethylene films. These new products will gain market share from the current face stocks – coated, uncoated and saturated papers, polyester (PET), PVC, and orientated polypropylene (OPP).

Future market needs include greater productivity from faster line speeds, compliance with VOC requirements, and reduced worker exposure to hazardous solvents. The last two will reduce the volume of solvent-borne adhesives in this market.

⌘⌘⌘

Flexible laminating, packaging, and converting – These markets include: heat seal applications, paper/film and paper/foil laminators, and flexible packaging and laminators.

Heat seal adhesives are tack free, non-blocking adhesive films used to bond substrates under heat and pressure. They may be coated on paper, film and nonwoven products such as duPont's Tyvek® polyolefin. Uses include food, industrial and medical packaging. Food packaging include individual packets of coffee creamers, tea bags, and orange juice containers. Medical applications include breathable Tyvek lidding for clear plastic trays or pouches containing surgical supplies. The porous lid allows entry of ethylene oxide gas to sterilize the surgical supplies but the narrow pores prevent subsequent entry of microbes into the sterile package. Heat seal coated paper, film and Tyvek can also be used as lidding material for toner cartridges for computer printers and copying machines.

Heat seals are based on hot melts and emulsion acrylics. The adhesives are coated onto the paper, film or synthetic nonwoven substrate using roll, pattern or extrusion coating equipment.

Paper/foil and paper/film laminations are used for many applications including candy wrappers, microwave packaging, ice cream carton lids, and reflective building insulation. Flexible laminations are multilayered composites, with each each component providing special properties. Examples include snack food bags, aseptic containers sterilized by radiation or heat, meat packages, solar screen laminations, and flexible air ducts.

The paper/film and paper/foil laminations use a wide range of adhesive systems including neoprene-casein, SBR-casein, SBR, polyvinyl acetate, and EVA. The adhesive is roll coated onto the nonporous substrate and the lamination is achieved by nipping the second substrate to the adhesive-coated one. Flexible packaging and laminating products use mainly solvent-borne adhesives including one and two part polyurethanes and polyesters. VOC compliant adhesives are being developed based on one and two-part water-borne urethanes, polyvinylidene chloride lattices, and 100% solids moisture-curing urethanes. The adhesives are gravure coated and slot die coated using wire rods to apply thin adhesive layers.

Construction (Back, 1990; Prane,1990)

Plywood, particle board and timber laminations – Interior grade plywood and particle board are made using urea formaldehyde adhesives due to their good bonding to wood and low cost. Their disadvantage is poor water resistance. Exterior grades of plywood use phenol formaldehyde adhesives due to their good water resistance and low cost.

Laminated beams and other wooden structures, especially for outdoor applications, use resorcinol-formaldehyde adhesives. These adhesives give joints which have very high structural strength, often stronger than the wood

⌘⌘⌘

itself. The joints have excellent weatherability and resistance to extreme temperature and humidity cycles.

Furniture – There are many different adhesives use in the manufacture of furniture, and cabinetry. Animal glue was the traditional adhesive for furniture making. This has been superseded with a range of new products from the polyvinyl acetate (PVA) "white glue"" popular with the DIY market because of easy application and clean-up, to strong bonding resorcinol formaldehyde adhesives, to fast setting ethylene vinyl acetate (EVA) hot melts. Phenol formaldehyde resins are used as contact adhesives. PVA is also used to laminate PVC film to wood and particle board used in the furniture industry. Polyamide hot melt resins and polyurethane adhesives are also used in this market.

Floor coverings – Wooden floors are glued to plywood subflooring using SBR, natural rubber and neoprene based adhesives. PVC plastic floor tiles use acrylate and nonvulcanized natural rubber adhesives.

Wall coverings – Starch, polyvinyl acetate and acrylic emulsions are all used to glue wallpaper to interior walls of buildings.

Bathrooms – Polyvinyl acetate sealants, acrylic emulsion sealants and silicone rubber sealants are all used as bath-tub caulks and for wall tile and wallboard joints. Polyvinyl acetate and acrylic emulsion sealants are easier to apply and clean up, and are lower cost than silicone sealants. The silicone sealants form stronger adhesive bonds and have better long term durability.

Glazing – Linseed oil based putties were extensively used in the past for glazing applications. Newer glazing sealants include polysulfide, silicone, and acrylic emulsion. Butyl tape, polysulfide and silicone sealants are used in the manufacture of insulated glass. Safety glass door and window panels can be made by sandwiching a layer of polyvinyl butyral film between two layers of glass.

DIY adhesives and sealants – Adhesives for consumer use include polyvinyl acetate, "white glue", for large wooden joints; rubber solution and latex systems for gluing rubberized fabrics, cloth and carpets; silicone rubber for flexible adhesive joints to a wide variety of substrates from metal to cloth; epoxy adhesives for high strength rigid bonds to metals, glass, ceramic, wood and fabrics; and cyanoacrylates for very fast cure and strong rigid bonds to a similar range of substrates as for epoxy resin systems. The cost of the adhesive increases dramatically from white glue to the cyanoacrylates, so that epoxy

⌘⌘⌘

and cyanoacrylate product are typically reserved for very small specialty applications.

There is a similarly wide range of sealants for the DIY household market, varying from oil based putties for glazing; polyvinyl acetate and acrylic emulsion sealants, with easy application and clean-up properties; and butyl, polyurethane and silicone sealants for durable, flexible, long lasting external seals. Again the cost increases progressively from oil based putties to silicone rubber based sealants.

Fire stops – Fire can travel through a building, especially a modern multistory office block, not only through readily visible areas such as offices or hallways, but through invisible openings between curtain walls, floors, and other areas that are inaccessible to conventional sealants. Two component silicone sealant systems have been developed for injection into these spaces. The silicone sealant forms a foam which fills voids and acts as a fire retardant barrier preventing the spread of flames between curtain wall and floors. The silicone foam has an additional benefit in that it can sound proof walls. Conventional one part silicone sealants can be gunned into small joints, cable ways, and other small apertures. to seal them and reduce water damage if a fire does occur.

Intumescents – Intumescent caulks and moldable putties are elastomers under normal conditions but expand up to ten times their normal volume when heated at very high temperatures to form a hard thermal insulating barrier. A major advantage is their ability to expand and fill voids left by combustible fixtures such as plastic pipe, and cables. Water-based latex caulks are also available which emit water to smother the flames when exposed to high heat.

Curtain wall and perimeter joints – Two component polysulfide and polyurethane and one component silicone sealants are used in these applications because they tolerate high joint movement. This ranges up to ±25% for polysulfide and from ±20 to 50% for the polyurethane and silicone sealants.

Concrete pipe seals – Butyl sealant tapes and neoprene rubber are frequently used as concrete pipe seals. The neoprene seals have very good chemical resistance.

Roadways, parking decks, and bridges – Neoprene, two part polyurethane and two part polysulfide sealants are excellent sealants for concrete expansion and paving joints. They offer excellent weatherability and 20 year life expectancy. Neoprene is especially resistant to oil and chemical attack and is

⌘⌘⌘

used in chemical plants. All three sealant systems have good water resistance and can be used for bridge, canal, and marine applications.

Transportation (Clemens, 1990; Prane, 1990)
The transportation industry represents a large market for a wide variety of adhesives and sealants. In 1989, 585 million wet pounds of these materials were used, 92% of which was in autos, trucks, buses and recreational vehicles. The aircraft and aerospace market consumed 6% and the boat industry 2%.

Auto, truck, buses, and recreational vehicles – The breakdown of adhesives and sealants for motor vehicles in the U.S. in 1989 was 25% epoxy, 23% PVC, 20% polyurethane, 14% elastomers, 10% acrylic, 2% silicone, and 6% miscellaneous. One part systems (PVC plastisol, epoxy, polyurethane, silicone) represented 58% of the products used. Solvent-borne and water-borne products accounted respectively for 18% and 11% of the market. Hot melts, two part (epoxy and urethanes) and miscellaneous products represented 6%, 5% and 2% of the market. Consumption in motor vehicles can be further subdivided between structural adhesives and sealants.

STRUCTURAL ADHESIVES
Metal body panels – Car bodies are now made of galvanized steel or plastic SMC (sheet molded compounds) to reduce the corrosion problems of cold rolled steel. Car panels, such as doors composed of inner and outer panels, were conventionally spot welded together. The switch to galvanized steel, which is difficult to spot weld, has resulted in the use of hem flange joints bonded with adhesives, as shown in Figure 6. The adhesive provides a very strong moisture resistant bond and a smoother panel surface than the earlier welded doors. Robot applied one part epoxy adhesives are the prime bonding agents, although some two part epoxies are also used.

SMC panels – Some automobiles now use door, hood and trunk lids made from inner and outer panels of SMC. Two part epoxies and polyurethane adhesives are robotically applied to the inside faces of the two panels which are then held together under heat for 1 to 2 minutes to initiate cure. Cure is later completed in the paint oven. Two part polyurethane adhesives are used to adhere SMC panels to painted space frame parts.

Bumpers – Automobile bumpers are now made from a range of polymers including polycarbonate, polyester thermoplastic elastomers, and reaction injection molded urethanes. These are being bonded to the automobile using a variety of adhesives such as two part acrylics and polyurethanes as well as RTV silicones.

⌘⌘⌘

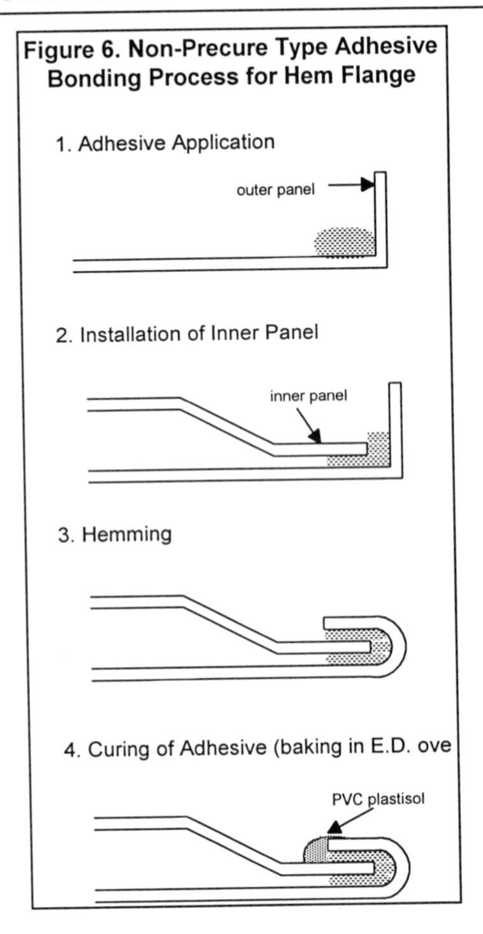

Head lights and tail lights – Glass headlights have been replaced by halogen lamp bulbs inserted into plastic subassemblies that include the lens and reflector. This change has enabled weight reduction and changes in body styling. The lens, reflectors, and housing materials include polycarbonate, nylon and thermoset polyester/glass. The lens can be bonded to the housing with any of several adhesives which can withstand the high temperatures (up to 325°F/162°C) generated by the high intensity halogen bulb. These include one part heat curing epoxies, two part polyurethanes and RTV silicone rubber. Tail light assemblies are also made of plastic materials, although the temperatures generated by the light bulbs are much lower. Butyl sealants with a mechanical clip or silicone adhesives are used.

Trim adhesives – Neoprene adhesives are used to adhere vinyl trim and tops as well as in laminating plastic to wood and steel. Nonvulcanizing natural rubber

⌘⌘⌘

and SBR based adhesives are used to join car interior ceilings (headliners), while hot melts are used in carpet applications.

Miscellaneous – Polyvinyl butyral is used as a clear adhesive film to laminate two sheets of glass to form safety glass for car windshields. Phenol formaldehyde resins are used to make brake lining composites, and blends of nitrile rubber and phenolic resin are used to bond brake linings to brake shoes.

SEALANTS

Formed-in-place gaskets – Silicone RTV sealants are widely used in this application for valve covers, oil pans, thermostats, rear view mirrors. Neoprene sealants have excellent oil and chemical resistance and SBR sealants give good bonding to oily surfaces. Gunnable butyl rubber mastics are used for vibration dampening of panels and as general sealants for frame and body parts.

Windshield sealants – One part polyurethane sealants are the main product used in sealing auto windshields at the factory. They are used in conjunction with a silane primer to increase their adhesion to the metal frame. Butyl rubber tapes are used by repair shops to replace broken windshields.

Underbody sealants – PVC plastisols remain the dominant sealant coating to protect the underbody of cars, although some sprayable waterborne acrylic resins are also used.

Aerospace – The aerospace market is comprised of three sections, commercial aircraft, military aircraft and space vehicles. The performance and long term reliability of the adhesives and sealants is very important in this high technology market. Adhesives have been used for the past twenty years to replace mechanical fasteners (rivets) or welding, especially in the construction of the exterior body and wing members. This change was prompted by the finding that corrosion and metal fatigue crack initiation occurred at the site of the mechanical fasteners and the fastener holes. The advantages of adhesive bonding in aircraft include smoother external surfaces (less drag), use of lighter gauge materials (higher fuel efficiency), and superior bonds in which stress forces are spread uniformly over the whole length of the joint. The usage by weight of adhesives and sealants in this market is 37% polysulfide, 30% epoxy, 20% elastomers (butyl, SBR and silicones) and 13% others (acrylic, polyimide and urethane).

Structural applications – Exterior structural applications include the manufacture of the exterior fuselage, some floor beams and flooring substructures, and wing members. These involve bonding aluminum to aluminum skin, aluminum to honeycomb, or composite to metal. In commercial aircraft these bonds must withstand a performance temperature range of -60 to 250°F (-51 to 121°C). Epoxies which cure at 350°F (177°C) and are toughened with nitrile rubber are preferred. These require a solvent based chromate primer for good bonds. The adhesive for bonding aluminum skins and honeycomb is supplied as a fully formulated film on a release liner because of the precise requirements of the bond line.

Interior structural and laminating applications for civilian aircraft include adhesives for bonding cabin floors, sides and overhead storage bins, and adhesive films for cabin dividers and liners. These interior structural applications use epoxies curing at 250°F (121°C). Laminating adhesives for decorative facing and carpeting are switching from solvent based systems to waterborne and hot melt adhesives with fire retardant properties.

Exterior structural applications in military aircraft are similar to civilian aircraft except that the adhesive bonds must be able to withstand humid jungle conditions, and higher surface temperatures for fighter planes. Adhesives for humid jungle conditions are tested at 95% relative humidity and 200°F (93°C) for extended periods. Most of the current fighter aircraft see skin temperatures of 300 to 350°F (149 to 177°C) and their adhesive needs can be met with high performance epoxies. Special high temperature fluorine based epoxy resins and polyimide adhesives are used in high speed fighter planes which reach speeds up to Mach 2.5 and can have skin temperatures of 400°F (204°C).

Fuel tank sealing – The major applications for sealants in aircraft are in fuel tank sealing and door and window gaskets. These two applications represent 70 – 90% of total sealant usage. Fuel tank sealants must be fuel resistant. This market is dominated by two part polysulfide sealants using lead dioxide or manganese dioxide for crosslinking. These can be used at temperatures up to 300 to 350°F (149 to 177°C). Military aircraft require sealants that can operate at temperatures of 400°F (204°C), which is possible with more stable polysulfide prepolymers and urethane and polythioether polysulfides.

Door and window gaskets – Silicone sealants are widely used for aircraft door and window gaskets. They have excellent UV, ozone, chemical and water resistance and can retain excellent rubber properties (extension, compression and high recovery) over the wide range of -85 to 400°F (-65 to 204°C). The silicone elastomer also does not support combustion.

⌘⌘⌘

ADHESIVE AND SEALANT PROCESSING

Mixing

Complete and uniform dispersion of the ingredients in an adhesive or sealant is critical to product performance. The choice of mixer, from among the many available, depends on the viscosity of the product, any special needs (e.g. vacuum) and how the mixed product is removed form the mixer to the filling line. Table 5 shows the viscosity of everyday items (Ross, 1989) in comparison to typical viscosities for several different types of adhesives and sealants (Flick, 1988).

Table 5. Comparison of the Viscosity of Adhesives and Sealants With Other Materials

Familiar Materials[1] (mPa.s x 10^3)	Adhesives and Sealants[2] (mPa.s x 10^3)
Water = (1 mPa.s)	Cyanoacrylate = 2 – 4
Castor Oil = 1	Epoxy adhesive = 1.2 – 90
Corn Syrup = 10	Urethane epoxy = 5- 9.5
Heavy Printing Ink = 50	Self-leveling urethane = 10
Chewing Gum = 100	Silicone sealant = 45 – 90
Heavy Plastisol = 250	Latex floortile adhesive = 40 – 60
Caulking Compound = 50 0	Acrylic latex caulk = 250
Putty = 2,000	Butyl gutter caulk = 450
Rubber Compound = 5,000	Latex glazing caulk = 750
	Hot melt adhesive at 300°F = 4 – 40
	Polyurethane construction sealant = 1,000

1mPa.s = 1cps
[1](Ross & Sons Co.)
[2](Flick, 1988)

A list of mixer types used in the manufacture of adhesives and sealants is shown in Table 6.

The three most commonly used mixers are:

1) The dispersator, which is good for mixing systems with viscosities from 1,000 to 50,000 mPa.s.
2) Planetary mixers, for systems with viscosities from 2,000 to 3,000,000 mPa.s.
3) Kneaders, for systems with vicosities from 100,000 to 5,000,000 mPa.s. (Ross,1989).

⌘⌘⌘

Table 6. Mixers Used in Manufacture of Adhesives and Sealants

	Adhesives	Sealants
Increasing Dispersion Capabilities	Change can	Dough mixer
↓	Dispersator blade	Low shear change can
↓		
↓	3 roll mill	High shear change can
↓		
		Twin screw extruder mixer

The mixer may have jacketed walls to heat or cool the contents during mixing. In addition, the mixer may be blanketed with dry nitrogen during the addition of the adhesive or sealant ingredients to protect moisture-cure systems, such as one component silicones and polyurethanes. Products of this type also have a vacuum cycle for drying or degassing the mixture. The size of the mixers can vary from 50 to 500 gallons. Scraper blades on the dispersator and planetary mixers give a better mix. For high volume sealant manufacture twin screw extruder mixers are very effective, but very expensive.

Figure 7 is a flow diagram showing a typical process for manufacturing a butyl sealant using dispersion, sigma blade and change-can mixers (Stucker, 1977). This involves first breaking the uncured rubber stock into manageable pieces. These are then dissolved in a solvent using a covered and jacketed dispersator or sigma blade mixer. The butyl rubber solution is pumped from the bottom of the mixer into a covered change-can mixer or covered sigma blade mixer using a positive displacement pump. The filler and other compounding ingredients are then added in small increments. After mixing is complete the sigma blade mixer can be emptied from the bottom using a positive displacement pump or a extrusion screw. For the change-can mixer, the blades are raised and the change-can slid on wheels to the filling room, where a follower plate connected to a hydraulic ram is placed on top of the sealant. As the follower plate moves downward the sealant is forced out of the change can to the filling machine.

At the filling machine, the sealant can be filled into caulking tubes or 1 and 5 gallon pails. If the sealant is a moisture cure system the 1 and 5 gallon pails are blanketed with dry nitrogen gas before sealing.

⌘⌘⌘

Figure 7. Typical Flow Plan for Butyl Caulk Manufacture

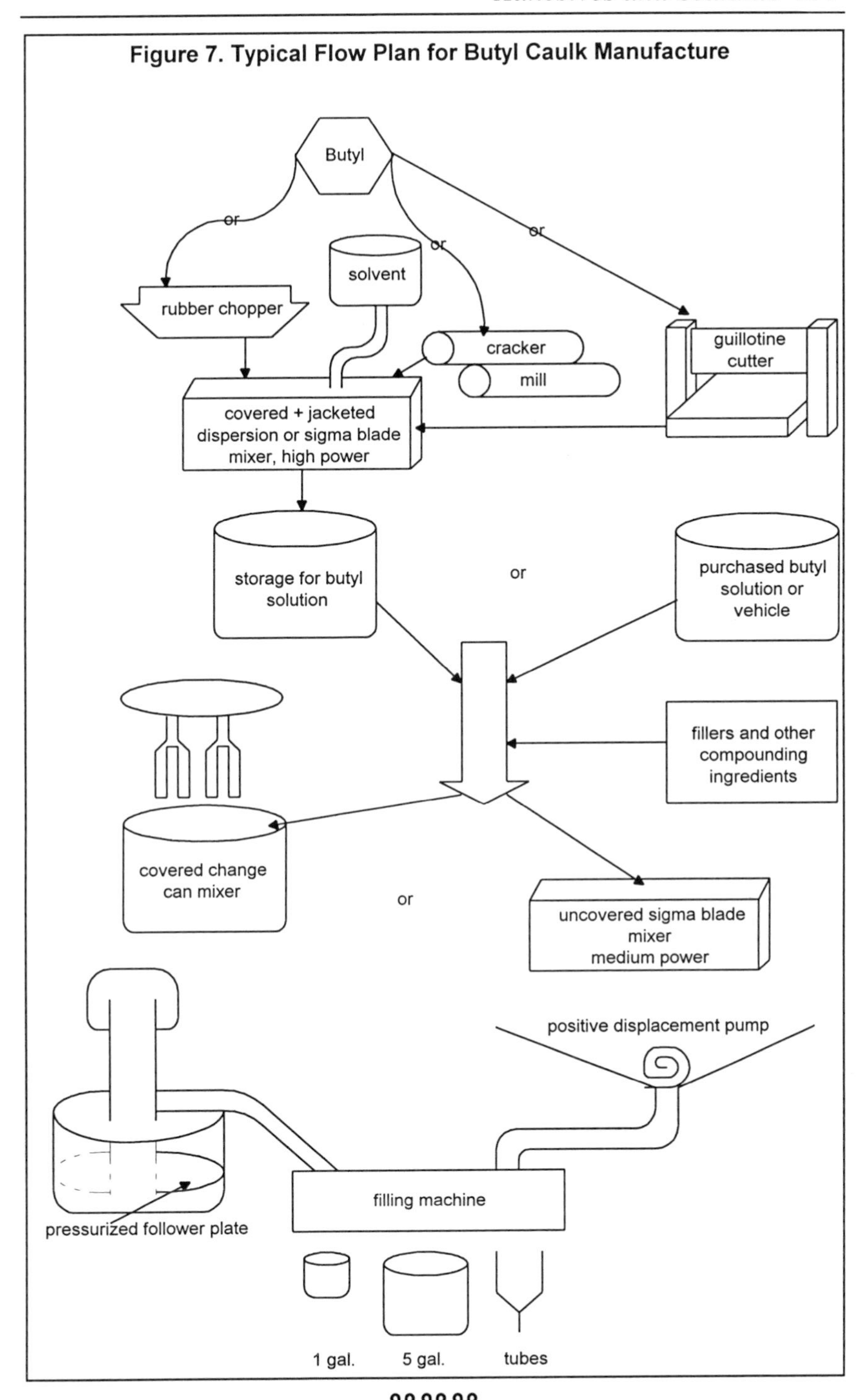

⌘⌘⌘

Application

Two component systems must be blended in correct proportions (usually 1:1) before applying to a substrate. This can be achieved by:

a) Simply adding part B to part A in a pail and mixing with a paddle or motorized propeller stirrer.

b) Pumping components A and B out of separate containers through an application gun with a mixing head.

The method of application depends on the viscosity of the adhesive or sealant and the end-use. This can vary from DIY home projects to fast highly automated automobile production lines. Low viscosity adhesives can be brushed, sprayed, or pumped through automated coating equipment, as in the encapsulation of computer chips with epoxy resin. High viscosity caulks and sealants can be applied with a hand caulking gun or an automated caulking system or a simple putty knife.

ADHESIVE AND SEALANT TESTING

There are a multitude of tests which have been established by various organizations to describe the properties and performance of adhesives and sealants. For construction sealants, these include Federal Specification (TT), Military Specifications (MIL), and American Society of Testing and Materials (ASTM). Major work has been carried out by the ASTM D-14 Committee on Adhesives and ASTM C-24 Committee on Building Seals and Sealants.

Table 7 lists the major tests of adhesive properties (Petronio, 1977). Bear in mind that bond strengths will decrease if the adhesive film is too thin or too thick. The optimum film thickness to give maximum bond strength will vary with the modulus of the adhesive. The optimum film thickness varies from 2 mils for high modulus materials to 6 mils for low modulus materials. Table 8 lists the major tests for sealants (Toporcer, 1986;Prane, 1990). Among the many other important sealant properties are: application temperature range, service temperature range, curing temperature range, life expectancy, adhesion, compression set, shelf life, UV resistance, ozone resistance, water pick-up, solvent resistance, color stability, stain resistance, and dirt pick-up.

CHEMICAL RAW MATERIALS – POLYMERS

Natural Base Polymers (Dick, 1987)

Until the initial development of synthetic polymers in the early 1900s, and certainly before the acceleration of their development from the 1940s to the current time, all adhesives and sealants (putties) were based on naturally

⌘⌘⌘

occurring polymers. Currently less than 10% of total adhesive/sealant sales are made with natural base polymers. The switch to the use of synthetic polymers is related to many factors including better uniformity, higher cured physicals, faster cure, and wider operating temperature.

Table 7. Adhesive Tests

	ASTM	Units
Uncured Properties		
Viscosity		
Rotational viscometer	D2556	cps
Flow through orifice	D1084	secs
Storage life	D1337	cps and/or psi
Coverage	D898, D899	lbs/1000 sq/ft
Blocking	D1146	
Tack		
Penetration	D1916	in/mm
Sag – Leneta chart	D 4400	
Curing Properties		
Pot (working life)	D1338	min
Cure rate		
Tensile,	D1144	psi/time
lap shear as function of time		
Cured Physicals		
Tensile strength	D897	psi
Lap Shear		
metal/metal	D1002	psi
wood/wood	D90G, D905	psi
Peel strength		
180 degree	D903	lb/in
T peel	D1876	lb/in
Impact strength	D950	ft lb/in
Cleavage	D1062	lbs/in
Creep (cold flow)	D2293	
Fatigue	D1061	psi/time

Animal glue – This is obtained from the hydrolysis of the protein collagen which is extracted from animal hides and bones. It is still used as an adhesive in wood working and as a packaging adhesive for tape, fiber cans, and paper tubes.

⌘⌘⌘

Table 8. Sealant Tests

	ASTM	Units
Uncured Properties		
Extrusion rate	C603	g/min
Extrusion rate – shelf aged	C731	g/min
Flow	C639	sec
Sag	C713, D2202 D 2376	ins
Curing Properties		
Skin over time		min
Tack free time		hours
Cure rate		psi/time
Cured Properties		
Durometer	D2240, C661	Shore A units
Modulus	D412	psi
Tensile at break	D412	psi
Elongation at break	D412	percent
Tear		lbs/in
Peel	C794	lb/in

Fish glue – This glue is prepared from fish bones in a similar fashion to animal glue; it is very water soluble.

Dried animal blood – This proteinaceous glue is made from dried blood. It is less water soluble than animal and fish glues and was used in the manufacture of plywood before the development of phenol formaldehyde adhesives.

Casein – This is a proteinaceous adhesive made from milk. Casein was used in the manufacture of plywood and packaging adhesives.

Starch – Starch can be derived from many plants including corn, wheat, rice, and potatoes. It is composed of two polysaccharide forms – amylose (20%) and amylopectin (80%). Starch is water dispersable and widely used as an adhesive in paper packaging and wall papering.

Dextrin – Dextrin is formed by the partial hydrolysis of polysaccharides to give lower molecular weight products. It is used for envelope seals and postage stamps.

⌘⌘⌘

Cellulosic adhesives – Cellulose is a naturally occurring water insoluble polymer which is a major component of plants – wood is 60% cellulose. It is a polysaccharide very similar to starch except that the saccharide units of the macromolecules are connected by beta-linkages in contrast to the alpha linkage for starch. Cellulose can be reacted with various acids to produce water soluble adhesives such as methylcellulose, hydroxyethylcellulose, and carboxymethylcellulose, or organic solvent soluble polymers such as hydroxypropylcellulose and cellulose acetate. The polymers have molecular weights of about 25,000 and are used for adhesives in paper, textile, plastic, and leather shoes.

Oil Based Caulks (Kilchesty, 1986; Prane, 1990)

Until the second half of the twentieth century, oil based caulking compounds or putties were used almost exclusively for sealing joints in buildings against the infiltration of air, dust, and water. They were adequate for sealing the stone and brick structures with small windows framed in wood or metal. Due to building design there was limited temperature induced building/joint movement. This made the oil based sealants, which have a very low movement capacity of $\pm 5\%$, suitable. Maintenance to the caulk seals was carried out during the repainting cycle of the building.

The change to high rise curtain wall buildings with large metal panels and glass windows resulted in the need for sealants with greater movement capacity. This led to the replacement of the oil based putties with elastomeric sealants. The elastomeric sealants also have a much longer life, significantly reducing maintenance costs, which is now primarily labor rather than sealant cost. Oil based sealants are very inexpensive and are still used in residential and light industrial construction and for the maintenance of older buildings.

Oil based caulking compounds are supplied in knife and gun grades. Gun grades are viscous semiliquids whereas knife grades are more viscous and are similar to mortar in consistency. A simple linseed oil putty has a composition of 12 parts oil and 88 parts calcium carbonate filler. The sealants can be applied at temperatures between 40 to 100°F (4 to 38°C). Below 40°F, the viscosity becomes too high for easy working, and above 100°F, the viscosity drops sufficiently for sagging to occur.

The oil base can be blown peanut oil, vegetable oil, soya oil, or fish oil, among others. These oils are 700 to 900 molecular weight triglyceride esters of fatty acids with long chain hydrocarbons. The chains have some unsaturation which slowly react with atmospheric oxygen increasing their molecular weight and effectively drying (curing) the sealant. Blown or oxidized peanut oil is used most frequently. This is prepared by partially oxidizing the heated oil with catalysts and bubbled air.

Elastic glazing compounds can be formulated using calcium carbonate and minor amounts of tremolitic talc, and small quantities of fatty acids for easier working (knifing) properties. Architectural caulking compounds can be made by adding 5 to 8% fatty acids and polybutenes as plasticizers to 20% blown vegetable oil, 56% calcium carbonate and 14% tremolitic talc, plus 4% mineral spirits to improve the caulk workability.

Polymers for Evaporative Adhesives and Sealants

Polyvinyl Acetate (Newman, 1986; Corey, 1977) – Polyvinyl acetate has been commercially available in the USA since the 1930s, and became widely used in the 1940s as a synthetic substitute for hide glue. It is commonly used in the packaging field for paper bags, envelopes, labels, cardboard boxes, and book binding, and as a wood adhesive. It is found in most households as a multi-purpose "white glue". Sealants and caulks based on vinyl acetate homopolymers and copolymers have been on the market since the late 1950s. These are cheap low performance materials which are mostly used in residential applications for bathtub caulking, wall tile joints, and wall board joints where there is little joint movement (±7.5%). A major advantage of polyvinyl acetate latex emulsions is their easy clean-up.

The polyvinyl acetate emulsions were designed for adhesive and coating use rather than for sealants. The homopolymers have little flexibility and are water sensitive. High levels of solvent type plasticizers (e.g. dibutyl phthalate) are used to obtain the required flexibility, but after application the plasticizer volatilizes out on aging. This results in a loss of flexibility in the sealant and failure through cracking if used in areas subject to joint movement.

Polyvinyl acetate emulsions are made by polymerizing vinyl acetate monomer to make a homopolymer or with other monomers to make copolymers which can build flexibility into the polymer chain.

$$\underset{\text{vinyl acetate}}{HC{=}CH_2-O-C({=}O)-CH_3} \longrightarrow \underset{\text{poly(vinyl acetate)}}{\left[-HC(-O-C({=}O)-CH_3)-CH_2- \right]_n}$$

The polyvinyl acetate emulsion is made by adding water to the reaction vessel together with emulsifiers, surfactants, or colloids as stabilizers. Vinyl acetate is added under shear and emulsified. An initiator or catalyst may be added together with the comonomer.

Most of the comonomers used with vinyl acetate are liquids such as n-butyl acrylate, dibutyl maleate, or dioctyl maleate. Gaseous comonomers such as ethylene can also be used but the addition and reactions must be done at pressures up to 2000 psi.

⌘⌘⌘

$$CH_2{=}CH_2 \;+\; \underset{\substack{| \\ O \\ | \\ O=C-CH_3}}{HC{=}CH_2} \;\longrightarrow\; \left[-CH_2-CH_2-\underset{\substack{| \\ O \\ | \\ O=C-CH_3}}{HC}-CH_2- \right]_n$$

ethylene vinyl acetate polyethylene vinyl acetate

The emulsions used for general purpose coatings, adhesive and caulks do not crosslink chemically but dry through water loss. At a critical point in the drying process the emulsion particles are forced close enough to coalesce. Complete evaporation of the remaining water and other volatile components may take hours or days depending on film thickness, drying conditions, and permeability of the resin system.

Caulk formulation typically consist of 30 to 60% emulsion, 30 to 60% filler, and 0 to 10% plasticizer. Small amounts of additives, such as dispersants, surfactants, preservatives, defoamers, thixotropes, adhesion promoters and solvents for tooling are added to optimize performance.

Calcium carbonate is the most widely used filler for polyvinyl acetate caulks. Typically the material has been ground and screened to give a product with >99.5% through a 325 mesh screen (<44 mm diameter) with most particles in the 1 to 30 mm size range. Water ground and dried grades give whiter products. The particle size range of the calcium carbonate filler is important, as substitution of a fine particle size grade, 1 to 10 mm for example, will improve sag resistance but decrease gunnability.

Talcs are sometimes used as fillers in polyvinyl acetate caulks, but the platy particles are not as easily dispersed as calcium carbonate. They do not have the same degree of whiteness and have higher oil absorption values with higher vehicle demand. They impart higher viscosities and thixotropy to the caulk.

Small quantities of attapulgite clays and synthetic silicas are sometimes added to improve the sag resistance and extrudability.

Acrylates (Kilchesty, 1986; Eisentrager, 1977) – The use of acrylic adhesives was first reported in 1928 for making pressure sensitive adhesives, adhesive labels, and price tags. In 1954 grades were developed for laminating fabrics and later for bonding PVC floor coverings. Acrylic sealants were developed in 1958.

Due to the high reactivity of the vinyl groups, acrylic compounds are capable of forming linear polymers of high molecular weight, up to 80,000, by reacting with themselves or to other vinyl compounds such as n-butyl acrylate and methyl methacrylate.

⌘⌘⌘

$$\underset{\text{acrylic ester}}{\begin{array}{c} HC{=}CH_2 \\ | \\ C{=}O \\ | \\ O{-}R \end{array}} \longrightarrow \underset{\text{polyacrylate}}{\left[\begin{array}{c} -HC{-}CH_2- \\ | \\ C{=}O \\ | \\ O{-}R \end{array} \right]_n}$$

Synthesis can be by solvent or emulsion polymerization. Solvent polymerization involves dissolving the monomer or monomer mixture in a solvent such as ethyl acetate, benzene, or toluene. Polymerization occurs through the addition of a free radical initiator such as a soluble organic peroxide. Emulsion polymerization involves adding water, emulsifier, and acrylic acid (neutralized to pH 7 to 8), any comonomers, and a catalyst. A typical latex dispersion is 50% solids, polymer particle size 0.3 mm and a viscosity of 20 mPa.s.

Acrylics have good adhesion to many substrates. This can be further enhanced by the addition of small amounts of unsaturated carboxylic acid during the polymerization process. Acrylic adhesives have replaced animal and vegetable adhesives on modern paper converting machines because of their faster bonding speeds. Acrylic ester adhesives are used for paper, metal foils, plastics, and laminating aluminum foil to paper. Acrylic latex adhesives are used to bond fabric to fabric, fabric to foam, and textiles to PVC polymer film. They are also used to laminate PVC film to wood and particle board in the furniture industry, and in the building industry for bonding flexible flooring and wall paper.

Solvent acrylic sealants are one component mixtures. A typical formula is 40 to 60% polymer; 1 to 4% tremolitic talc, for improved cohesive strength and antisag; 15 to 40% calcium carbonate, talc or clay; 0.05 to 3% tinting color; 2 to 7% plasticizer, to improve low temperature flexibility; and 2 to 5% solvent, such as xylene or toluene. These sealants are gunnable at temperatures above 60°F(16°C), but must be preheated if used at lower temperatures. They "cure" through solvent evaporation and develop a Shore hardness of 5 to 10 in two weeks and gradually increase to a final range of 45 to 60. If the temperature drops to 0°F(-18°C), the durometer will increase by 15 to 20 points.

Acrylic sealants are rubbery low recovery materials and not true elastomers. They have excellent adhesion peel strength, 2.1 to 2.6 kN/m, on a wide range of substrates and can be applied to wet, oily, or dusty surfaces. They stretch to 200 to 250% before failure, which is normally cohesive. Acrylics are best for small joints with limited movement, ±7.5 to ±12.5%. In wider joints they tend to wrinkle under cyclic joint movement. Acrylic sealants show good resistance to a wide range of chemical solvents and to water, but are not suitable for water immersion applications. Acrylics age and weather well, retaining color with very little surface chalking or crazing. They exhibit excellent UV resistance and are a good glazing sealant. Solvent acrylics are used for perimeter caulking of windows, door panels, and precast concrete joints. Their slow cure makes them susceptible to dirt pick-up and unsuitable for homeowners or small contractors.

Acrylic emulsion sealants have many of the same properties as solvent sealants but are easy to apply, adhere to most construction surfaces, clean up easily, and dry rapidly so that they can be painted soon after application. They are used indoors for bathtubs, trim and baseboard seams, and outdoors for masonry joints, roof and siding joints as well as for glazing. They are widely used by homeowners.

Natural Rubber (Wake, 1977) – The two major types of rubber based adhesives are those made from rubber latex and those made from dissolved rubber. In both cases there are nonvulcanizing and vulcanizing products.

The advantage of rubber latex products is lower viscosity and better wettability of surfaces at a given total solids level. The latex adhesives do not emit flammable vapors which require special ventilation for use. The solvent rubber adhesives are faster drying, however. Some latex and solvent rubber cements are used without curatives as one part systems with long pot lives. Others have curatives such as sulphur and zinc oxide added to crosslink the rubber molecules. These are two part systems which are blended just before use. One part of the system contains half of the rubber content plus sulphur and the second part contains the other half of the rubber content plus the accelerator. The blended components normally have a working life of 2 to 4 hours. Vulcanization increases the molecular weight of the polymer, increasing its cohesive strength and the adesive strength of the glue line.

Generally, natural rubber and synthetic rubber elastomers do not have enough natural stick and tack to adhere to many substrates. Tackify resins are added to improve the adhesive strength. Rubber adhesive formulations also contain plasticizers, antioxidants, carbon black, fillers, and colorants.

Natural rubber and synthetic elastomers are viscoelastic and can be used to make pressure sensitive adhesives. These are viscous and can flow under limited pressure to wet the adherend surface, and have enough cohesive strength to prevent ready separation under stress.

Natural and synthetic elastomers are used extensively in caulks and sealants because of their ability to expand and contract easily as the temperature changes. Less flexible polymers would fracture under wide temperature swings.

Nonvulcanizing natural rubber latex adhesives with tackifiers have been used for self-adhesive envelopes, floor tiles, leather applications, and the internal trim of autos. Vulcanizing latex adhesives are used in textile and carpet applications. Solution adhesives are used in surgical tapes, industrial pressure sensitive tapes, and do-it-yourself adhesives. Vulcanizing solvent adhesives are used for joining rubberized fabrics.

Natural rubber is used in mastics, which are high viscosity and become immobile on loss of solvent or water. Often mastics contain drying oils or bituminous components which harden and form a skin over the exposed mastic surface. Mastics are used as fillets in metal or brickwork corners and angles.

Styrene Butadiene Rubber (SBR) (Stricharczuk, 1977; Dick, 1987) – SBR was developed by the U.S. government during World War II when natural rubber was unavailable. The polymer found only limited use in adhesives before 1950 as it imparted poorer tack and cohesive properties than natural rubber. The poor tack has been overcome by adding tackifiers such as rosin and rosin esters, adding plasticizers, reducing the average polymer molecular weight, and widening the molecular weight distribution. SBR adhesives show good heat aging and water resistance, dissolve readily in hydrocarbon solvents, and have a lower cost than natural rubber adhesives. SBR adhesive sales have overtaken those of natural rubber.

SBR is made by emulsion copolymerization of butadiene and styrene under two different temperature conditions, at 122°F (50°C; Hot SBR) and at 41°F (5°C; Cold SBR). The lower molecular weight and broader molecular weight range of hot SBR favors its use in adhesives. The low molecular weight fraction provides quick stick, while the high molecular weight fraction provides shear strength.

SBR based adhesives are typically produced by first masticating the polymer on a two roll mill or Banbury mixer. Antioxidants, fillers, and other additives are then introduced. The processed rubber can be either calendared directly onto the adherend, dispersed and dissolved in a solvent with tackifying resins to form adhesive solutions, or dispersed in water with emulsifiers to form a latex.

SBR solvent adhesives are used as pressure sensitive adhesive, sprayable adhesives for paper, cardboard, wood, plastic, and cloth, tiretread adhesives, plastic tile adhesives, and as panel and flooring adhesives in the construction industry.

SBR latex adhesives are used for tire cord adhesives, packaging adhesives, tufted carpet adhesives in combination with natural rubber, and as a fabric adhesive for combinations of fabric to fabric, fabric to paper, and fabric to leather.

SBR is soluble in hydrocarbon solvents such as toluene, heptane, and hexane to give adhesives with long term stability and resistance to gelling. Solvent based SBR sealants and caulks are easy to compound , have good

⌘⌘⌘

tack, low cost, and bond well to oily surfaces. They require high dilution, however, as SBR has a high solution viscosity. This resulrs in substantial shrinkage on curing. They also tend to harden (crosslink) on long term aging due to some chain unsaturation in the SBR.

Butyl Rubber and Polyisobutylene (Stucker, 1977; Newton, 1986) – High molecular weight butyl rubber was developed in the 1930's and became commercially available after World War II. Butyl rubber and polyisobutylene are elastomeric polymers which are widely used in adhesives and sealants as primary binders, tackifiers, and modifiers. Butyl rubber is a copolymer of 97 to 98% isobutylene and 2 to 3% isoprene. The slight amount of unsaturation can be used to crosslink the polymer chains.

$$\underset{\text{isobutylene}}{(CH_3)_2C{=}CH_2} + \underset{\text{isoprene}}{CH_2{=}C(CH_3){-}CH{=}CH_2} \longrightarrow \underset{\text{butyl rubber}}{\left[{-}(CH_2{-}C(CH_3)_2)_{50}{-}(CH_2{-}C(CH_3){=}CH{-}CH_2){-} \right]_n}$$

Polyisobutylene is a homopolymer of isobutylene.

$$\underset{\text{isobutylene}}{(CH_3)_2C{=}CH_2} \longrightarrow \underset{\text{polyisobutylene}}{\left[{-}C(CH_3)_2{-}CH_2{-} \right]_n}$$

Butyl rubber has very little unsaturation and polyisoprene has none, making both polymers very stable to oxidative degradation from aging, weathering, or heat. They are superior in this respect to SBR or NBR (nitrile) polymers. Both polymers are resistant to chemicals, vegetable and animal oils, and water, but are soluble in hydrocarbon solvents such as hexane and toluene. The polymers are amorphous, flexible chain hydrocarbons with no polarity. Even though their tack is good, they have to be mixed with resins and other materials to impart some polar character to give good adhesion to polar surfaces.

Butyl rubbers are normally high molecular weight polymers but they are also available in several grades including low molecular weight, partially crosslinked, and as latex. Polyisobutylene is available in several forms from low molecular weight soft and tacky semi-liquids to strong tough high molecular weight elastomers.

Butyl and polyisobutylene are used for paper adhesives, laminating polymer film, pressure sensitive adhesives, pipe wrap, and electrical tape. Butyl latex adhesives can be formulated for a variety of special use foil/paper laminates, labels, plywood/polyolefin laminates and pressure sensitive adhesives.

Butyl rubber is used as the binder in a wide range of commercially available caulking and sealing compounds. One component solvent release caulks find use as general purpose construction sealants for channel glazing, do-it-yourself home repair, and joints with movement of ±10 to ±15%. Architectural sealing tapes (usually vulcanized) are extruded semisolid sections with pressure-sensitive adhesion. They are 100% solid and have no problem with solvent evaporation and subsequent shrinkage. The product is sold in rolls interlined with release paper. They vary from low cost highly polybutene-extended tapes for trailers and mobile homes to fully vulcanized butyl rubber for glazing applications in high rise buildings.

Butyl rubber tapes are also used in automobiles for the glazing of windshields and back lites. Other automotive applications for butyl tapes and gunnable mastics are as after-seals for windows, as vibration dampening materials on panels, and as general sealants on frame and body parts.

Recently hot melt extrudable sealants tapes have been developed. These are premasticated, high viscosity thermoplastic materials that can be fed as ribbons coated with talc into a sealant applicator. The compound is heated within the applicator, extruded into the joint opening, and on cooling gives a smooth joint with excellent adhesion. Hot applied, post-formed butyl tapes have been used to seal insulated glass units, curtain walls, and concrete pipes.

Nitrile Rubber (NBR) (Morrill, 1977; Dick, 1987) – Nitrile rubber was first commercialized in Germany in the late 1930's and in the U.S. in the early 1940's. Nitrile rubbers are copolymers of an unsaturated nitrile, usually acrylonitrile, and dienes, usually butadiene. Nitrile rubbers are much more polar than other elastomers and hence have good oil resistance. Polarity and oil resistance increase with the acrylonitrile (VCN) content. Nitrile rubber with 25% VCN has medium oil resistance and at 40% VCN level has excellent oil resistance. The higher polarity of the nitrile rubber compared to other elastomers has another advantage in that it is compatible at relatively high loading with a variety of different resins including phenol formaldehyde, alkyd resins, natural resins, polyindene, epoxies, and PVC (polyvinyl chloride). High VCN polymers with high resin contents have high cohesive strength and good adhesion to polar substrates such as metals, paper, wood, and textiles.

NBR products are available as solvent based adhesives, mastics, latexes, and tapes. Solvent based adhesives normally contain 20 – 35% nitrile rubber

⌘⌘⌘

and are dissolved in methyl ethyl ketone or chlorinated solvents. Other solvents commonly used for rapid evaporation are acetone and ethyl acetate. Slower evaporating solvents include nitroethane, chlorotoluene, and butyl acetate. Care must be taken in mixing and applying these adhesives due to the flammability and toxicity of these solvents.

NBR adhesives have excellent shelf stability and high temperature properties. They are used in aircraft, automobile, paper, and electronics applications. Blended with phenolic resins, they have been used to laminate aluminum and stainless steel, fabricate airplane structures, bond abrasives to metal, brake linings to brake shoes, laminate leather, attach soles in shoe manufacturing and bond cardboard, polymeric films, masonite, wood and metal to each other and themselves. Nitrile rubber-phenolic resin blends are poor adhesives for non polar elastomers such as natural rubber, butyl rubber, polyethylene unless their surfaces have been activated to improve adhesion.

NBR has little use in sealants due to the very high solvent levels required to produce gunnable products.

Polychloroprene (Neoprene) (Steinfink, 1977; Dick, 1987; Prane, 1990) – Polychloroprene polymers were developed in 1932 but not widely used because of high price until natural rubber became unavailable during World War II. Neoprene is manufactured by the polymerization of 2 chloro-1,3-butadiene. During polymerization the chloroprene molecule is added to the polymer chain in four different isomeric forms.

$$-CH_2-C(Cl)=CH-CH_2-$$

trans-1,4

$$-CH_2-C(Cl)=CH-CH_2-$$

cis-1,4

$$-CH_2-C(Cl)(CH=CH_2)-$$

1,2

$$-CH_2-CH(C(Cl)=CH_2)-$$

3,4

The proportions of these configurations determines the reactivity to vulcanization (1, 2 addition) and the amount of crystallization (trans-1, 4). Neoprene is more expensive than natural rubber or SBR but it is used more frequently for several reasons. Neoprene adhesives based on polymers with high trans-1,4, content crystallize quickly giving high bond strengths.

⌘⌘⌘

Neoprene adhesives show excellent resistance to attack by oils, solvents, water, acids, chemicals, heat, and weathering. There are many grades of Neoprene polymer available for use in both solvent based and latex adhesives. Neoprene adhesives, because of their high bond strengths, are used in the manufacture of shoes (gluing soles), for adhering automobile vinyl trim and tops, and in laminating plastic to wood and steel.

Neoprene sealants are available as one or two part systems with good elastomeric properties, modulus, exterior durability and abrasion resistance, but poor color stability and extended cure times when compounded for gunning. Many of their applications, such as chemical plant use, automotive, marine, and concrete pipe seals are due to their excellent oil and chemical resistance.

Thermoplastic Polymers

Hot melt adhesives and sealants are based on thermoplastic polymers. These polymers melt at elevated temperatures and can wet an adherend surface. On cooling the polymer can resolidfy forming a hard glue line, often reforming crystalline regions in the polymer matrix. Additives such as tackifying resins, waxes, plasticizers, fillers, and antioxidants are usually added to the thermoplastic polymer to improve the hot melt adhesive and sealant properties The use of hot melt adhesives dates back several hundred years as the waxes used to seal letters and official documents are simple thermoplastic polymers.

Polyethylene (Dick, 1987) – Polyethylene is a homopolymer of ethylene.

$$\underset{\text{ethylene}}{CH_2{=}CH_2} \longrightarrow \underset{\text{polyethylene}}{CH_3\text{-}(CH_2)_n\text{-}CH_3}$$

Polyethylene hot melts are low cost and adhere well to most porous easy to bond substrates and in applications where low temperature flexibility is not needed. Polyethylene hot melts are used primarily for sealing paper products, especially multiwall bags, cartons, and cases. They are not used more widely due to low adhesive strength on many substrates such as film, metals, and fabrics. This is due to the absence of strong polar groups, such as acetate, acrylate, or hydroxyl on the backbone of the polyethylene polymer chain.

High density polyethylenes have higher cohesive strength than low density polyethylenes but poorer wettability of surfaces when melted. These differences in properties are related to the linear structure and high crystallinity of the high density polyethylene and the branched chain structure of the low density polyethylene.

⌘⌘⌘

Ethylene Vinyl Acetate (EVA) (Dick, 1987) – EVA is the most useful and widely used thermoplastic polymer in hot melt adhesive systems.

$$CH_2{=}CH_2 + HC{=}CH_2\text{–O–C(=O)–}CH_3 \longrightarrow \left[CH_2\text{-}CH_2\text{-}HC(\text{–O–C(=O)–}CH_3)\text{-}CH_2 \right]_n$$

ethylene vinyl acetate polyethylene vinyl acetate

The copolymer is more expensive than polyethylene but is commercially available in several different grades varying primarily in % vinyl acetate content (18 to 40 wt% VA) and molecular weight (0.4 to 500 Melt Index). Grades with high vinyl acetate content give good adhesion to polar surfaces such as aluminum, steel, and vinyl, while grades with low vinyl acetate content give good adhesion to nonpolar surfaces such as polyolefins. Increasing the molecular weight of the EVA generally improves the low and high temperature adhesive strength. Another very significant advantage of EVA over other thermoplastic polymers is its ability to be formulated with a wide range of plasticizers, tackifying resins, and waxes. The end uses vary from soft tacky pressure sensitive adhesives to tough rigid hot melt furniture adhesives.

Ethylene Ethyl Acrylate (EEA) (Dick, 1987) – A much narrower range of EEA polymers are available than for EVA. This has reduced the number of hot melt adhesives developed using EEA to a few specialty formulations in which EEA imparts superior properties. These include superior adhesion to polyethylene and polypropylene, improved thermal stability and lower price.

$$CH_2{=}CH_2 + HC{=}CH_2\text{–C(=O)–O–}C_2H_5 \longrightarrow \left[CH_2\text{-}CH_2\text{-}CH(\text{–C(=O)–O–}C_2H_5)\text{-}CH_2 \right]_n$$

ethylene ethyl acrylate polyethylene ethyl acrylate

Ethylene Acrylic Acid (EAA) – EAA copolymers have poorer heat stability and are more expensive than EVA but have superior adhesion to metals and glass.

⌘⌘⌘

SBS and SIS Thermoplastic Rubbers (Harlan, 1977; Chu, 1986) – Styrene-butadiene-styrene and styrene-isoprene-styrene are thermoplastic rubber block copolymers. They were first marketed commercially in 1965. The polymers have rubbery midblocks of butadiene or isoprene molecules and two plastic end blocks of styrene molecules. The polymers have the modulus and resilience of vulcanized butadiene and isoprene at room temperature and act as thermoplastics at higher temperatures. When SBS or SIS molecules are combined in the solid phase, a two-phase structure is formed by the clustering of the styrene endblocks. The plastic endblock regions are called domains which act as crosslinks between the ends of the rubber chains (butadiene or isoprene) locking them in place. The block copolymers act like a typical vulcanized rubber that is filled with dispersed reactive filler particles.

Thermoplastic rubber polymers can dissolve quickly in many low cost hydrocarbon solvents to form high solid low viscosity solutions. They can be mixed easily with resins, fillers and other additives to give solution adhesives or hot melt adhesives and sealants that are tailor made. The final properties of the thermoplastic rubber in these systems depends on the compatibility of the added materials in the rubber and plastic phases of the polymer.

Thermoplastic rubber polymers show good resistance to most aqueous reagents, good electrical insulating properties and excellent low temperature properties. Improvements can be made in solvent resistance and high temperature strength by chemical crosslinking or addition of insoluble polypropylene polymer.

Polyamide Resins (Dexheimer, 1977; Dick, 1987) – These resins are widely used in adhesives primarily because of their excellent adhesion to a wide range of substrates. They are prepared by the polycondensation of diamines and diacids. Early resins used single diacids and diamines; more sophisticated resins are now available which use combinations of diacids and diamines to develop optimum properties. They vary from low molecule weight branched chain fluids to high molecular weight linear chain solids. The resins are used in four different types of adhesives: liquid thermoset adhesives, nylon epoxy adhesives, thermoplastic hot melt adhesives, and thermoplastic resins which interact when molten to form thermoset resins. The branched chain polyamide fluids are used in thermoset adhesives where unreacted amino groups on the polymer react with epoxy resins. Nylon epoxy adhesives are alloys of epoxy resins and polyamide resins. Polyamide hot melt adhesives are linear high molecular weight polymers which are nonreactive with epoxies. They have good adhesion to paper, wood, metals, and film due to the amide and amine groups on the polymer chain. In addition, these same groups impart good cohesive strength to the polymer as soon as it is cooled below its melt point.

⌘⌘⌘

Polyamide resins are used in a wide variety of adhesives for aircraft, automotive, construction, woodworking, household repair, electrical and electronic, and shoes.

Polyvinyl Butyral (PVB) (Dick,1987) – Formulations based on polyvinyl butyral resins make excellent hot melt adhesives that form tough clear films with good adhesives strength. Polyacetals such as PVB are prepared by reacting aldehydes and polyvinyl alcohol. The structure for PVB is shown below.

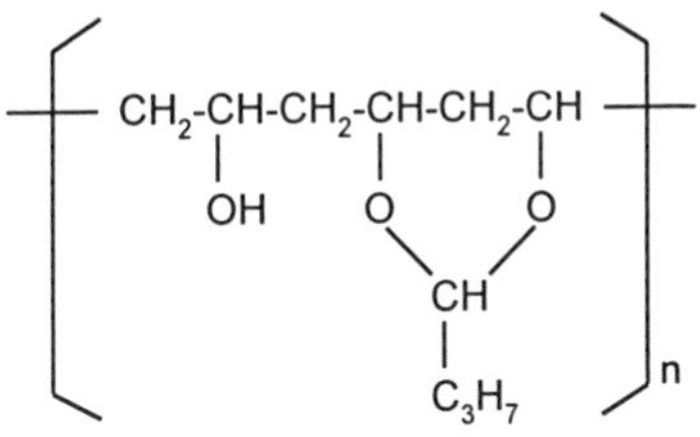

$$\left[CH_2\text{-}\underset{OH}{\underset{|}{CH}}\text{-}CH_2\text{-}\underset{O}{\underset{|}{CH}}\text{-}CH_2\text{-}\underset{O}{\underset{|}{CH}} \right]_n \quad \text{(the two O atoms joined through } \underset{C_3H_7}{\underset{|}{CH}}\text{)}$$

polyvinyl butyral

The largest use of PVB is as the interlayer in laminated safety glass for automobiles and architectural uses such as glass doors and windows. The laminate is composed of two pieces of glass permanently bonded together under heat and pressure with a plasticized polyvinyl butyral interlayer. If the glass is suddenly cracked or broken, the fragments of glass are prevented from flying off by the excellent adhesive properties of the film. The PVB film retains its excellent adhesive, optical, tensile, and chemical properties on long term environmental exposure.

Reactive Base Polymers

Phenol-Formaldehyde Resins (Barth, 1977; Dick, 1987) – Phenol-formaldehyde resins are prepared by the controlled condensation reaction of phenol and formaldehyde releasing water as a by-product.

There are two basic types of resin, one is called a two step novolac and the other is called a resole resin. The two step novolac is made using an acid catalyst with a stoichiometric excess of phenol to prevent the polymer from reaching a gel point where it crosslinks three-dimensionally. The novolac resins are cured by adding a methylene donor such as hexamethylene tetramine and then reheating the resin in its adhesive application.

A resole resin is made by condensing phenol and formaldehyde in the presence of an alkaline catalyst using a higher level of formaldehyde than used with novolac. The reaction is stopped by cooling before the gel point is reached. The adhesive is activated by heating just before use so gellation occurs. Resoles are frequently used in alcoholic or aqueous dispersions which allows good penetration in plywood applications.

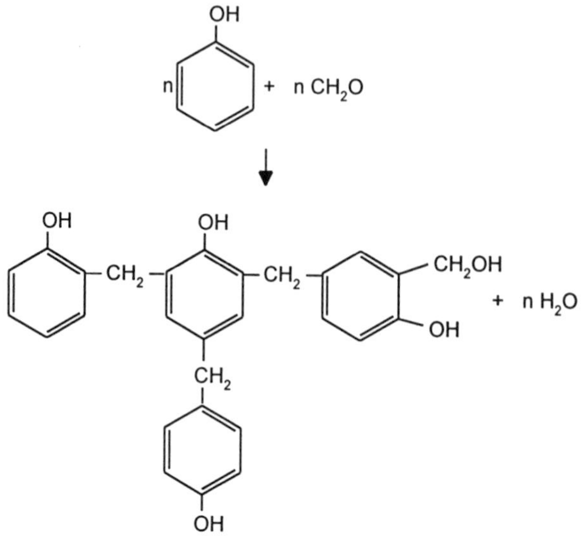

Phenol formaldehyde resins have the largest volume use of any synthetic adhesive. The major market is for adhesives for exterior grade plywood. This is due to their low cost and good water resistance. Other applications are as contact adhesives for furniture and shoes, in foundry sand applications, and brake lining composites.

On curing, phenolic resin adhesives form strong rigid brittle bonds which are unsuitable for some construction application. Flexibility and toughness can be introduced by compounding the resin with elastomers such as nitrile or neoprene. These "alloys" are tougher and stronger than the elastomers or phenolic adhesives alone. The nitrile/phenolic alloy gives high peel strength and oil resistance.

Resorcinol-Formaldehyde Resins (Moult, 1977; Dick, 1987) – Resorcinol adhesives are condensation products of resorcinol with formaldehyde or various phenol-formaldehyde resoles. Most of the adhesives manufactured are of the second type due to their lower cost. Resorcinol is much more reactive with formaldehyde than phenol since it has two meta hydroxyl groups. The two groups reinforce each other in their activation of the ortho and para positions of the benzene ring. To make stable resorcinol/formaldehyde resins which do not gel on aging, 0.5 to 0.7 moles of formaldehyde is added per mole of resorcinol. At the time of use, some additional formaldehyde is added and the resin is quickly converted to a very highly crosslinked adhesive. High strength bonds are possible at curing temperatures close to 32°F (0°C).

Resorcinolic adhesives are useful in bonding many different materials including wood, paper, glass, ceramic, concrete, textiles (including cotton, rayon and nylon), thermoset resins, rubber and primed metals. They are fast effective adhesives for wood, giving joints which have high structural strength often stronger than the wood itself. The joints have excellent weatherability and resistance to extreme temperature and humidity cycling, freezing and thawing, and many solvents except for very caustic solutions. Aged joints under heavy loads show no evidence of creep. A major market for resorcinolic adhesives is in the production of laminated beams and other wooden structures especially those used for exterior construction.

Urea-Formaldehyde Resins (Dick, 1987) – These resins are formed by the polycondensation of urea and formaldehyde with water as a by-product.

$$NH_2CONH_2 + 2CH_2O \longrightarrow HOCH_2NHCONHCH_2OH$$

$$RNHCH_2OH + HOCH_2NHR^* \longrightarrow RNHCH_2OCH_2NHR^* + H_2O$$

Urea formaldehyde resins are water soluble and usually sold as aqueous solutions. They are low cost and have a clean colorless appearance. The resin can cure at room temperature using catalysts such as ammonium chloride and sulphur, but the cured adhesives have poor water resistance compared with phenol formaldehyde resin products.

Major uses for urea formaldehyde adhesives are in the manufacture of particleboard and interior grade plywood.

Melamine-Formaldehyde Resins (Dick, 1987) - These resins are produced by the polycondensation of melamine and formaldehyde with water as a by-product.

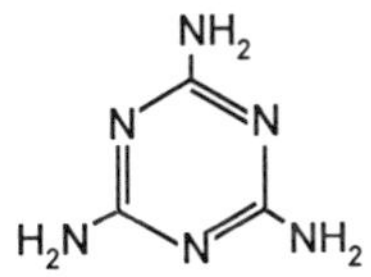

melamine

Melamine resins are more expensive than urea formaldehyde resins and have the same end uses. The melamine resins give a higher crosslink density

⌘⌘⌘

as melamine is hexafunctional whereas urea is tetrafunctional. Melamine resins also have better heat stability and water resistance.

Cyanoacrylates (Coover, 1977; Dick, 1987) – Cyanoacrylate adhesives are one part moisture cure systems which can rapidly form very strong bonds with a wide range of substrates. The adhesives was first commercialized in 1958 and most of those in use today are based on methyl 2-cyanoacrylate and ethyl 2-cyanoacrylate. The extremely fast cure is related to the presence of highly electronegative alkoxy (-COOR) and nitrile (-CN) groups in the alkyl 2-cyanoacrylate molecules. This allows initiation of anionic polymerization by weak bases such as alcohol or water at room temperature. The polymerization is propagated by the resultant carbanion as shown below.

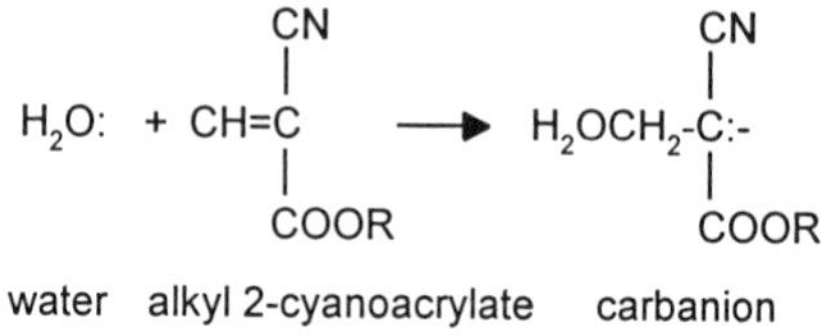

$$\underset{\text{water}}{H_2O:} + \underset{\text{alkyl 2-cyanoacrylate}}{CH{=}\overset{\overset{CN}{|}}{\underset{\underset{COOR}{|}}{C}}} \longrightarrow \underset{\text{carbanion}}{H_2OCH_2\text{-}\overset{\overset{CN}{|}}{\underset{\underset{COOR}{|}}{C}}\text{:-}}$$

The extremely fast cure (seconds to minutes), low viscosity (mPas), and ability to form strong bonds with many substrates, including glass, ceramic, aluminum, steel, rubber, plastic, wood, and their combinations, make the cyanoacrylates excellent for assembly line bonding. These adhesives have good solvent resistance but show some loss in bond strength on aging in water and in air at temperatures above 212°F (100°C).

Disadvantages include high cost, fast setting times, and need to avoid contact with skin and eyes, and prolonged breathing of the vapor. Despite these disadvantages the cyanoacrylate adhesives are widely used by homeowners for special repair and do-it-yourself projects.

Epoxy Resins (Salva, 1977; Dick, 1987) – Adhesives based on epoxy resins are very versatile, have high strength and give excellent adhesion to a wide range of substrates. Most of the adhesives are based on resins made by the reaction of epichlorohydrin and the diol, bisphenol A.

Epoxy resins are usually cured with reactive hardeners containing primary and/or secondary amines. Each hydrogen atom attached to the nitrogen can open an epoxy ring.

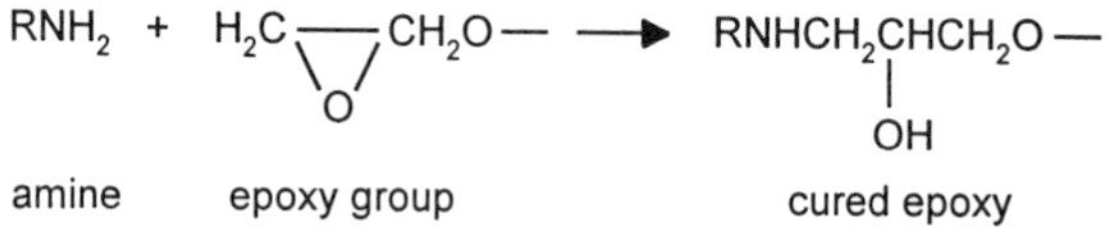

$$\underset{\text{amine}}{RNH_2} + \underset{\text{epoxy group}}{H_2C\underset{O}{\diagdown\!\diagup}CH_2O-} \longrightarrow \underset{\text{cured epoxy}}{RNHCH_2\underset{\underset{OH}{|}}{C}HCH_2O-}$$

⌘⌘⌘

Aliphatic primary and secondary amines such as diethylenetriamine (DETA) and diethylaminopropylamine (DEAPA) react rapidly with epoxy groups to give room temperature cures. These cure systems are the basis for the typical two part epoxy adhesives.

Other crosslinkers such as aromatic amines, dicyandiamide, dihydrazides, and boron trifluoride amine complexes give stable formulations at room temperature and only cure when heated at elevated temperatures. These cure systems are the basis for one part epoxy adhesives. Crosslinkers such as polysulfides and polyamides can be used to increase the flexibility of the adhesive bonds. Other components that may be added to modify the properties of the adhesives are diluents, flexibilizers, modifiers, fillers, and reinforcing agents.

Commercial adhesives are sold as one or two part systems. One component adhesives may be 100% non volatile liquids, solvent solutions, pastes, sticks, powders, supported or unsupported films. Two part adhesives have the resin and hardener (curative) in two separate containers as liquids or paste; these are mixed just before use. These systems normally cure at room temperatures.

Epoxy adhesives can bond to glass, ceramics, metal, wood, concrete, many plastics, fabrics, and any combination of these materials. The excellent adhesion is due to their low viscosity, which allows good surface wetting, and the presence of polar epoxide, hydroxyl, and amine groups. Epoxies also have good cohesive strength and often stress failure occurs in the adherend rather than in the adhesive. As most epoxy adhesives are 100% nonvolatile liquids, there is very little bond line shrinkage on cure, especially compared with polyesters and acrylics. Bond line shrinkage can be reduced to < 1% by the addition of inorganic fillers. Epoxies are thermoset resins and show very little creep on aging compared with thermoplastic adhesives such PVA or polyvinyl butyral. Epoxy adhesives can be formulated for long term use at cryogenic or high temperature (>500°F/260°C) conditions. They show excellent resistance to water and solvents, and are good electrical and thermal insulators. Epoxies can be formulated to have very short cure times (minutes) making them suitable for use on production assembly lines.

The disadvantages of epoxy adhesives are the need for care in handling, as some components may be toxic, the low pot and shelf-life, and their moderate to high cost. Major uses for epoxy adhesives include aircraft manufacture, automotive, and do-it-yourself markets.

Polysulfide Polymers (Panek, 1977; Peterson, 1986; Prane, 1990) – Liquid polysulfide polymers have been used in commercial adhesives and sealants since the early 1950's. The polymer is formed by the reaction of bi-chloroethylformal and sodium polysulfide.

⌘⌘⌘

$$ClC_2H_4OCH_2OC_2H_4Cl + NaS_x$$

$$\downarrow$$

$$Cl\text{-}(C_2H_4OCH_2OC_2H_4SS)_n\text{-}C_2H_4OCH_2OC_2H_4Cl + NaCl$$

The polymer is then split and terminated with mercaptan groups by replacement of the chlorine by NaSH. Small quantities of 1,2,3-trichloropropane are used to produce branching of the liquid polysulfide (LP) chains.

$$HS(C_2H_4OCH_2OC_2H_4SS)_n\text{-}C_2H_4OCH_2OC_2H_4SH$$

The mercaptan-terminated LP is readily polymerized to an elastomeric solid by the use of active oxidizing agents. The most widely used curing agent is lead dioxide which cures the LP as shown below. LP polymers are available with average molecular weights ranging from 1,000 to 80,000.

$$2RSH + PbO_2 \longrightarrow \text{-RSSR-} + PbO + H_2O$$

The lower molecular weight LP-3 and LP-33 are used as epoxy resin flexibilizers for adhesives, coatings, potting compounds, and plastic tooling compounds.

The actual change in the cured epoxy properties is dependent on the level of LP used. In general, elongation, impact resistance, and tensile strength increase with increasing LP level, while brittleness decreases.

Polysulfide sealants are widely available as one or two part systems and as gunnable products packaged in cartridges or in bulk. The elastomeric sealants are widely used in construction, glazing, aircraft, and marine applications.

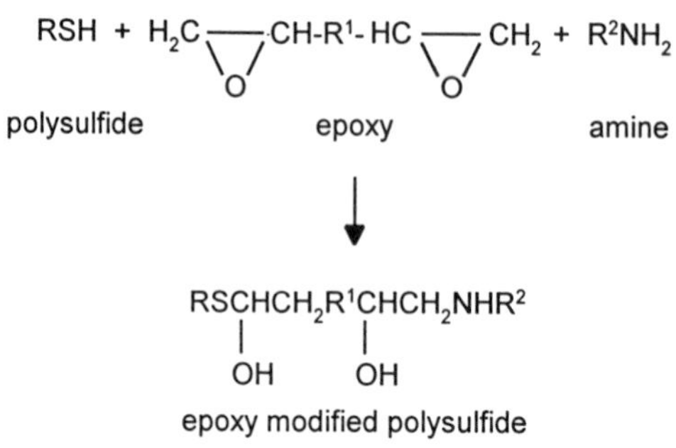

$$RSH + H_2C\underset{O}{\diagdown\!\diagup}CH\text{-}R^1\text{-}HC\underset{O}{\diagdown\!\diagup}CH_2 + R^2NH_2$$

polysulfide   epoxy   amine

$$\downarrow$$

$$RS\underset{OH}{\underset{|}{C}}HCH_2R^1\underset{OH}{\underset{|}{C}}HCH_2NHR^2$$

epoxy modified polysulfide

⌘⌘⌘

The two component sealants consist of a LP polymer base and an activator containing the curing agent. Construction sealants normally contain 35 to 55% LP polymer extended with 10% phthalate plasticizer. Two to three percent of phenolic or epoxy resin is added to improve adhesion, as LP polymers alone do not adhere well to solid surfaces. Twenty parts of surface treated precipitated calcium carbonate are added for reinforcement, and 2 to 4% of inorganic thixotropes, such as fumed silica or amine modified bentonite clay may be added to prevent slumping during application and cure. The second component contains the activator system of lead dioxide, stearic acid to retard the cure rate, and additional plasticizer. Ninety percent of final cure properties in strength, adhesion, and elastomeric properties are usually achieved within 24 hours. The cure rate is accelerated by high humidity and high temperature and retarded by low humidity and low temperature. It is necessary to ensure that all joint surfaces are very clean or have a primer coat.

Polysulfide sealants have good elasticity and can accommodate joint movements of ±25%. They have good UV and water resistance, fast cure, good adhesion, and a performance life of up to 20 years. Two component polysulfides are very widely used in construction, for curtain wall sealing, expansion and contraction joints, precast concrete joints, glazing, and insulated glass sealing; as a joint sealant for highways, bridges, and canals; for aircraft construction and fuel tank sealing; for marine construction; and for automobile windshields.

One component polysulfide sealants are moisture activated systems. Their compositions are similar to those of two part polysulfide sealants except that the curing agent is an alkali oxide such as calcium peroxide at a loading of about 2 wt%. The calcium peroxide is activated by atmospheric moisture to initiate the LP polymer chain extension and crosslinking reaction. For this reason the sealant must be formulated with moisture free ingredients and mixed and packaged under very dry conditions. Normally moisture scavengers such as molecular sieves are added to the formula to prevent premature crosslinking in the sealant package during storage. Gunnable grades contain about 30% LP polymer with 5% toluol added to improve extrusion properties.

The sealant skins quickly at relative humidities above 50%. The rate of cure is dependent on the temperature, relative humidity and seal depth, and area of exposure of the seal surface to the atmosphere. Full cure may take several days or weeks if the humidity is low.

Polyurethane (Schollenberger, 1977; Regan, 1986; Prane, 1990) – Polyurethane adhesives were first developed in Germany during World War II. Polyurethanes are formed by the reaction a polyhydroxyl compound with a di or polyisocyanate as shown below.

⌘⌘⌘

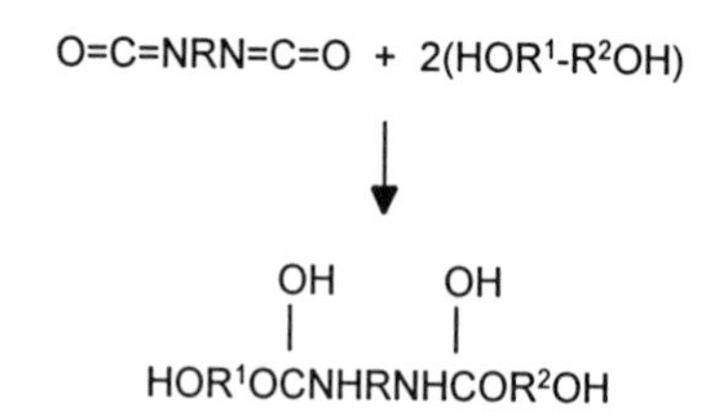

Due to the polar groups on the polymer chain, polyurethane adhesives will adhere to many different surfaces, including metals, textiles, wood, plastics, and concrete. An advantage of polyurethane adhesives over epoxies is in their much greater flexibility on curing due to their relatively low crosslink density.

Amines and water can react with unreacted isocyanate groups on the polyurethane chain to cause chain extension and cure.

$$RN{=}C{=}O + R^*NH_2 \longrightarrow RNHCONHR^*$$

$$2RN{=}C{=}O + H_2O \longrightarrow RNHCONHR + CO_2$$

The latter reaction is the curing mechanism used in one component polyurethane sealants. Some isocyanates are available in a blocked structure in which the reactive isocyanate group is only accessible for crosslinking at elevated temperature. This technology can be used to prepare one-component adhesives by blending a polyhydroxy material with a blocked isocyanate which is stable until the blend is heated to about 300°F (150°C). Polyurethane adhesives have a wide range of bonding applications including shoes, metal laminates, film, furniture, textiles, garments, and packaging. To date their usage is much lower than for the epoxy adhesives with which they compete.

In contrast, polyurethane sealants are widely used in many different applications. They are sold as two component and one-component gun-grade formulations, packaged in cartridges or in bulk. Pourable self levelling grades are also available. They compete directly with polysulfide and silicone sealants in many applications.

The polyhydroxy components of urethane sealants are usually hydroxyl terminated saturated polyesters or polyethers. Polyester based urethane sealants are hard and tough, give good adhesion but have poor water resistance and weatherability. Polyether urethanes have better water resistance, lower moduli but inferior adhesion.

Toluene diisocyanate (TDI) is the most commonly used polyisocyanate in urethane sealant production. All aromatic polyisocyanates cause urethanes to yellow on exposure to light. The yellowing problem can be eliminated by the use of aliphatic or cycloaliphatic polyisocyanates, but these are more expensive.

⌘⌘⌘

The isocyanate terminated prepolymer used in the manufacture of polyurethane sealants can be cured in two ways: in two component sealants by the addition of hydroxyl containing coreactant such as polyester or polyether; or in one component sealant by water absorption followed by polyurea formation with the production of carbon dioxide as a by-product.

There are many different formulations for two component polyurethane sealants dependent of the sealant properties required. The usual NCO/OH equivalent ratio is 1.05 to 1.10. One component is a liquid isocyanate-terminated prepolymer containing pigments, fillers such as calcium carbonate or talc, and an antisag agent such as fumed silica. The second component is a hydroxyl terminated polymer, pigment, and a catalyst such as methylene dianiline. The individual components are mixed at elevated temperatures under dry conditions so that premature crosslinking does not occur. Prior to use, the two components are mixed well and have a pot life of about 1 to 4 hours. Cure time to develop full sealant properties is 1 to 2 days.

One component gunnable sealants typically contain 30 to 60% prepolymer to which is added predried pigment, fillers, and fumed silica. Again the mixing is carried out at elevated temperatures under vacuum to remove any traces of moisture which could cause premature gelling in the cartridge during storage. Drying agents such as molecular sieves may also be used in the formulation. The cure rate is much slower than for two component polyurethanes as it is dependent on the relative humidity, temperature, and seal thickness. One component polyurethanes cure faster than one component polysulfides because of their greater moisture permeability.

Polyurethane sealants have excellent UV, ozone and chemical resistance, excellent abrasion and indentation resistance, good elongation and recovery, good tear, and inservice low temperature (-40°F/-40°C) properties. Major markets are in construction for high performance joints (movement capacity ±25%) in curtain walls and perimeter walls; in expansion and paving joints for parking decks, bridges, highways, and airfields; and in automobiles for glazing windshields and back lights. Polyurethane sealants sometimes need a metal primer to improve adhesion. Although polyurethane sealants are very versatile for household needs, such as window repair, bath and shower seals, gutter repair, sidewalk and automobile uses, they have seen limited sales in the do-it-yourself market dominated by acrylic, butyl, oil caulks and silicone sealants.

Silicone (Beers, 1977; Elias, 1986; Prane, 1990) – Silicone sealants became commercially available in the late 1950's. Since that time they have found widespread use in many markets including the construction, industrial, automotive, and consumer due to their unique combination of properties and versatility. These include excellent UV, ozone, chemical and water resistance; excellent elastomeric properties (extension, compression, high recovery) over a wide temperature range (-85 to 400°F/-65 to 200°C); ease of application (even at 0°F/-18°C) with a wide range of rheological properties (non sag to self levelling); rapid cure with low shrinkage; good release and thermal and electrical insulation properties; and moderate cost.

Silicone sealants are available as both one and two component systems. The main component of both systems is silanol terminated polydimethylsiloxane polymers. In one component sealants, cure is achieved through the use of trifunctional silane such as methyltriacetoxysilane, methyltrimethoxysilane, or methylethylketoximesilane using small quantities of metal soap catalysts such as dibutyl tin dilaurate. When the sealant is extruded from the tube, atmospheric moisture diffuses into the polymer hydrolyzing the reactive groups on the crosslinker. The unstable silanols (SiOH) on the crosslinking molecule condense with the silanol end groups of the silicone chain forming stable siloxane bonds (Si-O-Si) in the cured sealant. Methyltriacetoxysilane is not used as the crosslinking agent when the sealant application involves metals, concrete or limestone surfaces as the by-product acetic acid causes corrosion. In these applications the ketoxime cross linkers are used. One part sealants normally have tack free times of less than an hour and full cure may require several days depending on the relative humidity, temperature, and thickness of the sealant.

High modulus silicone sealants normally contain 80 to 85 parts silanol polymer, 10 parts fumed silica for reinforcement and slump control, about 5 parts crosslinker and 0.01% tin catalyst. Lower modulus sealants reduce the silanol polymer level, add silicone fluids as a plasticizer and replace part or all of the reinforcing fumed silica with much higher levels of low reinforcing calcium carbonate. To prevent premature gelling of the sealant in the tube during storage, great care must be taken to use dry ingredients and mix and package the product in a moisture-free environment.

Silicone sealants are used extensively in construction applications such as expansion joints (silicone sealants have movement capacities of ±25 to ±50%), perimeter sealing control joints and structural and nonstructural glazing. They have wide use in the automobile industry for formed in place gaskets for valve covers, oil pans, and thermostats, and as adhesives for rear view mirrors. They are used as well in aerospace applications for door and window gaskets, in household appliances as gaskets for washing machines, dishwashers, and refrigerators and for encapsulation of electronic components, such as connectors and terminals. Silicone sealants are widely used in the consumer market as a bathtub/shower caulk, auto sealant, window sealant and general purpose adhesive sealant.

Two component silicone sealants have as one component the silanol terminated silicone polymer, plus tetraethylorthosilicate as the crosslinker, plasticizers, reinforcing fillers, and other additives. The minor component contains the activator, an organometallic catalyst such as dibutyl tin dilaurate, in a paste of silicone fluid, and filler.

⌘⌘⌘

$$4-\overset{\overset{\displaystyle CH_3}{|}}{\underset{\underset{\displaystyle CH_3}{|}}{Si}}\text{-OH} \;+\; Si(OC_2H_5)_4 \longrightarrow -\text{O-}\overset{\overset{\displaystyle CH_3}{|}}{\underset{\underset{\displaystyle CH_3}{|}}{Si}}\text{-O-}\overset{\overset{\displaystyle CH_3\text{-}Si\text{-}CH_3}{|}\atop{\displaystyle O \atop |}}{\underset{{| \atop \displaystyle O}\atop{| \atop \displaystyle CH_3\text{-}Si\text{-}CH_3}}{Si}}\text{-O-}\overset{\overset{\displaystyle CH_3}{|}}{\underset{\underset{\displaystyle CH_3}{|}}{Si}}\text{-O}-$$

silanol polymer   tetraethyl orthosilicate   crosslinked polymer

After mixing the two components, the sealant cures as shown above with the formation of ethanol as a byproduct. No special requirements are needed to keep components one and two dry, in fact the presence of moisture within the formulation, from the fillers for example, is necessary for the crosslinking process to occur quickly.

Two component silicone sealants are used in the electronics industry for potting and encapsulation, in construction for sealing high performance insulating glass, and for structural glazing.

CHEMICAL ADDITIVES

Plasticizers (A.S.C., 1986; Dick, 1987) – Plasticizers are widely used in elastomer based adhesives and sealants and hot melt systems to improve processability and final cured properties. Typical use levels are 10 – 20 weight percent. Plasticizers are low molecular weight, low volatility molecules which are soluble in the base polymer. They have several functions:

a) They act as a processing aid in the breakdown of uncured rubber on the mill so that it can be dissolved in solvent.
b) They lower the viscosity of the base polymer allowing easier filler in corporation and dispersion.
c) They reduce modulus and lower hardness at the adhesive\sealant adher end interface.
d) They increase % elongation.
e) They improve low temperature flexibility for reduced cracking. Solvent acrylic sealants have poor low temperature properties. The addition of 2 to 7 weight % plasticizer can improve the low temperature flexibility by 40 to 50%.
f) They can reduce the cost of formulation by allowing the use of higher levels of low cost extender filler.

⌘⌘⌘

In hot melt systems, plasticizers lower the melt viscosity and can improve the wetting of the adherend, but may lower adhesion if the plasticizer level is too high. When added to emulsion adhesives or sealants the plasticizer migrates to the polymer droplets, swelling them and hence increasing the viscosity of the emulsion and helping to reduce slump.

Plasticizers are used in butyl rubber, polyurethane, polysulfide, solvent acrylics, polyvinyl acetate, and silicone rubber adhesives and sealants. The compatibility of the plasticizer depends on base polymer chemistry. Plasticizers for silicone systems are low molecular weight polydimethylsiloxane fluids, which have a wide viscosity range, from 1to10,000 mPas. There is a broad selection of plasticizers for other systems. The most commonly used are dioctyl phthalate (DOP), dibutyl phthalate (DBP), butyl benzyl phthalate (BBP), triaryl phthalate and dioctyl adipate.

Hot melt adhesives and sealants can use the same plasticizers as elastomers, as well as microcrystalline wax and paraffin wax, which are solids at room temperature.

Solvents (A.S.C., 1986; Dick, 1987) – Solvents are used in evaporation type adhesives and sealants. They act as the carrier in which the base polymer is dissolved to provide low viscosity solutions that can accommodate fillers and otheradditives. Butyl rubber solutions are made with 50 to 70% rubber and 30 to 50% solvent. These are the bases of products which vary from unfilled viscous liquids to semi-solid filled sealants. Solvent release sealants generally range from 70 to 80% total solids. The solvent level is chosen so that adhesive or sealant is easy to apply and wets the adherend surface. After application, the solvent evaporates from the glue-line leaving the polymer bonded to the adherend.

There are several factors to consider when choosing a solvent for a given polymer system. These include solubility of the polymer, viscosity stability, cost, drying time, flammability, toxicity and environmental factors. The solubility of a polymer in an organic solvent is dependent on the chemistry of the two materials. Generally, nonpolar polymers dissolve easily in nonpolar solvents and vice-versa. For example, nonpolar butyl rubber dissolves well in hydrocarbons such as hexane, whereas polar acrylic polymers are usually compounded with xylene or toluene. Solvents can be selected based on their Hildebrand Solubility Parameters and that of the polymer. If their solubility parameters are similar, within 2 units, the polymer will dissolve readily in the solvent.

The choice of solvent is very important in that the better the solvent, the less is required for polymer dissolution. Less shrinkage will occur at the polymer/adherend interface on drying, resulting in less built-in bond stress.

The rate of solvent evaporation is important as it affects the working-time of the adhesive or sealant, and more importantly the cure time to develop good

⌘⌘⌘

bond strength. Using VM&P naphtha rather than kerosene in butyl sealants will shorten both tack-free and cure times. The rate of solvent evaporation depends on molecular weight, polarity and degree of hydrogen bonding of the solvent. Most organic solvents are flammable and some are considered toxic. They must be handled carefully in well ventilated areas. The halogenated solvents are less flammable than other solvents. Hydrocarbon solvents are less expensive than aromatic and halogenated solvents. Frequently combinations of solvents are used to optimize cost/performance considerations.

The wide range of possible solvents for adhesives and sealants include:

1) hydrocarbons – gasoline, hexane, heptane, cydohexane, VM&P naptha.
2) aromatics – toluene, xylene
3) aliphatic alcohols – methanol, ethanol, butanol
4) aliphatic ketones – methyl ethyl ketone, acetone
5) aliphatic esters – ethyl acetate, butyl acetate, amyl acetate
6) halogenated hydrocarbons – 111-trichloroethane, trichloroethylene, perchloroethylene.

Tackifiers (A.S.C., 1986; Dick, 1987) – Tackifiers are used in certain types of adhesives and sealants to greatly improve initial adhesive strength (tack strength) on contact with an adherend surface before a stronger bond is formed later during the cure process. Elastomers such as natural rubber and butyl rubber, and thermoplastic hot melt systems all have low tack and require tackifiers to adhere well to polar substrates such as glass, ceramics, masonry and metal surfaces.

Tackifiers function in several ways to increase tack strength to surfaces. Incorporation of the correct tackifier type and loading will improve the compatibility of the polymer with the adherend. The small tackifier molecules will also reduce the viscosity and elasticity of the polymer molecules allowing better wetting of the adherend surface. Frequently this improvement in tack also improves the permanent bond strength on cure.

There are many different types of tackifiers, these include :

Rosins – These are abietic acid resins obtained from wood sources. There are several different types: wood rosins obtained from the naptha extraction of pine stumps, gum rosins obtained from the destructive distillation of turpentine, and tall oil rosins obtained by the destructive steam distillation of tall oil from the manufacture of paper. Rosins are widely used in the adhesive and sealant industry. In hot melts they oxidize and discolor easily at high temperatures due to the unsaturation in the resin. Hydrogenated grades are available which have much better thermal stability.

⌘⌘⌘

Hydrocarbon Resins – There are two general types, aliphatic and aromatic resins. Both resins are obtained from the polymerization of olefins from the petroleum industry. Aromatic resins are based on coumarone-indene and polyindene whereas the aliphatic resins are from polymerized mixed olefins which are not based on indene.

Polyterpene Resins – These resins are widely used in elastomeric and hot melt systems for their better temperature resistance. They are formed by the polymerization of beta-pinene extracted from wood.

Phenolic Resins – Phenol-formaldehyde tackifying resins are linear short chain polymers which have limited heat stability, and so may cause discoloration if used in hot melt systems.

Curing Agents (Hardeners) (A.S.C., 1986; Dick, 1987) – Curing agents are incorporated into a chemically reactive adhesive or sealant to increase the adherend bond strength by crosslinking the base polymer into a three dimensional matrix. The curing agents are usually small molecules used at low levels, <10 weight %, and are specific to the chemistry of each type of chemically reactive polymer system.

Silicone adhesives and sealants use tri or tetrafunctional organosilanes, $RSiX_3$ and SiX_4 respectively, as room temperature curing agents. The X groups may be acetoxy, methoxy, methylethyl ketoxime, or n-methyl benzamide. These groups readily hydrolyze in the presence of moisture to generate the reactive molecules $RSi(OH)_3$ and $Si(OH)_4$. These molecules can then condense with the terminal hydroxyl groups of the polydimethylsiloxane base polymer in the presence of a catalyst to give a cured elastomer.

Liquid polysulfide polymers are converted to high molecular weight elastomers through the reactive hydrogen of the mercaptan groups. The curing mechanism for polysulfide sealants is based on oxidizing agents such as lead dioxide, manganese dioxide, calcium peroxide, zinc peroxide, and cumene hydroperoxide.

Phenol-formaldehyde novolac resins are formed by the incomplete polymerization of phenol and formaldehyde. Insufficient formaldehyde is present to allow complete cure. Complete crosslinking can be achieved by adding a powdered methylene donor such as hexamethylenetetramine and heat. Resorcinol-formaldehyde novolac liquid resins can be cured by the addition of powdered paraformaldehyde.

To cure an epoxy adhesive, the epoxy ring must first be opened before polymerization can occur. The epoxy ring may be opened by an acid anhydride or an amine. The four general types of curing agents are aliphatic polyamine, aromatic polyamine, acid anhydride, and catalytic.

⌘⌘⌘

Aliphatic amines such diethylenetriamine (DETA), triethylenetetramine (TETA) and tetraethylenepentamine (TEPA) give fast cures at room temperature. They also have the disadvantages of short pot-lives, poor peel strength, high volatility and high toxicity. Aromatic amines such as m-phenylenediamine (MPDA) have lower volatility, but are still toxic, cure more slowly and have a longer pot-life. They impart better heat resistance but may require heat to activate cure. Acid anhydrides such as phthalic anhydride also provide long pot-lives at room temperature, but only cure at elevated temperatures.

Boron trifluoride amine complexes can be used in one part epoxy adhesives as they are stable at room temperature and must be heated to initiate cure. These cures are termed latent curing systems and the boron triflouride acts as a catalyst. Dicyandiamide (DICY) is frequently used in these cure systems for construction and metal to metal adhesion.

Catalysts and Retarders (Elias, 1986; Petersen, 1986; Regan, 1986) – Catalysts (accelerators) speed up the chemical curing reaction without being consumed themselves. Retarders (inhibitors) slow down such reactions. They are always used in conjunction with a curing agent which is itself consumed in the polymerization process. The catalysts or retarders in the adhesive or sealant are usually used at very low concentrations ~0.1 weight percent.

Some silicone adhesives and sealants require condensation catalysts to accelerate curing. These are usually organotin and organotitanium compounds such as dibutyl tin dilaurate , dibutyl tin diacetate, dibutyl tin dioctoate, or tetrabutyl titanate.

Catalysts are also used in polyurethane systems, particularly dibutyl tin dilaurate and tertiary amines.

The room temperature cure of two component polysulfide sealants can be controlled from a few minutes to several days by catalysts and inhibitors in conjunction with metal oxide curing agents as shown below. The pH of the system is the controlling factor, with bases catalyzing and acids retarding the cure.

Curing Agent	Catalyst	Retarder
PbO_2	Sulfur	Stearic acid
CaO_2	H_2O	Metallic stearate
ZnO_2	Amine	Sulfur
MnO_2	Bases	Stearic acid

Process Aids (A.S.C., 1986) – Process aids (wetting agents) are added to elastomeric adhesive and sealant formulations to aid in dispersing dry filler and pigment particles into the matrix of rubber, plasticizers and tackifiers. Although plasticizers are added to the formula to improve low temperature

⌘⌘⌘

flexibility and tackifiers for improved adhesion, they both are solubilized within the polymer and lower its viscosity. This reduction in polymer viscosity further assists in the dispersion of the filler and pigment particles.

Wetting agents can also act as process aids by reducing surface tension and allowing the organic matrix to better wet the surfaces of pigment and filler particles for easier dispersion. The more commonly used wetting agents are long chain fatty acids such as stearic acid, oleic acid , fatty acid esters such as zinc stearate, long chain hydrocarbon glycol ethers and carbonyl derivatives.

In emulsion adhesives and sealants, ionic and nonionic surfactants are used to disperse pigments and fillers. In polyvinyl acetate emulsion sealants, most grades of calcium carbonate can be dispersed using anionic or non ionic surfactants. Titanium dioxide is more difficult to disperse due to its finer particle size (~0.2µm) and anionic dispersants such as potassium tripolyphosphate and polymeric salts of acrylic acid are used.

Adhesion Promoters (Newton, 1986; Petersen, 1986) – Adhesion promoters are incorporated into an adhesive or sealant to improve their ability to adhere to a surface. Systems composed of nonpolar polymers, such as butyl rubber and polysulfides, especially require adhesion promoters to allow them to adhere to metal, glass, and other polar surfaces. Adhesion promoters must be bifunctional molecules. They must be miscible and have a strong attraction to the adhesive or sealant, as well as having a portion of the molecule with a strong affinity for the adherend surface. The most common adhesion promoters are the silanes ($RSiX_3$) and titanates (R_3TiX), where R is an organic group which has a strong attraction for the adhesive/sealant, and X is a hydrolyzable group, such as methoxy or ethoxy. On hydrolysis, the resulting hydroxyl groups – $Si(OH)_3$ and $-Ti(OH)_3$ can react with polar surfaces such as glass, ceramic, stone and metal. Other adhesive promoters include phenolics and epoxy resins. The level of adhesion promoter may vary from approximately 0.5 to 3 weight %.

Colored Pigments (Newton, 1986; Newman, 1986) – Adhesives and especially sealants are frequently sold as colored products. The common colors for construction sealants are black, white, grey, and neutral. The main colored pigments used are, carbon black, titanium dioxide (rutile) and red and yellow iron oxides. Clean durable colored products are made by adding enough titanium dioxide to mask the natural formula color and then adding other colored pigments to tint the white base.

Many pigments, such as calcium carbonate, talc, and titanium dioxide, look white in the dry state. When wet out by polymers and plasticizers, however, only titanium dioxide retains its brightness and ability to opacify. To function as an opacifying agent there must be a significant difference in the refractive index of the pigment and its surroundings. The refractive index of

⌘⌘⌘

calcium carbonate and talc are each about 1.55, which is much higher than the refractive index of air, 1.00. When these fillers are wet out into a sealant formulation they lose their opacifying ability since the matrix usually has a similar refractive index. In contrast, rutile titanium dioxide has a refractive index of 2.55, so that even when wet by polymers the fine rutile particles can diffract and scatter light allowing the pigment to retain its hiding power and brightness. The typical level of color pigments in a sealant ranges from 0.05 to 3.0 weight percent.

UV Stabilizers (Newton, 1986; Regan, 1986) – Adhesives and sealants can suffer polymer degradation when exposed to ultraviolet light for extended periods of time. This may involve polymer breakdown and discoloration. Carbon black filled systems show good UV resistance as carbon black absorbs UV light. Light colored sealants can use highly reflective pigments such as titanium dioxide and zinc oxide to reflect the UV light. Inorganic UV stabilizers can be added as well. The stabilizers function by absorbing the UV radiation, which they convert into molecular vibrations, rotations and eventually heat.

Antioxidants (Newton, 1986; Regan, 1986) – Antioxidants inhibit polymer oxidation. They are normally used with UV absorbers since they often they exhibit a synergistic effect. Antioxidants can be divided in primary and secondary types. The primary antioxidants are able to trap free radicals formed by oxidation or UV reaction processes. Aromatic amines and substituted phenols are of this type. Care should be taken when using aromatic amines as these can stain light colored sealants. Secondary antioxidants can break down any hyperoxides that may be formed, preventing carbonyl group formation and free radical initiation. Thioesters and organophosphites function as secondary antioxidants.

Surface Primers (Dick, 1987) – A surface primer is a coating applied to an adherend to improve an adhesive bond or subsequent coating. It is typically used where the adhesive and adherend have widely differing polarities, a butyl rubber adhesive on a steel surface for example. Primers are of two general types:

1) The primer has a polarity intermediate to that of the adhesive and adherend surface.
2) The primer contains a bifunctional compound with both a polar and nonpolar substituent.

Surface primers can be based on silanes, epoxies, methylacrylates, polyurethanes, and resins.

⌘⌘⌘

Thickening Agents (Podlas, 1988) – These materials are added to adhesive and sealant systems for several reasons:

1) To prevent settling of pigments and fillers during storage.
2) To control flow during application.
3) To ensure that the adhesive or sealant remains in the desired location during cure. For example, to prevent absorption by a porous adherend and consequent adhesive starvation at the glueline, and to prevent sag when applied to vertical surfaces.

There are a variety of thickening agents (thixotropes) designed for different applications. In aqueous and latex adhesives and sealants, such as acrylics, polyvinyl acetate, butyl, and natural rubber, frequently used types include:

1) cellulosics – carboxymethylcellulose (CMC), hydroxyethylcellulose (HEC), hydroxypropylcellulose (HPC), and methyl cellulose (MC).
2) natural gums – gum arabic and gum karaya.
3) starches – modified potato and corn starches.
4) hydrophilic clays – bentonite, hectorite, smectite, and attapulgite clays.

The types of thickening agents used in solvent and non aqueous adhesives and sealants such as acrylics, epoxies, polysulfides, silicones, polyurethanes, and natural rubber include:

1) fumed silicas – hydrophilic and surface modified grades
2) precipitated silicas – hydrophilic and surface modified grades
3) carbon black
4) organophilic clays
5) organic castor oil derivatives.

The mechanisms by which the individual types of inorganic thickeners function will be reviewed in the following section.

FILLERS

The general properties of industrial minerals in adhesives and sealants are summarized in Table 9.

Calcium Carbonate (Tapper, 1992; Kummer, 1978) – Calcium carbonates are the most widely used fillers in adhesives and sealants. There are two types

⌘⌘⌘

of calcium carbonate fillers – extender fillers derived from natural mineral sources, and rheology control and reinforcing grades produced synthetically.

Table 9. Mineral Properties in Adhesives and Sealants

	Adhesives	**Sealants/Caulks/ Sealant Adhesives**
Mineral particle size	fine	wide range of sizes
Loading	< 10phr	10 to 100phr
Functions		
In Storage	antisettling agent	antisettling agent
During Premix (2 part systems)	optimize rheology of A and B	optimize rheology of A and B
On Application	-flow control -spreading control on horizontal surfaces -control penetration into porous surfaces -antisag properties on vertical surfaces	-extrusion control -bead shape control on horizontal surfaces - -antisag properties on vertical surfaces
On Curing		
Reactive systems	control flow/sag for exothermic cures	control flow/sag for exothermic cures
Evaporative systems	reduce glueline shrinkage better adhesive bond	reduce bondline shrinkage better adhesive bond
Cured Properties	optimize durometer tensile, elongation	optimize durometer tensile, elongation
Elastomeric Systems	tear, modulus	tear,modulus
Other	-	reduce cost

Naturally occurring forms of calcium carbonate include limestone, chalk, marble, and calcite. After mining, the calcium carbonate is ground, screened and air classified. Some grades are water washed, water ground and redried. Wet ground grades have smaller particles with a narrower particle size distribution. They are also more expensive. Typical properties are shown in

⌘⌘⌘

Table 10. Ground calcium carbonate is an ideal extender filler. It is cheap, white, and readily available, and it has a low refractive index and low oil absorption (i.e. low binder demand). The preferred extender grades have average particle size of 3 to 9 µm. Oil based putties and caulks use ground limestone extensively at high levels to reduce costs.

Table 10. Commercial Grades of Ground Limestone

	Coarse	Medium	Fine
Avg. Particle Size, µm	7-9	5-6	3-1
Top Size, µm	31	25	15
Surface Area, m^2/g	1.6	2.8	3.7
Oil Absorption, g/100g	8-10	10-15	15-20
Dry Brightness	93-94	95-96	96-97
Trade Example – Vicron	31-6	25-11	15-15

(Tapper, 1992)

Precipitated calcium carbonates are made by first calcining limestone to calcium oxide and carbon dioxide. The calcium oxide is slaked with water to form calcium hydroxide, and this is treated with the carbon dioxide liberated on calcining to precipitate calcium carbonate. By varying the process conditions, grades with average particle sizes from 0.07 to 3 µm and distinct crystal forms are obtained. The latter include rhombic, acicular, and scalenohedral (rosette) crystals. The rosette shape enhances the paintability of certain sealants, such as silicones, by providing tooth to the sealant surface. The finer particle size, higher surface area, and higher oil absorption of the precipitated calcium carbonate grades, as shown in Table 11, make them useful in many adhesive and sealant systems as rheology control and reinforcing agents.

Table 11. Commercial Grades of Precipitated Calcium Carbonate

	Rosette	Medium*	Fine*
Avg. Particle Size, µm	2	0.6	0.07
Surface Area, m^2/g	9	7	19
Oil Absorption, g/100g	50	25	30
Dry Brightness	96	98	98
Trade Examples	Albacar	SuperPflex	Ultra-Pflex

* surface treated.
(Tapper, 1992)

⌘⌘⌘

In general, the effect of decreasing the filler particle size at a constant loading affects sealant properties by :

1) Increasing viscosity.
2) Reducing slump.
3) Reducing gunnability.
4) Slight decreasing elongation.
5) Increasing tensile.
6) Improving extrudate surface smoothness.
7) Improving brightness.
8) Improving tear strength.
9) Increasing Shore hardness and modulus.

Good dispersion of the precipitated calcium carbonate in the adhesive or sealant is essential to optimize the reinforcement and rheological properties. Frequently blends of ground and precipitated calcium carbonates are used to optimize cost/performance. The ratio of ground to precipitated typically ranges from 1:1 to 3:1.

Ground grades of limestone are used in ready mix tile cements. The limestone particles are easily dispersed in water, have low oil absorption values, are only very slightly soluble in water giving slight alkalinity, have excellent weathering properties, and do not stain or contribute to mildew growth. They are used in both polyvinyl acetate tile adhesives for dry conditions and styrene acrylate water resistant tile adhesives. Particle size distribution is important in tile compounds. Fine particles give smooth easy to apply adhesives whereas large particle size grades can give gritty, poor flowing pastes.

Properties of the natural ground and precipitated grades of calcium carbonates can be further enhanced by treatment with stearates to make the particles hydrophobic. Stearate treated precipitated calcium carbonates are widely used in sealants to achieve good precure properties such as ease of extrusion with good sag resistance, and low hardness and modulus with high elongation and high tensile. The sealant systems include acrylics, butyl rubber, polysulfides, polyurethanes and silicones. The treated precipitated calcium carbonates are also used as thixotropes to impart nonsag properties to PVC plastisol underbody coatings and thick bead sealants used in automobiles.

Typical ground calcium carbonate levels in adhesives are 5 to 65%, and in sealants 30 to 60%. Precipitated grades use levels in adhesives are 10 to 35%, and in sealants 10 to 50%.

Talc (Washabaugh, 1988; Radosta, 1978) – Talc is a naturally occurring magnesium silicate mineral. The material is very soft (Mohs hardness of one) and most talcs are white with a bluish undertone. Talc is commercially available in two forms – platy and tremolitic. Tremolitic talcs are lower in oil absorption, enabling higher loading levels, and easier to disperse. The platy talcs impart generally better reinforcement and barrier properties and are more commonly used. The platy talcs are used in many adhesive and sealant formulations in combination with mica and attapulgite as a rheology control replacement for asbestos. Because the surface of talc particles is hydrophobic (organophilic), talc imparts better water and chemical resistance to formulations than do hydrophilic fillers.

Platy talcs which have been ground and air classified to particle size ranges from 0.75 to 15μm are used at loadings from 5 to 30 weight % in adhesive and sealant formulations. They are used to:

a) Reduce cost as an extender.
b) Impart rheology control.
c) Act as a semi-reinforcing filler.
d) Impart water and acid resistance.
e) Improve sandability.

Talc applications include caulking compounds, sealants, auto putties, mastics, and joint compounds based on polybutene, polyisoprene, butyl rubber, polyurethane, and polyvinyl acrylate.

Kaolin (O'Driscoll, 1988; Washabaugh, 1988) – Kaolin clay is an aluminum silicate mineral widely used in adhesive and sealants for a variety of reasons. The usage depends on the particle size, particle size distribution and the form of the kaolin.

Kaolin has an off-white color with a slight yellowish cast, is slightly acid, and is inert with excellent chemical and acid resistance. The particles have a hexagonal shape. Coarse particles are composed of stacks (booklets) of platelets and the finer particles are composed of individual platelets. After mining and pulverizing, the kaolin is classified by air flotation or water washing. Air floated grades are less expensive than water-washed grades but have a poorer color, and more impurities and water soluble salts. Kaolin contains about 14% crystalline water; grades are available where this has been partially or totally removed by calcination. Calcination gives harder more porous particles (higher binder demand), and better color. Many surface modified versions of each type are available, offering good reinforcement, water resistance, and electrical properties. Typical grades range in average particle sizes from 0.3 to 5.0 μm.

The various types of kaolin are used in adhesives and sealants to:

1) Lower cost as an extender filler.
2) Improve and accelerate wet tack.

⌘⌘⌘

3) Increase rate and strength of adhesive bond formation.
4) Control adhesive penetration into cellulosic substrates and reduce excessive loss from the bond surface (larger particle size grades).
5) Function as a rheology modifier to control viscosity, prevent sag and improve surface smoothness (finer particle size grades).

Air floated and water-washed grades of kaolin are widely used in paper, packaging and construction adhesives based on starch, emulsion and latex based waterborne systems. They are also used in fabric adhesives, mastics and caulking compounds. Calcined kaolin is used as a low moisture reinforcing agent in one or two part construction and automotive sealants based on butyl, urethane and silicone elastomers. Calcined kaolin is less acidic and so can be used with precipitated calcium carbonate in polysulfide sealants. Typical loading of kaolin fillers are 5 to 25% in adhesives, 5 to 20 % in sealants, and 25 to 50 % in mastics.

Bentonite (O'Driscoll, 1988; Hahn 1992) – Bentonite, sodium aluminum magnesium silicate, is a smectite clay that is an excellent thixotrope for water based emulsion adhesives and sealants. Smectite clays are composed of small platelets (<0.1μm). Each macroscopic smectite particle is composed of thousands of these platelets stacked in sandwich fashion with exchangeable cations and oriented water between each. The platelet faces carry a negative charge from lattice substitutions, while edges have a slight positive charge from broken bonds and cation adsorption. When smectite and water are mixed together, water penetrates between the platelets forcing them further apart. When sodium is the predominant exchange ion, the platelets can further separate and the exchange ions begin to diffuse away from the platelet faces. Further penetration of water betwen the platelets then proceeds in an osmotic process until complete separation. The presence of dissolved substances in the water prolongs hydration time by inhibiting this osmotic swelling. Once the smectite platelets are separated, the weakly positive platelet edges are attracted to the negatively charged platelet faces. The resulting three dimensional structure is often referred to as the "house of cards". This loose association of platelets causes a viscosity increase or gelation of water based systems. The viscosity is thixotropic, since the gel structure can break down under shear (mixing) and reform with time after the shear forces are removed. Smectites build rheology mainly in the low and medium shear rate range and have very little effect on the high shear viscosity of a system.

Bentonites can improve water-based adhesives and sealants by:

1) Increasing viscosity.
2) Suspend fillers and pigments.

⌘⌘⌘

3) Reduce penetration in porous substrates such as paper.
4) Control sag.
5) Improve emulsion stabilization.
6) Increase water release, accelerating drying.
7) Act as a blocking agent.

Care must be taken to properly disperse the bentonite so that the individual platelets are separated and then allowed to fully hydrate so that they can develop the optimum gel structure. Heat can accelerate this process.

Bentonite thickening agents are used in various water based adhesives, caulks and sealants including those based on proteins and starch, polyvinyl acetate emulsions, natural rubber latexes, acrylic emulsions, neoprene emulsions, butyl emulsions, nitrile emulsions, and SBR emulsions.

Organophilic bentonite clays can be made by treating the clay with quaternary ammonium compounds such as tetraalkyl or alkylaryl ammonium chlorides. The clay's exchangable cations are replaced by organophilic quaternary amine groups. A range of organoclays can be made with different compatibilities, dispersability and thickening efficiencies in different polymer based adhesive and sealant systems by varying the exact quaternary amine used.

In normal use, the dry organoclay, or a pregel in a solvent, is added to the base solvent/polymer blend. Polar activators such as propylene carbonate or ethanol/water blends are added to help disperse the agglomerated organophilic platelets and provide hydrogen bonded bridges between platelet edges. High shear mixing maximizes the gel network thus formed throughout the adhesive or sealant base. The organoclays imparts similar rheology improvements to solvent based adhesives and sealants as the hydrophilic clays do to water based products.

Attapulgite (Washabaugh, 1988; O'Driscoll, 1988) – Attapulgite is a naturally occurring hydrated magnesium aluminum silicate. The mineral is mined, ground, and air classified to yield colloidal grades with very fine, 0.1μm acicular particles. The colloidal grades are excellent cost effective thixotropes and rheology modifiers for paints, coatings, animal feeds, drilling muds, adhesives, and sealants. It's thickening properties are due to interactions between the needle-like particles rather than to the swelling as with platy smectite clays. Also unlike most other mineral thickeners, attapulgite is stable in the presence of electrolytes.

Attapulgite can be used in both water and solvent based adhesives and sealants to impart thixotropy, and good filler suspension. In water based polyvinyl acetate and vinyl emulsion systems, levels of 2 to 8 % are used. The attapulgite is easy to disperse and is usually added to the formula with the other dry components. The thixotrope also imparts good shelf stability,

⌘⌘⌘

improved texture, ease of application and less sag. Attapulgite is also used in solvent based systems, vinyl plastisols and some epoxy systems. Typical use levels in sealants are 10 to 20 %. Cationic surfactants are used with attapulgite to improve thickening efficiency and viscosity stability in solvent based systems. In many systems it is a partial replacement for asbestos thixotropes.

Natural Silica (Flick, 1989; O'Driscoll, 1988) – Ground natural silica is used as a low cost extender filler in rubber based adhesives and sealants. It is available in particle sizes ranging from 1.2 to 4.2μm, equivalent surface areas of approximately 0.54 to1.89 m^2/g, and low oil absorption (resin demands) of about 27 to 44 g./100 g of silica. The silicas have low surface moisture and do not require predrying prior to use in moisture sensitive systems. They impart good weather resistance and are chemically inert to acids and bases. The silicas have a low refractive index (1.54), and can be used in transparent systems. Silica has good dielectric and thermal insulation properties and is used as an inert, low cost extender filler for silicone rubber. Special very high purity grades are used at high loadings to improve the thermal insulation properties of epoxy sealants for encapsulating computer chips In hot melt adhesives, ground silica imparts heat resistance and reduces adhesive penetration into porous adherends.

Diatomaceous Silica (Moreland, 1978; O'Driscoll, 1988) – Diatomaceous silica is a soft chalky rock of very low density. It is a form of natural silica derived from the siliceous remains of diatom fossils. After mining, the ore is crushed, milled and dried, air classified and calcined in rotary kilns. Grades of average mean diameter from 2.4 to 6.0μm are available. The materials have high oil absorption (resin demand) of 160 to 180g\100g of filler, and high water absorption capacity of 190 to 250g/100g of filler. The particles are neutral with a pH of 6.5 to 8.5.

Diatomaceous silica is a low cost extender filler which increases the hardness and stiffness of elastomers without increasing tensile or tear strength. It also inhibits the expansion of sealant bead diameter during extrusion.

Precipitated Silica (Statt, 1988; O'Driscoll, 1988; Degussa,1992) – Precipitated silicas are synthetically produced amorphous silicas. They are prepared by the reaction of an alkaline silicate solution, usually sodium silicate ("waterglass") with a mineral acid, usually sulfuric acid. By controlling the reaction conditions precipitated silicas can be made with surface areas ranging from 60 to 700m^2/g.. The silica slurry is pumped to filter presses where the precipitated silica is separated out and washed to remove salts formed during the reaction. The silica filter cake is then dried, milled and classified to yield different grades with average agglomerate sizes ranging from 3 to 100μm. The grades used in adhesive and sealant applications usually have average agglomerates sizes from 3 to 6μm, and BET surface areas from 140 to 475m²/g, DBP oil absorption values 275 to 330g/100g silica, and a moisture content of 3 to 6 %. The silicas are fine white powders with a refractive index of 1.46, which is similar to that of many resins and elastomers. The surface of precipitated silica is very hydrophilic due to the presence of hydroxyl groups attached to the surface silicon atoms.

Hydrophobic grades of precipitated silica are available in which part of the surface hydroxyl groups have been reacted with silicone oil or dimethyldichlorosilane.

Precipitated silica can function as a thickener/thixotrope where a balance of viscosity increase and filling is required. Precipitated silicas are less efficient than fumed silicas but are more effective than other fillers such as talcs, clays, diatomaceous silica and calcium carbonate. The thickening and thixotropic properties of precipitated silicas are due to the surface hydroxyl groups on adjacent silica agglomerates being able to hydrogen bond with each other to form a loose silica network throughout the liquid. Hence the thickening mechanism for fumed silica and precipitated silicas are similar. However, fumed silica is a much more effective thickener/thixotrope due to the much more open structure of their agglomerates compared to those of the precipitated silicas.

Precipitated silica is an effective rubber reinforcing filler and can be used to replace carbon black when non-black rubber compounds are required. Precipitated silica is used as a reinforcing aid in butyl caulks and tapes

Precipitated silicas are used in adhesives and sealants to:

1) Impart thickening/thixotropy, flow control, anti-sag and reinforcement to acrylic, polyurethane, polyvinyl plastisols, styrene butadiene rubber and latexes
2) Prevent resin separation and the settling of pigments and heavy fillers.
3) Prevent excessive penetration of adhesive into porous adherends.
4) Improve flow control and heat resistance of hot melts in application.
5) Serve as a free flow agent for spray dried resins including urea formaldehyde and melamine formaldehyde.

Hydrophobic precipitated silicas are used in polysulfide sealants to give improved weather resistance; improved reinforcement, as higher loadings can be used without increasing process viscosity; and faster cure rates, as the silicas have higher pH values.

Fumed Silica (Cochrane, 1988; Cabot, 1994) – Fumed silica is an amorphous synthetic silica which is widely used in adhesives and sealants as a rheology control and reinforcing agent. The silica is produced by the high temperature hydrolysis of chlorosilanes in a hydrogen/oxygen flame.

⌘⌘⌘

$$SiCl_4 + 2H_2 + O_2 \rightarrow SiO_2\downarrow + 4HCl\uparrow$$

In the flame process, molten spheres of silica are formed. Their diameters can be varied by adjusting the process conditions to average values of from 0.007 to 0.0270μm resulting in grades whose B.E.T. nitrogen surface areas range from 380 to 100 m^2/g. The molten spheres, termed primary particles, collide and fuse to form three dimensional branched chain aggregates ranging in average size from 0.1 to 0.3μm. As the aggregates cool below the fusion temperature of silica (~1710°C) further collisions result in reversible mechanical entanglement or agglomeration. Further agglomeration takes place in the collection system.

Residual adsorbed byproduct hydrogen chloride is removed by high temperature steam calcination. The resultant silica has a hydrophilic surface with 2.5 to 3.5 hydroxyls (OH)/nm^2. Hydrophobic grades are produced by treating hydrophilic grades with hexamethyldisilazane, dimethyldichlorosilane, or with silicone fluids

The important properties of fumed silica for adhesive and sealant applications include:

1. Fine particle size, giving very high surface area for reinforcement.
2. Branched chain aggregates giving high structure level with high DOP oil absorption values of 300 to 500g/100g silica for rheology control.
3. High chemical purity for good electrical and thermal insulation.
4. Surface hydroxyl groups used in thickening and silicone rubber reinforcement.
5. Refractive index of 1.46, similar to many polymers, making transparent compounds possible.
6. White powder
7. Hydrophobic grades available with very low absorbed moisture content which requires no predrying for use in moisture sensitive systems such as polyurethane and silicone sealants.
8. Ease of dispersion.

Fumed silica is used in adhesives and sealants primarily to:

1. Impart flow control and thixotropy even at elevated temperatures.
2. Prevent pigment and filler settling during storage.
3. Provide extrusion and flow control during application.
4. Prevent sagging during cure.
5. Provide reinforcement by increasing modulus, tear, tensile, elongation, and peel strength of elastomeric systems.

⌘⌘⌘

Hydrophilic grades of fumed silica are effective thickening agents in non polar liquids and polymers but are ineffective thixotropes in water, latex, amines, and other polar systems. Fumed silica functions as a thixotrope in non polar media by forming a loose network of silica aggregates throughout the media. The silica aggregates interact with one another through their surface hydroxyl groups. When the media is subject to shear, the hydrogen bonds are broken, reducing viscosity. On removal of the shear force, the hydrogen bonds reform and rebuild viscosity.

If the silica network reforms quickly and is strong enough, the adhesive or sealant will not sag during application and cure. The strength of the silica network is measured by its yield value, the force required to initiate flow in the adhesive or sealant. The higher the yield value, the greater the thickness of adhesive or sealant that can be applied before sagging occurs.

Hydrophilic grades of fumed silica are ineffective thixotropes for use in polar liquids as the polar liquids adsorb on the silica surface hydroxyls, preventing the formation of the hydrogen bonded silica network. Treated grades of fumed silica can act as effective thickening agents for polar organic systems. In this case, the silica forms a loose network of aggregates throughout the system via weaker Van der Waal forces. The thickening efficiency of the treated silicas depends on the type of treatment and treatment level.

The reinforcement properties of fumed silicas are dependent on their primary particle size (BET N_2 surface area), structure level, and surface chemistry. At a given surface area and structure level, the order of reinforcement ability is untreated hydrophilic grades > partially treated grades > fully treated grades.

The general trends observed for hydrophilic fumed silicas as their surface areas increase are:

1. Moisture content increases
2. Rate of incorporation decreases
3. Ease of dispersion decreases
4. Anti-sag and anti-slump properties increase
5. Viscosity increases, extrusion rate decreases but anti-settling properties increase.
6. Rheological properties are more stable on aging.
7. Reinforcement properties increase.
8. Transparency increases under good shear conditions.

As the fumed silica treatment level increases:

1. Moisture content decreases.
2. Rate of incorporation may decrease.
3. Ease of dispersion decreases.

⌘⌘⌘

4. Slump or sag properties may increase or decrease depending on the chemistry of the system.
5. Viscosity normally decreases and extrusion rate increases.
6. Rheological properties may or may not be more stable on aging.
7. Reinforcement properties usually decrease.

Fumed silicas are used in a wide range of adhesives and sealants for thixotropy, antisettling, antisag and reinforcement properties.

Mica (Flick, 1989; O'Driscoll, 1988) – Muscovite mica is a naturally occurring potassium aluminum silicate mineral. The mica can be easily cleaved to give transparent, colorless, thin, highly flexible, resilient platelets with high aspect ratios of 100 (flake diameter/thickness). Mica is available in various particle sizes from 100 to 325 mesh (143 to <44 μm). The filler has very low moisture content < 0.25% and oil absorption values (resin demand) of about 54g/100g mica. Mica has good thermal and electrical insulation properties and is opaque to UV light. The material is used as an extender filler in joint cements, caulks and sealants. At high loadings the mica platelets can overlap, similar to fish scales, and act as a barrier to moisture and to prevent penetration into porous substrates.

Wollastonite (O'Driscoll, 1988; Nyco, 1994) – Wollastonite, calcium metasilicate, is a naturally occurring very bright white acicular filler and reinforcement. After mining, the material is dry ground and then air classified to give a series of different particle size grades. Surface modified grades are also available to give easier processing at higher filled loadings. The acicular crystal structure reinforces thermoplastics, increasing impact, tensile and flexural strengths. Wollastonite can be used as a low cost filler and asbestos replacement. It is used in epoxy structural adhesives and encapsulants.

Alumina Trihydrate (Gibbsite) (Woycheshin, 1978) – Fine particle size 0.5 to 1.7μm) grades of this white filler can be used at high loadings in adhesives and sealants as a fire retardant. The filler is stable at temperatures below 205°C, decomposes slowly at 205 to 220°C and rapidly above 220°C to form aluminum oxide and water vapor.

$$2Al(OH)_3 \rightarrow Al_2O_3 + 3H_2O$$

This endothermic process allows the alumina trihydrate to behave as a heat sink to provide flame retardance and smoke suppression to polymeric systems.

The disadvantages of alumina trihydrate are that it can not be processed at more than 205°C, and that the high loadings required for effective fire

⌘⌘⌘

retardancy of this non-reinforcing filler may result in high process viscosity and a significant loss in mechanical properties. Major markets are in rubber carpet backings and adhesives.

Antimony Oxide (Waddell,1978) – Fine particle (1.25μm) antimony oxide can be used at levels of 1 to 15% in conjunction with halogenated compounds to form an effective fine retardant. The antimony oxide by itself imparts very little flame retardancy to adhesives and sealants. Halogen compounds by themselves are only effective as fire retardants at relatively high loadings. However, the two compounds together show great synergy and are very effective, especially at the optimum concentration of 3 moles of halogen for each mole of antimony oxide.

The source of the halogen can be from the base polymer (e.g. neoprene, PVC) of the adhesive or sealant, or organic halogen compounds, especially long chain halogenated aliphatic compounds which can decompose readily to release reactive halogen.

Antimony oxide is also a brilliant white opacifying pigment with a pink undertone. However, it is much more expensive than titanium dioxide and finds little use for this purpose.

Carbon Black (Queen, 1978) – Carbon black is the most widely used filler for reinforcing rubber compounds and, as such, is used in the manufacture of black rubber based adhesives and sealants. Carbon black reinforcing fillers and pigments are made by the incomplete combustion of organic hydrocarbons under controlled conditions.

The two major commercial processes for manufacturing carbon black are the thermal black process which uses natural gas as the hydrocarbon feed stock, and the furnace black process which petroleum oil as the feedstock. The thermal process produces the largest particle size (0.1 to 0.5μm) and lowest structure blacks. The furnace process accounts for over 95% of the total carbon black produced and can be controlled to produce both small and large particle size grades in the range of 0.01 to 0.10μm.

Reinforcing grades have BET surface areas of 70 to 150m²/g and DBP oil absorption values of 0.70 to 1.3 ml/g carbon black. The higher the BET and oil absorption values, the greater the reinforcing properties of the carbon black.

Titanium Dioxide (O'Driscoll, 1988) – Titanium dioxide is used primarily as a opacifying agent and a white pigment. The material is available in two different crystalline forms, anatase and rutile. Anatase is manufactured from ilmenite ore by treatment with sulfuric acid, rutile by the high temperature flame hydrolysis of titanium tetrachloride, $TiCL_4$. Pigment grades are brilliant white powders with 0.2 – 0.3 μm particles. The advantage of these materials over other white pigments has been discussed above under Colored Pigments.

⌘⌘⌘

COMPARISON OF MINERAL FILLERS IN ADHESIVES AND SEALANTS

PVC Plastisol (Armstead, 1994)
The data in Table 12 compares the rheological properties of a PVC plastisol, unfilled, containing ground limestone, and finally a blend of ground limestone and coated precipitated calcium carbonate. The unfilled plastisol had a low apparent viscosity, no yield value and severe sagging when applied to a vertical surface. Introducing a high loading of ground limestone raised the viscosity and imparted a slight yield value but insuffient to prevent severe sagging. Replacing 140 parts of the ground limestone with a high surface area treated precipitated calcium carbonate increased the apparent viscosity of the plastisol to 3200 mPas and raised the yield value to 271 Pa. This was suffient to prevent the sealant sagging when applied to a vertical surface. The high surface area precipitated calcium carbonate imparted a stronger particle-particle network in the PVC plastisol.

Table 12. Ground Limestone vs. Precipitated Calcium Carbonate In a PVC Plastisol

Formulation	A	B	C
Phthalate Plasticizer	180	180	180
PVC Dispersion Resin	100	100	100
Ground Limestone[1]	–	216	76
Coated Ppt. $CaCO_3$[2]	–	–	140
Calcium Oxide	12	12	12
Lubricant	22	22	22
Adhesion Promoter	3	3	3
Flow Properties			
Yield Stress, Pa	0	15	271
Apparent Visc., mPas	166	400	3200
Sag Value, cm	3	3	0

[1]VICRON® 25-11 (Specialty Minerals)
[2]UltraPlex® (Specialty Minerals)

Epoxy Sealant (Cabot, 1994) – Hydrophilic grades of fumed silica are widely used as thickening agents for epoxy resin adhesives and sealants. They impart excellent initial viscosity and sag resistance. After shelf aging for several months, however, the epoxy systems lose part or all of their sag resistance. A simple two part epoxy sealant formulation to test the utility of a series of mineral thixotropes is shown in Table 13.

⌘⌘⌘

Table 13. Two-Part Epoxy Sealant

	Parts by Weight
Part A	
EPON 828, epoxy resin	100
Calcium Carbonate, filler	100
Thixotrope	6
Part B	
Diethylenetriamine (DETA)	
Mixture	
Part A	100
Part B	11

The thixotropes were a 200m^2/g hydrophilic fumed silica (M-5), three surface treated fumed silicas – a 120m^2/g product partially treated with dimethyldichlorosilane (TS-610), a 100m^2/g product fully treated with polydimethylsiloxane fluid (TS-720), and a 215m^2/g product fully treated with hexamethyldisilazane (TS-530). The four silicas were compared with a high surface area defibrillated asbestos and a treated clay thixotrope. The sag resistance of the formulations after mixing Parts A and B were measured initially and after accelerated aging of Part A for periods of 1, 2 and 4 weeks at 60°C. After each time period samples of Part A were allowed to cool to room temperature prior to mixing with part B and measuring the sag values.

The results in Figure 8 show that only the silicone fluid treated thixotrope prevented sagging during the accelerated aging test. The defibrillated asbestos and the treated clay imparted poor initial sag resistance which remained unchanged during the test. The hexamethyldisilazane treated fumed silica imparted poor initial sag resistance which further deteriorated during the accelerated aging. Both the hydrophilic silica and the partially dimethyldichlorosilane treated silicas imparted good initial sag resistance which degraded over time. All four silicas have generally similar surface areas within the range of 100 to 215m^2/g and similar aggregate structures. Their main differences are related to differences in surface chemistry.

All mineral thixotropes function in adhesives and sealants by building up a three dimensional network of interacting particles. This network can be partially or fully destroyed if other components of the formulation preferentially interact physically or chemically with the mineral particles to prevent particle-particle interaction. In the case of epoxy adhesives and sealants, it is believed that the epoxide groups on the epoxy molecules gradually preferentially adsorb on the hydroxyl groups at the silica surface destroying the silica networks, resulting in an increase in sag. In the above comparison, the long chain silicone molecules are the most effective in preventing epoxy disruption of the silica network..

⌘⌘⌘

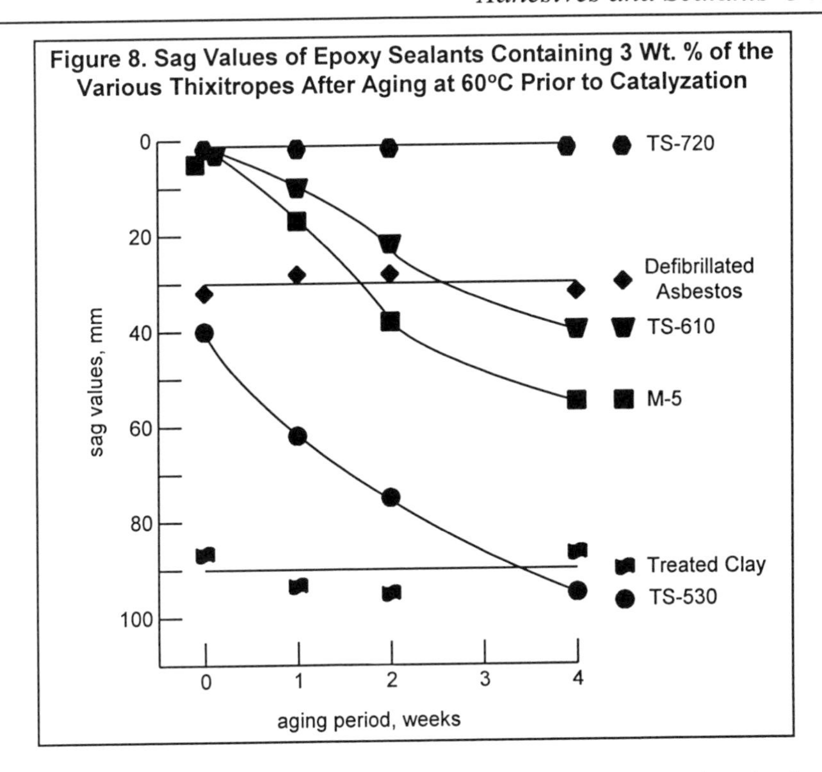

Figure 8. Sag Values of Epoxy Sealants Containing 3 Wt. % of the Various Thixitropes After Aging at 60°C Prior to Catalyzation

Silicone Rubber RTV-1 Sealants (Cabot, 1994) – The superior rheology control and reinforcement properties of fumed silica compared to a similar surface area grade of precipitated silica has already been shown in Figure 5 and Table 4. The more open chained structure of the fumed silica allows the formation of a much more effective silica network with more of the silica surface available for polymer/filler interaction, resulting in superior cured physical properties.

The influence of varying the surface area of hydrophilic fumed silica in the simple sealant has been studied. The results in Table 14 show that the lower surface area silicas were easier to disperse but gave lower yield values, viscosities and shear thinning indices, and faster extrusion rates. These results indicate that the the silica network strength increases as the silica surface area increases up to about 200 m^2/g, and then remains fairly constant. All of the networks were strong enough to prevent sagging. The cured physical properties show a similar trend.

⌘⌘⌘

Table 14. Effect of Fumed Silica Surface Area In a Silicone Rubber RTV-1 Sealant

Formula: (parts by weight)

Baysilone® E-18 Gum	100
Baysilone M-1000 Non-Reactive Diluent	50
Baysilone Croslinking Agent 3187	7
Fumed Silica	14
Dibutyl Tin Dilaurate (DBTDL)	0.03

Fumed Silica	**L-90**	**LM-130**	**LM-150**	**M-5**	**EH-5**
Surface Area, m^2/g	108	136	164	192	402
Processing and Rheological Properties					
Dispersion level	←←→← Excellent →→→→			Good	Fair
Yield value, Pa	410	560	700	700	530
Viscosity @ 1 rpm, Pa.s	45	65.3	68.7	81.6	90.6
STI	1.5	2.1	2.1	2.5	2.8
Slump value, cm	0	0	0	0	0
Extrusion rate, 0.28 MPa, g/min	250	110	100	80	70
Cured Physical Properties					
Shore A Durometer	20	20	21	23	23
Die B Tear, kN/m(ppi)	4.4(25)	5.1(29)	5.4(31)	5.8(33)	6.0(34)
Peel Adhesion, kN/m (ppi)	1.1(6.0)	1.3(7.3)	1.1*(6.5*)	1.5(8.8)	1.6*(9.3*)
Modulus @50%, MPa (psi)	0.29(40)	0.30(45)	0.34(50)	0.36(50)	0.37(55)
Tensile, MPa (psi)	1.6(240)	1.4(210)	1.6(235)	1.7(250)	1.6(230)
Lap Shear, MPa (psi)	1.9(275)	1.0(150)	0.8(110)	1.2(170)	0.6(85)
% Elongation	450	460	430	440	410
Clarity,% T@ 500nm	47	57	63	68	62

All samples showed cohesive failure, except * where 40% adhesive failure occured.
(Cabot)

Table 15 shows the effect of varying the surface treatment of the fumed silica in the same sealant formula. The partially silane treated silica (TS-610) had a moderate yield value, moderate extrusion rate and no sagging, indicating a moderately strong silica network. The fully surface treated silicas (TS-530, TS-720) gave low yield values, lower viscosities, higher extrusion rates, and high slump values of >5cm. These results indicate that the fully treated silicas

⌘⌘⌘

provide weak silica networks. The cured physical properties imparted by the three silicas were generally similar.

Table 15. Evaluation of Treated Fumed Silicas in A Silicone Rubber RTV-1 Sealant

Fumed Silica	**TS-610**	**TS-530**	**TS-720**
Treatment Agent	DIMEDI	HMDZ	PDMS
Incorporation time	3.5	2.5	3
Dispersion level	←←←	Good →→→	
Yield value, Pa	410	195	220
Viscosity @ 1 RPM, Pa.s	43.8	37.4	31.8
STI	1.39	1.02	1.26
Slump value, cm	0	> 5	> 5
Extrusion rate, 0.28 MPa, g/min.	160	175	260
Cured Physical Properties			
Shore A Durometer	20	16	17
Die B Tear, kNm-1(ppi)	4.2(24)	3.5(20)	3.5(20)
Modulus @ 50% MPa(psi)	0.31(45)	0.31(40)	0.28(40)
Peel Adhesion, kNm^{-1}(ppi)	1.0(5.5)	1.2(6.8)	0.9(5.0)
Lap Shear, Mpa(psi)	1.9(280)	1.9(280)	1.9(280)
Tensile, MPa (psi)	1.4(205)	0.79(115)	1.0(150)
% Elongation	405	305	350
Clarity, % T @ 500nm	66	62	55

DIMEDI = dimethyldichlorosilane
HMDZ = hexamethyldisilazane
PMDS = polydimethylsiloxane
(Cabot)

The preceeding three studies reflect the general findings that, at a given loading, the effect of the filler on product properties is determined mainly by three filler characteristics: surface area, which is generally related to particle size; structure level, as determined by particle or aggregate shape; and surface modification, according to chemical type and level.

REFERENCES

Adamson, A.W., 1967, *Physical Chemistry of Surfaces*, Interscience, pp 57.

Armstead, J.C., 1994, "Use of Calcium Carbonate in Sealant Applications", *Supplier Short Course 1*, Adhesive and Sealant Council, Washington, DC.

⌘⌘⌘

A.S.C., 1986, *Caulk and Sealant Short Course*, Adhesive and Sealant Council, Washington, DC

Asselin,D.A., et al, 1990, *Journal of the Adhesive and Sealant Council*, Vol.XIX, No.3, Fall, pp 181-198 .

Back,J., et al, ibid, pp 79-89.

Barnes,H.A.,et al,1989, *An Introduction to Rheology*, Elsevier, New York.

Barth, B.R., 1977, "Phenolic Resin Adhesives", *Handbook of Adhesives*, 2nd ed., I. Skeist(ed.), Van Nostrand Reinhold, New York, pp 382-416.

Beers, M.D., "Silicone Adhesive Sealants", ibid, pp. 628-639.

Cabot Corporation, 1990, *CAB-O-SIL Fumed Silica in Adhesives and Sealants.*

Chu,S.G., 1986, "Sealants Based on Block Copolymers", *Caulks and Sealants Short Course*, Adhesive and Sealant Council, Washington, DC, pp 269-339.

Clemens, L.M. ,et al, 1990, *Journal of Adhesive and Sealant Council*, Vol.XIX, No.3, Fall, pp 55-89.

Cochrane, H.and Lin, C.S., 1988, "The Use of Fumed Silica in Adhesives and Sealants", *Supplier Short Course I,* Adhesive and Sealant Council, Washington, DC, Dec., pp 45-78.

Coover, H.W.Jr., McIntire, J.M., 1977, "Cyanoacrylate Adhesives", *Handbook of Adhesives*, 2nd ed., I. Skeist(ed.), Van Nostrand Reinhold, New York, pp 569-580.

Corey, et al, 1977, "Polvinyl Acetate Emulsions and Polyvinyl Alcohol for Adhesives", ibid, pp 465-483.

Degussa Corporation, 1981, *Application of Highly Dispersed Synthetic Products in Adhesives*, Brochure No.44.

Degussa Corporation, 1979, *Synthetic Silicas for Joint Sealing Compounds,* Brochure No.63.

Degussa Corporation, 1992, *Precipitated Silicas, Manufacturing / Properties / Applications.*

Dexheimer, R.D.,and Vertinik, L.R., 1977, " Polyamide Adhesives", *Handbook of Adhesives*, 2nd ed., I. Skeist(ed.), Van Nostrand Reinhold, New York, pp 581-591.

Dick, J.S., 1987, "The Adhesive Industry", *Compounding Materials for the Polymer Industries,* Noyes Publications, Park Ridge, NJ, pp 165-204.

Eisentrager, K., and Druschke, W., 1977,"Acrylic Adhesives and Sealants", *Handbook of Adhesives*, 2nd ed., I. Skeist(ed.), Van Nostrand Reinhold, New York, pp. 528-559.

Elias, M.G., 1986, "Formulating Silicone Sealants", *Caulks and Sealants Short Course,* Adhesive and Sealant Council, Washington, DC, Dec., pp. 241-267.

Ferrigno, T.H.,and Taranto, M., 1978, "Kaolin, Hydrous and Anhydrous", *Handbook of Fillers and Reinforcements for Plastics*, H.S.Katz and J.V. Milewski (eds.), Van Nostrand Reinhold, New York, pp 119-126.

Flick, W.F., 1989, *Handbook of Adhesives Raw Materials,* 2nd Ed., Noyes Publications.

Gilbertson, T.J., 1991, "Mixing Water with Electrical Energy: Successful Printing with Water Based Inks," *Polymers, Laminations and Coatings*, TAPPI Press, Atlanta.

Hahn, J., 1992, "Clay and Castor-Based Rheological Additives for Adhesives and Sealants", *Supplier Short Course on Additives,* , Adhesive and Sealant Council, Washington, DC, April. pp vii, 1-27

⌘⌘⌘

Harlan, J.T. and Petershagen, L.A., 1977, "Thermoplastic Rubber in Adhesives", *Handbook of Adhesives*, 2nd ed., I. Skeist(ed.), Van Nostrand Reinhold, New York, pp 304-330.

Kilchesty, A.A., 1986, "Oleo Resinous and Oil-Base Caulking Compounds", *Caulks and Sealants Short Course*, Adhesive and Sealant Council, Washington, DC, Dec., pp 13-23.

Kilchesty, A.A., 1986A, "Solvent Acrylics", ibid, pp 77-84.

Kummer, P.,and Crowe, G., 1978, "Calcium Carbonate", *Handbook of Fillers and Reinforcements for Plastics*, H.S.Katz and J.V. Milewski (eds.), Van Nostrand Reinhold, New York, pp 81-118.

Kirn, W.A., 1986, "Waterborne Acrylic Caulks", *Caulks and Sealants Short Course*, Adhesive and Sealant Council, Washington, DC, Dec., pp 183-203.

Moult, R.H., 1977,"Resorcinolic Adhesives", *Handbook of Adhesives*, 2nd ed., I. Skeist(ed.), Van Nostrand Reinhold, New York, pp 417-423,.

Moreland, J.E., 1978, "Silica Fillers, Extenders And Reinforcements," *Handbook of Fillers and Reinforcements for Plastics*, H.S.Katz and J.V. Milewski (eds.), Van Nostrand Reinhold, New York, pp 136-159.

Morrill, J.P. and Marguglio, L.A., 1977, "Nitrile Rubber Adhesive," *Handbook of Adhesives*, 2nd ed., I. Skeist(ed.), Van Nostrand Reinhold, New York, pp 273-292.

Newman, L.E., 1986, "Polyvinyl Acetate-Based Caulks and Sealants", *Caulks and Sealants Short Course*, Adhesive and Sealant Council, Washington, DC, Dec., pp. 205-240.

Newton, M.V. et al, 1986, "Butyl Sealants: Formulating, Developing, Processing", ibid, pp 25-75.

Nyco Minerals, Inc., 1994, *Wollastocoat in Thermoplastics, Thermosets and Elastomers*.

O'Driscoll, M., 1988," Minerals in Adhesives and Sealants", *Industrial Minerals*, Feb., pp 32-51.

Panek, J.R, 1977,"Polysulfide Sealants and Adhesives", *Handbook of Adhesives*, 2nd ed., I. Skeist(ed.), Van Nostrand Reinhold, New York, pp 368-381.

Petersen, E.A., 1986, "Formulating Sealants with Mercaptan Terminated Polymers", *Caulks and Sealants Short Course*, Adhesive and Sealant Council, Washington, DC, Dec., pp 87-132.

Petronio, M., 1977, “Properties, Testing Specification and Design of Adhesives", *Handbook of Adhesives*, 2nd ed., I. Skeist(ed.), Van Nostrand Reinhold, New York, pp 92-114.

Podlas, T.J., 1988, "Cellulosics and Natural Gums: Their Chemistry and Utility in Adhesives", *Supplier Short Course I,* Adhesive and Sealant Council, Washington, DC, Dec., pp 79-103.

Prane,J .W. et al, 1990, " Sealants And Caulks", *Handbook of Adhesives*, 3rd ed., I. Skeist(ed.), Van Nostrand Reinhold, New York, pp.611-640.

Queen, E.J., 1978, "Carbon Black", *Handbook of Fillers and Reinforcements for Plastics*, H.S.Katz and J.V. Milewski (eds.), Van Nostrand Reinhold, New York, pp 277-291.

Radosta, J.A and Trivedi,N.C., 1978, " Talc", ibid, pp 160-171.

Regan, J.F., 1986, "Urethane Formulations", *Caulks and Sealants Short Course*, Adhesive and Sealant Council, Washington, DC, Dec., pp 133-182.

Ross & Son Company, "Double Planetary Mixers", Brochure edition D386.

⌘⌘⌘

Savla, M., 1977, "Epoxy Resin Adhesives", *Handbook of Adhesives*, 2nd ed., I. Skeist(ed.), Van Nostrand Reinhold, New York, pp 434-445.

Schollenberger, C.S., 1977, "Polyurethane and Isocyanate-Based Adhesives", ibid, pp 446-464.

Statt, B.K. et al, 1992, "Precipitated Silicas in Adhesives and Sealants", *Supplier Short Course on Additives, ,* Adhesive and Sealant Council, Washington, DC, April, pp IV, 1-30.

Steinfink, M., 1977, "Neoprene Adhesives: Solvent and Latex", *Handbook of Adhesives*, 2nd ed., I. Skeist(ed.), Van Nostrand Reinhold, New York, pp 343-367.

Stucker, N.E.,and Higgins,J.J, 1977, " Butyl Rubber and Polyisobutylene", ibid, pp 255-272.

Stricharczuk, P.T., and Wright, D.E., 1977, "Styrene Butadiene Rubber Adhesives", ibid, pp 293-30.

Tapper, M., and Mathur, K., 1992, " Use of Calcium Carbonates in Sealants", *Supplier Short Course on Additives, ,* Adhesive and Sealant Council, Washington, DC, April, pp. II, 1-19.

Toporcer, L.H., 1986, " Sealant Specifications", *Caulks and Sealants Short Course*, Adhesive and Sealant Council, Washington, DC, Dec., pp 1-10.

Waddell, H.H.,and Touval,I., 1978, "Antimony Oxide", *Handbook of Fillers and Reinforcements for Plastics*, H.S.Katz and J.V. Milewski (eds.), Van Nostrand Reinhold, New York, pp 219-236.

Wake, W.C., 1977, " Natural Rubber and Reclaimed Rubber Adhesives", *Handbook of Adhesives*, 2nd ed., I. Skeist(ed.), Van Nostrand Reinhold, New York, pp 242-254.

Washabaugh, F.J., 1992, "Kaolin, Talc and Attapulgite", *Supplier Short Course on Additives, ,* Adhesive and Sealant Council, Washington, DC, April, pp III, 1-19.

Woycheshin, E.A. and Sobolev, I., 1978, "Alumina Trihydrate", *Handbook of Fillers and Reinforcements for Plastics*, H.S.Katz and J.V. Milewski (eds.), Van Nostrand Reinhold, New York, pp 237-249.

⌘⌘⌘

EIGHT

PLASTICS

George Hawley
George C. Hawley & Associates, Ltd.
Saranac, NY

Minerals today are primarily used for the property enhancements they bring to polymers, not for their low cost. Although they are commonly called fillers, they are often actually more expensive than the polymer itself when specific gravity, volume replacement and compounding costs are taken into consideration. The major improvements imparted by minerals include increases in stiffness, strength, impact and temperature resistance, dimensional stability, surface hardness and scratch resistance. These properties are in ever-increasing demand as the service requirements for plastics grow more stringent than can be met by unfilled polymers. These demands have largely originated from the automobile industry, where fleet mileage regulations have required extensive reduction in vehicle weight. This has been acheived through downsizing and by the substitution of low density plastics for high density metals.

PLASTIC POLYMERS

Polymer Types

Most polymers are formed by either addition or condensation reactions, and are conventionally differentiated as thermoplastic or thermosetting.

Thermoplastic addition polymers – The common addition polymers are those based on a polyethylene backbone. Their synthesis starts with the catalytic activation of monomers such as ethylene, propylene or vinyl chloride to free radicals. These combine sequentially in an addition or chain reaction. The polyethylene backbone consists of a chain of methylene groups. The other common addition polymers, as illustrated in Figures 1 and 2, are essentially polyethylenes with one or both of the hydrogens on the methylene group substituted.

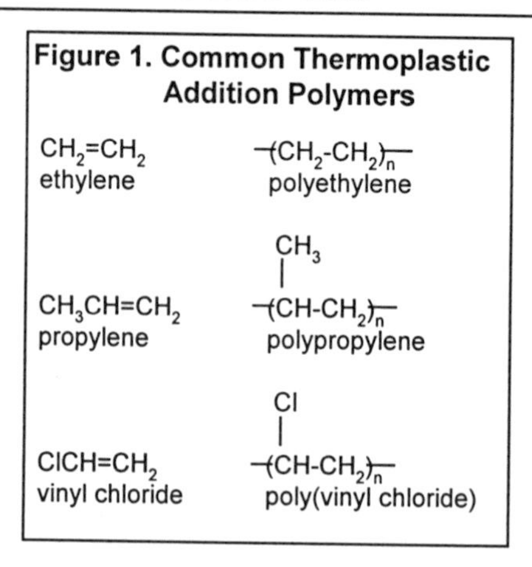

The addition polymers most often used in plastics compounding are thermoplastic solids at room temperature. They melt and flow on heating, and resolidify on cooling. This cycle can be repeated indefinitely, except that some polymer breakdown occurs during each heat cycle, usually by oxidation and scission of the long chain into smaller segments.

Thermoplastic condensation polymers – Condensation polymerization relies essentially on acid-base reactions, generally with the elimination of one small molecule of waste product (e.g. water) per repeating unit. Thus, in the formation of the polyamide Nylon 66, adipic acid ($C_6H_{10}O_4$) reacts with hexamethylene diamine ($C_6H_{16}N_2$) with the liberation of water. Similarly, in the condensation of thermoplastic polyesters, a dibasic acid reacts with a difunctional alcohol with the elimination of water to form a chain that consists of units linked by ester groups. Thermoplastic condensation polymers are synthesized from saturated monomers, as shown in Figure 3.

Thermoset condensation polymers – Thermosetting polymers contain unsaturated backbones and/or reactive substituents. With heat, these polymers crosslink into an irreversible network, forming a hard infusible plastic. The major thermoset polymers are polyesters and polyurethanes. For polyesters, at least one of the reactants is unsaturated, as shown in Figure 4.

The major applications for thermosetting polyurethanes are rigid and flexible foams, and reaction injection molded (RIM) motor vehicle fenders and fascias. The formation of thermoset polyurethane, as represented by RIM polyurethane, is described in Figure 5. Curing (crosslinking) can make thermoset polymers shrink and warp, problems which are routinely overcome through the use of reinforcing minerals. Acicular wollastonite and

⌘⌘⌘

mica, for example, are routinely used for reinforcement in RIM polyurethane.

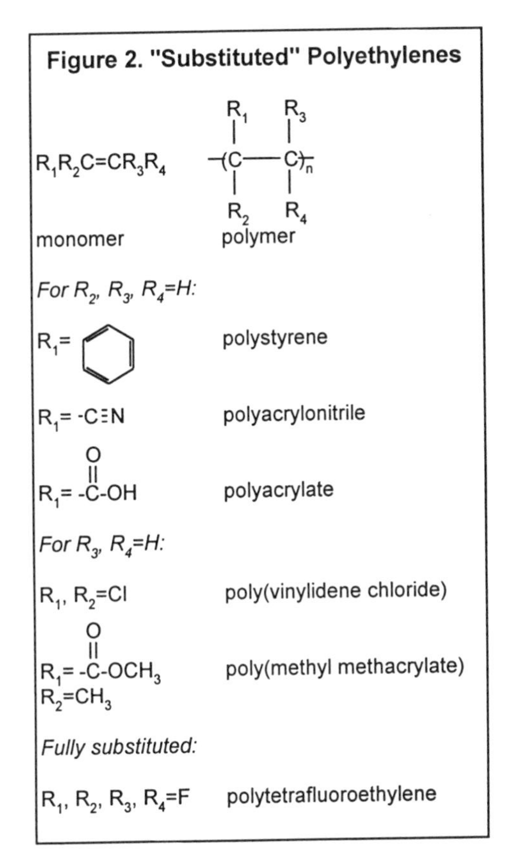

Figure 2. "Substituted" Polyethylenes

Special Considerations

Crystallinity in thermoplastics – Long unbranched chains with very regular structure – typical of several thermoplastic polymers – are able to line up with each other to give a solid that has localized domains with properties resembling crystals. These domains, called crystallites, are separated by amorphous polymer zones. Polymers with a high proportion of crystallites are regarded as crystalline or semi-crystalline. Polymers with side chains or bulky substituents are oriented in random fashion, so that the chains are hindered from aligning. These polymers have no oriented domains and are considered amorphous.

Crystalline polymers tend to be harder and denser than the equivalent amorphous polymers, and have a narrower melting range. The crystallites

⌘⌘⌘

also make them opaque or translucent, whereas amorphous polymers may be clear and water white.

Crystalline polymers show the greatest interaction with minerals. In amorphous polymers, minerals merely increase stiffness, with little or no reinforcing effect. Polypropylene, high density polyethylene, thermoplastic polyester and polyamides are crystalline polymers. Low density polyethylene, polystyrene and polycarbonate are amorphous.

Figure 3. Common Thermoplastic Condensation Polymers

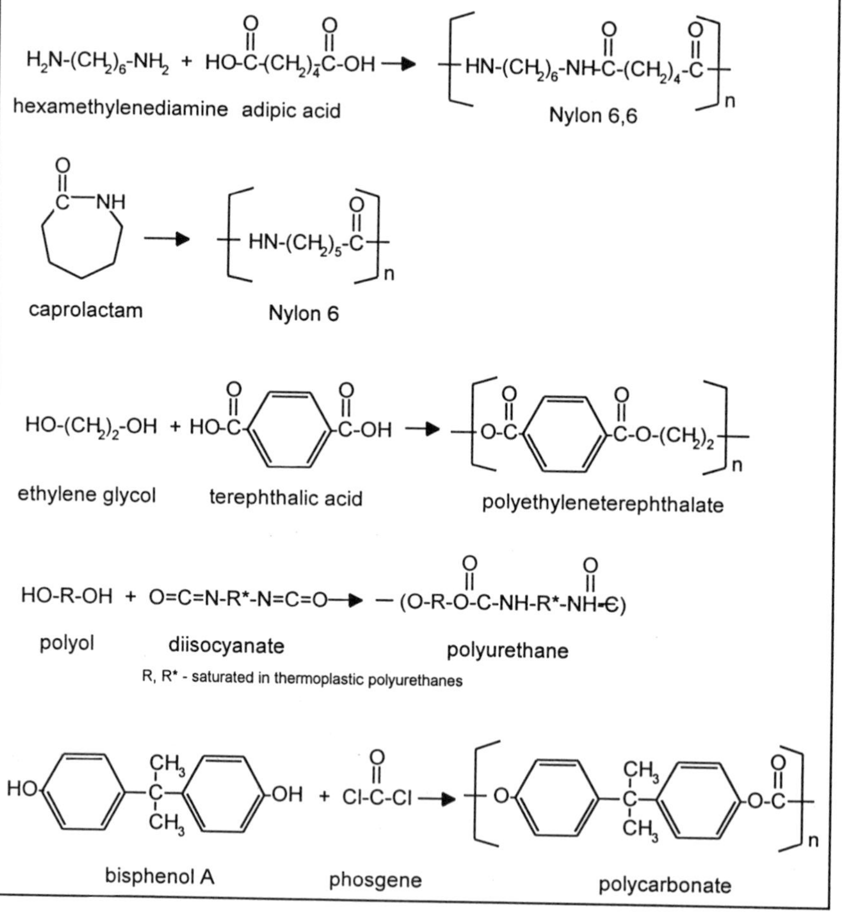

$H_2N\text{-}(CH_2)_6\text{-}NH_2$ + $HO\text{-}C(=O)\text{-}(CH_2)_4\text{-}C(=O)\text{-}OH$ → $[\text{-}HN\text{-}(CH_2)_6\text{-}NH\text{-}C(=O)\text{-}(CH_2)_4\text{-}C(=O)\text{-}]_n$

hexamethylenediamine adipic acid Nylon 6,6

caprolactam → $[\text{-}HN\text{-}(CH_2)_5\text{-}C(=O)\text{-}]_n$

caprolactam Nylon 6

$HO\text{-}(CH_2)_2\text{-}OH$ + $HO\text{-}C(=O)\text{-}C_6H_4\text{-}C(=O)\text{-}OH$ → $[\text{-}O\text{-}C(=O)\text{-}C_6H_4\text{-}C(=O)\text{-}O\text{-}(CH_2)_2\text{-}]_n$

ethylene glycol terephthalic acid polyethyleneterephthalate

HO-R-OH + O=C=N-R*-N=C=O → —(O-R-O-C(=O)-NH-R*-NH-C(=O))

polyol diisocyanate polyurethane

R, R* - saturated in thermoplastic polyurethanes

$HO\text{-}C_6H_4\text{-}C(CH_3)_2\text{-}C_6H_4\text{-}OH$ + $Cl\text{-}C(=O)\text{-}Cl$ → $[\text{-}O\text{-}C_6H_4\text{-}C(CH_3)_2\text{-}C_6H_4\text{-}O\text{-}C(=O)\text{-}]_n$

bisphenol A phosgene polycarbonate

⌘⌘⌘

Figure 4. Thermoset Polyesters

glycol + saturated anhydride or saturated dibasic acid + unsaturated anhydride or unsaturated dibasic acid = unsaturated polyester

unsaturated polyester + reactive diluent + promoter = liquid polyester resin

liquid polyester + catalyst + heat (optional) = cured polyester

Reactant	**Example**	
glycol	ethylene glycol	$HO\text{-}CH_2CH_2\text{-}OH$
saturated anhydride	phthalic anhydride	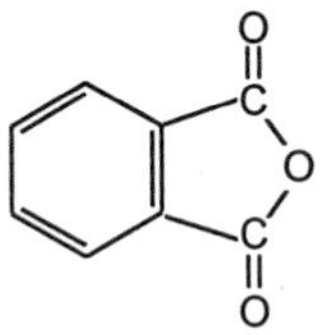 [structure]
saturated dibasic acid	orthophthalic acid	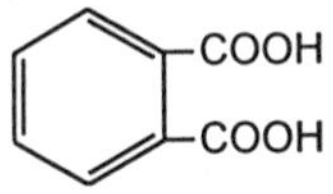 [structure: benzene ring with two COOH]
	fumaric acid	HOOCCH=CHCOOH
unsaturated anhydride	maleic anhydride	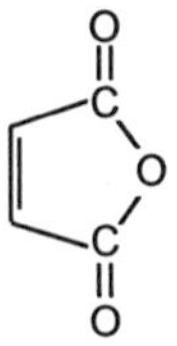 [structure]
reactive diluent	styrene	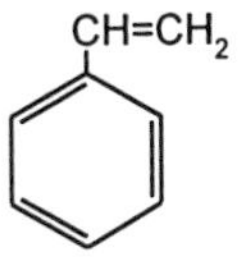 [structure: benzene ring with $CH{=}CH_2$]
promoter	cobalt naphthenate	
catalyst	methylethylketone peroxide	

⌘⌘⌘

Figure 5. Thermoset RIM Polyurethane

polyol + chain extender + isocyanate = crosslinked polyurethane

Reactant	Example	
polyol	linear low MW polyether	$HO\text{-}(CH_2CH\text{-}O)_n\text{-}CH_2CH\text{-}OH$ [poly(propylene glycol)]
isocyanate	aromatic diisocyanate	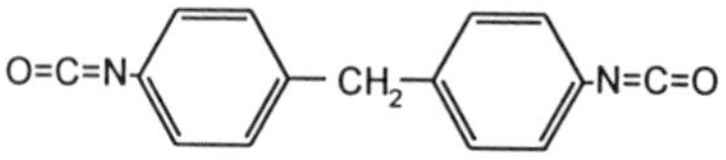 O=C=N-C_6H_4-CH_2-C_6H_4-N=C=O [diphenylmethane-4,4'-diisocyanate]
chain extender	diol	$HO\text{-}(CH_2)_4\text{-}OH$ [butanediol]
	or	
	diamine	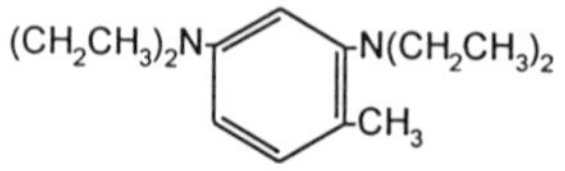 $(CH_2CH_3)_2N$-C_6H(CH_3)-$N(CH_2CH_3)_2$ [diethyltoluenediamine]

Thermal expansion – The requirements of rigidity, low coefficient of thermal expansion and surface smoothness are problems with plastic automobile body parts. Thermoplastics, unless exotic and expensive, cannot meet all these requirements without the addition of functional fillers. Certain thermosetting plastics, such as polyester sheet molding compound (SMC) reinforced by glass fibers and calcium carbonate, and reaction injection molded polyurethane or polyurea reinforced with mica and wollastonite (RRIM) are therefore commonly used for automotive applications.

Table 1 compares the thermal properties of metals, minerals and plastics. The thermal expansion of plastics is up to nine times greater than that of metals. This becomes important when plastic parts are used in combination with metals. For example, the chassis and framework of an automobile are made from steel. If a large part – a door, trunk lid or hood – made of plastic is set in a metal frame, problems will occur as the automobile undergoes its service temperature cycle of 140°F in summer in the Texas sun to -60°F in Alaskan winters. Between these temperature extremes, the differential expansion can be as high as 0.01 inch per inch. In a part that is 50 inches wide, this corresponds to one-half inch. If the door fits its opening in winter, it will jam in summer; if it fits in summer, it will rattle in winter. As Table 1 shows, minerals have one-half to one-third the thermal expansion of metals. Compounding plastics with minerals brings their thermal expansion closer to that of metals and reduces the problem.

Thermal expansion is not isometric, it tends to be smaller along the polymer chain than across it. This becomes important when a crystalline

⌘⌘⌘

polymer cools after it has been molded. The injection or extrusion molding process tends to align the polymer chains in the direction of flow of the molten plastic. For economic reasons, to keep the molding cycle short, the part is always released from the mold while it is still hot, just as soon as it has cooled enough to keep its shape. On cooling to room temperature, the part will tend to warp as it shrinks more in the cross-flow direction than in the direction of flow.

Table 1. Relative Thermal Properties of Materials

	Resistance °F	**Linear Expansion** in./in.°F x 10^{-6}	**Conductivity** BTU/ft^2,hr.,°F/in
Metals	**melting**		
Iron, gray cast	2150	6.7	310
Steel, carbon	2760	6.7	360
Zinc, die cast	728	15.2	784
Aluminum, die cast	1100	11.6	670
Titanium	3000	5.3	114
Magnesium, die cast AZ91A	1105	14.5	370
Minerals	**decomposition**		
Calcium carbonate	1515	5.5 - 6.5	4.3
Talc	1650	4.4	3.8
Mica, muscovite	932 - 986	4.4 - 5	4.6
Mica, phlogopite	1562 - 1832	7 - 8	4.6
Wollastonite	2804	3.6	na
	deflection temp. under load (66 psi), °F		
Polymers, unfilled			
PP homopolymer	225 - 250	45 - 55	0.8
HDPE homoploymer	175 - 196	33 - 61	3.5
PVC, rigid	135 - 180	28 - 55	1 - 1.5
PBT	240 - 375	33 - 53	1.2 - 2
Polymers, filled			
PP/40% $CaCO_3$	270	15	2
PP/40% talc	273	23	2.2
PP/40% mica	280 - 295	12	n.a.
PP/40% glass fiber	290 - 320	12	n.a.
Nylon 6,6/40% glass fiber	260 - 500	8.3 - 30	1.5 - 3.4
PBT/15% glass fiber/ 25% mica	426	1	n.a.

⌘⌘⌘

Fibrous reinforcements, such as glass and asbestos, and acicular reinforcements, such as wollastonite, also tend to orient parallel to the flow direction. Since they all have lower thermal expansions than plastics, use of these minerals tend to promote warping, especially in crystalline polymers. Fillers such as mica, talc and kaolin, also orient in the flow direction, but their platy, more two-dimensional shape does not contribute to warping. High aspect ratio mica, because of its extreme platiness, is used to counteract the natural warping tendency of highly crystalline thermoplastic polyesters.

Advantages and Disadvantages of Plastics

Unfilled plastics have certain features that can promote or prevent their use in place of metals or wood.

Advantages of Plastics – The main advantages of plastics are :

Plasticity – Plastics can be molded into complex shapes that need no further machining or finishing. Many metal pieces can be consolidated into one plastic piece, reducing assembly costs.

Low temperature processability – Plastics are processable at much lower temperature than metals, with consequent saving.

Repeatability – Multiple identical parts can be molded in rapid cycles.

Colorability – Colorants may be incorporated in the plastic so that no painting or other decoration is necessary.

Flexibility – Plastics are resilient so that they will withstand impact without breakage and will spring back to the original shape when deflected.

Low conductivity – Most plastics are poor conductors of heat and electricity.

Chemical resistance – Plastics do not corrode and resist most inorganic chemicals.

Weatherability – Plastics can be compounded for long-term weathering resistance.

Low cost – On a volume basis, plastics are generally much less expensive than metals.

⌘⌘⌘

Light weight – Plastics are less dense than metals, slightly heavier than wood.

Abrasion/erosion resistance – Plastics are routinely compounded for good wear resistance.

Disadvantages of Plastics – The main disadvantages of plastics are:

Low strength – The strength of plastic components is often much lower than those of metal or wood.

Low stiffness – Stiffness is much lower than metal and wood.

Low hardness – Plastics tend to scratch and scuff more than metals.

Poor high temperature resistance – High temperature tolerance is much lower than metals, but superior to wood.

Poor low temperature resistance – Plastics tend to embrittle below their glass transition temperature.

Inferior decorability – Plastics are more difficult to paint and plate than metals. Adhesion failure is common.

Low thermal conductivity – Plastic parts cool slowly, restricting the thicknesses that can be molded and requiring the part be left in the mold a relatively long time before it can be demolded without warping.

Creep – Plastics tend to cold-flow under small continuous loads until they fail at much lower values than the initial conditions.

Flammability – Most plastics are highly inflammable, although this can be improved by additives.

Low solvent resistance – Most plastics are swelled by solvents; some are weakened by boiling water.

Stress cracking – Most plastics crack along stress lines.

The science of compounding polymer-mineral composites consists of retaining the desired features of the plastic, while overcoming its inherent

⌘⌘⌘

deficiencies. The mechanical properties of selected polymers, unfilled and filled, and metals are compared in Table 2.

Table 2. Relative Mechanical Properties of Materials

	Tensile Strength Kpsi	Elongation %	Modulus Kpsi
Metals			(tensile)
Iron, gray cast	25	0.5	30,000
Steel, carbon	38	30	30-60,000
Zinc, die cast	41	10	12,035
Aluminum, die cast	43	2	10,600
Titanium	127	12	16,000
Magnesium, die cast AZ91A	33	3	6,525
Polymers, unfilled			
PP homopolymer	4.7	350	180
HDPE homoploymer	4.1	600	160
Nylon 6,6	3.5	200	n.a.
PVC, rigid	13.7	80	500
PBT	8.7	50	435
Polymers, filled			
PP/40% $CaCO_3$	2.8	10	500
PP/40% talc	4.3	3	575
PP/40% mica	7.3	3	575
PP/40% glass fiber	6.3	1.8	1000
PVC/40% $CaCO_3$	3.5	200	n.a.
Nylon 6,6/40% glass fiber	28	22.5	1700
PBT/15% glass fiber/ 25% mica	12	5	1350

Why Plastics Are Used

Plasticity – Plastics are mainly used for the reason implied by the name – plasticity. They can be molded into extremely complex forms with accurately controlled dimensions, and with no need for post-machining. Shrinkage does occur, on cooling or curing, so that the final dimensions are less than that of the mold, but the shrinkage is predictable and can usually be compensated in the mold dimensions and design. This is not true in the case of cast metals, which usually require casting to a larger size, followed by machining to the final dimensions.

⌘⌘⌘

Plastic parts can be molded with much greater complexity than is achievable with cast, forged or stamped metals. Thus, components that were assemblies of many metal parts can be molded as one plastic piece. This saves assembly, which is costly in fasteners and man-power. This is a major reason for the conversion from metals to plastics by the automobile industry.

Table 3. Relative Raw Material Costs

	Density g/cc	**Density** lb./in.^{3}	**Cost** cents per lb.	**Cost** cents per in.^{3}
Metals				
Iron, gray cast	7.8	0.282	30	8.46
Steel, carbon	7.86	0.284	30	8.52
Zinc, die cast	6.6	0.238	46	10.95
Aluminum, Al 380 ingot	2.72	0.098	78	8.13
Magnesium, die cast AZ91A	1.81	0.065	190	12.35
Minerals				
Calcium carbonate	2.7	0.097	5 - 16	0.48 - 1.55
Talc	2.8	0.101	10 - 25	1.01 - 2.53
Kaolin	2.6	0.094	5 - 25	0.47 - 2.35
Barite	4.3	0.155	10 - 15	1.55 - 2.33
Alumina trihydrate	2.4	0.087	14 - 36	1.21 - 3.13
Mica, muscovite	2.9	0.105	4.3 - 48	0.45 - 5.04
Mica, phlogopite	2.7	0.097	13.5 - 35	1.31 - 3.39
Wollastonite	2.9	0.104	15 - 100	1.58 - 10.5
Polymers, unfilled				
PVC	1.4	0.051	31 - 32	1.58 - 1.63
PP	0.9	0.032	40	1.28
HDPE	0.96	0.035	36 - 38	1.26 - 1.33
PBT	1.3	0.047	143 - 150	6.72 - 7.05
Nylon 6 & 6,6	1.13	0.041	129 - 133	5.29 - 5.45
Thermoset polyester	1.25	0.045	57 - 65	2.56 - 2.92
Polymers, filled				
PVC/40% $CaCO_3$	1.74	0.063	33-49	2.08-3.09
PP/40% $CaCO_3$	1.23	0.044	54	2.38
PP/40% talc	1.23	0.044	55	2.42
PP/40% mica	1.24	0.045	63	2.83
SMC	1.8	0.065	110	7.15
BMC	1.9	0.069	90	6.21
Nylon 6 or 6,6/ 40% glass fiber	1.45	0.052	122 - 130	6.34 - 6.76
PBT/15% glass fiber/25% mica	1.6	0.058	125-150	7.25-8.70

⌘⌘⌘

Low Density – Plastics are considerably lighter than metals, as seen in Table 3. For many applications this is not important. One case where density has caused a great swing from metals towards plastics has been in the manufacture of automobiles. Regulations concerning fleet average gas mileage have caused the automobile industry to switch to plastics wherever possible. Engine size is a major factor in gas utilisation, and the most effective method to reduce gas usage is to downsize the motor. A smaller motor can only pull a lighter car. A lighter car is either a smaller car or a car made with lighter components. A weight saving of one kilogram reduces fuel consumption by 0.0025 US gallons per 100 miles (Holland, 1980). Reducing vehicle weight by 30% also leads to a 20% reduction in production of carbon dioxide, a contributor to global warming.

Historically, small cars have not been as popular with the American motorist as they are in Europe and Japan. The goal, therefore, is to reduce weight without reducing size. The switch to plastics first took place in non-structural passenger compartment parts, but has been moving toward parts with more stringent requirements in the drive train and engine compartment.

A direct switch from stamped or cast metal parts is not possible since unfilled plastics are much weaker and more flexible than metals, as shown in Table 2. This means that plastic parts must be much thicker than those of metal to give the same strength and stiffness. Thus, the weight reduction factor is not as good as would be envisioned from the density figures alone. However, these problems of low strength and stiffness can be partially addressed by clever design. Further gains in stiffness and strength are obtained by the reinforcement of the plastics with minerals and glass fibers. These latter materials are almost as strong and stiff as ferrous metals and are intermediate in density between these and plastics. The properties of plastic-mineral composites are based on the volume concentration of the components, and can be tailored to a great extent.

Cost – As Table 3 shows, the raw materials cost of plastics are of the same order as the metals on a weight basis, but are much lower on a volume basis. This latter is the more important measure, since manufactured goods are usually bought on an area or volume basis. For example, a box is bought to hold a certain volume. The volume held is a function of the area of the sides of the box. Thickness of the sides is only important in that they must be thick enough to support the load of the box itself and its contents. A metal box will be able to have thinner sides than an equivalent plastic box, but in most cases, a box of equal strength in virgin or mineral-reinforced plastics will be considerably lighter and cheaper.

⌘⌘⌘

Cost per cubic inch is a first measure of true cost. Strength, stiffness and overall fabrication cost must also be taken into consideration. Yard-sticks like cost per eqivalent strength or stiffness are used. Low cost per cubic inch and design plasticity, with its savings in post-fabrication costs and in parts consolidation, have made plastics much cheaper than metals.

The last metal parts to be replaced in automobiles are exterior body parts, and engine parts. For body parts, sheet metal can be stamped cheaply into relatively simple forms much more rapidly than plastics can be molded. Very large area plastic parts require molding machines that are very large and expensive. The high pressure required necessitates clamps or presses capable of holding more than 100 tons force. Foam molding or injection/compression molding alternatives operate at substantially lower pressures. For engine parts, the service temperatures are too high for conventional plastics. Engineering resins, nevertheless, are being used with and without fillers and reinforcements in intake manifolds and radiator headers.

MINERALS IN PLASTICS

The primary mineral fillers and reinforcements in plastics are calcium carbonate, kaolin, talc, mica, wollastonite and silica. Calcium carbonate is by far the mineral most commonly used to fill plastics. This is mainly because it is low in cost, widely available and provides a good balance of properties. Calcium carbonate may be pure or in combination with magnesium carbonate as a dolomitic limestone. Dolomitic limestone is harder than pure calcium carbonate and is preferred when abrasion resistance is needed, as in floor tiles.

Natural calcium carbonate is processed by dry grinding to coarse and medium particle sizes, but is wet ground for the finest grades. Surface treatment with stearic acid and calcium stearate is very common and is used to improve dispersion and to act as a processing aid. Coarse and medium grades are used in cultured marble. Blends are often used to give the best packing of particles so that the maximum loading can be incorporated. Fine and stearate coated grades are used in injection molding and extruded plastics. Precipitated calcium carbonate (PCC) is more expensive and accordingly most often used where the highest reinforcing effects are needed, mainly in wire and cable compounds.

The major use of kaolin in plastics is for calcined grades in PVC wire insulation to improve electrical resistivity. Air floated and water washed grades are used in thermosets to ensure a smooth surface finish, and

⌘⌘⌘

delaminated clay is used in thermoplastics to improve physical properties and, when silane treated, impact resistance.

Talc is unusual among the mineral fillers in that it is naturally oleophilic, making it highly compatible with olefinic polymers. Its single largest application is in polypropylene to increase stiffness and resistance to high temperature creep. Talc is not amenable to treatment with silanes, but proprietary surface modified grades are available which improve tensile strength and impact properties.

Micas are hydrophilic and are thus compatible with polar polymers without surface treatment. But for best performance in polyolefins they are surface treated with silanes or surface active agents. Mica is considered the most effective mineral for reducing warpage and improving stiffness and heat deflection temperature. Its primary use is in polyolefins.

The aspect ratio of wollastonite grades used in plastics range from approximately 5 (powder grades) to 20 (acicular grades). The needles are strong and stiff and are excellent reinforcements. They are hydrophilic and work well with polar polymers, particularly nylons. Acicular grades are used to supplement or replace short milled glass fibers in thermoplastic and thermoset polymers. Wollastonite is much more abrasive than most of the commonly used industrial minerals, but is no more so than glass fibers or titanium dioxide pigment. Wollastonite reacts well with silane surface treatments.

The amorphous synthetic silicas, fumed and precipitated, are widely used as thixotropes in thermoset polyester resins and gel coats and in PVC plastisols. They are also used as anti-blocking, anti-slip, anti-plateout and flatting agents. Fine ground natural silicas find some use in thermosets to improve dimensional stability, thermal conductivity and electrical resistance at low cost.

Surface Treatment of Minerals Used in Plastics

Most fillers – calcium carbonate, kaolin, mica and wollastonite – have polar surfaces. Conversely, many polymers, such as the polyolefins, are hydrophobic. These will not readily wet hydrophilic fillers. It is therefore necessary to treat the filler surface to facilitate intimate polymer-mineral contact. Surface treatments also act as internal lubricants and improve the dispersion of the filler in the plastic matrix and the flow characteristics of the filled polymer. A further effect of some surface treatments is to improve the mechanical properties when exposed to water in vapor or liquid form, especially at high temperatures. These topics are discussed in detail in the chapter *Mineral Surface Modification.*

⌘⌘⌘

Effects of Mineral Addition on Plastics

Industrial minerals are considerably harder, stronger and heavier than plastics. Because of these factors, they change the mechanical properties of plastics into which they are incorporated in proportion to the volume they displace. In general, the amount of change in any property is given by the Rule of Mixtures:

$$F_c = F_m V_m + F_f V_f$$

where

F = the property under consideration
V = volume fractions of matrix and filler.
subscripts $_{c, m, f}$ = composite, matrix polymer, filler.

However, this relationship does not give the full picture. In order for stresses to be transferred from the weak, flexible plastic to the strong, stiff mineral, the mineral must be wetted by the plastic and dispersed evenly throughout the composite. Most minerals are hydrophilic and are not easily wetted by most plastics unless coupling agents are used. Also, when compounding minerals into plastics, it is important that the minerals be totally dry and that no air is entrained. This is because air and steam have no tensile or flexural strength and reduce these properties in a composite in proportion to their volume fraction.

Special considerations come into play when the minerals have a shape factor, as with flakes and fibers. A factor that is commonly used to describe the shape is the aspect ratio.

Aspect ratio (fiber) = Ratio of mean length to mean diameter.

Aspect ratio (flake) = Ratio of mean diameter of a circle of the same area as the face of the flake to the mean thickness of the flake.

For tensile and flexural stresses to be transferred from the polymer matrix into the fiber or flake, the aspect ratio must have a minimum (critical) value, or else the stresses pass around the particle. Once this critical ratio is exceeded, the efficiency of stress transfer increases with increasing aspect ratio. For the same volume fraction of flake or fiber, increasing the aspect ratio increases the composite's strength and stiffness. In practice, there are limitations. As flakes and fibers become thinner, they become fragile and tend to be broken by handling and processing. This reduces the aspect ratio. To achieve the best mechanical properties, then, it is important to have a high initial aspect ratio for the reinforcing flake or fiber, and to maintain this ratio as high as possible in the final composite after all the processes of conveying, mixing, compounding and molding.

General Effects of Industrial Minerals on Plastics Properties

The general effects of minerals additions to plastics are to:

Increase stiffness (tensile and flexural moduli)
Increase strength (compressive, flexural; tensile in some cases)
Reduce elongation at break
Decrease impact resistance
Reduce creep
Increase opacity
Reduce gloss
Improve weather resistance
Improve water resistance
Improve solvent resistance
Increase surface hardness
Increase abrasion resistance
Increase density
Reduce raw material cost
Reduce flammability
Increase thermal conductivity
Reduce coefficient of thermal expansion
Reduce shrinkage out of the mold
Increase dielectric strength
Increase arc resistance.
Increase heat resistance (deflection temperature)

All industrial mineral fillers provide the above-mentioned effects to a greater or lesser degree. Table 4 demonstrates the effects on cost of the addition of various fillers to polypropylene and PVC, while Table 5 shows the beneficial and undesirable effects contributed to plastics by calcium carbonate. Some fillers are naturally more effective than others in enhancing specific properties. Others have special effects not listed above.

Special Effects

Tensile strength of plastics is generally reduced by the incorporation of industrial minerals, but anisotropic minerals, such as mica and wollastonite, are used to increase the tensile strength of polypropylene, polyethylene and polyamides. The degree of enhancement is dependent on such variables as aspect ratio, surface treatment, particle size and degree of dispersion.

Permeability of plastics to gases and liquids depends on the affinity of the gas or liquid to the polymer. For example, blow molded polyethylene

⌘⌘⌘

gas tanks will allow gasoline vapor to pass through at a fairly rapid rate unless the interior is treated to make it polar. The introduction of zero emission regulations together with the increased addition of alcohol to gasolines has made this hydrophilising treatment less effective. High aspect ratio platy minerals are impermeable to gases and liquids, regardless of polarity, and may find uses in such applications.

Table 4. Volume Costs of 40% Mineral-Filled Plastics

	Density lb./in.^{3}	Cost cents/lb.	Cost cents/in.^{3}
PP homopolymer	0.0324	40	1.3
Phlogopite mica[a]	0.1044	31.75	3.31
60% PP/40% mica	**0.0448**	**36.7**	**1.64**
PP homopolymer	0.0324	40	1.3
Muscovite mica[b]	0.1044	4.5	0.47
60% PP/40% mica	**0.0448**	**25.8**	**1.16**
PP homopolymer	0.0324	40	1.3
$CaCO_3$[c]	0.0972	8.54	0.83
60% PP/40% $CaCO_3$	**0.0442**	**27.4**	**1.21**
PVC homopolymer	0.0504	38	1.91
$CaCO_3$[d]	0.0972	16	1.55
60% PVC/40% $CaCO_3$	**0.0625**	**29.2**	**1.82**

[a]high aspect ratio, silane treated [b]low aspect ratio, untreated
[c]dry ground, 3 μm, uncoated [d]dry ground, 1 μm, coated

Some polymers are sensitive to ozone and their resistance is improved by the use of platy fillers.

High density is desirable in plastics which must sink in water, such as the devices for dispensing chemicals in toilet tanks. High density is also regarded as a mark of quality, as in cultured marble articles. Barite is the preferred heavy mineral filler for these applications.

Lowered density, where required, is obtained by the use of perlite and vermiculite.

Low smoke generation is imparted to plastics by the incorporation of minerals that decompose at plastics processing temperatures. The volatiles lost are usually combined water but may also be carbon dioxide or

⌘⌘⌘

ammonia. Alumina trihydrate is used at lower temperatures, while magnesium hydroxide is preferred for higher temperatures. Hydrated calcium and magnesium carbonates, such as huntite and nesquehonite also emit water vapor and carbon dioxide at higher temperatures.

Anti-stick properties are desirable in polymers that are used to make thin films. Most fine particle sized minerals are acceptable for this purpose, but those most generally used are silicas, talc , mica and calcium carbonate.

Table 5. Calcium Carbonate Fillers

Desirable Properties:
Non-toxic, non-irritating, odorless, tasteless.
High brightness, low refractive index – easy to color.
Soft, Mohs hardness 3.
Dry, no water of crystallization.
Stable under normal plastics processing conditions.
Low cost
Available in a wide range of particle sizes.
Easy to disperse.
Can be incorporated at higher loadings than other fillers without deleterious effects.
Enables low stiffness at high loadings.
Improves PP impact resistance.
Reduces shrinkage in liquid polymers.
Reduces coefficient of thermal expansion.
Increases thermal conductivity.
Acid scavenger in PVC.
Anti-plate-out agent.
Easy to coat with low cost additives which improve performance.
Disadvantages
High specific gravity, 1.9 to 3 times that of polymers, so increases weight.
Increases liquid and molten plastic viscosity (but less than other fillers).
More abrasive to tooling than polymers.
Increases thermal conductivity (undesirable in window & door frames).
Decreases compresive strength.
Decreases tensile and flexural strength (but less with ultrafine, stearate coated grades).
Acid soluble.
Poorer weatherability than other fillers.

⌘⌘⌘

MAJOR END-USES

Industrial minerals are used in both thermoplastic and thermosetting polymer systems.

Commodity Thermoplastics

Poly(Vinyl Chloride)

Poly(vinyl chloride) is one of the least expensive polymers and its price can be further reduced by the incorporation of fillers. PVC does in fact use the largest amount of industrial mineral filler, primarily calcium carbonate. This is mostly ground calcium carbonate, although precipitated calcium carbonate is used where the highest reinforcing effects are needed.

PVC has inherently poor stability to heat, even the heat of processing. When thermally decomposed it releases HCl gas and becomes brown and brittle. The acid gas acts as a catalyst to accelerate further breakdown. Calcium carbonate absorbs and neutralizes this acid and therefore acts as a low cost stabilizer. This is a major reason for using this mineral in preference to others. Ultrafine ground grades, coated or uncoated, give the best stabilization. Table 6 shows calcium carbonate usage in PVC.

Table 6. Calcium Carbonate Usage in PVC (USA/1994)

Market	PVC lbs. x 10^6	phr $CaCO_3$	*$CaCO_3$ tons x 10^3
Plasticized			
Flooring, rigid	228	150 - 250	155 - 259
Flooring, flexible	204	150 - 250	139 - 231
Wire & cable	411	100	186
Rigid (unplasticized)			
Pipe & conduit	4294	15 - 20	292 - 390
Siding	1271	10	58
Window/door profiles	282	10	12.8
Pipe fittings	272	10	12.3
Total PVC	**11,213**		
Total $CaCO_3$ in PVC			**855 - 1149**

*estimate

Vinyl roll flooring and vinyl tiles use the greatest amount of calcium carbonate, followed by pipe and conduit, and wire and cable. Vinyl siding is a fast growing sector of the vinyl industry, but the filler content is low.

⌘⌘⌘

Talc can provide higher strength and stiffness than calcium carbonate in rigid PVC, but is not commonly used since calcium carbonate is whiter and has the additional benefit of serving as an acid scavenger. Mica has been evaluated in vinyl siding extrusions to reduce the coefficient of thermal expansion down to the same level as aluminum, but the flow characteristics were poor resulting in a rough surface. Starve feeding and special die design may overcome this problem. Calcined kaolin is used in PVC wire and cable because it imparts good electrical characteristics, especially under humid conditions.

Plasticized PVC – Poly(vinyl chloride) polymer is a hard, somewhat brittle plastic. The addition of a plasticizer makes PVC soft, flexible, and extensible. PVC is the thermoplastic most often compounded with fillers, and the polymer used with the highest filler levels due, in part, to compounding with plasticizer. This lowers polymer viscosity and enables high filler loadings. In addition, the high specific gravity of PVC makes the use of fillers economical, compared with lower specific gravity polymers like polyethylene and polypropylene.

Plasticized PVC is used in flooring and in wire and cable insulation. These compounds are highly filled, mainly to reduce cost. The calcium carbonate grades used are the cheapest possible commensurate with acceptable properties of the finished product. In hard vinyl flooring, for example, inexpensive coarse ground grades with particles mostly larger than 150 μm are used. Where better gloss, surface finish, scratch resistance and reduced stress whitening are required, the more costly ultrafine grades are prefered. In general, stiffness and strength increase slightly and softness decreases slightly as filler particle size is reduced. Stiffness is often not desirable in plasticized PVC compounds as it usually means poor impact resistance. Of all normal filler types, calcium carbonate increases viscosity and stiffness the least and lowers impact resistance the least.

Stearate coated ultrafine ground calcium carbonate usually provides the best overall balance of properties and cost, and is the preferred grade for wire and cable insulation, footwear and sheeting. Use levels of ground natural calcium carbonates in plasticized PVC range from 80 to 400 parts per hundred of resin (phr).

PVC plastisols are used in autobody undercoatings and gap sealants, using mainly precipitated calcium carbonate (PCC) to add slump resistance while maintaining ease of pressure injection. Stearate coated PCC with mean particle size of 0.075 μm is the preferred product. It is used in underbody coatings at 25% by weight and at 50% in gap sealants.

⌘⌘⌘

Rigid PVC – Rigid PVC contains little or no plasticizer. End-uses include pipe (drain, water and vent) and their fittings; rainwater gutters, downspouts and fittings; and window and door frame extrusions. The absence of plasticizer means that only a small amount of filler is tolerable before viscosity is increased too high for proper processing. Calcium carbonate in unplasticised PVC is regarded more as a processing aid than as a filler. It aids extrusion and improves impact properties. In the US, levels up to 30 phr are used. In Europe, levels as high as 65 phr are used in pipe. The preferred grades are stearate coated ultrafine ground calcium carbonate and PCC. The ultrafine ground products can be used at a higher loading than PCC without unacceptable loss in impact resistance. Table 7 compares the properties imparted by various fillers in unplasticized PVC.

Table 7. Filler Effects in Unplasticized PVC, Type S

Filler	Wt. %	Density g./cc	Tensile Strength psi	Elong. @ Break %	Modulus Elasticity Kpsi	Impact Strength* ft-lb./in. of notch
None		1.36	8680	6 - 10	390	2.2
Talc	20	1.46	4920	6	506	n.a.
Silica	20	1.45	5500	6	448	n.a.
Wollastonite	20	1.47	3620	5.4	n.a.	2.2
$CaCO_3$, grnd	30	1.53	6660	8	463	6.6
$CaCO_3$, ppt	15	1.45	4340 - 6800	6	448	10.3

*Charpy Notched

Polyolefins

Polyolefins are the workhorses among the thermoplastic polymers, accounting for more than 40% of resin sales. They are produced from wellhead and refinery gases and can be regarded as high molecular weight waxes. They are divided into two main classes – polyethylenes and polypropylenes – although copolymers (alloys of the two) are used. Polyethylenes are divided into low density types, which are mainly used for films and bottles, and high density grades, which are stronger and stiffer and which are used for more general packaging uses. Polypropylene competes with high density polyethylene, as it is available in a wider range of properties, and is lighter and cheaper. It consequently has a very high growth rate.

The mechanical effects of fillers are shown in Table 8 for polypropylene, Table 9 for high density polyethylene and Table 10 for copolymer.

⌘⌘⌘

The vast majority of polyolefin end-uses are without filler, since they are inexpensive and of such low density that it has been difficult to produce fillers that are as cheap on a volume replacement basis. The most widely used minerals in filled polyolefins are calcium carbonate and talc.

Ground calcium carbonate is used mainly in polyethylene and polypropylene wire and cable compounds as a low cost pigment extender for titanium dioxide; in non-structural polypropylene automotive parts as a filler and pigment extender; and in outdoor and restaurant furniture to increase rigidity, temperature resistance and ultraviolet light resistance. The use of calcium carbonate in garden furniture has mushroomed in the past few years. Stearate coated fine ground calcium carbonate provides rigidity in this application, while retaining impact resistance and a high gloss.

Table 8. Mechanical Properties of 40% Mineral-Filled Polypropylene

Filler	None	Wollastonite[1]	Wollastonite[2]	Talc	$CaCO_3$
Tensile Str.[a]	4.7	4.5	6.6	4.3	2.8
Flexural Str.[a]	4.5	8.3	11.8	6.4	4.7
Flex. Modulus[a]	180	535	639	680	420
Impact Resist.[b]					
notched	0.45	0.9	0.7	0.45	0.75
unnotched	no brk.	n.a.	n.a.	4.5	23
Heat Distortion[c]	56	81	93	78	84
Mold Shrinkage[d]	0.02	n.a.	n.a.	0.012	0.014

Filler	Mica[3]	Mica[4]	Mica[5]	Mica[6]	Mica[7]
Tensile Str.[a]	4.57	7.32	4.21	4.1	6.2
Flexural Str.[a]	8.24	13.22	8	6.5	9.5
Flex. Modulus[a]	1100	1279	900	930	1100
Impact Resist.[b]					
notched	0.38	0.39	0.49	0.6	0.65
unnotched	1.82	3.37	1.9	3.8	4.4
Heat Distortion[c]	91	118	113	89	108
Mold Shrinkage[d]	0.006	0.004	0.009	0.008	0.008

[a]Kpsi [b]Izod; ft.lb./in. [c]264 psi, °C [d]in./in.

[1]untreated [2]treated [3]wet ground muscovite, untreated [4]wet ground muscovite, treated

[5]dry ground muscovite, untreated [6]dry ground phlogopite, untreated [7]dry ground phlogopite, treated

Calcium carbonate adds high brightness, improves stiffness, does not reduce impact resistance like talc and does not have the severe

⌘⌘⌘

stress-whitening problem associated with talc. Thus, as more appearance parts are seen in polypropylene, calcium carbonate is likely to gain at the expense of talc. Calcium carbonate usage in polyolefins is summarized in Table 11.

Coated ultrafine ground grades increase notched Izod impact resistance over the unfilled polypropylene when used up to 40 phr. Tensile strength is retained up to 100 phr. Maximum impact resistance is obtained using ultrafine ground grades with 80% minus 2 μm and 90% minus 2.6 μm treated with 1% stearic acid .

Ground calcium carbonate, fine stearate coated grades, are used in small amounts in pigmented plastic films and containers, and as blocking agents to prevent the 'static cling' of linear low density polyethylene blown films. The carbonate increases the stiffness, dimensional stability, tensile strength and impact resistance of the film, and reduces plate-out (deposition of a thin smear of polymer) on the die lips by acting as a mild internal abrasive.

Table 9. Mechanical Properties of Mica Reinforced HDPE

Filler	None	Mica[1]	Mica[2]	Mica[3]	Mica[4]
Filler Wt. %	0	25	25	25	25
Tensile Str.[a]	4.06	4.21	6.73	4.18	6.93
Flexural Str.[a]	4.78	5.83	7.31	6.25	7.99
Flex. Modulus[a]	160	340	380	470	500
Impact Resist.[b]					
notched	16.7	7.4	n.a.	3.4	n.a.
unnotched	no break	32.1	18.7	n.a.	n.a.
Heat Distortion[c]	44	n.a,	55	56	66
Mold Shrinkage[d]	0.028	0.023	0.02	0.018	0.011

[a]Kpsi [b]Izod; ft.lb./in. [c]264 psi, °C [d]in./in.

[1]wet ground muscovite, untreated [2]wet ground muscovite, treated

[3]dry ground muscovite, untreated [4]dry ground phlogopite, untreated

The calcium carbonate used in film must have a particle diameter 1/4 to 1/3 the thickness of the film. Since films are blown as thin as 0.25 mil, a particle 1 to 2 μm may be necessary, and a top cut of 10 μm is usually required. Such particles must not agglomerate and are usually treated to make the surface hydrophobic. High purity, with no taste or odor, is required to meet FDA requirements. Low content of silica and other high hardness

⌘⌘⌘

minerals is required to avoid machine wear. High brightness is desired for colored films.

Recent work has shown that the use of up to 20% stearate coated ultrafine ground calcium carbonate in linear low density polyethylene (LLDPE) garbage bags increased the production rate by 50%, increased dart impact resistance fivefold, reduced cost and improved blocking, with no change in tensile or tear strength.

Table 10. Acicular Wollastonite vs. Talc in Ethylene-Propylene Copolymer

		Wollastonite A[1]	Wollastonite B[2]	Talc
Filler Wt.%		20	20	2o
Melt Index, g/10 min.		3.7	3.7	3.8
Flexural Modulus, Kpsi		283.6	272	283.6
Izod Impact Resistance				
ft.lb./in.	0°C	1.2	4.2	3
	20°C	0.7	1.4	1.2
Heat Distortion Temp., °C		61	60	63
Scratch/Mar Resist. (PSA)		0.9	0.5	2.6

[1]Acicular grade, silane A [2]Acicular grade, silane B

A total of 1180 million pounds of LLDPE garbage bags were produced in 1994. At the 20% level this corresponds to 107,000 tons of stearate coated ultrafine ground calcium carbonate. If this technology is applicable to LDPE and HDPE trash liners, it would add a further 50,000 ton to the demand for this grade.

In what may become a major market for ultrafine ground calcium carbonate, work has been done in Germany on filled HDPE plastic paper. Notched Izod impact resistance with 40% untreated filler was better than stearate coated filler at the 30 to 40% level, while the Unnotched Izod was slightly lower than the unfilled. The Germans have also developed a biodegradable agricultural mulch, using calcium carbonate filled HDPE.

Talc has been the second mineral of choice to fill polypropylene. Such applications have usually been automotive and industrial. Talc is a general purpose filler mineral and is highly compatible with polyolefins. It is widely used to increase stiffness and heat resistance in polypropylene automobile trim parts that do not require high reinforcement.

Talc gives higher stiffness and a lower reduction in tensile strength than calcium carbonate in polypropylene, but does not give as high values as mica or wollastonite. The use of surface treated talc brings the tensile

⌘⌘⌘

strength back up to the value of the unfilled polypropylene, but at a cost premium. A major problem with talc-filled plastics, however, is poor mar resistance, so that scratches and scuff marks show.

Talc is also used at 0.5% as an antiblocking agent in low density polyethylene and linear low density polyethylene blown and cast films .

Kaolins are used primarily in wire and cable insulation where the calcined clay gives good electrical characteristics, a smooth surface and good resistance to humidity. Air-floated kaolin clay is used in polypropylene in automotive applications to increase viscosity and thereby improve processability.

Table 11. Calcium Carbonate Usage in Polyolefins (USA/1994)

Market	**Resin lbs. x 10^6**	**phr $CaCO_3$**	***$CaCO_3$ tons x 10^3**
Polypropylene			
Wire & cable	30	50 - 100	5 - 15
Consumer	1080	35 - 65	170 - 315
Appliances	242	25 - 50	27 - 54
Transportation	352	25 - 50	40 - 80
Rigid packaging	1008	10	45
Total Polypropylene	**9752**		
Total $CaCO_3$ in PP			**287 - 509**
Polyethylene			
Wire & cable			
LDPE	255	50 - 100	55 - 110
LLDPE	206	50 - 100	45 - 90
HDPE	128	50 - 100	30 - 60
Total Polyethylene	**25,683**		
Total $CaCO_3$ in PE			**130 - 260**

*estimate

Mica is used mainly in automotive applications. It gives the greatest stiffening and anti-warping effects of all the industrial minerals. Mica-reinforced polypropylene panels can have the same strength and stiffness as a steel panel, but weigh 45% less. Mica is used in polypropylene parts, such as passenger compartment trim, wheel covers, instrument panels, glove boxes, crash pad retainers, air conditioner/heater housings, wheel arch liners, battery support trays and fan shrouds. It adds strength, stiffness, high

⌘⌘⌘

temperature resistance and warp resistance. When talc does not give sufficient reinforcement, mica is the next choice, although at a higher cost. Where mica is insufficient, more costly glass fibers must be used.

Mica is also used in high density polyethylene to make blow molded rear seat backs and load floors used in intermediate size cars and vans to replace much heavier steel assemblies. The mica imparts stiffness and heat resistance. Mica is very effective in foamed plastics, especially polyethylene copolymers, and modified poly(phenylene oxide). It promotes a very uniform cell structure with improved strength and stiffness, and unexpectedly increased impact resistance. The high stiffness of mica gives improved sound transmission, so that it is used in loudspeaker cones, cabinets and earphones. Paradoxically, mica is the preferred mineral in coatings based on polyurethane or asphalt binders to reduce road and motor noise and to damp vibrations.

The main problems with mica are poor knit line strength and impact resistance. When mica-reinforced plastic flows around inserts and rejoins on the other side, the knit line may only have 60% of the strength of the rest of the part. Impact resistance of mica-reinforced plastics is poor, as it is with talc and kaolin. These problems are often overcome by combinations of mica with glass fibers, acicular wollastonite, calcium carbonate or glass microspheres, and by special molding machinery.

Wollastonite is also used in polypropylene, high density polyethylene and ethylene-propylene copolymer in applications similar to mica's. Most end-uses are in transportation.

Styrenics

The strength of amorphous styrenics – polystyrene and acrylonitrile butadiene styrene – is reduced by mineral addition. Calcium carbonate and talcs, especially surface treated grades, are used, nevertheless, to increase the impact resistance of polystyrene modified with thermoplastic elastomer. The more crystalline styrenics – styrene acrylonitrile and styrene maleic anhydride – respond to reinforcement with glass fibers, mica or wollastonite. Table 12 compares the effects of various minerals in styrenics.

Engineering Thermoplastics

Engineering thermoplastics are those which have higher strength and stiffness and higher temperature stability than the commodity plastics. They are also more often used in mineral-polymer composites. Glass fiber is the most common filler, but increasingly mica, wollastonite and kaolin are being used, alone or in combination with glass fiber.

⌘⌘⌘

The main filled engineering plastics are polyamides, thermoplastic polyesters, poly(phenylene oxide)/polystyrene alloy, poly(phenylene sulfide), polyphthalamide, and polysulfones.

Polyamides

The two most widely used polyamides are Nylon 6 and Nylon 66. Nylons are tough strong polymers. Their main fault is sensitivity to water. They can lose half their strength after prolonged immersion in water or steam. The effects of reinforcing minerals in nylons are shown in Tables 13 and 14.

Table 12. Filled and Reinforced Styrenic Thermoplastics

	POLYSTYRENE				
Filler	**None**	**Mica[1]**	**Mica[2]**	**$CaCO_3$**	**Talc**
Filler Wt.%	0	40	40	30	40
Tensile Str.[a]	4.84	4.2	5.1	2.2	5.6
Tensile Modulus[a]	n.a.	n.a.	n.a.	298	810
Flexural Str.[a]	8.04	8.59	9.9	n.a.	n.a.
Flex. Modulus[a]	460	1910	2130	n.a.	n.a.
Heat Distortion[b]	n.a.	n.a.	n.a.	n.a.	n.a.

	ABS		SAN	
Filler	**None**	**Mica[1]**	**None**	**Mica[1]**
Fuiller Wt.%	0	50	0	30
Tensile Str.[a]	6.2	9.4	10.1	12
Tensile Modulus[a]	376	2113	521	2359
Flexural Str.[a]	n.a.	n.a.	n.a.	n.a.
Flex. Modulus[a]	n.a.	n.a.	n.a.	n.a.
Heat Distortion[b]	83	103	90	110

[a]Kpsi [b]264 psi, °C

[1]phlogopite, untreated [2]phlogopite, treated

Wollastonite is the reinforcing filler of choice for polyamides. Levels of 40 to 45% by weight are typically used to increase flexural strength, modulus and heat distortion temperature. Mica and kaolin are also effective. Mica in combination with glass fibers gives an excellent combination of high strength and stiffness, together with reduced warping and improved resistance to water immersion. Surface modified talc can substitute for part of the glass fiber reinforcement in Nylon 6, with only slight loss of properties.

⌘⌘⌘

Thermoplastic Polyesters

Thermoplastic polyesters are linear polymers made by the condensation of terephthalic acid and polyethylene or polybutylene glycols. They are highly crystalline and have a pronounced tendency to warp, especially when reinforced by glass fibers. As noted above, these fibers tend to line up in the flow direction producing a major difference in thermal expansion characteristics in the flow versus the cross-flow directions. Warpage control is particularly important in exterior automobile parts which have to withstand the high temperatures of the paint oven. Mica is the mineral of choice to overcome warping. Mica is usually used in combination with glass fibers, which impart improved impact resistance.

Table 13. Mechanical Properties of Mineral-Filled Nylon 6

Filler	None	Muscovite Mica[1]	Muscovite Mica[2]
Filler Wt.%	0	40	40
Tensile Str.[a]	9	14	12.9
Flexural Str.[a]	12.5	22.4	20.5
Flex. Modulus[a]	270	1910	1610
Impact Resist.[b]			
notched	0.6	0.5	0.4
unnotched	n.a.	3.4	3.3
Heat Distortion[c]	56	177	173
Mold Shrinkage[d]	0.15	0.002	0.003

Filler	Wollastonite[3]	Wollastonite[4]	Wollastonite[3]	Wollastonite[4]
Filler Wt.%	50	50	70	70
Tensile Str.[a]	8.9	10.5	7.6	11.9
Flexural Str.[a]	15.4	18.6	12.2	21
Flex. Modulus[a]	830	770	960	990
Heat Distortion[c]	150	140	177	180

[a]Kpsi [b]Izod; ft.lb./in. [c]264 psi, °C [d]in./in.

[1]wet ground, untreated [2]dry ground, untreated [3]untreated [4]treated

Polybutyleneterephthalate and polyethyleneterephthalate reinforced with glass fibers and mica are used in exterior automobile panels, such as cowl/vent hoods, rear quarter panels and headlight housings, and in under-the -hood parts, such as high performance distributor caps and rotor arms and the E coil on the Ford Lynx and Escort, where mica improves electrical characteristics. Similar compounds are used in computer keyboards and key facings, solder-side circuit board covers, glue gun

⌘⌘⌘

housings and vacuum cleaner heads. The properties of mica- and wollastonite-reinforced PBT are shown in Table 15.

Poly(Phenylene Oxide)/Polystyrene Alloy

The high strength, stiffness and heat distortion temperature of poly(phenylene oxide)/polystyrene alloy is improved by reinforcement with mica or wollastonite. These alloys are used in appliances, and in electrical and electronic products.

Table 14. Mechanical Properties of 40% Mineral-Filled Nylon 6,6

Filler	None	Wollastonite	Kaolin	ATH	$CaCO_3$
Tensile Str.[a]	12	4.8	10.9	9.2	10.5
Flexural Str.[a]	18.7	7.7	23.7	14.7	16.5
Flex. Modulus[a]	440	790	1010	645	660
Impact Resist.[b]					
notched	0.33	0.6	0.3	0.5	0.5
unnotched	36.1	9.4	12.3	6.4	9.6
Heat Distortion[c]	58	221	199	202	198
Mold Shrinkage[d]	0.016	0.009	0.004	0.008	0.012

Filler	Mica[1]	Mica[2]	Mica[3]	Mica[4]
Tensile Str.[a]	14.3	15.3	9.9	14.2
Flexural Str.[a]	24.4	26.2	16.5	19.7
Flex. Modulus[a]	1520	1430	1200	1420
Impact Resist.[b]				
notched	0.54	0.55	0.6	n.a.
unnotched	7.1	8.4	4	4.2
Heat Distortion[c]	179	175	182	215
Mold Shrinkage[d]	0.004	0.003	0.008	0.008

[a]Kpsi [b]Izod; ft.lb./in. [c]264 psi, °C [d]in./in.

[1]wet ground muscovite, untreated [2]wet ground muscovite, treated

[3]dry ground phlogopite, untreated [4]dry ground phlogopite, treated

Polycarbonate

Historically, polycarbonate has had its greatest use in clear parts, such as glazing. But its excellent properties are leading to its use in small kitchen appliances. It is usually not reinforced, but wollastonite imparts some benefits as shown in Table 16.

⌘⌘⌘

Thermosetting Polymers

Thermosetting polymers are usually liquids at room temperature. They are converted to solids by combining with other liquid reactants in the presence of liquid or paste catalysts. Heat may be applied to hasten the setting reaction. Unsaturated polyesters constitute the largest volume, followed by polyurethanes and polyureas. Phenol-, urea- and melamine-formaldehyde resins are used in large volumes in laminates, but these contain little or no mineral filler. Epoxies are a relatively low volume market for fillers.

Table 15. Mechanical Properties of Mineral-Reinforced Thermoplastic PBT

Filler	None	Wollastonite[1]	Wollastonite[2]
Filler Wt.%	0	50	50
Tensile Str.[a]	7.3	6.1	8.2
Flexural Str.[a]	9.1	12.3	15.3
Flex. Modulus[a]	347	n.a.	n.a.
Impact Resist.[b]			
notched	0.33	n.a.	n.a.
unnotched	no break	3	3.9
Heat Distortion[c]	53	155	174
Mold Shrinkage[d]	0.02	n.a.	n.a.

	Mica[3]	Mica[4]	Mica[5]	Mica[3]	Mica[4]
Filler Wt.%	30	30	30	40	40
Tensile Str.[a]	10.4	7.9	8.7	11.0	9.9
Flexural Str.[a]	16.3	12.8	14.6	16.7	16
Flex. Modulus[a]	1300	1210	1290	1770	1680
Impact Resist.[b]					
notched	0.29	n.a.	n.a.	0.32	n.a.
unnotched	3.2	n.a.	n.a.	2.6	3.1
Heat Distortion[c]	152	n.a.	n.a.	168	169
Mold Shrinkage[d]	0.006	n.a.	n.a.	0.004	0.006

[a]Kpsi [b]Izod; ft.lb./in. [c]264 psi, °C [d]in./in.

[1]acicular, untreated [2]acicular, treated [3]wet ground muscovite, untreated

[4]dry ground phlogopite, untreated [5]dry ground phlogopite, treated

⌘⌘⌘

Unsaturated Polyesters

These are similar in chemical composition to the thermoplastic polyesters, but contain reactive unsaturated sites so that they can be cross-linked. Fillers, mainly calcium carbonate but also alumina trihydrate and talc, are added to increase the viscosity, strength, stiffness and impact resistance. The major thermoset applications for calcium carbonate are in cultured marble for sink units, counter tops, spas and bowling balls.

The automobile industry uses reactive polyester in the form of sheet molding compound (SMC) to mold exterior panels – lift gates, doors, side panels – for the General Motors mini-vans (Lumina APV, Silhouette and Transport). Exterior panels are also made for the Camaro, Firebird and Corvette. SMC uses 5 to 10 μm ground calcium carbonate at about 200 phr and oriented glass fiber to produce a composite offering stiffness, resistance to temperature and a smooth paintable surface.

Table 16. Wollastonite-Filled Polycarbonate

Filler	None	Wollastonite, treated	Wollastonite, untreated
Filler Wt.%	0	50	50
Tensile Str.[a]	7.8	6.4	8.1
Flexural Str.[a]	12.6	11.9	14.4
Flex. Modulus[a]	300	920	940
Heat Distortion[b]	131	133	130

[a]Kpsi [b]264 psi, °C

A similar compound, but without the strength conferred by the orientation of the fiberglass, is bulk molding compound (BMC). This is used for molding electrical switches and boxes, plumbing fixtures and appliance parts.

Polyurethanes, Polyureas

Micronized and surface treated micas and wollastonites are used in automobile and van fascias and fenders made by Reinforced Reaction Injection Molding (RRIM) of polyurethanes and polyureas. These minerals add strength, stiffness and heat resistance, and lower the coefficient of thermal expansion so that the plastic part mates better with the steel frame of the vehicle. Combinations of wollastonite with mica are used, since the mica gives a smooth surface (improved Distinctness of Image), an essential characteristic for aesthetics after painting. Properties of mica- and wollastonite-reinforced RRIM are shown in Tables 17 to 20.

⌘⌘⌘

The high modulus of mica, which confers excellent sound deadening characteristics, accounts for its favored use in sound and vibration damping composites, based on polyurethanes.

Large amounts of calcium carbonate, alumina trihydrate, and barite are used to fill foam underlay for carpets. The base polymers are rubber, synthetic thermoplastic and polyurethane in latex form. The fillers serve two functions – to add weight so that the carpet lies flat on the floor and does not shift with traffic, and to reduce flammability.

Table 17. Mica & Glass Fiber Reinforcement Of Polyurethane and Polyurea RRIM Automobile Facia

	Polyurethane	High Performance PU	Polyurea
Reinforcement Wt.%			
Wet-ground mica	15	15	15
Glass Fibers	14.2	14.8	14.8
Density, g/cc	1.173	1.111	1.145
Dimensional Stability, avg. % changed			
after 24 hours	0.16	0.07	0.28
after 144 hours	0.46	0.27	0.43
after 240 hours	0.55	0.51	0.49
Tensile Strength, Kpsi	3.5	2.4	2.85
flow direction	2.7	2.4	2.6
cross-flow direction	2.5	2.2	2.2
Elongation, %	140	100	116
flow direction	177	126	198
cross-flow direction	164	114	170
Flexural Modulus, Kpsi			
flow direction	86.6	64.5	70.7
cross-flow direction	76.5	56.2	61.7
Tear Strength, lb./in.			
flow direction	495	386	430
cross-flow direction	468	363	391
Impact Resistance			
Gardner			
in.lbs., 70°F	216	144	216
in.lbs., -20°F	16	160	168
Izod Notched			
in.lbs., 70°F	109.6	78.3	104.8
in.lbs., -20°F	35.2	38.7	61.7
Heat Sag, flow direction 1 hr. @ 250°F, mm	14	9	8

⌘⌘⌘

Table 18. Treated Acicular Wollastonite in Polyurethane/ Polyurea RRIM Automobile Panels

	On-Line Top Coat or Off-Line Assembly	Full On-Line
Reinforcement[1] Wt.%	30	30
Density, g/cc	1.28	1.34
% Shrinkage		
flow direction	0.6	0.5
cross-flow direction	1.5	1.3
Tensile Strength, Kpsi	4.35	7.05
Elongation, %	58	20
Flexural Modulus, Kpsi		
flow direction	280	420
cross-flow direction	140	202
Impact Resistance, Izod, in.lbs., -30°C		
flow direction	29	30
cross-flow direction	37	32
Heat Sag, 152mm overhang 1 hr. @ 121°C	3	1
Thermal Expansion Coefficient in./in.°Cx10^{-6}, -40 to 120°C		
flow direction	34	30
cross-flow direction	106	85
Distinctness of Image[2], ASTM E430	91	95

[1]RRIMGLOSS 10013 (NYCO) [2]D.O.I. of steel = 95

COMPOUNDING METHODS

Thermoplastics

In order for thermoplastics to be molded into useful articles, they must first be plasticated, i.e. melted, and then held in the mold until resolidified. For ease of handling and to avoid excessive air entrapment, thermoplastics are usually supplied as cylindrical pellets about 1/8 inch in diameter and 1/8 to1/4 inch in length. Powdered polymers are used in the processing of PVC and in rotational molding. The incorporation of minerals as fillers and reinforcements creates certain problems, since they are much harder and more abrasive than plastics. Generally, minerals are extrusion compounded into pellets or dice before being used in the injection molding or extrusion of the final products, but they can be dry blended with powdered polymers or even pellets.

Table 19. Treated Acicular Wollastonite vs. Glass Fiber in Polyurethane RRIM Automobile Fascia

	Wollastonite[1]	Wollastonite[2]	Glass Fiber
Reinforcement Wt.%	15	17	15
Density, g/cc	1.16	1.19	1.15
% Shrinkage, flow direction	0.5	0.6	0.6
Tensile Strength, Kpsi	3.5	2.4	2.85
Elongation, %	140	100	116
Flexural Modulus[a], Kpsi			
flow direction	82	59	63
cross-flow direction	42	34	38
Tear Strength, lb./in.	500	450	450
Impact Resistance, Gardner in.lbs., -29°C	190	150	190
Heat Sag, 152mm overhang 1 hr. @ 121°C	6	7	7
Thermal Expansion Coefficient in./in.°Cx10^{-6}	50	65	58
Distinctness of Image[b], ASTM E430	91	77	75

[a]unfilled = 30 kpsi [b]D.O.I. of steel = 95

[1]RRIMGLOSS 10013 (NYCO) [2]G-RRIM Wollastokup 10013 (NYCO)

Primary Processing

Plastication – The injection molder or extruder feeds pelletized plastic into a hopper from which it falls by gravity into the barrel. The barrel is heated from the outside by bolted-on electric heaters. A rotating screw runs down the central axis of the barrel with a small clearance. This screw pulls in the plastic pellets, shears them and smears them against the heated barrel so that they are heated by a combination of frictional and conductive heat. As the pellets are pushed toward the front of the machine they melt.

When a filler is added directly at the hopper and passes into the barrel, there is wear from the dry filler rubbing against the screw and barrel lining until the plastic melts and encapsulates it. All fillers are much harder than plastics and most plastics machines are designed to use pure plastics only.

⌘⌘⌘

To reduce wear, fillers are often added at an ancillary hopper at a point downstream where the plastic is already melted. In some machines, specially hardened barrels and screws are used.

Table 20. Treated Acicular Wollastonite vs. Glass Fiber in 3.3mm Low Density RRIM Polyurethane Composite Panel

	Wollastonite[a]	1/16" Milled Glass Fiber	Glass Mat
Reinforcement Wt.%	15	15	15
Density, g/cc	0.5	0.53	0.5
Flexural Modulus[a], Kpsi	122.3	124.6	135.6
Heat Distortion Temp., °C	95	93	n.a.
Dimensional Stabilty % change, 24 hours	0.04	0.21	0.13
Composite Material Cost, $/lb. (unfilled PU = $1.25/lb.)	1.11	1.19	1.26

[a]G-RRIM Wollastokup 10013 (NYCO)

Extrusion – In extrusion, the melt is forced forward by the turning screw and is pushed under considerable pressure through a die to form the shape of the extruded part. The die constricts the plastic so that it is accelerated. Wear occurs from the rubbing of the molten plastic on the die surfaces. In compounding mineral fillers and glass fibers into plastics, the die consists of a series of circular holes often about 1/8 inch in diameter. The filled plastic emerges as spaghetti-like strands, is water- or air-cooled and is cut into cylindrical pellets about 1/8 to 1/4 inch long. Alternatively, the pellets are cut hot, under water, and are spherical or lens-shaped.

Other compounding methods – Compounding of plastics is also done in sigma blade mixers (Banbury), high intensity blade mixers (Henschel), special designs such as the K Mixer, and on roll mills. In these processes, the thermoplastic is melted by heat developed by the high shear imparted by large slowly turning blades, smaller high speed rotors, or by heated rolls. The molten mass is usually passed onto a roll mill which further homogenises the batch and forms it into sheets which are then diced as a feed to injection molding machines. The mass may also be fed into an accumulator cylinder, which feeds an extruder, or directly to calender rolls to produce the final plastic product.

Compounding is an expensive process and means that an additional heat cycle is added to the process. This causes degradation of the plastic and

⌘⌘⌘

some loss in mechanical properties. Machinery to achieve direct feeding of plastics and fillers (powder or pellets) is under development. This is based on reciprocating extrusion mixing screws with or without accumulators.

Injection molding – In injection molding machines, the end of the machine is closed by a valved nozzle. A screw pushes the melt forward and is itself pushed backwards. When the barrel ahead of the screw is full, the screw is forced forward, acting as a piston. The valve at the front of the barrel opens, and the melt is pushed through the nozzle into a series of runners (channels). These direct the flow to one or more gates where it enters the female cavity of the mold itself. These gates may be pinhole sized, or arranged as thin sheets when the plastic part must be of high quality with no visible flaws. Small area gates also help break the part off from the runner system without the need for manual separation.

The flow through a small gate is accelerated and this is a major site for wear. Severe wear also occurs on the face of the mold itself where the flow from the gate first impinges. Wear on other parts of the mold is less since the molten plastic flows against, but does not drag against, the faces of the mold.

Blow molding – Blow molding takes the plastics molding process one stage further. A parison (preform) is made by conventional injection molding or by pinching off the end of an extruded tube. The parison is placed in a mold. Pressurized gas is introduced into this parison while it is still hot and malleable, and it is blown like a balloon to fill the mold. Formation of the parison is a high pressure process and therefore needs a large clamping device, but the blowing stage is at comparatively low pressure. This means that large parts can be made with less expensive machines than by injection molding.

Blow molding is most commonly used to make bottles, but can also make complex shapes like automobile air-conditioner ducting and large parts such as seat backs and load floors.

Rotational molding – Rotational molding is another low pressure process. Unfilled or filled powdered plastic, usually high density polyethylene but sometimes polypropylene, is placed inside a hollow mold that can be rotated about one or two axes. This distributes the powder evenly against the surface of the mold. The entire rotating mold is placed in an oven to fuse the powder and form a layer of plastic against the mold. The mold is then rotated to a cooling station to solidify the plastic. It is then removed from around the plastic part and the cycle repeats. This process is used to make

⌘⌘⌘

very large parts including septic tanks and reservoirs up to 6000 gallons capacity.

If filled polymers are used, they are usually first extrusion compounded into pellets which are then pulverized, often cryogenically. Blends of powdered polymer and filler mineral do not work well since there is no means to force the wetting of the filler powder by the plastic as it melts.

Plastisols and organosols – PVC powders are dissolved in solvent (organosol) or dispersed in plasticizer (plastisol), and fillers are added in conventional slow speed mixers. The resultant viscous liquids are used to rotationally cast, slush mold or dip articles. Mineral fillers are used to increase stiffness and hardness, reduce tackiness and reduce costs, although they may increase viscosity and reduce shelf life.

Feeding Injection Molding and Extrusion Machines

Delivery, storage and supply systems – Incoming plastics are usually delivered in bulk, for large operations; in mini-bulk containers made from cardboard (gaylords) or plastic, for medium sized processors; and less often, in 50 lb. bags for in small operations,.

With mini-bulk containers, the plastic is suctioned, into the hoppers that feed the machines. This transfer system works well with pellets of uniform size, but less so with powders, which tend to stratify. Medium-sized injection molders prefer to use the suction transfer system because it is inexpensive. In contrast, a single silo bulk system feeding one hopper bin costs about $100,000, plus $20,000 for each additional receiver hopper bin.

Dry mineral powders must be fed using some type of bulk handling system, preferably totally enclosed to prevent contamination of non-filled parts. If a molder, for example, is making filled parts and also polystyrene chandelier lights, one or two specks of filler are enough to ruin the crystal-clear parts.

Dry mixing of powders – Powdered minerals and polymers may be mixed by such simple devices as a drum tumbler, or by paddle or ribbon mixers. V-shell and double cone mixers are more efficient. The most efficient and fastest mixing devices are the high intensity mixers such as the Henschel, Pappenmeier or Welex mixers. Their high speed impellers are run for sufficient time to blend and heat the mix without fusing it to a mass. The mix then is dumped into a cooler, often integral with the mixer, to prevent further coagulation of the mass. This is then fed to the extruder. These devices are used with powder PVC for feeding pipe and other profile lines.

The transfer of fine powders causes air entrapment and fluffing to a low bulk density. All of this air must be removed during plastication or bubbles will develop in the plastic as the pressure is released after extrusion or on demolding. This is done by venting the injection or extrusion barrel and sometimes the hopper, either under natural pressure or with the assistance of vacuum. The barrel length in injection molding is often shorter than that in extrusion, so there is less opportunity to remove entrained air. For this reason, injection molders prefer to purchase compounds in which the fillers are already fully mixed, de-aerated and coated with plastic.

Some mineral suppliers are producing pelletized fillers to overcome these handling problems. For injection molding of such pellets, special venting may not be necessary. For example, only about 36% air is entrained in pelletised mica, which is not much more than in a pelletized plastic. In contrast, the air entrained by high aspect ratio mica powder is 2 1/2 times greater. Because of its low bulk density, high aspect ratio mica entraps much more air than a granular filler mineral like calcium carbonate.

Hopper feeding systems – The actual feeding into the machine hopper must be controlled to match the rate that plastic is plasticated and extruded or molded. This may be by hand or volumetric feeder, or best by a weigh feeder. It is becoming common to use complex weigh feeder/mixers. These weigh out individually into a separate hopper the various ingredients (plastic pellets, pigments, fillers, additives), then mix these and meter them into the machine feed hopper, overcoming segregation problems. Small users tend to mix plastic with fillers in a drum tumbler, but as soon as the tumbling stops, segregation starts. When transferred, the material in the machine hopper segregates so that the mineral powder sifts down between the plastic pellets and the first shots are mineral-rich. This is much less of a problem if the plastic is in powder form, but only some grades of polypropylene and most PVC grades are available as powders. Pelletized polymers can be pulverized, but at added expensive.

Some relief from segregation can be obtained by coating the plastic pellets with 1 to 3% mineral oil so that the filler sticks to them, but this is only successful at low filler levels.

Thermosetting Polymers

Open Mold /Hand Lay-up Process

In hand lay-up of thermosets, fillers are not normally used but can be added to modify viscosity, add some stiffness and reduce cost. A 10 to 20 mil thick gel coat (unreinforced hard surfacing resin, sometimes pigmented) is applied to a waxed mold and is allowed to partially cure to a rubbery state. Promoted and catalyzed resin mix is then applied to the gel coat by brush or spray at about 3 fluid ounces per sqare foot. Reinforcement in the form of a

⌘⌘⌘

sheet of fibreglass is then pressed into the wet resin, and entrapped air is removed with a disc roller.

The process is repeated to build up the desired thickness. The catalyzed resin may contain fumed or precipitated silica to add thixotropy, so that the resin will not sag when applied to vertical surfaces. Other fillers are not normally used since they tend to interfere with the wet-out of the fiberglass and make the laminate opaque so that flaws such as entrapped air bubbles are not visible. Alumina trihydrate (ATH) may be added, however, to add flame retardancy. ATH and the synthetic fillers make the resin translucent but not opaque. Mica in the first layer will prevent the glass fibres from telegraphing (showing through the gel coat) after the resin has shrunk during curing. Only the mold side has a good finish.

Hand lay-up is used to make one-off and large articles that cannot be made by pressure processes. Commonly, boats, kit automobiles, swimming pools, spas, van bodies, slides, large tanks, piping and aircraft are made by this process. Polyester resin is used to the largest extent, followed by epoxies, polyurethanes and phenolics.

Open Mold /Spray-up Process

Spray-up is similar to hand lay-up except that the glass reinforcement and resin are applied to the gel coat simultaneously by a gun. The gun mixes catalyst and resin and sprays this onto the surface. At the same time, a chopper fitted with rotating blades cuts continuous rovings (strands) of glass fibers into lengths of 1/2 to 1 inch. These are impelled into the stream of resin which carries them onto the mold, wetting them in the process. The layers are rolled to remove air as in hand lay-up. This process is much faster and less labor-intensive than the hand process.

Mineral fillers are not often used in spray-up apart from ATH and synthetic silicas. Mica may be used in the first layer to avoid telegraphing. A machine is available to spray dry fillers such as mica and ATH into the resin spray.

Both lay-up and spray-up processes use only one-sided molds. Female molds are preferred for ease of demolding, since the resin shrinkage on curing tend to lock the part onto a male mold.

Spray-up is generally used only with polyester resin. It is not uncommon for the molding to be reinforced by a backing of chipboard or polyurethane foam. Parts manufactured by this process are essentially the same as those made by hand lay-up. A common variation is the manufacture of tubs, showers, spas and small boats by using a thermoplastic acrylic facing instead of the gel coat. The facing is first made by thermoforming and is then set in a holding jig while the polyester is sprayed onto the reverse side.

⌘⌘⌘

Both open-mold processes allow styrene vapor (the reactive diluent; see Figure 4) to pass into the work place atmosphere. Regulations limit the amount of permissible styrene exposure to as low as 20 parts per million of air.

Resin Transfer Molding

A mold with male and female faces is used in resin transfer molding to provide articles with two good faces. The calculated amount of glass fiber reinforcement in the form of mats is placed in the cavity of the mold, which is locked closed. Catalyzed resin is forced under pressure into the cavity, wetting out the reinforcement. Care is taken to allow the displaced air to escape; vacuum is used in some cases.

Resin transfer molding is used mainly for polyesters and epoxies. The content of glass reinforcement may be as high as 80%, so the strength of such parts is very high Apart from ATH, fillers are not often used because they increase resin viscosity and interfere with the wetting of the fiber reinforcement. When used with polyester, this process contains the styrene so that much less escapes into the workplace atmosphere.

Small boats, electrical housings, chairs and exterior automobile parts are made by this process. It is also used to encapsulate electrical components such as coils, semiconductors, resistors, capacitors, transformers, chokes and delay lines. Encapsulation is typically with epoxies, but polyesters, silicones, polyurethanes or phenolics are used for certain applications.

Casting

In casting, open single or closed double face molds may be used. The liquid polymer, catalyst, fillers, reinforcements and pigments are mixed in a paddle or similar low speed low shear mixer to the consistency of a pourable paste. High speeds are avoided to reduce air entrapment. In some systems, vacuum is used to remove entrapped air. Reinforcing high aspect ratio fillers are best added late in the mixing cycle to avoid their breakdown and loss of reinforcing properties.

Once mixed, the paste must be cast into molds quickly if the catalyst is of the room temperature curing type. Alternatively, a premix or masterbatch of all ingredients other than catalyst is made first to ensure a long shelf life. Portions of this masterbatch are weighed out and the prescribed amount of catalyst is added just prior to casting.

Fillers are used in considerable amounts in this process. The most common filler is calcium carbonate used at levels of 35 to 75%. No glass fiber reinforcement is used since it builds excessive viscosity.

The quality of the casting may be improved by centrifugal casting, where a cylindrical mold is spun to align reinforcing fillers and force air to

⌘⌘⌘

the interior. Rotational molding is also used, where the mold is spun in three axes. Molds may also be heated to speed up curing.

Casting is commonly used for large, and especially thick items. Resins used are polyesters, thermosetting acrylics, epoxies and polyurethanes, plus some thermosetting polyamides. Common items made by casting are cultured marble counter tops, sink and bath units, and bowling balls. A gel coat may or may not be sprayed onto the mold surface before casting.

Polymer concrete, made from polyester reinforced with up to 80% aggregates and fillers is used for machine bases, and for simulated granite countertops and flooring.

Bulk Molding Compound

Bulk Molding Compound, BMC, (Dough Molding Compound in Europe) is produced by first mixing pre-catalyzed liquid resin with fillers, mainly calcium carbonate and talc, in a heavy duty low speed sigma blade mixer. This is compression molded at 500 psi and 300 to 400°F. The resin most commonly used is unsaturated styrene-diluted polyester. Other BMC resins are alkyds, phenolics, urea, melamine, diallyl phthallate, silicones and epoxy. All are highly filled with calcium carbonate, talc, mica or alumina to improve mechanical properties and reduce shrinkage.

BMC may also be transfer molded or injection molded. In transfer molding, the filled resin is heated and liquefied in a 'pot' and is then pushed by a plunger, via a series of runners, into a heated mold. There it is held under heat and pressure until polymerized. The injection molding process is similar to that used for thermoplastics, except that the mold is heated, not cooled.

End-uses for BMC include washtubs, trays, tote boxes, and electrical parts.

Sheet Molding Compound

Sheet Molding Compound, SMC, is a variant of BMC. SMC achieves higher strength and stiffness of the molded part through reinforcement with long glass fibers, a process which minimizes fiber breakdown, and because the glass fibers are partially oriented in the SMC machine and may be further oriented in the compression mold.

In forming SMC, a paste is made of styrenated polyester resin with calcium carbonate or alumina hydrate, catalyst and thickener. This is deposited in an even layer on plastic film laid on a moving conveyor. Glass fibers chopped to the desired length are deposited on the paste layer, and a top layer of plastic film is applied. This sandwich passes through rollers that knead it to wet out the fibers. It is then rolled up and stored until its viscosity builds to a leather-like consistency. When ready for molding, the sandwich

⌘⌘⌘

is cut into various shapes to place the reinforcement in the desired orientation. The films are stripped off, and the pieces of 'leather' are placed in the mold. They are placed in layers as necessary to build up thickness in high stress areas. The mold is closed and the mass is heated to 300 to 400°F and pressed at 1000 to 2000 psi.

SMC parts are used for such automotive components as trunk deck and hood lids, front-end panels, headlight housings, and rear wheel opening covers. Non-automotive uses include bathtubs, shower pans, laundry tubs, and electrical goods.

Reinforced Reaction Injection Molding (RRIM)

Reaction Injection Molding (RIM) is a process wherein two reactive liquid components are injected at approximately equal proportions under 2000 to 3000 psi pressure and high volumes into a mixing head (chamber). There they mix and react very quickly. The mixture then passes into a mold where it quickly gels. The two components are usually a polyisocyanate and a polyol, but some caprolactam- and cyclopentadiene-based polymers use this process. The technology for high volume pumping and impingement mixing came from the mixing of liquid propellants used by the aerospace industry.

Unfilled RIM polymers have low rigidity and high coefficients of thermal expansion, which is a problem when they are mated with metal parts. To overcome these defects fillers, mainly mica and wollastonite, are added to the polyol components by the molder to create RRIM. These fillers tend to settle out, especially since the viscosity of the components is reduced when they are heated to accelerate gelation. To overcome this, the filled polyols are continuously pumped from the mixing/storage tank through the lines to the mixing head and back.

Automotive applications for SRIM include front and rear fascias and bumper covers, and exterior trim panels. Non-automotive end-uses include air filters and play equipment.

TEST METHODS

The testing of mineral filled plastics follows the protocols from the American Society for Testing & Materials (ASTM), although ISO specifications are starting to be used. There are in excess of 700 tests and specifications for plastics in ASTM Volume 8. Relatively few are routinely used in the processing and evaluation of filled plastics.

⌘⌘⌘

Filler or Reinforcement Content

ASTM D2584 Ignition Loss – This is commonly called the "burnout" test, whereby the filled or reinforced plastic is ignited to 565°C until the polymer is carbonized and the carbon oxidized to carbon dioxide gas. The residue is the filler and reinforcement. If a combination of mineral filler and glass fiber reinforcement are present, they may be weighed separately by screening the cooled residue through 60 mesh. The glass fibers are usually much longer than this while the fillers are usually finer than 100 mesh. The burnout test will give some misleading data when there are fillers such as alumina trihydrate or magnesium carbonate since these decompose below 565 degrees. In such cases, the residue may be analyzed by X-ray diffraction, X-ray emission, or atomic absorption spectrophotometry, with the probable composition of the filler back-calculated.

Alternatively, the polymer may be extracted in a Soxhlet apparatus if there is a suitable solvent or if the polymer can be decomposed by chemicals that have no effect on the fillers

Thermoplastics Processing Tests

Certain common tests are used to determine how the filled polymer will process. The most important processing tests for thermoplastics are melting temperature and melt viscosity.

ASTM D1525 Vicat Softening Temperature – A test specimen of the compound is heated until a needle under a load of one kilogram penetrates the plastic by one millimeter.

ASTM D2117 – This test measures the melting point of semi-crystalline polymers by means of a hot stage microscope.

ASTM D1238 Flow Rate of Thermoplastics by Extrusion Plastometer (Melt Index) – Compound melted at a predetermined temperature is extruded by a piston loaded with a specified weight under conditions that are determined by the composition. The weight in grams of compound extruded in 10 minutes is the Melt Flow Index. A higher value means a lower melt viscosity.

Thermoset Processing Tests

Important parameters for the processing of thermosets are viscosity, gel time, exotherm and shrinkage during cure.

ASTM D2393 – Viscosity as measured by Brookfield viscometer is used for liquid systems; a special head can be used for pastes.

⌘⌘⌘

ASTM D3795 – Thermal Flow and Cure Properties by Brabender Torque Rheometer – This test give a simulation of what will happen during injection or transfer molding.

ASTM D731 Molding Index – This test is used as a guide to required molding pressure and time. The prepared and weighed compound is loaded into a standard cup mold. The mold is closed using enough pressure to form the cup. The pressure is reduced step by step until the mold cannot close. The next higher pressure and time, those that would be just sufficient to cause mold closure, are reported as the molding index.

ASTM D3123 Spiral Flow – This test measures how a filled molding compound will behave when injection or transfer molded. It enables the molder to set the optimum mold temperature. The compound is forced into a heated spiral mold under controlled conditions. Once cure is complete the part is removed and the flow length is measured directly from the specimen.

ASTM D2471 – Gel time and peak exotherm.

ASTM D 2566 – Shrinkage during cure.

Physical Properties

ASTM D792 Specific Gravity – This is used to determine the weight of the final part and its volume cost.

ASTM D570 Water Absorption – This indicates how the compound will behave in humid and wet environments.

Appearance

Color, gloss, light transmittance, haze, and absence of defects, such as sink marks, flow and weld lines are all important appearance factors.

ASTM D2244 – Color by Tristimulus Colorimeter.
ASTM D523 – Specular gloss of nonmetallic specimens.
ASTM D2457 – Specular gloss of plastic films.
ASTM D2103 – Luminous transmittance of polyethylene film.
ASTM D2562 – Classification of defects in molded parts.
ASTM D1492 – Diffuse transmittance, reinforced thermoset building panels.

⌘⌘⌘

Short Term Mechanical Properties

The short term properties of most interest are tensile (resistance to pulling), flexural (resistance to bending) and compressive strengths and moduli, stiffness and elongation. These properties are often measured at the yield point, when the polymer starts to cold-flow, and at the breaking point. Compressive impact resistance is also of major importance. This depends to a great extent on the method of test – whether by side impact on a bar (Izod or Charpy tests) or in the center of a surface (Gardner or falling dart impact). Another factor is the notch-sensitivity of the filled polymer. Some polymers fail more readily in impact when there is a flaw in the polymer, such as a crack or scratch. This is measured using a notched specimen. Surface hardness is also measured, since minerals are often added to increase this property.

ASTM D638 – Tensile strength, modulus and elongation.
ASTM D790 – Flexural strength, modulus and elongation.
ASTM D695 – Compressive strength.
ASTM D256 – Izod and Charpy impact resistance.
ASTM D1709 – Free falling dart impact resistance.
ASTM D3029 – Falling (weight) impact resistance.
ASTM D785 – Rockwell hardness.
ASTM D2240 – Shore hardness.
ASTM D2583 – Barcol hardness.

Electrical Properties

Filled compounds are often used to manufacture electrical cables, switches, and housings and to encapsulate electrical parts. Resistivity, arc resistance and dielectric properties are thus important. The main test methods are as follow.

ASTM D149 – Dielectric strength.
ASTM D150 – Dielectric constant.
ASTM D229 – Electrical resistance.
ASTM D495 – Arc resistance.

Thermal Properties

Important thermal characteristics determined for a plastic compound are:

–how it will behave under the action of heat and cold
–its temperatures of glass transition, melting and decomposition
–at what temperature it will start to yield under various loads

⌘⌘⌘

–how it will expand and contract with heat cycling, especially when in contact with metals with low thermal expansivity
–how well it will conduct heat when molded
–how brittle it may be at low temperatures.

ASTM D648 – Deflection temperature under flexural load at 66 psi and 264 psi.
ASTM D696 – Coefficient of linear thermal expansion.
ASTM C177 – Thermal conductivity.
ASTM D256 – Izod and Charpy impact resistance at low temperatures.
ASTM D746 – Brittleness temperature.
ASTM D3418 –Transition temperatures by thermal analysis.

Fire Resistance

Most plastics will ignite on heating and will continue to burn once the source of heat is removed. The products of combustion may be toxic gases, like formaldehyde from phenolics; corrosive, like hydrochloric acid from PVC; or sooty, like polystyrene. Tests are available to measure the ease with which they ignite, if they continue to burn once the source of ignition is removed and the smokiness of the products of combustion. The topic of flammability is complex and there are many tests related to various plastics, plastic products and end-use scenarios. None of the laboratory scale tests are wholly acceptable, and must be backed by full scale simulated fire tests.

ASTM D1929 – Flash ignition temperature using a hot-air ignition furnace.
ASTM D3801 – Comparative extinguishing characteristics.
ASTM D2843 – Smoke density.
ASTM D2863 Oxygen Index – Minimum oxygen concentration to support candle-like combustion.

Aging Properties

It is important to know how the filled plastics will behave after a long period in service; for example under long-time exposure to high temperature under the hood of a car, immersed in hot water in a clothes or dishwasher, on exposure to the elements, on exposure to chemicals such as motor oil and brake fluid, and to fatigue by cyclical loads. Plastics tend to creep (cold-flow) when exposed to continuous loads for long periods of time. They typically have poor resistance to heat and UV light, and require additives to combat these defects.

ASTM D1299 – Dimensional change due to heat aging.
ASTM D1693 – Environmental stress crack resistance.

⌘⌘⌘

ASTM D671 – Fatigue resistance.
ASTM D2990 – Creep rupture of plastics - tensile and compressive.
ASTM D3045 – Heat aging.
ASTM D42 – Accelerated weathering by Weatherometer.
ASTM D1435 – Outdoor weathering resistance.

BIBLIOGRAPHY

Anon., 1994, 1995, *Extender and Filler Pigments,* Charles H. Kline & Co., NJ

Anon., 1963, *Bayer Pocket Book for the Plastics Industry*, Farbenfabriken Bayer AG

Anon., 1971, *Crystic Monograph No. 2, Polyester Handbook*, Scott Bader Co.

Anon., 1961, *Cellobond Polyester Resins, Technical Manual No. 12*, British Resin Products Ltd.

Anon., *Stock List and Reference List, Ed. 14*, Drummond McCall & Co.,

Dick, J.S., 1987, *Compounding Materials for the Polymer Industries: A Concise Guide to Polymers, Rubbers, Adhesives and Coatings*, Noyes, Park Ridge, NJ

Doyle, E.N., 1971, *The Development and Use of Polyurethane Products*, McGraw-Hill, NewYork

Edenraum, J. Ed., 1992, *Plastics Additives & Modifiers Handbook,* Nostrand Reinhold, New York

  Sekutowski, D., Chapter 36: *Calcium Carbonate*

  idem, Chapter 37: *Kaolin*

  idem, Chapter 39: *Talc*

  idem, Chapter 40: *Mica*

  idem, Chapter 41: Wollastonite

Gachter R., Muller H., 1983, Plastics Additives Handbook, Hanser

Griffith, J.B, Ed., 1994-1996, *Industrial Minerals; 11th Industrial Minerals International Congress*, Berlin , April, 1994; *Industrial Minerals Meeting: Pigments and Extenders- Filling a Colorful Market*, Atlanta, 1984.

Harper, C.A.., 1961, *Electronic Packaging with Resins,* McGraw-Hill, New York

Hawley, G.C., 1995, *Expanding Markets for Calcium Carbonate Fillers,* Seventh Annual Canadian Conference on Markets for Industrial Minerals, Vancouver, October 1995

Hawley, G.C., 1994, *The Future of Industrial Mineral Enhancers in Plastics,* Eleventh Industrial Minerals International Congress, Berlin, April 1994

Hawley, G.C., 1992, *Industrial Minerals and Environmental Regulation: Problems and Opportunities*, Fourth Annual Canadian Conference on Markets for Industrial Minerals, Toronto, October 1992

Hawley, G.C., 1984, *Composites-Alloys of Polymers and Minerals,* Industrial Minerals Meeting: Pigments and Extenders- Filling a Colorful Market, Atlanta, November 1984

Kaplan, W.A, Managing Ed., *Modern Plastics Encyclopedia*, McGraw - Hill, New York, 1996, 1995, 1994, 1985-86,1984-85, and *Modern Plastics*, 1994 - 1996

Lange, A.D., Forker, G.S.; 1961, *Handbook of Chemistry*, McGraw Hill, New York

Lewis, P.A.Ed., 1988, *Pigment Handbook, Vol. 1. Section B- Extender Pigments* Wiley-Interscience, New York

  North, R.B., Chapter b.1: *Natural Calcium Carbonate*

⌘⌘⌘

Thieme, C. , Aumann G., Chapter b.2: *Precipitated Calcium Carbonate*
Berry, H. K., Chapter d.1: *Aluminum Silicate [Kaolin]*
Copeland, J.R., Chapter d.2: *Natural Calcium Silicate*
Harvey, A.M., Chapter d.5: *Magnesium Silicate [Talc]*
Hawley, G.C., Chapter d.6: *Aluminum Potassium Silicate [Mica]*

Milewski, J.V., Katz , Harry S., Eds., 1987, *Handbook of Reinforcements for Plastics*, Van Nostrand Reinhold , New York
Hawley, G.C., Chapter 4: *Flakes.*
Copeland, J.R., Chapter 8: *Wollastonite*

Milewski,J.V., Katz , Harry S. Eds. 1987, *Handbook of Fillers for Plastics*, Van Nostrand Reinhold , New York
Baker, R.A., Koller, L.L., Kummer, P.E., Chapter 6: *Calcium Carbonate*
Ferrigno, T.H., Florea, T.G., Chapter 7: *Kaolin*
Radosta,J.A., Trivedi, N.C., Chapter 11: *Talc*

Naitove, M.H., 1994 - 1996, *Plastics Technology*, Bill Communications, New York

Penn, W.S., 1962, *PVC Technology*, MacLaren & Sons, New York

Perry, R.H., Chilton, C.H., Eds., 1973, *Chemical Engineers' Handbook*, McGraw Hill, New York
Norden,R.B., Chapter 23: *Materials of Construction*

Seymour, R.B., 1978, *Additives for Plastics, Vol.1. State of the Art,* Academic Press, New York

⌘⌘⌘

NINE

PHARMACEUTICALS

James W. McGinity
Drug Dynamics Institute
University of Texas at Austin

Patrick B. O'Donnell
University of Texas at Austin

The pharmaceutical industry is a multi-billion dollar enterprise which affects and improves the lives of people world wide. In the broad sense, the pharmaceutical industry encompasses all aspects of drug design, testing, and delivery. Although the industry is highly competitive, the research and development of a new drug into a final dosage form is expensive and prolonged. Due to stringent testing and strict governmental controls, a newly discovered drug may take up to twelve years to reach the marketplace. From the thousands of new chemical entities identified yearly, only a small number are actually brought to the patient.

The largest market in the world for pharmaceutical products is the United States. Tables 1 and 2 respectively illustrate the leading pharmaceutical companies in the United States and their ranking in the world market.

Table 1. Leading American Pharmaceutical Companies

	1994 Sales (millions), Adjusted
Johnson & Johnson	$15,734
Merck	$14,970
Bristol-Myers Squibb	$11,984
Pfizer	$ 9,946
Abbott Laboratories	$ 9,156
American Home Products	$ 8,966

(Fortune, 1995)

Table 2. World's Largest Pharmaceutical Companies

	1994 Sales (millions), Adjusted
Johnson & Johnson (US)	$15,734
Merck (US)	$14,970
Bristol-Myers Squibb (US)	$11,984
Sandoz (Switzerland)	$11,661
Roche (Switzerland)	$10,790
Pfizer (US)	$ 9,946
Abbott Laboratories (US)	$ 9,156
American Home Products (US)	$ 8,966
Glaxo (UK)	$ 8,466
SmithKline Beecham (UK)	$ 8,281

(Fortune, 1995)

Figure 1 is a schematic representation of the new drug development process. When a new drug compound is discovered, a series of preclinical studies are performed to establish the safety and efficacy of the drug. Under the Food Drug and Cosmetic Act , the sponsor of a new drug is required to submit to the Food and Drug Administration (FDA) a "Notice of Claimed Investigational Exemption for a New Drug" (IND) before the drug can be clinically tested in human beings. After completion of preclinical and clinical trials, the pharmaceutical company may file a New Drug Application (NDA), and if approved, the drug may be marketed.

Clinical trials are performed in three stages, which can take from two to six years. After successful completion of the clinical trials, a new drug application (NDA) is submitted to the FDA. The NDA review typically takes two years, after which the NDA is approved. Figure 2 illustrates the time frame for development of a new drug.

The United States Pharmacopeia (USP) contains detailed monographs on drug substances and dosage forms. Monographs on pharmaceutical excipients are outlined in the National Formulary (NF). Pharmaceuticals and excipients must meet the specifications as stated in the United States Pharmacopeia or the National Formulary to be listed as a USP/NF ingredient. For international markets, an excipient or drug must also meet the compendial standards of other countries, such as the European Pharmacopeia (Ph. Eur.) and the Japanese Pharmacopeia (JP). In 1989 these two organizations joined with the United States Pharmacopeia to form the Pharmacopeial Discussion Group (PDG) to work on the harmonization of excipient standards and test methods. This standardization of compendial

⌘⌘⌘

requirements will help proliferate world-wide trade of pharmaceutical excipients (Chowhan, 1994).

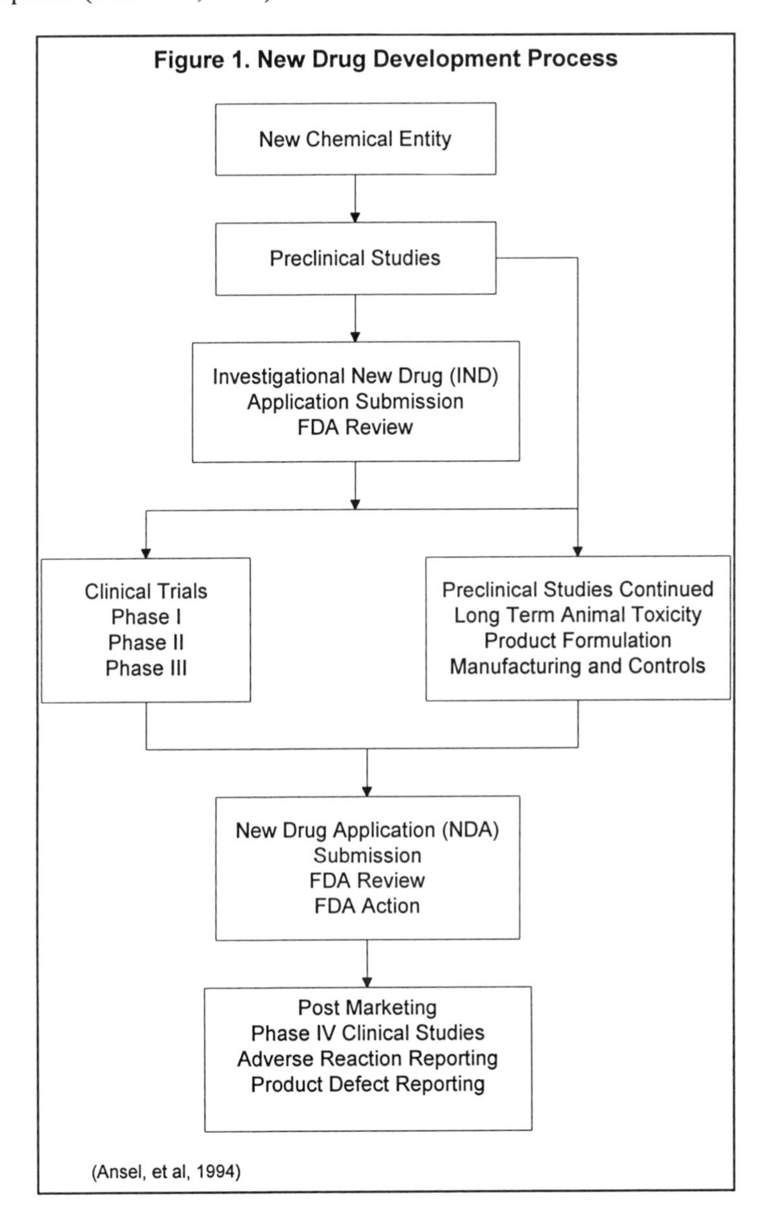

Figure 1. New Drug Development Process

⌘⌘⌘

Figure 2. Time Frame for the Development of a New Drug

Pre-Clinical R & D	Clinical R & D	NDA Review	Post-Marketing Surveillance
Initial Synthesis and Characterization			Adverse Reaction Reporting
	Phase 1		
	Phase 2		
	Phase 3		Surveys Sampling Testing
Animal testing			
Short Term Long Term			Inspections
Avg. 3 1/2 Year	Avg. 6 Years	Avg. 2 1/2 years	

(Ansel., *et al*, 1994)

MINERALS AS ACTIVE PHARMACEUTICAL INGREDIENTS

Several minerals are included in pharmaceutical preparations as the medicament or active ingredient. These minerals are generally as gastric antacids, adsorbents, laxatives, or topical drugs.

Gastric Antacids

Gastric antacids are drugs which, on ingestion, react with the hydrochloric acid in the stomach to reduce acidity. They are prescribed for the systematic relief of hyperacidity, irritation caused when gastric fluids are forced into the esophagus (gastroesophageal reflux), gastrointestinal bleeding, kidney stones (renal calculi), abnormally high presence of phosphates in the blood (hyperphosphatemia), and abnormally low blood concerntration of calcium (hypocalcemia). In order for a product to be labeled an antacid, according to the FDA, it must contain one or more approved active ingredients. The finished product must contain at least 5mEq/dose unit neutralizing capacity and raise the pH of gastric secretions to 3.5 or greater within 10 minutes. Table 3 illustrates the onset, duration, sodium content and acid neutralizing capacity of some common single entity antacids.

Antacids are usually used in combination for three reasons: first to combine fast and slow reacting antacids in order to obtain a product with rapid and even action; second, to lower the dose of each component and reduce the possibility of adverse side effects, and third; to use one component to reduce or prevent a side effect caused by another component. Virtually all antacid combinations are sold in either suspension or tablet form, with only a few exceptions.

Table 3. Some Common Single Entity Antacids

	Onset	Duration	Sodium (mg/unit)	ANC (mEq)
Aluminum Carbonate gel	slow	short	0.12	12
Aluminum Hydroxide Gel	slow	prolonged	<2.5	16
Calcium Carbonate	fast	prolonged	<2.3	10
Magnesium Carbonate	intermediate	prolonged		20/g
Magnesium Hydroxide	fast	prolonged	0.12	14
Magnesium Oxide	fast	short		21
Sodium Bicarbonate	fast	short	88	12/g

(Gennaro, 1990)

Laxatives

Laxatives are drugs that affect fecal consistency and accelerate the elimination of feces. A number of magnesium salts as well as other sulfates are used as saline laxatives. Some of the more common laxatives include magnesium carbonate, magnesium oxide, magnesium sulfate, and sodium phosphate.

Magnesium carbonate is prepared from dolomite, and is available in light or heavy forms, with the light being about two and one half times as bulky as the heavy. It is used as an antacid and a cathartic, with pharmacologic properties similar to magnesium oxide except that carbon dioxide is released during neutralization. It is often used with calcium carbonate to reduce the constipating action of that salt. Magnesium carbonate is used by the pharmacist in the preparation of Magnesium Citrate Oral Solution USP, a pleasant saline laxative. The preparation as described in the USP XXII is as follows:

Magnesium Citrate Oral Solution USP

Magnesium carbonate	15g
Anhydrous citric acid	27.4g
Syrup	60ml
Talc	5g
Lemon oil	0.1ml
Potassium bicarbonate	2.5g
Purified water sufficient to make	350ml

⌘⌘⌘

Magnesium oxide occurs in nature as the mineral periclase, and commercial preparations are made from magnesite (magnesium carbonate) ores. As with magnesium carbonate, it is also available in light and heavy forms. In water it is converted to the hydroxide, and does not release carbon dioxide.

Magnesium sulfate can be prepared by neutralizing sulfuric acid with magnesium carbonate, or from natural sources as the mineral kieserite, for the monohydrate form, and epsomite, for the heptahydrate form. It is an effective and widely used saline laxative. The laxative action probably results from two factors. The first is that the magnesium sulfate is not absorbed from the intestinal tract and thus retains sufficient water within the lumen of the bowel to make an isotonic solution. The second is that the magnesium ion stimulates the release of cholecystokinin-pancreozymin which causes an accumulation of fluid and electrolytes within the small intestine. Magnesium Sulfate Injection USP is also used as a parenteral anticonvulsant.

Sodium phosphate is prepared from the mineral phosphorite which is a tribasic calcium phosphate. The phosphatic material is digested with sulfuric acid, the mixture is then leached with hot water, neutralized with sodium carbonate, and the sodium phosphate is crystallized from the filtrate. It is the active ingredient in Effervescent Sodium Phosphate, one of the most pleasant of the saline laxatives. It is also used in the preparation of Sodium Phosphate Oral Solution. This solution, which corresponds to 0.75g of the official hydrated salt in each ml. provides an easy form for the administration of a saline laxative and antihypercalcemic (Gennaro, 1990).

Adsorbents

Adsorbents are chemically inert powders that have the ability to adsorb gasses, toxins, and bacteria. The fine state of subdivision of these powders facilitates their high adsorptive capacity. Only certain materials that possess chemical adsorptive properties lend themselves to gastrointestinal detoxification. These substances include kaolin, or kaolinite, and magnesium trisilicate.

Kaolin is distributed worldwide in nature, but is frequently contaminated with impurities, and must be purified for use in pharmaceuticals. It is used alone or as a mixture with pectin. It is of value primarily in the treatment of diarrhea caused by agents capable of being adsorbed, such as diarrhea caused by dysentery or food poisoning.

Magnesium trisilicate occurs in nature as the minerals meerschaum, parasepiolite, and sepiolite. It has been used as a nonsystemic antacid and as an adsorbent. It is claimed to exert a protective action in the stomach by

⌘⌘⌘

virtue of released silicic acid which acts more as a demulcent or protective agent rather than as an adsorbent.

Topicals

Astringents are locally applied protein precipitants which have such a low cell permeability that the action is essentially limited to the cell surface. Astringents are used to coagulate blood, reduce inflammation of mucus membranes, promote healing, and toughen the skin. The principle astringents are the salts of aluminum, zinc, manganese, iron, or bismuth, in addition to certain other salts which contain these metals and tannins.

Alum (aluminum potassium sulfate dodecahydrate) is prepared from the mineral bauxite, a hydrated aluminum oxide, and sulfuric acid. Ammonium alum is predominant in the market due to its low cost. It is a powerful astringent in acidic solutions, and is slightly antiseptic. As an astringent it is used in concentrations ranging from 0.5% to 5.0%.

Aluminum sulfate occurs in nature as the mineral alunogenite, and is a powerful astringent acting much like alum. It is also used as an antiperspirant and is the effective ingredient in some commercial antiperspirant products. It is also a pharmaceutical necessity for the preparation of Aluminum Subacetate Topical Solution USP.

Calamine is prepared by mixing zinc oxide with sufficient ferric oxide to obtain a product of desired color. The name is also applied to a native form of zinc silicate which is not suitable for making medicinal calamine. Calamine is used as an astringent and in protective lotions and ointments for treatment of sunburn and poison ivy.

Zinc oxide occurs in nature as the mineral zincite, and can be prepared from zinc metal by vaporization and oxidation. It can also be prepared from franklinite or zinc sulfide. It is a very fine, odorless, amorphous, white or yellowish white powder. Zinc oxide has a mild astringent, protective and antiseptic action, and is usually formulated into ointment, paste or lotion vehicles. It is used widely in the treatment of dry skin and disorders such as acne vulgaris, prickly heat, insect bites, ivy poisioning, diaper rash, seborreha, impetigo, and psoriasis. It is also contained in some sunscreens.

Zinc sulfate is prepared by reacting metallic zinc or zinc oxide with diluted sulfuric acid. It has use as an astringent, emetic, and a weak antiseptic. Its antiseptic and astringint properties made it a valuable agent for use as an aqueous based eyewash for the treatment of conjunctivitis caused by *Morax-Axenfeld bacillus*, although it has been replaced by antibiotics for this purpose. It may be applied to the skin in lotion form, as in White Lotion USP, for the treatment of acne, poison ivy, and impetigo.

Minerals have also been developed or investigated to treat other health problems not included in the catagories above. For example, a number of substances are currently being investigated as treatments for osteoporosis. These substances include calcium, a combination of phosphates and diphosphonate, magnesium, manganese. and sodium fluoride (Ratafia, 1989).

MINERAL EXCIPIENTS IN PHARMACEUTICAL APPLICATIONS

An excipient is defined as an inert material used as a formulation aid in pharmaceutical products. Most minerals used by the pharmaceutical industry are found in this category. Minerals find use as suspending agents for drug suspensions, adsorbents, and excipients in tablet formulations. Since tableted dosage forms comprise approximately 80% of the pharmaceutical trade, minerals find most frequent use in this type of solid dosage form. Typical tablet formulations may contain a disintegrant (5-10%) to aid in the dissolution of the tablet, a binder (5-10%) to facilitate good bonding of the tablet components, a lubricant (0.5-1%) to ensure that the tableting machinery operates smoothly, and a glidant (0.25-1%) to ensure that the granulation flows smoothly through the machinery. The remaining approximately 80% of the tablet weight is composed of drug plus filler, with minerals in some cases serving the latter function..

The choice of excipients impacts the physical properties of the tablet. They can accelerate or retard tablet disintegration, affect the hardness of the tablet, or impact the stability of the dosage form by adsorbing moisture from the atmosphere. Careful selection and testing of excipients is therefore of paramount importance.

It has been shown that certain excipients such as calcium carbonate, terra alba, and kaolin can interfere with the bioavailability of certain drug compounds (Naggar, 1981). For example, anticholinergic drugs such as atropine sulfate prevent acetylcholine from acting as a neurotransmitter, which results in relaxation of smooth muscle and decreased secretion of digestive juices. They are often administered in combination with antacids. However, atropine sulfate is readily adsorbed by calcium carbonate, thus reducing its bioavailability, or inhibiting its therapeutic effect. In the case of ambutonium bromide, another anticholinergic, no adsorption was observed (Blaug and Gross, 1965). The excipients used in the tablet formulation must be tested with the drug to insure that no adverse effect on drug stability or bioavailability occurs.

Certain minerals also have a therapeutic use, as discussed in the preceding section. For example, kaolin has found medicinal use in poultices, and has good adsorptive properties. Kaolin is frequently employed

⌘⌘⌘

to adsorb the enterotoxin of *pseudomonas aeruginosa* as a treatment for diarrhea (Said *et al,* 1980). Calcium carbonate is frequently employed as an antacid, as a source of calcium and as a phosphate binder for controlling abnormally high levels of phosphate in the blood (hyperphosphatemia) of patients on chronic dialysis (Ookubo *et al*, 1992). Boehmite (hydrous aluminum oxide) has also been investigated as a phosphate binder (Nishida *et al*, 1990). The therapeutic use attributed to the mineral product must be considered when selection of excipients is made for a pharmaceutical preparation.

Minerals Used in Pharmaceutical Suspensions

A suspension is a two phase system consisting of finely divided solids dispersed in a solid, liquid or gas. For the pharmacist however, the main emphasis is placed on solids dispersed in a liquid vehicle. There is a need for more specific terminology, so this type of pharmaceutical preparation may be classified as a suspension, mixture, magma, gel or lotion (Gennaro, 1990). The preparation of a stable suspension depends on the dispersion of the drug in a suspending medium. A surfactant may be utilized in order to effectively wet the drug by the dispersion medium. The suspending agent in the aqueous medium can then be added. An alternate method of preparation allows the dry suspending agent to be mixed thoroughly with the drug and then triturated with the aqueous media. Other methods include the formation of a flocculated suspension or a flocculated preparation in a suspending vehicle.

The ideal characteristics of a suspending agent have been defined; the material should:

1. Be readily and uniformly incorporated into the substance to be suspended.
2. Be readily dissolved or dispersed when water or an aqueous solution is added.
3. Ensure the distribution of the insoluble material uniformly in an easily redispersable loosely-packed system.
4. Not influence the dissolution or absorption rates *in vivo* of poorly soluble substances.
5. Be inert, non-toxic, and free of incompatabilities.
6. Be of acceptable color and odor, with a bland taste.
7. Be free of microbial contamination.
8. Be available and inexpensive.

When these criteria are applied to potential suspending agents, only a small number can be considered for pharmaceutical use (Farley, 1976).

⌘⌘⌘

Suspensions

Pharmaceutical suspensions are defined as preparations intended for oral or injectable (parenteral) use containing finely divided insoluble drug particles, referred to as the suspensoid, and distributed uniformly throughout the vehicle. Some drug suspensions are available in a ready-to-use form already dispersed in the suspending media, while others must be reconstituted prior to use with a suitable vehicle. Drugs that are unstable in the presence of an aqueous vehicle, as are many antibiotics, are usually supplied as dry powder mixtures for reconstitution by the pharmacist at the time of dispensing.

Certain drugs may be supplied as suspensions in order to facilitate ease of swallowing. This is particularly true in the case of infants and geriatric patients. Additionally, the unpleasant taste of certain drug compounds is markedly reduced in suspension form.

The reduction in the particle size of a suspensoid is beneficial to the stability of the suspension in that the rate of sedimentation is reduced as the particles decrease in size. Fine particles, however, may have a tendency to form a compact cake upon settling to the bottom of the container, making redispersion by shaking difficult or impossible. To avoid this problem, measures must be taken to prevent the agglomeration of the particles into larger masses or crystals. The most common method of preventing caking is by the intentional formation of loose agglomerates of particles held together by weak particle to particle bonding forces. This type of particle aggregation is called a floc or a floccule, with the flocculated particles forming a loose lattice structure that resists complete settling. The loose structure is easily disrupted by shaking and the product returns to a fine suspension (Heistand, 1964).

There are several methods for preparing flocculated suspensions, and the choice of method depends on the drug and product type. In the preparation of an oral suspension of a drug, clays such as bentonite magma are commonly used as the flocculating agent. The addition of small amounts of bentonite flocculate bismuth subnitrate suspensions and eliminate caking (Schott, 1976). This type of flocculation is attributed to charge neutralization of the bismuth subnitrate by the bentonite. The structure of the bentonite and other clays used for this purpose also aid the suspension by helping to support the floc once it is formed. Aqueous suspensions of kaolin can be flocculated with a variety of flocculant molecules, including pectin, tragacanth, carboxymethylcellulose, and polyvinylpyrollidone (Alexander, et al, 1989). Some common examples of oral suspensions are listed in Table 4. All contain natural smectite clays which comply with the NF monograph for Magnesium Aluminum Silicate (MAS).

⌘⌘⌘

Table 4. Oral Suspensions

Suspension	Name of Product	Mineral	Purpose of mineral
Grisofulvin Oral Suspension	Grifulvin® V	MAS	Suspending Agent
Primidone Oral Suspension	Mysoline® Suspension	MAS	Suspending Agent
Bismuth Subsalicylate Suspension	Pepto-Bismol ®	MAS	Suspending Agent
Erythromycin Ethylsuccinate and Sulfisoxazole Acetyl	Pediazole®	MAS	Suspending Agent
Chlorprothixene Oral Suspension	Taractan® Concentrate	MAS	Suspending Agent
Trisulfapyrimidines Oral Suspension	various	MAS	Suspending Agent

Redispersability is one of the major considerations in assessing the acceptability of a suspension. The sediment formed should be easily redispersed with moderate shaking. Many suspending agents may, however, create bioavailability problems due to the tendency of certain drugs to bind to the suspending agent. Suspensions containing either ampicillin or sulphadimidine and kaolin have shown reduced bioavailability due to their adsorption to the kaolin. Modification of the formulations with a variety of additives increased bioavailability and improved redispersity of the suspension (Hosny, *et al*, 1988).

Gels

The USP defines gels as semisolid systems of suspensions composed either of small inorganic particles or large organic molecules interpenetrated by a liquid. Where the gel mass consists of a network of small discrete particles in the colloidal region, less than 0.5μm, the suspension may be considered a gel consisting of a two phase system, such as aluminum hydroxide gel. In a two phase system, with the dispersed phase particle size greater than 0.5μm, the gel mass is sometimes referred to as a magma, such as bentonite magma. Both gels and magmas can be thixotropic, forming semisolids on standing and becoming liquid on agitation.

Aluminum hydroxide gel contains the equivalent of not less than 3.6g and not more than 4.4g of aluminum oxide (Al_2O_3), in the form of aluminum hydroxide and hydrated oxide. Aluminum oxide, also known as alumina, occurs in nature in the minerals bauxite, bayerite, boehmite, corundum, diaspore and gibbsite. Hydrated alumina, or aluminum hydroxide has applications in pharmaceutical products as a gel or as a dehydrated gel called algedrate. It can also be found in some antiperspirants and dentifrices. It has therapeutic applications as an antacid and as an antihyperphosphatmic utilized to counteract excessive phosphate levels in the blood. There have been clinical comparisons of aluminum hydroxide to calcium carbonate as a phosphate binder in chronic renal failure (Mak, *et al.* 1985). An example of a formulation utilizing aluminum hydroxide gel is listed below (Gennaro, 1990).

Aluminum Hydroxide and Belladonna Mixture PC

Belladonna Tincture	100ml
Chloroform Spirit	50ml
Aluminum Hydroxide Gel to	1000ml

Aluminum hydroxide gel is also used as the suspending agent for Theroxide® 5% and 10% benzoyl peroxide lotions.

Magmas and Milks

Magmas and milks are aqueous suspensions of insoluble inorganics and differ from gels mainly in that the suspended particles are larger, but still of small dimension. They are also termed fine dispersions. They are thick and viscous by nature and thus there is no need to add suspending agents. Magmas may also be prepared by chemical reaction. For example, Magnesium Hydroxide USP can be prepared by the hydration of magnesium oxide, which occurs in nature as the mineral brucite. Milk of magnesia, a common antacid suspension, is comprised of a 7.0 to 8.5% suspension of magnesium hydroxide.

Bentonite Magma NF is used as a suspending agent in pharmaceutical products, particularly for topical lotions. It is prepared by placing 500g of purified water in a blender and while the machine is running, 50g of bentonite is added. The mixture is blended for 10 minutes, and purified water is added to make 1000g.

Lotions

Lotions are liquid suspensions or dispersions intended for external application to the body. Most lotions require a suspending agent and may require a wetting agent or a surfactant. Two of the most common suspending agents are magmas made from bentonite, which is principally the smectite mineral montmorillonite, or magnesium aluminum silicate, which is the nomenclature use to indicate a blend of the smectite clays montmorillonite and saponite. These lotions are prepared by levigating the ingredients to a smooth paste and then adding the remaining liquid. Calamine lotion is a

⌘⌘⌘

classic example, consisting of finely powdered, insoluble solids held in permanent suspension by the suspending agent. The standard method for preparation of calamine lotion is to dilute bentonite magma (259ml) with an equal volume of calcium hydroxide solution. The calamine (80g) and zinc oxide (80g) are mixed thoroughly with 20ml of glycerin and 100ml of the diluted magma. This mixture is triturated until a smooth uniform paste is formed. The remainder of the dilute magma is slowly incorporated. Finally, calcium hydroxide solution is added to make 1000ml of the lotion. The bentonite magma is used to suspend the zinc oxide and the calamine. Other medicated lotions which contain minerals either as the suspending agent or the active ingredient are listed in Table 5.

Table 5. Medicated Lotions

Lotion	Commercial Product	Mineral	Purpose of Mineral
Ammonium Lactate Lotion	LAC-HYDRIN®	MAS	Suspending Agent
Benzoyl Peroxide	DESQUAM®-X5 and X10 cleanser	MAS	Suspending Agent
Chloroxine Shampoo	CAPITROL®	MAS	Suspending Agent
Calamine Lotion	Various	Bentonite	Suspending Agent
Selenium Sulfide Lotion	SELSUN®	Bentonite	Suspending Agent
Sodium Sulfacetamide and Sulfur Lotion	SULFACET-R® Acne Lotion	Sulfur	Keratolytic Agent

Adsorbents

Adsorbents are chemically inert powders that have the ability to adsorb moisture, gasses, toxins, and to some extent, bacteria. Some minerals have found extensive use as adsorbents, such as magnesium aluminum silicate, bentonite, attapulgite, talc, calcium carbonate, and kaolin. The activities of kaolin and calcium carbonate were discussed in a preceding section.

Extensive research has been done on the adsorptive properties of various excipients for specific drugs. This adsorption phenonemon can affect the bioavailability of the drug substance. When excipients with adsorptive properties are present in a dosage form or are administered concurrently with certain drugs, the activity or bioavailability of the drug may be compromised. The ingestion of digoxin with antacids possessing adsorptive properties causes a decrease in the bioavailability of the drug (Khalil, 1974). Adsorption of paracetamol and chloroquine phosphate by various antacids with adsorptive properties was studied and and the adsorptive capacity of each antacid was calculated (Iwuagwu and Aloko, 1992). It was found that magnesium trisilicate adsorbed 5.204 mg/g of paracetamol and 4.721 mg/g of chloroquine phosphate, magnesium oxide 4.342 mg/g and 4.175 mg/g, and edible clay 2.324 mg/g and 2.140 mg/g of each respective drug. The bioavailability of some antibiotics such as oxytetracycline hydrochloride, tetracycline hydrochloride, doxycycline hydrate, triacetyloleandomycin, chlorampenicol, ampicillin, and cloxacillin sodium is reduced by co-administration of most of the antacids. Ciprofloxacin hydrochloride, a wide spectrum antibiotic, was found to be more readily adsorbed by magnesium oxide (7.88 mg/g) or magnesium carbonate (1.68 mg/g) than by kaolin, which was found in this case to have an adsorptive capacity of 0.42 mg/g (Tuncel, 1992).

Clays have found use in pharmaceuticals both as therapeutic agents and as excipients. Because of their structure and unique rheological behavior, the identity and characterization of the clay is very important to the pharmacuetical scientist. The characterization and adsorptive properties of bentonite, hectorite, saponite, attapulgite, and kaolin were shown to be related to the clay structure and composition (Browne, et al, 1980). The clay structure, the nature of the exchangable cation, and the presence of non-clay components all were found to effect the clay interaction with tetracycline, while clay structures with a high surface charge led to greater interaction with the protonated form of tetracycline. Conversely, clay structures with a minimum surface charge interacted with the zwitterionic form of tetracycline. Dissolution and dialysis methods have been used to show that cationic drugs and some non-ionic drugs may bind to montmorillonite clay. It was demonstrated that the amount of drug released *in vitro* from the drug-clay complex was dependent on the amount of montmorillonite in the formulation (McGinity and Lach, 1976).

Drug-clay complexes have been examined as a method of controlling the release *in vivo* of the bound drug (McGinity and Lach, 1977). Amphetamine sulfate was the model drug studied, and drug concentrations *in vivo* were tracked by urinary recovery studies. It was found that montmorillonite significantly affects the initial therapeutic levels of amphetamine sulfate, and recovery profiles resembled those obtained from prolonged release dosage forms. In order to obtain sustained action in the administration of oral antihistamines, the interaction between montmorillonite and chlorpheniramine maleate was studied (Camazano, 1980). It was determined that the chlorphenirammonium cations are adsorbed into the interlayer space of the clay. The adsorption mechanism is cation exchange, and the maximum amount adsorbed by this mechanism was determined to be 76 mEq/100g.

⌘⌘⌘

The surface adsorption of griseofulvin, indomethacin, and prednisone to colloidal magnesium aluminum silicate has been shown to improve the dissolution rates of these poorly soluble drugs. The rapid release of the drug from the surface of the clay was attributed to the weak physical bonding between the two materials and to the swelling of the clay in aqueous media. The hydrophilic nature of the clay facilitated the wetting of hydrophobic drug substances (McGinity and Harris, 1980). Recently magnesium aluminum silicate has been investigated as an additive in tablet formulations to prevent sorption of bromhexine hydrochloride in solid dosage forms to polyethylene films. When bromohexine hydrochloride is stored in polyethylene packages, transfer of the drug to the film occurs and poor content uniformity in the dosage form occurs. The addition of 40% magnesium aluminum silicate substantially reduced the amount of bromohexine hydrochloride transferring to the polyethylene film (Kukita, *et al* 1992).

TABLETING APPLICATIONS OF MINERALS

Drugs are most frequently taken by oral administration in solid dosage form. Compared with other routes, the oral route is considered the most natural, uncomplicated, convenient and safe method of administering drugs. The tablet is the most common type of dosage form and its popularity stems from the fact that it is portable, easy to administer, stable, and less complex to manufacture. Tablets are composed of the active ingredient and a variety of excipients which give the tablet its physical characteristics. Excipients should be selected that will enable the production of a tablet which meets or exceeds the specifications established for the tablet. The selected excipients will have an impact on the tablet weight variation, hardness, friability, disintegration time, drug dissolution rate, water content, potency and content uniformity.

There are two major functional classifications of tablet excipients. The first class includes those additives which affect the compressional characteristics of the tablet, including diluents, binders and adhesives, lubricants, antiadherents, and glidants. The second class of excipients includes those which affect the biopharmaceutics, chemical and physical stability, and marketing considerations of the tablet. This category includes disintegrants, colors, flavors and sweeteners, and miscellaneous components such as buffers and adsorbents. All of these must meet certain criteria as follows:

⌘⌘⌘

1. Physiologically inert.
2. Acceptable to regulatory agencies.
3. Physically and chemically stable.
4. Free of bacteria considered to be pathogenic or objectionable.
5. Not interfere with the bioavailability of the drug.
6. Commercially available in form and purity specified by pharmaceutical standards.
7. For drug products that are classified as food, such as dietary aids, the excipients must be approved as food aditives.

The most common minerals in tableting include calcium carbonate, calcium phosphate, magnesium aluminum silicate, talc, and silicon dioxide. As mentioned previously, some of these chemical excipients also have physiological effects. In this section only the role of the chemical as an excipient will be discussed.

Tablet Fillers or Diluents

Tablet fillers or diluents often make up the bulk of a tablet formulation, and the selection of the type of filler is of prime importance

Calcium phosphates have a number of fundamental properties which make them ideal for solid dosage formulations with a wide range of actives. Such properties include white color, stability, insolubility, no case hardening, good action with disintegrants, compatable with a wide range of active ingredients, and a source of calcium and phosphorus.

Dibasic Calcium Phosphate Dihydrate NF is a common tablet diluent. It is generally used in as a diluent and binder in direct compression formulations where the active ingredient occupies less than 50% of the tablet weight. Disintegration properties of dibasic calcium phosphate tablet formulations using soluble and insoluble disintegrants has been studied (Khan and Rhodes,1975). The insoluble disintegrants showed a greater effect when compressional forces were varied than did soluble disintegrants. Typical properties of calcium phosphates in tableting are listed in Table 6.

Calcium Sulfate Dihydrate has been suggested as a diluent for granulated tablet systems containing up to 30% of active ingredient. Also known as terra alba (gypsum), it is an inexpensive, nonhydroscopic powder. It is the least expensive tablet filler and can be used for a wide variety of acidic, neutral, and basic drugs.

Calcium carbonate is a dense, white, insoluble powder. It is available in various degrees of fineness. Precipitated calcium carbonate of a very fine particle size is used as a tablet filler.

⌘⌘⌘

Table 6. Typical Properties of Calcium Phosphates in Tableting

	Dicalcium Phosphate Dihydrate $CaHPO_4 \cdot 2H_2O$	**Dicalcium Phosphate Anhydrous** $CaHPO_4$	**Tricalcium Phosphate** $Ca_5(OH)(PO_4)_3$
Water Solubility	No	No	No
Kso(25°C)	2.5 x 10-7	2.5 x 10-7	4.7 x 10-59
Kso(37°C)	1.9 x 10-7	1.9 x 10-7	2.3 x 10-59
HCl Solubility	Yes	Yes	Yes
pH (20% slurry)	7.4	5.1	6.8
Surface Area (M^2/g)	1 - 2	20 - 30	70 - 80
Disintegration:			
Disintegration w/o Disintegrant	No	No	No
Disintegration with Disintegrant	Rapid	Rapid	Rapid
Primary Bonding Mechanism	Brittle Fracture	Brittle Fracture	Plastic Deformation
Compressibility	Good	Good	Acceptable

(Rhone-Poulenc, 1992)

Disintegrants

The purpose of a disintegrant is to aid in the breakup of a tablet after ingestion. Disintegrants are materials that, when in contact with water, will swell, hydrate, change in volume or react chemically to disrupt the tablet structure. Clays such as bentonite and magnesium aluminum silicate have been investigated as potentially useful disintegrants (Granberg and Benton, 1949; Ward and Trachtneberg, 1962). Due to the natural color of most clays, their use is limited in white tablets due to the resulting discoloration.

Three different grades of magnesium aluminum silicate were investigated as a disintegrant for a calcium sulfate tablet. It was found that the clay with the smallest particle size distribution functioned as an excellent disintegrant at levels of 4% of tablet weight. The tablets disintegrated within 75 seconds, whereas tablets with 4% starch as the disintegrant displayed a disintegration time of 133 seconds (Wai, et al, 1966). The same study found that magnesium aluminum silicate functioned best as a disintegrant when added dry to the finished granulation. When added during the granulation process as an internal disintegrant, poor disintegration properties were observed.

⌘⌘⌘

Lubricants, Antiadherents, and Glidants

Most tablet formulations require the addition of lubricants, antiadherents, and glidants in order to produce consistent and properly formed tablets which are free of defects. Lubricants reduce the friction between the granulation and the die wall during compression and ejection. The addition of a lubricant will increase tool life of the punches and dies used to form the tablet. Antiadherents prevent sticking to the punch face and the die wall. A granulation with an insufficient or ineffective antiadherent will result in tablets with surface defects, usually referred to as picking. Glidants improve the flow characteristics of the granulation and ensure that uniform die fill occurs, thus reducing weight variations in the tablet formulation. The most common minerals used in these catagories are talc, silicon dioxide, and sodium chloride.

Usually a material that is a good glidant is a poor lubricant. One exception is talc, which can function as glidant at use levels of 5% of the formulation, or as a lubricant at use levels of 1 to 5 %, and even as an antiadherent at use levels of 1 to 5%. The physical properties of talc will influence its effectivness as a lubricant in a tablet formulation. In an analysis of nine different Talc USP samples, variations in density, particle size, surface area, and tendency of preferred orientation were compared to maximum ejection force of a compact (Lin and Peck, 1994). The results suggested that the talc sample with a higher tendency of preferred orientation has closer bulk packing, and talc with lower bulk density and higher surface area requires less peak force to break the tablet/die-wall adhesion. Differences in USP grades of talc from various suppliers have been reported by other investigators (Phadke, Keeney, and Norris, 1994). Variations in particle size and specific surface were observed in two types of Talc USP. These differences could have an effect on the glidant properties of the talc.

Excipients used for lubrication in tablet formulations can have an adverse effect on the dissolution characteristics of the final tablet. Hydrochlorothiazide tablets lubricated with talc have been shown to have improved dissolution properties as compared to similar tablets lubricated with magnesium stearate (Dawoodbhai *et al* ,1987). This same study also found talc to improve hardness, friability and appearance for some systems when compared to other lubricants. It was additionally noted that different sources of talc showed significant variation in their glidant and lubricant efficiencies.

⌘⌘⌘

PRINCIPAL MINERALS AND SELECTION CRITERIA

The following section lists the minerals most commonly used in the pharmaceutical industry, outlining the properties and compendial specifications for each.

Kaolin

Kaolin is a native hydrous aluminum silicate. Large deposits of kaolin with a degree of purity suitable for use in pharmaceuticals are mined in the state of Georgia and in Cornwall, England. It is powdered and freed of course gritty particles by elutriation or by screening. Although kaolin is distributed widely in nature, most deposits are associated with ferric oxide, calcium carbonate, magnesium carbonate and other such contaminants. To make this kaolin suitable for pharmaceutical use, it must be acid treated and then thoroughly washed to remove the contaminants. Kaolin USP must meet the following pharmacopeial specifications:

Test	USP
Identification	to pass test
Acid soluble substances	<2.0%
Loss on ignition	< 15.0%
Lead	< 0.001%
Iron	to pass test
Carbonates	to pass test
Microbial test limit	Negative to *E. Coli*

Kaolin is practically insoluble in water and organic solvents. It is also insoluble in mineral acids and solutions of alkali hydroxides. Unless sterilized, it may be heavily contaminated with pathogenic microorganisms, such as *bacillus anthracis, clostridium tetani* and *clostridium welchii.* It can be sterilized by maintaining the whole at a temperature not less than 160°C for not less than one hour. The primary applications of Kaolin USP in the pharmaceutical industry are as an adsorbent at concentrations of 7.5 to 55%, dusting powders at 25%, and poultices at a level of 53%. Kaolin has also been used as a diluent in tablet and capsule formulations. Kaolin Mixture with Pectin (Gennaro, 1990), an example of a pharmaceutical adsorbent formulation, is shown below.

The adsorbent properties of kaolin may influence the adsorption of other orally administered drugs, including cimetidine, lincomycin, tetracycline and digoxin. Since it is sometimes employed as a tablet diluent, it must never be used in tablets containing cardiac glycosides, alkaloids, estrogens and medicaments which may be adsorbed by kaolin.

⌘⌘⌘

Kaolin Mixture with Pectin

Kaolin	200g
Pectin	10g
Tragacanth	5g
Sodium saccharin	1g
Glycerin	20ml
Benzoic acid	2g
Peppermint oil	0.75ml
Purified water q.s.	1000ml

Bentonite

Bentonite is a native colloidal hydrated aluminum silicate or clay found in the midwest United States, Wyoming, and Canada. Consisting mainly of montmorillonite, a smectite clay, it is a very fine, odorless, pale buff or cream colored to grayish powder. It consists of particles in the range of 50 to 150 microns, but also has numerous particles in the 1 to 2 micron range.

The primary use for Bentonite NF in pharmaceutical applications is as a suspending agent or a viscosity building agent. It also finds uses as an adsorbent, a gelling agent, an emulsion stabilizer, and a clarifying agent. The typical amount of bentonite used in a suspension would be 0.5 to 5.0%, whereas a concentration of 1% would be common for use as an emulsion stabilizer. Use of bentonite as an adsorbent or clarifying agent would be at concentrations in the 1 to 2% range.

The Pharmacopeial specifications for Bentonite NF are as follows:

Test	NF
pH (4g in 200 ml of H_20)	9.5 - 10.5
Loss on drying	5.0 - 8.0% after 2 hours at 105°C
Gel formation	to pass test
Swelling power	to pass test
Fineness of powder (wet sieved)	must not be retained on 200 sieve
Microbial limit *E. Coli*	absence of *E.Coli* per microbial limit tests
Identification	to pass test
Alkalinity	not specified

Bentonite is insoluble in water, but swells to approximately twelve times its volume when added to water. It is insoluble and does not swell in organic solvents, including absolute alcohol, isopropanol, glycerin and fixed oils. Bentonite can be sterilized by maintaining the solid at 150°C for one hour after drying at 100°C. Aqueous suspensions can be sterilized by heating in an autoclave. It is hydroscopic and sorption of atmospheric moisture should be avoided. Bentonite is incompatible with strong

⌘⌘⌘

electrolytes, charged particles and solutions, sulfurated potash, and acriflavine hydrochloride.

Magnesium Aluminum Silicate

Magnesium aluminum silicate is obtained from the ores of the smectite clays montmorillonite and saponite. The ores are blended with water to form a slurry to remove impurities and separate out the purified colloidal clay fraction. The refined dispersion is drum dried to fine flakes or micronized to a powder. Magnesium aluminum silicate has a creamy off-white color and is odorless and tasteless. Magnesium Aluminum Silicate NF must meet the following specifications:

Test	NF
Arsenic	< 3ppm
Loss on drying	< 8.0%
Viscosity	to pass test
Microbial limits - total aerobic count	< 1000 per g
E. Coli	Absent
pH (5 in 100 suspension)	9.0 to 10.0
Acid demand	to pass test
Lead	< 0.0015%
Al/Mg content	to pass test

Magnesium aluminum silicate is insoluble in water or alcohol, but swells to many times its original volume to form colloidal dispersions. It is practically insoluble in organic solvents. Dispersions in water at the 1% to 2% level are thin colloidal dispersions. At 3% or above the dispersions are opaque, and as the concentration is increased above 3% the viscosity of the dispersions increases rapidly and exhibits thixotropic behavior. Magnesium aluminum silicate forms a gel at levels of 10% and above. Several different grades are available differentiated by Al:Mg ratio (montmorillonite: saponite ratio), particle size, dispersion viscosity, and degree of chemical activity.

Magnesium aluminum silicate is stable indefinitely when stored under dry conditions. Due to its inert nature it has few incompatibilities, however it is generally unsuitable for acidic drugs below pH 3.5. As with other clays, adsorption of active drugs may occur, resulting in low bioavailability if the drug is tightly bound or slowly desorbed. Typical pharmaceutical applications are:

⌘⌘⌘

Use	Concentration
Topical suspending agent	1 - 10%
Oral suspending agent	0.5 - 2.5%
Adsorbent	10 - 50%
Stabilizing agent	0.5 - 2.5%
Binding agent	2 - 10%
Disintegrating agent	2 - 10%
Emulsion stabilizer (topical)	2 - 5%
Emulsion stabilizer (oral)	1 - 5%
Viscosity modifier	2 - 10%

An example suspension formulation with magnesium aluminum silicate is as follows:

Trisulfapyrimidines Oral Suspension USP	
Magnesium aluminum silicate	1g
Syrup NF	90.6g
Sodium citrate	0.78g
Sulfadiazine	2.54g
Sulfamerazine	2.54g
Sulfamethazine	2.5%

Talc

Talc is a naturally occurring hydrous magnesium silicate which is pulverized and subjected to flotation processes to remove mineral impurities. The talc is then finely powdered and treated with dilute hydrochloric acid, washed with water and dried. It is very fine, white to grayish white, impalpable, and odorless. Talc is a crystalline powder, unctuous, and adheres readily to the skin. It is soft to the touch and free from grittiness. To be classified as Talc USP, the following specifications must be met:

Test	USP
Identification	To pass test
Loss on ignition	<6.5%
Acid soluble substances	< 10mg (2.0%)
Reaction and soluble substances	< 0.1%
Water soluble iron	To pass test
Microbial limit	< 500/g
Arsenic	< 3ppm
Heavy metals	< 0.004%
Lead	< 0.001%

⌘⌘⌘

Talc is insoluble in water, organic solvents, cold acids and dilute alkalis. It may contain bacteria, but it can be sterilized by dry heat at 160°C for not less than 1 hour, or by exposure to ethylene oxide. Depending on the source from which it is mined, talc displays different surface characteristics. The Montana talcs are hydrophobic in nature, whereas the California talcs are predominantly hydrophilic.

Talc has many varied applications in the pharmaceutical industry. Its medicinal use as a dusting powder depends on its desiccant and lubricant effects. Dusting Powder USP contains 2% cornstarch to further promote adsorption of moisture from the skin, and is used as a lubricant for rubber or latex gloves. Talc should not be used on surgical gloves because even the smallest amount deposited in an organ or healing wound may cause granuloma formation. When perfumed and in some cases medicated, it is used extensively for toilet purposes under the name talcum powder. In this type of application the talc should be in the form of an impalpable powder. A more coarse variety of talc is often used as a filtration aid, as in the in-pharmacy preparation of Aromatic Elixir USP. Talc is often used as a lubricant or glidant in tablet and capsule manufacture, and as a filler for these types of dosage forms. It has the disadvantage of retarding disintegration.

Precipitated Calcium Carbonate

Precipitate calcium carbonate is a fine white odorless, tasteless powder or crystal prepared by the double decomposition of calcium chloride and sodium carbonate in aqueous solution. Calcium carbonate exists in nature as the minerals aragonite, calcite, and vaterite, and is a dense, fine, white insoluble powder.

The primary uses of precipitated calcium carbonate in the pharmaceutical industry are as a fast acting antacid, and as a tablet filler. It is inexpensive, very white, nonhydroscopic, and inert. Calcium carbonate, like the calcium phosphates, can serve as a dietary source of calcium. Precipitated calcium carbonate must meet the pharmacopeial specifications shown below.

Precipitated calcium carbonate is insoluble in alcohol and its solubility in water is less than 1 in 50,000. It dissolves with effervescence in dilute acetic, hydrochloric, or nitric acids. It is stable in air, and should be kept in a well closed container. Precipitated calcium carbonate adsorbs less than 1% moisture at 25°C at relative humidities up to 90%.

Calcium carbonate is classified as a nonsystemic antacid in that it does not tend to cause a systemic alkalosis, a condition in which the alkalinity of body fluids and tissues becomes abnormally high. However, long term therapy with large doses taken with milk or other sources of phosphate will

⌘⌘⌘

cause renal pathology (milk-alkali syndrome) and some systemic alkalosis. For this reason, calcium and magnesium antacids are often alternated in therapy.

Test	USP
Identification	To pass test
Loss on drying	< 2.0%
Acid insoluble substances	< 0.2%
Fluoride	< 0.005%
Arsenic	< 3ppm
Barium	to pass test
Lead	0.001%
Heavy metals	< 0.003%
Magnesium and alkali metals	< 1.0%
Assay	98 - 100.5%

Calcium Sulfate

Calcium sulfate or calcium sulfate dihydrate is a white to yellowish powder which is odorless and tasteless. It is manufactured by carefully heating a pure variety of native gypsum until about three-fourths of the water has been expelled forming a powder containing about 95% calcium sulfate and 5% water. A purer variety of calcium sulfate may also be made by reacting calcium carbonate with sulfuric acid. Calcium sulfate, also known as terra alba, snow white, or mineral white, has the empirical formula of $CaSO_4$, and the dihydrate $CaSO_4 \cdot H_2O$. Classification as Calcium Sulfate NF requires that the following standards be met:

Test	NF
Identification	to pass test
Loss on drying	Anhydrous, <1.5%
	Dihydrate, 19.0 - 23.0%
Iron	<0.01%
Assay range	98.0 - 101.0%
Heavy metals	<0.001%

Calcium sulfate is chemically stable, but may adsorb moisture and cake. It must be stored in a well closed moisture resistant container. It is the least expensive tablet filler and can be used for a wide variety of acidic, neutral, or basic drugs. It has a high absorptive capacity for oils and has few incompatibilities. It the presence of moisture, the ionic calcium may be incompatible with amines, amino acids, peptides and proteins, which may form complexes. It must be used with caution in the formulation of moisture

⌘⌘⌘

sensitive drugs. An example of a formulation using calcium sulfate as the diluent is shown below.

Phenylpropanolamine HCl Tablets	
Phenylpropanolamine Hydrochloride	50mg
Calcium sulfate, dihydrate	175mg
10% PVP in ethanol	q.s.
Starch 1500	12mg
Magnesium stearate	6mg

Dibasic Calcium Phosphate

Dibasic Calcium Phosphate is a white odorless, tasteless powder or crystalline solid prepared from a phosphate mineral such as apatite. The apatite is dissolved in sulfuric acid and then filtered. The addition of calcium hydroxide precipitates dibasic calcium phosphate. The USP lists two types of calcium phosphate: dibasic calcium phosphate, anhydrous, which has the empirical formula of $CaHPO_4$, and dibasic calcium phosphate dihydrate which has the empirical formula $CaHPO_4 \cdot 2H_2O$. Synonyms for dibasic calcium phosphate include calcium hydrogen orthophosphate, dicalcium orthophosphate, and secondary calcium phosphate.

The primary use for dibasic calcium phosphate in pharmaceutical applications is as a tablet or capsule diluent. It is also used as an excipient for direct compression formulations of hard tablets requiring a good disintegrant and an effective lubricant. It has limited use as a mineral supplement for calcium replenishment. Diluents often comprise the bulk of the tablet formulation, with use levels between 40 to 90 % of the tablet weight. It has also found use as an adsorbent and as a thickening agent in creams and ointments. The pharmacopeial specifications for Dibasic Calcium Phosphate USP are as follows:

Test	Dihydrate USP
Identification	to pass test
Loss on ignition	24.5 to 26.5%
Acid insoluble substances	<0.20%
Carbonate	to pass test
Chloride	<0.25%
Floride	<0.005%
Sulfate	<0.5%
Arsenic	<3 pp.
Barium	to pass test
Heavy metals	0.003%
Assay	30.0 to 31.7% of Ca^{2+}

⌘⌘⌘

Dibasic calcium phosphate is only very slightly soluble in water. It is soluble in dilute hydrochloric acid and nitric acid, and slightly soluble in acetic acid. It is relatively stable, but should be stored in well closed containers in a cool dry place. It is incompatible with acids and interferes with the adsorption of tetracyclines. If used in tablet formulations containing inorganic acetate salts, the tablets are likely to develop an acetic odor on aging. An example of a typical tablet formulation containing dibasic calcium phosphate is as follows:

Diphenylhydramine Tablets

Diphenhydramine Hydrochloride	30mg
Calcium Phosphate, Dibasic	175mg
Starch 1500	25mg
10% PVP in 50% alcohol	q.s.
Magnesium Stearate	75mg
Microcrystalline cellulose	25mg

Tribasic Calcium Phosphate

Tribasic calcium phosphate occurs naturally as the minerals oxyapatite, voclicherite, and whitlockite. It is commercially prepared by treating phosphate-containing rock with sulfuric acid and then precipitated by the addition of calcium hydroxide. It is a white, odorless, tasteless powder which is stable in air. To be classified as Tribasic Calcium Phosphate NF the following standards must be met:

Test	NF
Identification	to pass test
Loss on ignition	<8.0%
Water soluble substances	<0.5%
Chloride	<0.14%
Carbonate	to pass test
Fluoride	<0.0075%
Nitrate	to pass test
Sulfate	<0.8%
Arsenic	<3 ppm
Barium	to pass test
Heavy metals	<0.003%
Dibasic salt and Ca oxide	to pass test
Acid insoluble substances	<0.2%
Assay range	34.0-40.0% Ca^{2+}

⌘⌘⌘

Tribasic calcium phosphate is soluble in dilute mineral acids, very slightly soluble in water, and insoluble in alcohol. It is used in the pharmaceutical industry primarily as a tablet and capsule diluent and as a flow and non-caking agent. Tribasic calcium phosphate provides a higher calcium load than dibasic calcium phosphate. It should not be used with strong acidic salts of weak organic bases or in the presence of acetate salts. It influences the adsorption of vitamin D, and should not be used with water soluble B vitamins or with certain esters such as vitamin E or vitamin A acetate or palmitate. It forms a calcium complex with tetracycline. A tablet formulation utilizing tricalcium phosphate as a filler is shown below.

Aminophylline Tablets

Aminophylline	100mg
Tricalcium Phosphate	50mg
Pregelatinized starch	15mg
Talc	30mg
Magnesium Stearate	10mg

Magnesium Carbonate

Magnesium carbonate is prepared from a suspension of dolomite which is saturated with CO_2 under pressure. On increasing temperature, $CaCO_3$ precipitates almost entirely, leaving a solution of magnesium bicarbonate. The filtered solution is heated to boiling, the magnesium bicarbonate loses CO_2 and H_2O, and magnesium carbonate precipitates. Magnesium carbonate is a light white friable mass or bulky white powder. It is odorless but will adsorb odor. The USP recognizes three types of magnesium carbonate: anhydrous magnesium carbonate, $MgCO_3$; the more common basic magnesium carbonate, a product that can vary between light magnesium carbonate $(MgCO_3)3{\cdot}Mg(OH)_2$ and magnesium carbonate hydroxide $(MgCO_3)_4{\cdot}Mg(OH_2){\cdot}5H_2O$; and normal magnesium carbonate, which is a hydrous magnesium carbonate with varying amounts of water, $MgCO_3{\cdot}Mg(OH)_2$. Magnesium Carbonate USP must meet the specifications given below.

Magnesium carbonate is practically insoluble in water and other solvents. Dilute acids dissolve it with effervescence. It is soluble in water containing CO_2 and insoluble in alcohol. It is stable in dry air and light, and should be stored in tightly closed containers. It is used by the pharmaceutical industry as a filtering aid for alkaline solutions, as a direct compression tableting excipient, and as an adsorbent in tablets of drug extracts which contain oily compounds. Magnesium carbonate also has therapeutic applications as an antacid. It is contraindicated as an antacid for those individuals whose stomachs cannot tolerate the evolution of CO_2.

⌘⌘⌘

Test	USP
Identification	to pass test
Soluble salts	< 1.0%
Acid insoluble substances	< 0.05%
Heavy metals	< 0.003%
Arsenic	< 4ppm
Iron	< 0.02%
Calcium	< 0.45%
Assay of MgO	40.0 - 43.5%
Microbial limit *(E. Coli)*	absent

Colloidal Silicon Dioxide

Silicon Dioxide occurs in nature as agate, amethyst, chalcedony, flint, quartz, sand, and tridymite. Silicon dioxide is obtained by acid precipitation from a sodium silicate solution to yield very fine particles. If silicon dioxide is obtained by the addition of sodium silicate to a mineral acid, the product is termed a silica gel. The pharmaceutical industry uses colloidal silicon dioxide, which is a submicroscopic fumed silica prepared by the vapor phase hydrolysis of a silica compound, such as silicon tetrachloride. Other names for colloidal silicon dioxide include light anhydrous silicic acid, silicic anhydride, and silicon dioxide fumed. Colloidal Silicon Dioxide NF must meet the following standards:

Test	NF
Identification	to pass test
pH (1 in 25 aqueous dispersion)	3.5 - 4.4
Loss on drying	<2.5%
Loss on ignition	<2.0%
Arsenic	<8ppm
Assay Limits (1000°C for 2 hours)	99.0 - 100.5%

Colloidal silicon dioxide is a submicroscopic, non-gritty amorphous powder. It is bluish-white in appearance and is odorless and tasteless. Colloidal silicon dioxide has a variety of applications in pharmaceutical formulations and technology. It is used as a drying agent for hygroscopic materials, and as an absorbent dispersing agent for liquids in powders or suppositories. In tableting processes, colloidal silicon dioxide is used as a glidant and an anti-adherent at levels ranging from 0.1% to 0.5% of the formulation. In gels and semi-solid preparations it is used as a thixotropic thickening agent and suspending agent with use levels ranging from 2% to 10%. The degree of viscosity increase depends on the polarity of the liquid, with polar liquids requiring a greater concentration than non-polar liquids. In

⌘⌘⌘

aerosols, colloidal silicon dioxide promotes particulate suspension, eliminates hard settling, and reduces clogging of the spray nozzle. The level of use for this type of preparation ranges from 0.5% to 2%.

Colloidal silicon dioxide is insoluble in purified water, but it does form a colloidal dispersion. It is soluble in hot solutions of alkali hydroxide, and in hydrofluoric acid. It is insoluble in other acids and in organic solvents.

Low dissolution rates of some drugs can be increased by their adsorption or deposition onto colloidal silicon dioxide. The dissolution rate of piroxicam deposited on silicon dioxide was found to be significantly increased in comparison to its micronized form (Vrecer, *et al*, 1992). However, the contact between the drug and carrier surface at the molecular level can be of great importance for the chemical stability of pharmaceutical preparations. Digoxin undergoes hydrolytic degradation to its molecular components when depositied on various grades of amorphous silicon dioxide. The extent of the degradation is dependent on the acidity, surface area, and pore size of the silicon dioxide used (Lau-Cam, *et al*, 1991).

Sodium Chloride

Sodium chloride occurs in nature as mineral halite, and is produced by mining, by evaporation of brine from underground salt deposits, and from sea water by solar evaporation. It is white, odorless and has a saline taste. the pharmacopeial specifications are as follows.

Test	USP
Identification	to pass test
Acidity or alkalinity	to pass test
Loss on drying	< 0.5%
Heavy metals	< 5 ppm
Arsenic	< 3 ppm
Barium	to pass test
Iodide and/or Bromide	to pass test
Calcium and Magnesium	, 0.005%
Iron	< 2ppm
Sulfate	< 0.015%
Assay range	99 - 101%

Solutions of sodium chloride more closely approximate the composition of the extracellular fluid of the body than solutions of any other single salt. A 0.9% solution has the same osmotic pressure as body fluids, and is thus isotonic with body fluids. An isotonic solution can be injected without affecting the osmotic pressure of the body fluids and without causing any appreciable distortion in chemical composition. An isotonic solution is the choice as a vehicle for many drugs which have to be administered parenterally. Sodium chloride is stable, and should be stored in well closed containers. Solutions are sterilized by autoclaving or by bacterial filtration. Sodium chloride reacts to form precipitates with silver, lead, and mercury salts. Strong oxidizing agents liberate chlorine from acidified solutions.

The primary applications in the field of pharmaceutics include:

Use	Concentration
Means to obtain isotonicity	up to 0.9%
Tablet diluent and adsorbent (direct compression)	10 - 80%
Water soluble tablet lubricant	5%
Capsule diluent	10 - 80%
Controlled flocculation of suspensions	up to 1%
Control of micellar properties of surfactants	up to 5%
Control of viscosity in detergent systems (shampoos)	up to 5%

Sodium Chloride Injection USP is a sterile isotonic solution of sodium chloride in Water for Injection USP. It contains 154 mEq of sodium and chloride ions per each liter, and may be used as a sterile vehicle in preparing solutions or suspensions of drugs for parenteral administration.

REFERENCES

Alexander, K.S., Azizi, J., Dollimore, D., and Patel, F.A., 1989, "An Interpretation of the Sedimentation Behavior of Pharmaceutical Kaolin and other Kaolin Preparations in Aqueous Environments", *Drug Development and Industrial Pharmacy*, **15**(14016), pp 2559-2582.

Blaug, S.M., and Gross, M.R., 1965, "In Vitro adsorption of Some Anticholinergic Drugs by Various Antacids", *Journal of Pharmaceutical Sciences*, **54**(2), pp 289-294.

Browne, J.E., Feldkamp, J.R., White, J.E., and Hem, S.L., 1980, "Characterization and Adsorptive Properties of Pharmaceutical Grade Clays", *Journal of Pharmaceutical Sciences,* **69**(7), pp 816-823.

Rhone-Poulenc, 1992, "Calcium Phosphate Pharmaceutical Ingredients", *Rhone-Poulenc Basic Chemicals Co. Technical Bulletin.*

Camazano, M.S., Sanchez, M.J., Vicente, M.T., and Dominguez-Gil, A., 1980, "Adsorption of Chlorpheniramine Maleate by Montmorillonite", *International Journal of Pharmaceutics*, **6**, pp 243-251.

Chowhan, Z.T., 1994, "Harmonization of Excipient Standards and test Methods: A Progress Update on Development of Test Methods and Standards Part II", *Pharmaceutical Technology,* **18**(12), pp 22-35.

⌘⌘⌘

Dawoodbhai, S.S., Chueh, H., Rhodes, C.T., 1987, "Glydants and Lubricant Properties of Several Types of Talcs", *Drug Development and Industrial Pharmacy*, **13**(13), pp 2441-2467.
Farley, C.A., 1976, "Suspending Agents for Extemporaneous Dispensing: Evaluation of Alternatives to Tragacanth", *The Pharmaceutical Journal,* June 26, pp 562-566.
Gennaro, A.R., 1990, *Remington's Pharmaceutical Sciences*, 18th Ed., Mach Publishing Co., Easton, PA, p 779

Ibid., p 787

Ibid., p 1538

Ibid., p 1540

Ibid., p 796

"Global 500", *Fortune*, August 7, 1995, F-26

Ibid., F-39

Granberg, C.B., and Benton, B.E., 1949, "The Use of Dried Bentonite as a Disintegrating Agent in Compressed Tablets of Thyroid", *Journal of the American Pharmaceutical Association, Scientific Edition*, 38, pp 648-651
Hiestand, E.N., 1964, "Theory of Course Suspension Formulation", *Journal of Pharmaceutical Sciences*, **53**(1), pp 1-18.
Hosny, E.A., Kassem, A., and El-Shattawy, H.H., 1988, "Availability of Anhydrous Ampicillin and Sulphadimidine from their Suspensions", *Drug Development and Industrial Pharmacy*, **14**(6), pp 779-789.
Iwuagwu, M.A., and Aloko, K.S., 1992, "Adsorption of Paracetamol and Chloroquine Phosphate by Some Antacids", *Journal of Pharmaceutical and Pharmacological Science*, **44**, pp 655-658.
Khalil, S.A.H., and El-Masry, S., 1974, "Adsorption of Atropine and Hyoscine on Magnesium Trisilicate", J*ournal of Pharmaceutics and Pharmacology*, **26**, pp 243-248.
Khalil, S.A.H., Mortada, L.M., Shams-Eldeen,M.A., and El-Khawas,M.M., 1987, "The in vitro Uptake of a Low Dose Drur (Riboflavine) by Some Adsorbents", *Drug Development and Industrial Pharmacy*, **13**(3), pp 547-563.
Khan, K. A., And Rhodes, C. T., 1975, "Disintegration Properties of Calcium Phosphate Dibasic Dihydrate Tablets", *Journal of Pharmaceutical Sciences*, **64**, pp 166-168.
Kukita,T., Yamaguchi, A. Okamoto, A., and Nemoto, M., 1992, "Interaction Between Polyethylene Films and Bromhexine Hydrochloride in Solid Dosage Form. IV. Prevention of the Sorption by Addition of Magnesium Aluminum Silicate", *Chemical Pharmaceutical Bulletin,* **40**(5), pp 1257-1260.
Lau-Cam, C.A., Pendharkar, C.M., Jarowski, C.I., and Yang, K.Y., 1991, "Instability of Digoxin-Amorphous Silicon Dioxide Triturates Prepared by Solvent Deposition and Ball Milling", *Pharmazie,* **46**, pp 522-525.
Lin, K., and Peck, G.E., 1994, "Characterization of Talc Samples from Different Sources", *Drug Development and Industrial Pharmacy*, **20**(19), pp 2993-3003.
Mak, R.H.K., 1985, "Aluminum Hydroxide and Calcium Carbonate as Phosphate Binders in Chronic Renal Failure", *British Medical Journal*, **291**, p 623.

⌘⌘⌘

McGinity, J.W., and Harris, M.R., 1980, "Influence of a Montmorillonite Clay on the Properties of Griseofulvin Tablets", *Drug Development and Industrial Pharmacy*, **6**(1), pp 49-59.

McGinity, J.W., and Lach, J.L., 1976, "In Vitro Adsorption of Various Pharmaceuticals to Montmorillonite", *Journal of Pharmaceutical Sciences*, **65**(6), pp 896-902.

McGinity, J.W., and Lach, J.L., 1977, "Sustained-Release Applications of Montmorillonite Interaction with Amphetamine Sulfate", *Journal of Pharmaceutical* Sciences, **66**, pp 63-66.

Naggar, V.F., 1981, "An In Vitro Study of the Interaction Between Diazepam and Some Antacids or Excipients", *Pharmazie,* **36** (2), pp 114-117.

Nishada, M., Yoshimura, Y., Kawada, J., Ookubo, A., Kagawa, T., Ikawa, A., Hashimura, Y., Suzuki, T., 1990, *Biochemistry International*, **22**, pp 913-920.

Ookubo, A., Ooi, K., and Hayashi, H., 1992, "Hydrotalcites as Potential Adsorbents of Intestinal Phosphate", *Journal of Pharmaceutical Sciences,* **81**(11), November, pp 1139-1140.

Phadke, D.S., Keeney, M.P. And Norris, D.A., 1994, "Evaluation of Batch-To-Batch and Manufacturer-To-Manufacturer Variability in the Physical Properties of Talc and Stearic Acid", *Drug Development and Industrial Pharmacy*, **20**(5), pp 859-871

Ratafia, M., 1989, "Big Business in Osteoporosis Products", *Medical Marketing and Media,* November, pp 96-110.

Russell, A., 1988, "Minerals in Pharmaceuticals", *Industrial Minerals,* August, pp 32-43.

Said, S.A., Shibl, A.M., and Abdullah., M.E., 1980, "Influence of Various Agents on Adsorption Capacity of Kaolin for Pseudomonas aerugiosa Toxin", J*ournal of Pharmaceutical Sciences,* **69**(10), October, pp 1238-1239.

Schott, H., 1976, "Controlled Flocculation of Coarse Suspensions by Colloidally Dispersed Solids I: Interaction of Bismuth Subnitrate with Bentonite", *Journal of Pharmaceutical Sciences,* **65**(6), pp 855-861.

Tuncel, T., and Bergisadi, N., 1992, "In Vitro Adsorption of Ciprofloxacin Hydrochloride on Various Antacids", *Pharmazie*, **47**, pp 304-305.

Vrecer, F., Kristl, J., Pecar, S., Rotar, A., 1992, "Study of the Physical State of Piroxicam Deposited on an SiO_2 Surface"., *6th International Conference on Pharmaceutical Technology*, June, pp 398-407.

Wai, K., DeKay, H.G., and Banker, G.S., 1966, "Applications of the Montmorillonites in Tablet Making", *Journal of Pharmaceutical Sciences*, **55**(11), pp 1244-1248.

Ward, J.B., and Trachtenberg, A., 1962, "Evaluation of Tablet Disintegrants:, *Drug and Cosmetics Industry*, 91, p 35

BIBLIOGRAPHY

Ansel, Popovich, and Allen., *Pharmaceutical Dosage Forms and Drug Delivery Systems, Sixth Edition*, 1995, Williams and Wilkins, Malvern, PA

Handbook of Pharmaceutical Excipients, 1994, American Pharmaceutical Association, Washington, DC

⌘⌘⌘

Lieberman, Lachman, and Schwartz., *Pharmaceutical Dosage Forms: Tablets, Volume 1, 2nd Edition*, 1990, Marcel Dekker, New York

Martin, Swarbrick, and Cammarata., *Physical Pharmacy, 4th Edition*, 1994, Lea & Febiger, Philadelphia, PA

The Merck Index, 11th edition, 1989, Merck & Co., Inc., Rahway, NJ

Physicians Desk Reference, 47th edition, 1993, Medical Economics Company Inc., Montvale, NJ

Remington's Pharmaceutical Sciences, 18th Edition, 1990, Alfonso R Gennaro, Mack Publishing Co., Easton, PA

The United States Pharmacopeia XXIII, The National Formulary XVIII, 1995, The United States Pharmacopeial Convention Inc., Rockville, MD

⌘⌘⌘

NOTES

⌘⌘⌘

TEN

AGRICULTURAL PESTICIDES

Bruce M. McKay, Ph.D.
CITE
Pennington, NJ

DEFINING AND REGULATING PESTICIDE PRODUCTS

United States – FIFRA

Pests – Congress delegates authority to regulate pesticide products in the United States to the Environmental Protection Agency (EPA) under FIFRA (Federal Insecticide, Fungicide, and Rodenticide Act). Any product, including its individual components, offered for sale to the public and claiming to control a pest must be approved and registered with EPA (Environmental Protection Agency, 1988). FIFRA generally defines a pest as any insect, rodent, nematode, fungus, weed, or any other form of terrestrial or aquatic plant or animal life, or virus, bacteria, and micro-organism. Excluded from this definition and scope of regulation are micro-organisms which live in or on man, and new animal drugs.

Inert Ingredients – Formulated pesticide products (end-use products) have two essential components – active and "inert" ingredients. An active ingredient is intended to destroy or mitigate the effect of some pest. Everything else in a formulated product is defined as inert. This definition does not mean that inert ingredients will always have no effect on pests, man, or the environment. Rather, the definition is designed to categorize materials based on their intended effect against some pest, and to establish the overall approach EPA will use to regulate a particular material.

Although inert ingredients are not regulated as pesticidal active ingredients they are nonetheless subject to extensive regulatory scrutiny before being approved in use in pesticide products. The approval process results in a list analogous to the FDA's GRAS (Generally Recognized As Safe) list for excipients (inert ingredients) used in food, drug, and cosmetic products. In the case of pesticide formulation inerts this list is specified in the Code of Federal Regulations (CFR), Title 40 (Protection of the

Environment), parts 180.1001(c), (d), and (e). Any component of a pesticide end-use product must appear in this list, and, additionally, its use in a particular product must be reported to, and approved by, EPA before a registration for production and sale of the product is granted by EPA.

In the three subsections of 180.1001 the EPA lists materials which are "exempt from the requirement of a tolerance" when used on growing crops, raw agricultural commodities after harvest, or animals. Exemption from tolerance means the material can be applied to a target site without the need to determine what residues remain in or on the edible portion of a crop or animal. All active ingredients have a strictly specified residue tolerance. Without the exemption from tolerance the cost for obtaining approval to use an inert ingredient would be in the tens of millions of dollars. As it is, under current EPA guidelines the cost to obtain an exemption from tolerance can require environmental and human health testing which totals one-quarter to one-half million dollars.

U.S. EPA's Inerts Lists – In addition to the primary listing at 40 CFR part 180.1001 the U.S. EPA has initiated a process for further classifying pesticide formulation inert ingredients according to their mammalian and environmental hazard potential. EPA has begun the process of placing each inert ingredient onto one of four inerts lists.

List 1 ingredients are considered highly hazardous and generally can no longer be used in formulation of pesticidal active ingredients. Most List 1 items were used as formulation solvents.

List 2 ingredients are in a sort of limbo. Placement on this list means EPA has some information available which suggests the potential for a hazard, but the information has not been fully evaluated. Over the course of time each of the items on List 2 will be reviewed and moved to List 1 or to List 4.

List 3 has been defined, but no one outside EPA is exactly certain what materials constitute the list. EPA's definition is that List 3 contains everything not found on one of the other Lists. In practice this means there is no information EPA has found to suggest a hazard potential, but properties of the individual materials have not been reviewed in enough detail to allow them to be declared as posing no hazard.

List 4 ingredients have been reviewed by EPA and found to pose no substantial hazard to mammals or the environment. They are presumptively safe and can be used by pesticide formulation chemists. This does not mean, however, they never undergo further scrutiny. Every pesticide product containing one, or more, of the ingredients on this list and ingredients from Lists 2 and 3 is subjected to a battery of EPA mandated toxicological and environmental effects tests before a company submits the product to EPA

⌘⌘⌘

for registration approval review. Being on List 4 only assures the formulator that there are no known hazard factors associated with the ingredient on its own.

Most of the mineral products reviewed here are on EPA's List 4; some are on List 3, with the expectation they will be moved to List 4 after their review is completed by EPA.

Europe – EC Guidelines

Regulation of pesticidal formulation ingredients in Europe is in a state of transition. With the advent of the European Community, regulation of pesticidal products is moving from country-by-country standards to a set of harmonized guidelines employed by every member of the EC. Completing the transition is expected to take anywhere from five to ten years. Under the EC guidelines there will be two categories of pesticide formulation additives: adjuvants and coformulants.

Adjuvants are defined as those materials which are not pesticidally active and are combined with a pesticidal product immediately before application. Some typical types of adjuvants would be surfactants added to an aqueous spray to improve wetting of plant foliage, or buffers added to an aqueous spray to change the pH of the spray water so as not to chemically degrade an active ingredient.

Coformulants are defined as any material other than the active ingredient which is added to a pesticidal active ingredient prior to packaging for sale to an end-user. This category includes all the items commonly employed by formulations chemists – solvents, solid diluents and carriers, surfactants, dyes, and aerosol propellants. Minerals in agricultural chemical applications will invariably be used as coformulants and not as adjuvants.

Harmonization Of Regulations – At the present time there is very little data generated for pesticide registration applications which can be used both in Europe and in the United States. Both sides are, however, discussing how their respective testing protocols might be revised so as to avoid duplication of testing by registrants and duplication of reviews by regulatory authorities.

With respect to pesticide formulation inerts and coformulants, as matters now stand a supplier must obtain separate approvals for use in Europe and the United States. Because formulation inerts and coformulants must receive separate approvals for use in Europe or the United States, the formulation chemist working in a multi-national company must keep two lists of acceptable ingredients – one for Europe and one for the United States.

⌘⌘⌘

Why Formulate Pesticidal Active Ingredients

Pesticidal active ingredients as manufactured are, with very few exceptions, not suitable for use in agricultural applications. The reason is quite simple – the physical form of the active ingredient obtained from the chemical manufacturing process.

With the advent of synthesized organic pesticidal active ingredients in about 1940 the application rate of active ingredients has continually been decreasing. From about 1940 to 1970 the average application rate was typically one pound of active ingredient per acre. Since 1970 the discovery of more effective compounds has resulted in application rates measured in hundredths of a pound (a few grams) per acre. A quick calculation shows why the typical active ingredient is unsuitable for direct application in undiluted form.

One acre of flat, bare ground has a surface area of about 44,000 square feet. Uniformly covering this area with one pound of material means the active ingredient must have a distribution of 0.0000227 pounds per square foot, or just about one one-hundredth of a gram per square foot. Such a distribution is equivalent to covering the walls, floor, and ceiling of an 8 foot x 8 foot x 8 foot room with a teaspoon of paint pigment. In many agricultural situations the distribution problem is increased because the acre of ground is not bare but is covered with a crop to be protected. The surface area to be treated, then, increases by factors of tens to hundreds.

The physical form of typical pesticidal active ingredients ranges from viscous liquid to tacky solid. Consequently, there is no efficient or cost effective method to uniformly apply the very small amounts of active ingredient required for pest control without doing something to the active ingredient. The answer to the problem for pesticidal active ingredients is the same as for paint pigments – dilute the active material with an inexpensive substance to achieve a volume which can be easily and uniformly applied to the desired target.

The dilution of pesticidal active ingredients necessarily involves at least one step, and will often involve two. The one-step process is, again, analogous to paint manufacture. Dilute the active material with something else and then apply the dilution to the desired target. For pesticidal materials this dilution is called the end-use product; it is the physical form in which a user such as a farmer purchases an active ingredient. The physical form for pesticides can be solid or liquid.

In the two-step dilution the end-use product is further diluted by the applicator. Almost universally this means the end-use product will be diluted with a large volume of water and then sprayed onto the target area to achieve uniform distribution and coverage. The solid forms of pesticide

⌘⌘⌘

formulations we will discuss below can be designed such that they can be applied as-is or after further dilution with water.

MINERAL USES

There are two primary reasons for using any mineral in a pesticide product formulation, and a limited number of subsidiary reasons. The two overwhelming uses of minerals are as a carrier or as a solid diluent. Regardless of the means by which the end-use product is applied to a target site by a grower or commercial applicator, products utilizing minerals as carrier or solid diluent will be sold in a dry form.

Carriers

Liquid Active Ingredients – Offering a liquid pesticidally active ingredient to the end user in a dry form requires that the liquid be absorbed or adsorbed by a dry carrier. In order to be efficient the carrier should have a relatively high liquid holding capacity. Manufacturing and transportation costs can be a significant portion of final sale price, so a difference of a few percent in active ingredient concentration of the end-use product can determine success or failure in the market.

Solid Active Ingredients – Some pesticide applications require a particular physical form, generally to conform with application equipment available to the grower. Granules, for example, are commonly used as the means to apply insecticides to corn. A solid active ingredient can be dissolved in a solvent and then absorbed by a carrier. This is, however, often not economical. The alternative used by the pesticide industry is to use a mineral as a substrate onto which the solid pesticide is adhered. High absorbency here is not required, and in fact can be detrimental.

Solid Diluents

Economic, environmental, and human health concerns require application of the smallest possible quantity of pesticide to target sites. Within the past two decades these concerns have led to the development of active ingredients which are applied at use rates of a few grams per hectare (1 hectare = 2.47 acre). While technology at the grower level has also increased, a working farm is not a laboratory. In particular farmers do not have pipettes and electronic balances readily available to them.

Rather than rely on the grower to accurately measure a few grams of dry end-use product, the end-use product is diluted by the manufacturer to allow convenient and accurate dose measurement. Already dry products are further

⌘⌘⌘

diluted at the manufacturing level with a mineral so that the quantity of end-use product to be dispensed by the grower is easily and accurately measured. Requirements for the mineral to be used are quite flexible, usually no more than that it be relatively dense (to reduce packaging cost) and not be chemically reactive toward the active ingredient.

Minor Uses

Flow Aid – Aesthetic and practical concerns require that dry products be free flowing and non-clumping. In many instances the selection of a carrier or diluent will provide the answer. When this does not occur a material will be added specifically to improve flowability and reduce clumping. While a mineral ordinarily used as a carrier or diluent will sometimes suffice, more usually the formulator must resort to a mineral which has been chemically derivatized. Minerals with hydrophobic surface treatments are particularly popular.

Thickening – Flowable concentrates are pesticide vehicles in which a solid active ingredient, or a liquid absorbed on a solid carrier, is suspended in a liquid, typically aqueous, medium. The density difference between the solid and liquid phases is invariably large enough to result in rapid settling and bottom caking of the suspended solids. As in many other fields, the answer to maintaining a homogeneous suspension is to compensate for the density difference by increasing the viscosity of the liquid vehicle with a thickener. Derivatized minerals, often in combination with a polysaccharide gum, are most commonly used by pesticide formulators.

PESTICIDE FORMULATION TYPES

As suggested by the discussion above, the typical pesticide formulator is not going to be an expert on minerals. Selection of a mineral for a pesticide formulation is fairly simple – can one be found that is on the EPA's list of exempt inerts, that has adequate absorbency, that has an appropriate density, that gives a free flowing end-use product, and is chemically compatible with the active ingredient? Any mineral that fits can be used. If there is no fit the pesticide formulator has the option to develop an entirely different formulation.

Therefore, pesticide formulations are most usefully categorized by formulation type rather than by mineral class. In Table 1 minerals are placed according to the pesticide formulation type and use with which they are most commonly associated. Keep in mind, however, that this scheme represents general use patterns. The variety of pesticide product end-use

⌘⌘⌘

patterns is diverse enough that almost every mineral has been used for almost every purpose at some time.

Table 1 - Minerals And Their Uses

Formulation	Mineral Use	Typical Minerals
dust	carrier	diatomite; kaolin
	diluent	calcium carbonate; diatomite; kaolin; pyrophyllite; talc
flowable	thickener	attapulgite; bentonite; montmorillonite
granule	carrier	attapulgite; gypsum; montmorillonite; quartz
water dispersible granule	carrier	attapulgite; montmorillonite
	diluent	kaolin
wettable powder	carrier	attapulgite; diatomite
	diluent	attapulgite; kaolin

Generalized compositions for each of the five formulation types are presented in Table 2.

Dusts

A dust is characterized by its small particle size and that it is provided to the applicator in ready-to-use form. The user adds nothing to the dust before application.

As the name dust implies, the average particle size is small enough to allow wide area dispersal with minimum energy input by the applicator. Dust product particles will range from 1 to 40 microns, with a 1 to 10 micron range being typical.

Ready-to-use means the applicator need do nothing more than spread the product where it is required. As a consequence there are only two necessary ingredients in a dust formulation – active ingredient and carrier or solid diluent.

Because dusts are generally intended to cover a large target area the concentration of active ingredient will be low, usually between 0.1 and 5% of the total product on a weight/weight basis.

While aerial crop dusting was once a typical method of insect and plant disease control, environmental concerns and advances in synthetic pesticide development have relegated this method of application to the history books. Dusts are now employed for restricted area treatments such as carpets, pets, and crop seeds in storage.

⌘⌘⌘

The low active ingredient concentration in a dust means that absorbency is often not a concern in mineral selection. Formulation criteria will be a small particle size for the added mineral, high bulk density to reduce package size, and chemical compatibility with the active ingredient. This last is usually the major factor affecting choice of a mineral. The very high ratio of carrier or diluent to active ingredient and the large surfaces areas present can result in chemical decomposition of active ingredient if the mineral has any catalytic surface sites.

Table 2. Generalized Pesticide Formulations

Dust Components	% w/w
active ingredient	0.1 - 5
carrier/solid diluent	95 - 99.9
Flowable Components	
active ingredient	20 - 50
wetting agent	0.25 - 1
dispersing agent	2 - 6
thickening mineral	1 - 3
water	qs 100
Granule Components	
active ingredient	1 - 25
carrier	75 - 99
Water Dispersible Granule Components	
active ingredient	50 - 90
wetting agent	0.5 - 1
dispersing agent	3 - 6
solid diluent	qs 100
Wettable Powder Components	
active ingredient	10 - 90
wetting agent	0.5 - 1
dispersing agent	3 - 6
carrier/solid diluent	qs 100

Flowables

This is the only liquid pesticide formulation type in which minerals play a significant part. The quantity of mineral used is small, but the physical

⌘⌘⌘

characteristics imparted by the mineral are absolutely necessary for commercial viability of the formulation.

A flowable is a concentrated formulation, intended to be diluted immediately before use. One part concentrate will be mixed with from ten to several hundred parts water and the dilution sprayed onto soil or foliage.

Flowables utilize solid active ingredients, or liquid actives absorbed onto a solid carrier, but are presented to the end user as a liquid. Micronized particles of active ingredient are suspended in a liquid carrier. Water is almost invariably the carrier of choice, since it is much less expensive than any organic liquid. As noted above the problem to be overcome with this formulation is the density difference between suspended solids and the fluid medium. The specific gravity of pesticide solids is typically on the order of 1.2, resulting in rapid settling of solids in water.

Since a concentrate might remain in storage for up to two years after production there must be some means to maintain solids in suspension. At the same time the viscosity of the concentrate should not be extremely high, as the concentrate must be diluted with water prior to application at the target site. High viscosity also results in substantial residues of concentrate in its container (usually a 2.5 gallon jug). Residues mean the grower is losing product for which he paid, and can result in environmental contamination when the container is discarded.

The solution is to prepare a concentrate which has viscosity characteristics similar to those of liquid antacids. The viscosity of a flowable concentrate will generally be in the range 2500-3500 cps. Either mineral thickeners alone or minerals in combination with polysaccharide gums can be employed as the thickening agent. While modified bentonites are used on occasion, attapulgites are most often the mineral of choice as the sole thickener (Sawyer, 1987). Combination thickeners are based on seaweed derived xanthan gum (a common food thickening additive) and montmorillonite (R.T. Vanderbilt).

Because pesticide active ingredients are hydrophobic and they are to be mixed with water in preparing the flowable and diluting for spray application, surface active agents must be incorporated into the formulation. Two types are commonly employed – a wetting agent to allow initial mixing with water, and a dispersing agent to overcome the tendency for hydrophobic solids to agglomerate in water.

Granules

Granules are also presented to the user as a ready-to-use product. Granules need only to be transferred to a mechanical applicator and then spread over soil. As with dusts, there are only two necessary components – active ingredient and carrier. Granules can be prepared either by impregnating the carrier with a liquid active ingredient, or by coating a solid active on the surface of the granule. Impregnated granules are the most common variety for pesticide formulations.

Again because of economic considerations, high liquid holding capacity is desirable in the carrier. The minerals of choice combining high absorptivity, low cost, and availability are attapulgite and montmorillonite.

The agricultural industry utilizes only a small subset of the possible size ranges for granular carriers. The overwhelming choice is a 24/48 (-24, +48) mesh size range. Granules in this size range are large enough to not be easily dispersed by wind during application, but are small enough to have a large number of particles per unit weight (up to 8 million particles per pound). Particle count is required in order to ensure that a soil insect or an emerging weed has a high probability of encountering pesticide.

Water Dispersible Granules and Wettable Powders

These formulations differ in the wetting and dispersing agents employed, but are essentially identical with respect to minerals usage. Each formulation is a concentrate designed to be diluted with water and sprayed onto soil or foliage.

The difference between the two is physical form. A wettable powder has the physical characteristics of a dust; a water dispersible granule can be thought of as a wettable powder which has been agglomerated. The agglomeration step is intended to reduce airborne dust contact during handling operations by human mixers preparing dilutions for spray application. Once diluted in water the water dispersible granule will, as the name implies, deagglomerate to give a suspension containing dispersed particles with the size range of a wettable powder, 1 to 40 microns.

Concentrates are prepared from either liquid or solid active ingredients. If liquid, the active ingredient must be impregnated onto a solid carrier. If solid, the active ingredient will be mixed with a solid diluent for convenience in processing and to dilute active ingredient to the desired end-use product concentration.

Active ingredient concentration in a product containing a liquid active will range from about 10% w/w to about 50% w/w. In order to have a freely flowing product the active ingredient concentration will be no more than about 25% if the carrier is a mineral. Above this concentration synthetic silica carriers will be utilized.

Active ingredient concentration in a product containing a solid active ingredient will range from about 10% w/w to about 90% w/w. The maximum concentration achievable is dictated by physical properties of the active ingredient. Product must be ground to reach the desired size range of 1 to 40 microns. High melting, crystalline, nontacky solids require only a small quantity of diluent in the grinding operation. As melting point and crystallinity decrease and tackiness increases more diluent must be added to allow efficient grinding.

Although the same carriers and solid diluents can in principle be used in either formulation, in practice the choices are reduced for water dispersible granules. The agglomeration step in water dispersible granule formulation adds enough cost that this formulation is almost exclusively reserved for products which contain at least 75% and preferably 90% w/w active ingredient. Thus, water dispersible granules are most usually prepared from high melting, crystalline active ingredients. Very absorptive solid diluents are not required. Kaolin is very often found as the diluent in water dispersible granules. Attapulgite is used occasionally.

Wettable powder formulations can utilize a variety of mineral carriers and solid diluents, either singly or in combination. When the active ingredient is a liquid the formulator will want a highly sorptive carrier. Attapulgite is the primary choice when the highest possible concentration is to be attained, with montmorillonite sometimes receiving consideration.

When maximum loading of the carrier is not required a combination of carrier and solid diluent may be employed. Often in these situations a first concentrate of liquid active on a sorptive carrier such as attapulgite will be prepared. The concentrate is then diluted with a solid diluent such as kaolin or diatomite. A solid diluent other than the carrier will be used to impart a particular physical property such as free flow, or will be used to reduce the possibility of chemical decomposition of active ingredient during storage.

PROCESSING TECHNIQUES

Three common unit operations suffice in describing the essential features for production of any of the formulation types discussed. These operations are – mixing/blending, liquid impregnation, and size reduction. Water dispersible granules require one additional operation – agglomeration.

Mixing/Blending

Solid Active Ingredient – Active, solid diluent, and wetting/dispersing agents, if used, must be homogeneously mixed. An effective and simple device is the ribbon blender. All ingredients in the final product are added to the blender, mixed, then delivered to the next operation. Order of addition is rarely important. Blender size and configuration are determined only by the batch size and preference of the formulator.

In the case of a simple dust the blending step will be the only operation required before packaging the end-use product. Some dusts, almost all

⌘⌘⌘

wettable powders, and all water dispersible granules will require size reduction, so this operation is an intermediate production step.

Liquid Active Ingredient – The first production step will be impregnation of active onto the carrier. Once this operation is complete the impregnated carrier can be treated as if it is a solid active ingredient.

Liquid Impregnation

Production of any of the formulations which contain a liquid active ingredient will involve a liquid impregnation operation. For impregnated granules this is the only major step; for the other formulations it is the first production step.

Impregnation for dust, water dispersible granule, and wettable powder formulations is accomplished using the same ribbon blender as for mixing, with the addition of a spray bar to the blender.

The blender may on occasion need to be heated. Viscous liquids can be inefficiently or incompletely absorbed at ambient temperatures. Some low melting solid technicals can be impregnated by heating them above their melting point. If heating is involved the temperatures will never be in excess of about 80°C, and preferably not in excess of about 50°C. This restriction is to reduce the possibility of thermal decomposition of active ingredient.

Order of addition now becomes important. If wetting/dispersing agents or other components are required they will be added after the impregnation step so as not to become coated with hydrophobic technical. Only carrier will be present in the blender until impregnation is completed.

Spray bars will be mounted on the long axis of the blender. The spray nozzles should deliver as fine a spray as possible consistent with maintaining an adequate production rate. The charge of carrier should be such as to minimize the area of ribbons uncovered during spraying, while allowing good agitation of the carrier bed.

Operating conditions which maximize throughput and uniformity of impregnation are determined by chemically analyzing impregnated carrier to determine percentage concentration of active ingredient.

Granule carrier impregnation can be accomplished in a ribbon blender, although physical attrition of the carrier is often unacceptably severe. Most formulation facilities employ a rotary blender such as the Munson Glass Batch Mixer®. Rotary blenders, analogous to closed horizontal cement mixers, contain no internal moving parts and mix by tumbling the batch as the shell rotates. As the batch tumbles a new face is continually exposed to the spray nozzle pattern. Physical attrition of granules is kept to a minimum.

⌘⌘⌘

Size Reduction
Particle size reduction is an essential part of formulation production when utilizing solid active ingredients. The only exceptions are for granules and a very few dusts. When the active ingredient is a liquid the formulator can select a carrier grade which has the desired size range. Impregnation properly carried out will yield a product with a size distribution the same as that of the carrier.

The particle size distribution of granules is fixed when the granule carrier size is selected. As we have mentioned above, carrier size is based on maximizing the number of individual granules applied per unit area, consistent with other considerations such as ease of application and resistance to dispersal by wind.

Dusts which do not require a size reduction operation are those containing a solid active ingredient which is already available in the required size range.

Biological efficacy of an active ingredient applied to soil or to foliage is often directly correlated with the density of the applied deposit; the higher the density the higher the efficacy (Hartley, 1980). Solid active ingredients are rarely manufactured to have the 1-20 micron particle size range required for optimal biological efficacy.

Size reduction of solid pesticide products can be accomplished with either mechanical energy or fluid energy mills (Perry, 1973). The choice depends on the physical behavior of the product, on the fineness of grind which must be achieved, and the concentration of active ingredient in the end-use product.

Mechanical energy mills such as hammer mills and ring-roller mills will be appropriate when the solid active ingredient is high melting, crystalline, and nontacky. High melting in this instance means a melting point above about 100°C. Below this temperature, or if the active ingredient is at all tacky, smearing and caking of the solid results in unacceptably high retention of active on mill parts, and in reduced particle size reduction efficiency.

Even the best behaved solid active ingredients are rarely milled without addition of at least some solid diluent as a conditioning and milling aid. At the least, solid diluent will be added to give better flow properties to the active ingredient allowing most efficient feeding from storage containers to the mill. An absorptive diluent can also assist by reducing smearing of technical inside the milling chamber. Addition of 3-10% w/w diluent to active ingredient feedstock is a common practice even with higher melting active ingredients.

When active ingredient is relatively low melting or particularly tacky, and a device such as a hammer mill is being used, the proportion of solid diluent to active ingredient may be increased to as high as 1:1. Should even this proportion not give acceptable results, mechanical energy milling is probably the wrong choice for particle size reduction.

When solid active ingredient is low melting or tacky, or when hammer milling does not produce a fine enough grind, fluid energy mills are utilized. The most common mill of this type in pesticide formulation facilities is the centrifugal jet air mill. High air flow through the mill reduces smearing and caking from melting or softening of active ingredient. Particle size range or average particle size of milled solids will be smaller than that obtained from the mechanical energy mill. Solid fungicidal active ingredients applied to foliage are known to have biological efficacy particularly sensitive to particle size, and so they are very commonly ground in air mills so as to obtain the finest possible grind.

Flowables, being liquid, must be ground using different equipment. All flowable production in the United States at one time utilized horizontal batch ball mills. The pesticide industry has slowly converted operations so that now both the United States and Europe employ stirred media mills (Perry, 1973). The Attritor® and Dyno-Mill®, and their variants, are now standard. In each either steel or glass beads are stirred while a slurry is pumped through the media bed. The primary advantage over ball mills is that the temperature of the product can be maintained at or near ambient, reducing undesirable physical or chemical changes which can occur at the higher operating temperatures inside ball mills.

During flowable grinding the mineral, or combination mineral and polysaccharide thickener will not be present. Increased viscosity imparted by the thickener appreciably reduces the operating efficiency of the mill. Typical production involves grinding all ingredients except thickener, then thickening under high shear agitation in a mixing tank.

Agglomeration

Production of water dispersible granules involves a step unique to this formulation type. Starting with either a solid or liquid active ingredient the formulator prepares a wettable powder. Processing wettable powder to water dispersible granules necessitates the additional operation of converting a finely sized, dusty material into a nondusting granule. In addition, the granule must disintegrate within a few minutes when added to water, to prepare a dilution for spraying, giving a dispersion that has essentially the particle size distribution of the powder from which it was prepared.

The agglomeration process of choice seems to depend on geographic location. In Europe the preferred method is to spray dry an aqueous slurry. In the United States pan agglomeration has been the method of choice. Within the past one to two years formulators in both areas have begun developing water dispersible granules/pellets based on low pressure extrusion. There is a mild consensus among formulators that either spray drying or extrusion will eventually replace pan agglomeration in the United States.

TESTING TECHNIQUES

Minerals

Pesticide formulators rarely, if ever, require an unusual specification for a mineral raw material. Shortly following the rise in formulation of synthetic pesticides during and immediately after World War II the industry settled on a set of readily available, relatively inexpensive mineral carriers and solid diluents. This set has been used with little modification for some fifty years.

Since the pesticide industry consumes only a minor fraction of mineral output in the United States pesticide formulators make use of standard grades of minerals produced for other industries. For the same reason, pesticide formulators apply industry standard tests when evaluation of minerals is required. Except in unusual circumstances only three properties of a mineral will be of significance – liquid holding capacity, bulk density, and surface pH (McKay, 1979; Weidhaas, 1955).

Evaluation of mineral products for pesticide formulation purposes will make use of tests designed and recommended by suppliers (Floridin, Engelhard, Edward Lowe).

Formulated Products

The pesticide formulator will spend the bulk of his evaluation time examining the end-use product to be sold to a grower or applicator. The pesticide industry has developed a set of procedures which can be applied to specific formulation types or to particular formulation types of a specific pesticidal active ingredient. ASTM (American Society for Testing and Materials) is the most active organization for this purpose in the United States. Because ASTM procedures do not cover all aspects of pesticide performance the world-wide sets of standards accepted by formulators are those issued by the World Health Organization and CIPAC (Collaborative International Pesticides Analytical Council). Many of the tests recommended are for a specific formulation of individual active ingredients, being designed to ensure acceptability of products tendered for use in public health projects (WHO, 1985; CIPAC, 1970).

Suspensibility – Flowable concentrates, water dispersible granules, and wettable powders are each intended to be diluted with water for spray application to a target site. The spray apparatus can range from a one or two liter hand-held tank used by a pest control operator treating a home for roach or flea control to an intensely agitated four-thousand liter mechanical sprayer used to treat fruit orchards. Spray applications to field crops are subject to the vagaries of weather, among other outside influences; spraying might be interrupted for up to twenty-four hours by, for example, rain.

Dilutions prepared for application, then, must be capable of remaining in suspension for extended periods and must be capable of being resuspended easily after settling when agitation has been discontinued for long periods. The concentrate should also, of course, form a good initial suspension even when the concentrate has been stored under uncontrolled conditions for up to several years.

Suspensibility, therefore, is one of the most important evaluations for any end-use product intended to be diluted with water for application. Regardless of the particular pesticide and formulation type the testing procedures to evaluate this property possess common features.

Suspensibility must be determined in a standard water. Dispersing surfactants are added to the formulation to assist in maintaining stable suspensions. These surfactants are most usually anionic sulfonate or carboxylate salts. Divalent metal ions, such as calcium and magnesium, commonly found in water can form insoluble salts with the anionic portion of surfactant (e.g., bathtub soap scum), leaving them ineffective and greatly reducing or destroying suspensibility of the pesticide product. Testing reproducibility depends on all formulators using the same dilution system.

The standard hard water recommended by WHO, and used in labs around the world, is made from a combination of calcium and magnesium salts in a set ratio. The hardness of this water is 342 ppm, when determined as calcium carbonate. The recipe for preparing this water is to dissolve 0.304 grams anhydrous calcium chloride and 0.139 grams magnesium chloride hexahydrate in distilled water and dilute to one liter. A typical evaluation program will use 342 ppm water as the standard, with other waters ranging from 34 ppm to 1000 ppm hardness added as demanded by end-use requirements.

The most reliable suspensibility tests make use of chemical analysis for active ingredient. Determinations by weighing settled residues are prone to error since water soluble fractions of a formulation are not taken into account. A generalized procedure is to prepare a slurry of pesticide concentrate in standard water and further dilute so the pesticide product is present at about 5% w/v in the final dilution. At some time, such as one-half, one, or twenty-four hours after preparing the dilution, the uppermost seventy-five percent of the dilution is carefully removed by pipette. The lower fraction is collected and chemically analyzed, typically by GC or

⌘⌘⌘

HPLC, for active ingredient. There will be a specification for the particular product for the quantity of pesticide active ingredient which must not have settled in order for the formulation to be commercially acceptable.

Flowability – Dusts, granules, water dispersible granules, and wettable powders must possess free flow characteristics. For granules this property is of course critical, since a granule will be applied as-is. Uneven or poor flow of the granule can translate to uneven coverage of a target site, with consequent reduced control of weed or insect pests. Poor flow of dusts, water dispersible granules, and wettable powders can result in problems when these products are transferred from storage containers prior to application. Poor flow can also result in unnecessary exposure of handlers to pesticides. Lumpy, nonflowing end-use product might need to be mechanically broken up before removal from a container is even possible. Under practical conditions this often is done by using a handy stick to break up lumps in a bag or box.

All methods in use or proposed for measurement of flow employ variations of a simple procedure. A funnel, either glass or metal, with specified dimensions is used to hold a powder or granules. A typical specification for powder product testing would be a glass funnel having a 51 mm top diameter and a 3.5 mm bottom opening, and with the walls of the funnel making a 60° angle. While product is loaded the mouth of the funnel is closed. To start the test the mouth is exposed. The test can be either a yes/no variety – product does or does not flow unaided, or can be a time-to-deliver – a stated weight or product must flow unaided in a certain time.

Chemical Stability – The most critical test for an end-use product is chemical stability of the pesticide active ingredient. Evaluation of chemical stability does not depend on any particular property of a mineral carrier or solid diluent. Evaluations are carried out by asking whether active ingredient is stable in the end-use formulation for a certain length of time under certain storage conditions. Answering the question makes use only of a chemical analysis for active ingredient, and is essentially independent of any property of the mineral.

⌘⌘⌘

EXAMPLE PESTICIDES

Dusts

Combination Fungicide-Insecticide

	% w/w
sodium bicarbonate	10
potassium carbonate	5
pelargonic acid	2
capric acid	2
talc	81

(Jones, 1994)

Carbonate salts function as fungicides and the fatty acids as insecticides. The dust is prepared by blending all ingredients and then milling to obtain a small particle size. The composition is intended to be dusted onto leaves of ornamental plants.

Termiticide

	% w/w
d-phenothrin	10
aromatic hydrocarbon solvent	15
silica gel	10
cellulose powder	10
diatomaceous earth	55

(Teijima, 1994)

The insecticide d-phenothrin is dissolved in aromatic solvent and sprayed onto a blend of the dry materials in a ribbon blender equipped with a spray bar. Silica gel is added as a flow control aid. Cellulose powder is added as an attractant for termites. Diatomaceous earth functions as a carrier for the solution of insecticide. The composition is spread near termite trails and is carried into the nest by workers.

⌘⌘⌘

Flowable Concentrate

Acid Stabilized Herbicide

	% w/w
linuron	8.7
monolinuron	8.7
silicone defoamer	1.0
propylene glycol	8.0
dispersant surfactants	4.5
xanthan gum thickener	0.2
Veegum® mineral thickener	1.0
citric acid	0.4
water	q.s. 100

(Frisch, 1993)

Linuron and monolinuron herbicides are prone to base-catalyzed decomposition in water. In order to stabilize the chemicals citric acid is added to bring the final pH to 5-6. Propylene glycol is added as a humectant, retarding evaporation of spray droplets on treated surfaces. The surfactants used are a combination of sulfonated naphthalene-formaldehyde condensate and sulfonated lignin with an ethoxylated nonylphenol. Suspension of the solid herbicides in water is achieved by thickening the mixture with a synergistic combination of polysaccharide xanthan and a montmorillonite thickener.

The flowable is prepared by mixing all ingredients except the thickeners and first milling with a toothed disk mill to reduce average particle size to under 200 microns. Ultimate particle size of less than about 5 microns is attained by milling the slurry in a ball mill loaded with 2 mm glass beads. After milling the slurry is thickened by mixing in, under high shear, the combination of polysaccharide and mineral thickeners.

⌘⌘⌘

Granules

Coated Granules

	% w/w
carbofuran	2.0
diazinon	3.0
polyurethane	1.0
polyvinyl alcohol	2.0
silicate powder	3.0
granular quartz	91.0

(Antfang, 1994)

A coated granule containing a solid (carbofuran) and a liquid (diazinon) insecticide is to be prepared. The solid insecticide is ground to a fine powder; the liquid insecticide is sprayed onto an absorbent silica. A combination of two adhesives is used. Granular (0.4-0.8 mm) quartz is first sprayed with a liquid polyurethane while tumbling the quartz in a shell blender. Once the mass is wetted the insecticide powders are added. After tumbling the mass to adhere the powders, an aqueous solution of polyvinyl alcohol is sprayed onto the surface. In order to set the adhesive the mass is transferred to a drier and heated to about 60°C.

Impregnated Granule

	% w/w
Dyfonate®	15.0
Igepal® DM970	5.0
attapulgite	80.0

(Scher, 1991)

In order to reduce the dermal toxicity of granules containing a phosphorodithioate insecticide (Dyfonate) a surfactant with an HLB (hydrophile-lipophile balance) of about 19 is added to the granules. Preparation involves simply adding the granular carrier (24/48 mesh for example) to a shell blender, tumbling the granules and spraying onto the granular carrier each of the liquid materials. The granules can be packaged immediately after the impregnation step.

⌘⌘⌘

Water Dispersible Granules

Spray Dried Fungicide

	lbs.
Water	490
Dispex® N40	44
Reax® 88B	53
sodium tripolyphosphate	30
cuprous hydoxide wetcake	2155
Volclay® HPM-75	167

(LeFiles, 1994)

The ingredients above are mixed to form a slurry and then spray dried to generate 130 micron average diameter granules which will disintegrate to form a fine particle size suspension in water. In order to make the spray drying slurry mobile and to stabilize the suspension of particles in crop spraying water, surfactants must be added. Dispex is a partially neutralized polyacrylic acid; Reax is a sulfonated lignin. Both surfactants function as dispersion stabilizing additives. Polyphosphate is added as a stabilizing agent for the copper hydroxide fungicide. Volclay (sodium bentonite) functions partially as a binder for the granules and as a water swellable mineral disintegrant. When granules are added to water the bentonite will assist in promoting rapid disintegration. The suspension of disintegrated granules will have an average particle size of about 2 microns.

The initial slurry is spray dried in a conventional tower equipped with a single fluid nozzle, and having an inlet temperature of about 190°C and an outlet temperature of about 90°C. The final granules will be dried to a moisture content of less than three percent. Granules will contain about 85% w/w copper hydroxide.

Pan Granulated Fungicide

	% w/w
triazole	25.0
surfactants	11.5
kaolin	63.5

(Katayama, 1993)

In order to obtain good disintegration of water dispersible granules in water for spray application solid diluents must be carefully selected. In this example the inventors claim that optimum breakup of granules in spray

⌘⌘⌘

water and good mechanical strength of granules during transportation and handling can be accomplished by selection of the grade of kaolin clay. Sulfonated surfactants are employed as wetting and dispersing agents. All ingredients are blended and then size reduced in an air mill. Powder is granulated on a pan agglomerator using a ratio of about 20 parts water to 100 parts powder. The wet granules are dried at low temperature in a fluid bed drier.

The grade of kaolin employed is selected so that the volume median diameter is greater than about 2 microns and less than about 10 microns. Depending on the physical and biological properties of the particular active ingredient, the final concentration of pesticide in granules can be as low as 10% w/w or as high as 80% w/w.

Wettable Powders

Impregnated Insecticide

	% w/w
diflubenzuron	15.0
N-methylpyrrolidone	20.0
surfactants	5.0
silicate	25.0
diatomaceous earth	15.0
kaolin	20.0

(Shibahara, 1993)

The insecticide diflubenzuron is a solid, but the inventors believe its biological activity is enhanced when it is in a liquid form. The particular application for the product requires that the end-use product used by the grower should be a solid. In order to meet each requirement the solid insecticide is dissolved in a nonvolatile solvent and then impregnated onto absorbent carriers, as if it were a liquid. The primary carrier is a synthetic silicate having high absorptivity. In order to reduce overall cost of the formulation and increase the powder's bulk density other materials will be added as solid diluents.

A concentrate is prepared by spraying the solution of insecticide in NMP onto silicate in a ribbon blended equipped with a spray bar. The concentrate is then diluted with minerals. Diatomaceous earth adds some absorptivity. Kaolin is added primarily for its bulk density, and partially to improve flowability of the final product. Since solid carriers and diluents are sufficiently fine there is no need to mill the mixture after the impregnation and blending steps.

⌘⌘⌘

Sustained Release Fungicide

	% w/w
dodine	30.0
polylactide	15.0
surfactants	10.0
montmorillonite	25.0
kaolin	20.0

(Friedrichs, 1994)

The fungicide dodine is incorporated into a polymer matrix in order to extend its duration of activity on leaf surfaces. One method of incorporation is to mix powdered fungicide with a molten polylactide and then extrude the mixture. Once cooled the matrix is roughly ground in a cutter-chopper mill. Sulfonated wetting and dispersing surfactants are added, as are clay diluents. Since the polymer matrix is susceptible to softening or melting during final milling the solid diluent blend is a combination of montmorillonite for absorption and kaolin for flowability. The mixture of solid ingredients is ground to its final particle size (20-40 microns) in a hammer mill.

REFERENCES

Antfang, Elmer; Kerimis, Dimitrios; Singer, Rolf-Jurgen, "Coated Granules Containing Liquid And Solid Active Compounds", United States Patent 5,326,573, issued July 5, 1994.

CIPAC Handbook, volume 1, *Analysis of Technical and Formulated Pesticides*, Collaborative International Pesticides Analytical Council Ltd., Cambridge, England (1970).

Engelhard Minerals & Chemicals Corp., "Flowability of Dry and Impregnated Dusts, Diluents, and Carriers", standard method No. 4252.

Floridin Co., "Minerals For Agrichemical Processing", brochure 900423-1M (1990).

Friedrichs, Edmund; Albert, Guido, "Fungicidal Composition", United States Patent 5,304,376, issued April 19, 1994.

Frisch, Gerhard, "Aqueous Herbicidal Dispersion Concentrate Containing Linuron And Monolinuron As Active Substances", United States Patent 5,226,945, issued July 13, 1993.

Hartley, G.S. and Graham-Bryce, I.J., *Physical Principles of Pesticide Behavior*, pp 781-790, Academic Press, New York, New York (1980).

Jones, Keith A., "Environmentally Safe Pesticide Compositions", United States Patent 5,342,630, issued August 30, 1994.

Katayama, Yasuyuki; Tsuda, Shigenori, "Water Dispersible Granules", United States Patent 5,180,420, issued January 19, 1993.

⌘⌘⌘

LeFiles, James H.; Taylor, Evelyn J.; Crawford, Mark A., "Copper Hydroxide Dry Flowable Bactericide/Fungicide And Method Of Making Same And Using Same", United States Patent 5,298,253, issued March 29, 1994.

Lowe Industries, Inc., "Standard Procedures for Clay Products Evaluation", (1987).

McKay, Bruce M., *Pesticide Solid Diluents And Carriers*, CITE, Middleport, New York (1979).

R.T. Vanderbilt Co., "Van Gel B/Rhodopol 23 Suspending System for Agricultural Flowables".

R.T. Vanderbilt Co., "Diluents, Dispersing Agents And Stickers for Agricultural Dusts and Sprays", Bulletin No. 23.

Sawyer, Edgar W., "Use of Gelling Clays to Stabilize Agricultural Pesticide Suspensions and Emulsions in Aqueous Systems", Special Technical Publication 943, pp 177-186, American Society for Testing and Materials, Philadelphia, Pennsylvania (1987).

Scher, Herbert B.; Rodson, Marius; Morgan, Ronald L., "Pesticide Compositions And Method", United States Patent 4,994,261, issued February 19, 1991.

Shibahara, Tetsuya; Kondo, Naohiko; Kato, Jun, "Process For Preparing Highly Active Water Dispersible Pesticides", United States Patent 5,264,213, issued November 23, 1993.

Snow, Richard H., et al, "Size Reduction And Size Enlargement", Chemical Engineer's Handbook, Fifth Edition, Section 8, R.H. Perry and C.H. Chilton editors, McGraw-Hill, New York, New York (1973).

Teijima, Isato; Aki, Seietsu; Ito, Takaaki, "Termiticidal Dust Containing Termite Attractant", Japanese Patent 06 87,797, issued March 29, 1994.

.D. and Brann, J.L., Handbook of Insecticide Dust Diluents and Carriers, second edition, Dorland Books, Caldwell, New Jersey (1955).

World Health Organization, *Specifications for Pesticides Used in Public Health*, WHO, Geneva, Switzerland (1985).

⌘⌘⌘

ELEVEN

CERAMICS & GLASS

John F. Mooney, Ph.D.
Muliakeramik
Cikarang, Bekasi, Indonesia

The broad term "ceramics" is comonly taken to mean any of a large family of materials, usually inorganic, requiring high temperatures in their processing or manufacture. In practice, these materials are generally divided into categories as follows:

1. Glass
2. Whitewares, including artware and structural ceramics.
3. Refractories

A brief description of these categories will follow to allow the non-ceramist a better understanding of the use of "Ceramic" materials.

MAJOR CLASSIFICATIONS OF CERAMICS

Glass

Glass has been variously described as a supercooled liquid, a randomly structured material, or a non-crystalline solid. In general the term "glass" refers to an amorphous solid with non-directional properties, characterized by its transparency, hardness and rigidity at ordinary temperatures, and capacity for plastic working at elevated temperatures

Major commercial uses of glass include:

Plate or "float" glass, as used for windows and windshields.
Glass tubing, or formed shapes, used for electric lighting envelopes.
Glass containers such as tumblers and bottles.
Electronic materials for cathode ray tubes, insulators, microchips.
Ophthalmic glasses for lenses, eyewear.
Decorative artware.
Glass-ceramics for everything from nose cones to cookware.
Various specialty glasses, particularly for fiber optics.

Whiteware

Whiteware is the earliest known form of ceramic, predating recorded history as evidenced by the "ceramic" artifacts uncovered from various anthropological digs in the form of earthenware vessels and figures representing early man. This family is characterized by a crystalline matrix held together by a glassy phase and usually covered by a glazed coating.

The major classifications of whiteware are:

1. Ceramic tiles, glazed and unglazed, for floors, walls and external use.
2. Sanitaryware in the form of toilets and lavatories.
3. Tableware, from earthenware to fine china.
4. Other specialty products.

Refractories

Refractories represent a broad classification of ceramics characterized by the ability of the materials to withstand high temperatures. These materials are used for steel making, glass melting, and in various chemical processing industries. They are used as containment materials, insulation, and heat and chemical barriers of all kinds.

Artware

The use of ceramic materials for art purposes is as varied as the imagination of the artist. Ceramic materials, and base minerals themselves, are used for everything from carvings (on such " soft" materials as talc and pyrophyllite) to glass art and traditional earthen clay pots.

Structural Ceramics

The traditional "heavy clay" industry encompasses a wide range of products from drainpipes, roofing tiles, road pavers and their like, to the cement industry, which is also a basic ceramic process of beneficiation through the use of high temperatures.

Other Ceramic Industries

New ceramic industries have emerged as a result of the imagination of the ceramist and increasingly inventive researchers. The catalytic converters which control automotive emissions, for example, are not easily classified in the traditional "ceramic" sense.

The major portion of emphasis in materials usage in this chapter will be placed on glass, whiteware (especially ceramic tiles) and refractories. These three areas encompass the majority of minerals used in the ceramic industry as well as the varied reasons for choices of these minerals for their different

⌘⌘⌘

roles in the manufacture and final properties of the ceramic objects under discussion.

GLASS

Since glass is an amorphous solid formed by fusing silica (SiO_2) with a basic oxide, the range of minerals available for use in the basic glass industries is, at first glance, limited. When one considers the wide range of specialty glasses manufactured today, however, the range of materials opens considerably.

There are a number of general families of glasses, some of which have many hundreds of variations in composition. It is estimated that there are over 50,000 glass formulas, depending on desired end properties. The soda-lime glasses are, however, the oldest and simplest of the glass families and will be discussed first.

Soda-Lime Glass

Soda-lime glasses are the lowest in cost, easiest to work, and most widely used of the many formulas available to the glassmaker. They account for 90% of the glass used in the world, and are composed of silica, sodium oxide (soda), and calcium oxide (lime). This is the glass of ordinary windows, bottles, and tumblers. The simplest combinations of materials to produce a soda-lime glass are silica sand, soda ash, and limestone (calcium carbonate). These materials are usually in the form of -30/+65 mesh particles, and are introduced into the glass melter together with various amounts of cullet (scrap glass from processing or reject product).

The glass melting process involves several stages:

1.) Melting, in which the glass raw materials, together with cullet, are introduced into the glass melter (or "tank") through a screw feeder. The materials melt to form a glassy phase with a large amount of bubbles, or "seeds", entrapped in the viscous glass.

2.) Refining, where the temperature is raised to the highest point in the process to lower the glass viscosity and allow the seeds to rise out of the melt. This refining process may be aided by chemical agents, called fining agents, to promote the fining process by releasing large bubbles by chemical breakdown or oxy-redox reactions. These lager bubbles "sweep" the smaller seeds from the melt.

⌘⌘⌘

3.) Conditioning, where the glass is cooled to the proper temperature and viscosity to prepare for the forming process.

4.) Forming, where the viscous material is shaped by blowing, pressing, drawing, rolling, or any number of additional forming processes.

Aluminosilicate Glass

Adding aluminum oxide to basic soda lime glasses increases the durability of the glass and opens the choices for raw material selection. Aluminosilicate glasses are useful at higher temperatures and have greater thermal shock resistance. Resistance to weathering, water, and chemicals is excellent, although acid resistance is only fair when compared with other glasses. Alumina may be obtained by the addition of albite ($NaAlSi_3O_8$) in the form of feldspar or nepheline syenite. Potash feldspars, either the orthoclase or microcline varieties, are added as a source of K_2O for fluxing.

Barium is added to some bottle glasses used for food or drug containers to increase brilliance or sparkle. The barium may be added as barium carbonate or barite in which case the sulfate content of the barite may be used to assist the refining process.

Borosilicate Glass

Borosilicate glasses, which contain boron oxide, are the most versatile of the glass families. They are noted for their excellent chemical durability, for resistance to heat and thermal shock, and for low coefficients of thermal expansion. The low expansion glasses are best represented by Pyrex glass, long known for its excellent thermal behavior. The most common source of boron oxide is borax, the hydrous sodium borate mineral also referred to as the tetraborate. This mineral was first obtained from Tibet under the Persian name borak, meaning white. Natural borax is $Na_2O \cdot B_2O_3 \cdot 10H_2O$. The borax deposits in California and Nevada may be in the form of colemanite or calcium borate in the form $2CaO \cdot 3B_2O_3 \cdot 5H_2O$. Boric oxide may also be added to glasses as boric acid generally derived by adding hydrochloric or sulfuric acid to borax and then crystallizing. Boric acid occurs naturally in volcanic fissures in Italy as the mineral sassolite, $B_2O_3 \cdot 3H_2O$. Anhydrous boric acid is generally preferred as the additive to glasses, since this form is much easier to handle as a powder in batching and silo storage.

Lithia Glass

A specialized family of glasses used for high thermal shock resistance derived from very low thermal expansion is based on lithium oxide, or lithia. The major source of lithia is lepidolite, the most widespread of the lithia-containing minerals with deposits in the United States, Canada, Rhodesia,

South Africa, India, China, Russia, Japan and Germany. The lithia oxides also form the basis for ion exchange chemistry in chemically strengthened glasses. The sanctions placed on Rhodesia in the past were major concerns for American glass manufacturers because the lepidolite deposits in that country were the purest and suitable for glass manufacture with little or no beneficiation. Lithia glasses have been used for ceramic stove tops, glass-ceramic cookware and the like.

Phosphate Glass

Phosphate glasses were developed for resistance to the action of hydrofluoric acid and fluorine chemicals. The glass structure is based on P_2O_5 with no silica, and some alumina and magnesia as network modifiers. Phosphate glasses are transparent and can be worked like silica glasses but are not highly resistant to water, the compromise for their fluorine resistance. The mineral apatite is a major source of phosphorus, with deposits in the Appalachian range, Idaho, Brazil and French Oceania. An additional source of phosphorus is phosphate rock, which occurs in the form of land pebbles. There are several deposits in the United States, but phosphate rock is most commonly found in Morocco, Tunisia, and the Pacific Islands of Christmas Island and Nauru.

Opal Glass

Opal glasses, commonly utilizing calcium fluoride as the source of the opalizing fluoride structure, have been used for many years to produce low to medium priced dinnerware and "milk" glass. One of the major problems of melting fluorine containing glasses has been the volatilization of fluorine from the surface of the hot glass melt and the resultant major pollution from the airborne fluoride. In combination with moisture, this volatile fluoride forms hydrofluoric acid which is especially destructive to windowpanes and evergreen trees in the neighborhood of the offending polluter. The high cost of stack scrubbers and bag collectors can be offset in some fluoride glass melting tanks by the use of large amounts of water vapor over the surface of the molten glass. The fluoride ion and the water molecule are close enough in atomic weight that the volatizing fluoride ions are " fooled" into believing the atmosphere is saturated with fluorine (rather than water vapor) and therefore the volatility of the fluorine ion is significantly reduced.

Lead Glass

Lead glasses or lead silicate glasses have been used as decorative or functional crystal ware and/or radiation shielding. The thermal expansion of these glasses increases as the lead content increases, thereby reducing the rigidity of the structure at very high lead contents. The presence of lead

⌘⌘⌘

oxide in crystalware produces a luster or sparkle to the glass which is extremely difficult to reproduce with any other material. The major lead minerals used in glass manufacture are lead oxide and lead silicate, both of which are refined from the naturally occurring mineral galena. Galena is lead sulfide, PbS, theoretically 86.6% lead. Southern Missouri is the chief source of galena in the United States.

WHITEWARE

Ceramic Tiles

The basic classes of ceramic tiles, as defined by the Tile Council of America, are: Wall Tiles, Glazed Floor Tiles, Unglazed Floor Tiles and Ceramic Mosaic Tile. The minerals used in the manufacture of ceramic tiles exhibit the entire range of properties and usefulness that exemplify the reasons for their use in almost all fields of ceramics.

Wall tile includes that family of glazed tile which exhibit relatively low mechanical strength and high water absorption. The process of manufacturing wall tiles may include tunnel kiln firing (14 hour cycle) or fast fire roller hearth kilns (40 to 60 minute cycle). The more modern roller kilns are replacing the older tunnel kilns throughout the world as the standard firing vehicle for all types of ceramic tile. Figure 1 illustrates a typical process flow diagram for the production of wall tile, exhibiting the "two-fire" process commonly used in today's wall tile industry.

Floor tiles have higher mechanical strength and lower water absorption, reflecting the tougher use criteria to which floor tiles are exposed. In today's state-of -the-art technology, floor tile are usually produced by the monocottura, or once-fired process. Figure 2 illustrates a commonly used once fired process. Some companies are firing wall tile in a single firing process referred to as monoporosa.

Generally, ceramic tiles are composed of three basic ingredients:

1.) A plastic component for strength and formability.
2.) A fluxing agent for strength at high temperatures.
3.) A filler or inert component.

In most ceramic systems one material may serve more than one of the above functions. For example, some clays are used as both plastic components and fluxes, depending on the alkali content of the clay.

The basic tile making process consists of the following steps:

1.) Hard materials grinding.
2.) Clay feeding / preparation.
3.) Weighing.
4.) Slip preparation via ball milling or blunging.

⌘⌘⌘

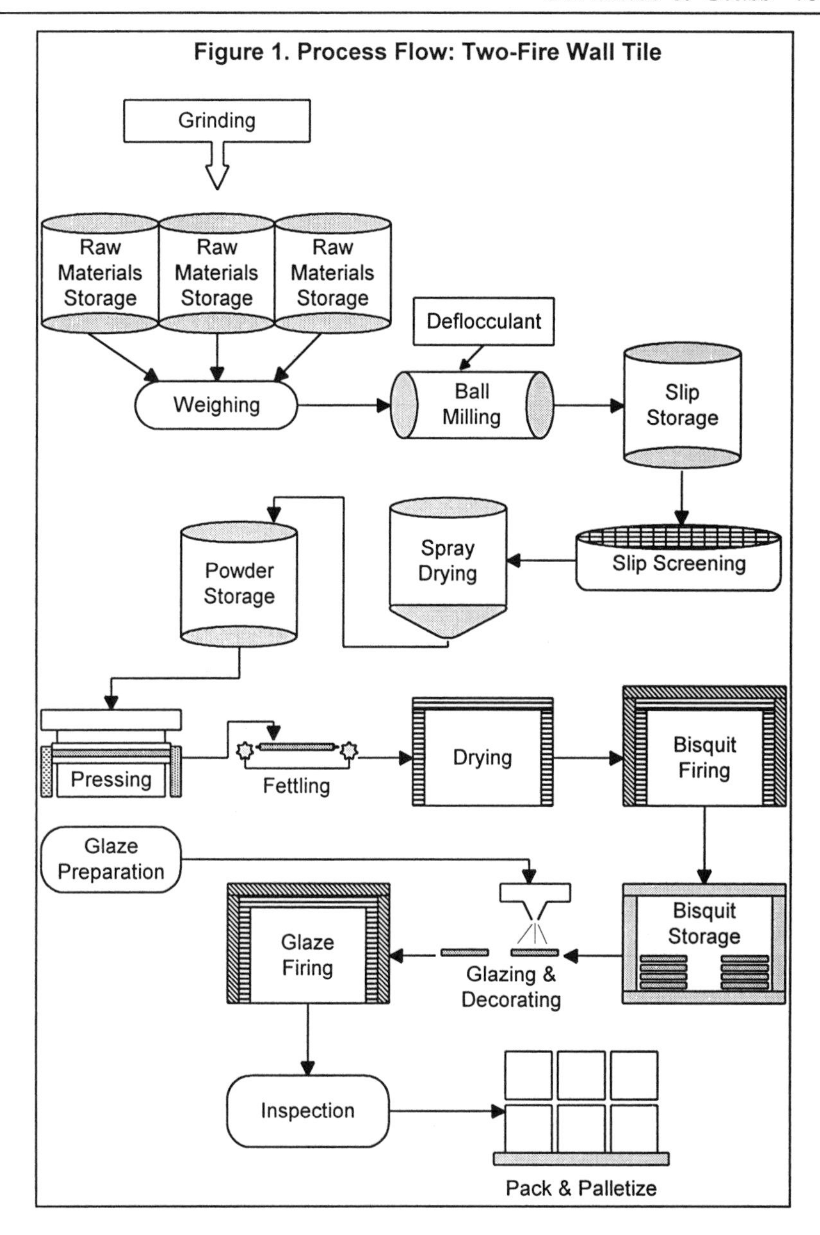

Figure 1. Process Flow: Two-Fire Wall Tile

5.) Slip storage to allow "aging" of the slurry, generally with respect to the organic components present in most clays.
6.) Spray drying of the slurry to form spherical powder granules with high flowability for today's modern hydraulic presses.
7.) Aging of the powder to allow redistribution of the moisture in

⌘⌘⌘

the spray dried granules.

8.) Pressing of the powder into tile blanks of the desired shape.
9.) Drying of the tiles.
10.) Bisquit firing (for wall tiles, especially).
11.) Storage of the fired tiles.
12.) Glazing by any one of several methods too numerous to mention here.
13.) Gloss firing of the glaze.
14.) Inspection, sorting, packaging, and palletizing.

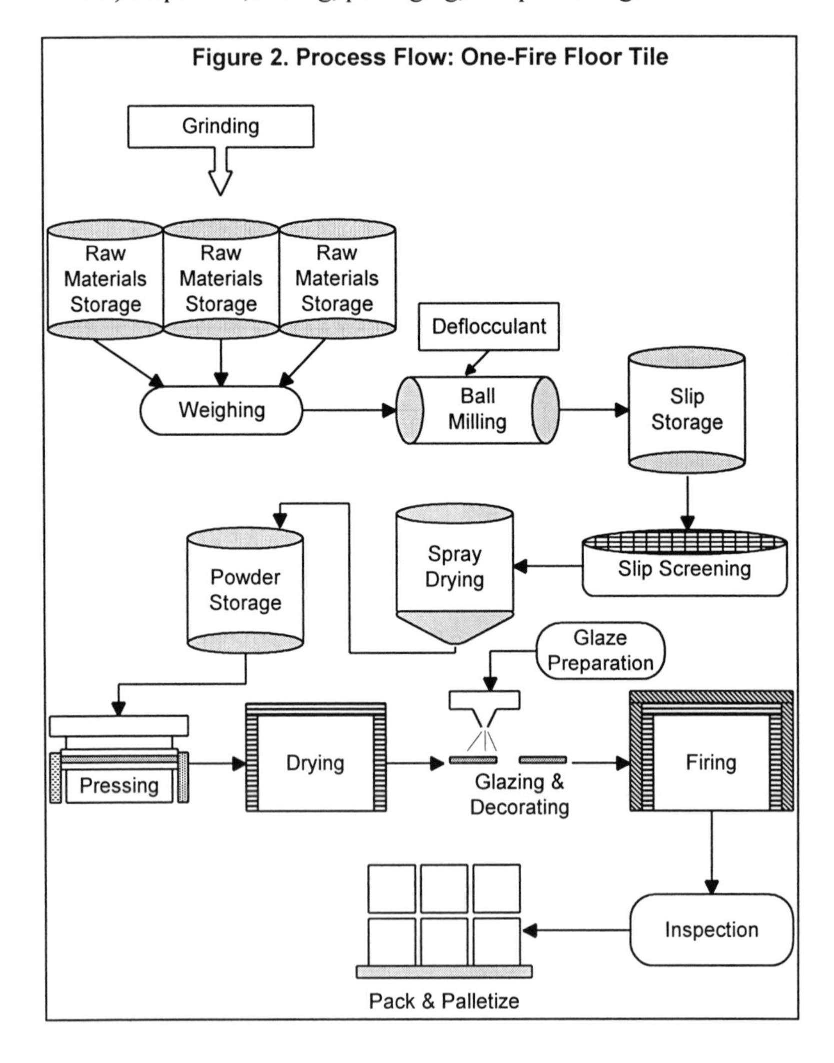

Figure 2. Process Flow: One-Fire Floor Tile

⌘⌘⌘

Tile Raw Materials

The desired color for a wall tile body, in most parts of the world, is white or buff. Some producers manufacture a pink or red body, depending on the availability of local materials. The iron content of the clays and/or other minerals used in tile manufacture is generally the determining factor in the tile body color. The higher the iron content, the redder the body.

Kaolin – The most common minerals used in the ceramic tile industry worldwide are the clays. Of these, the most widly used are the kaolin clays. The name kaolin originated from the Chinese, "kau-ling" meaning "high-ridge" in reference the place in Kiangsi province where the material was first obtained and used in ceramics. The most important kaolin deposits relative to the ceramic industry are located in Devon and Cornwall, England: Westerwald in Germany; the districts of Wiesau-Tirschenreuth and Hirschau- Schnaittenbach in Bavaria; Brittany in France; Karlovy Vary in Czechoslovakia; Georgia and South Carolina in the United States; and Foshan Province in China. All of these areas are, or at one time were , important ceramic tile manufacturing locations or sources of supply relative to tile manufacture.

Pure kaolin is characterized by its milky white color, low plasticity, greasy feeling and the tendency to leave a whitish powder film on the fingers after touching. The more plastic kaolins are the result of secondary weathering and are generally finer in particle size and associated with vegetable and carbonic elements. Plastic kaolins are generally darker in color than pure kaolins as a result of this contamination, but both types "burn" to a white or off-white color on firing. This characteristic makes kaolin extremely important to the tile industry.

The two major types of kaolin use in tile manufacture are commonly known as China Clay and Ball Clay. Ball clays are generally very plastic and composed of 50% kaolin, 24% micacious materials and 22 to 26% quartz. China clays are generally 80 to 85% Kaolin and 15 to 20% micacious. China clays are generally used for their whiter burning characteristics where little plasticity is needed, while ball clays are used for their plastic nature to provide high strength during the forming process.

Characteristics of these two types of clay relative to the tile making process are:

	Ball Clay	**China Clay**
Green and dry strength	High	Low
Shrinkage at 1020°C	0-2%	None
Apparent Porosity	18-22%	25-30%
Coefficient of thermal expansion	120-170	110-140

⌘⌘⌘

Combinations of these materials are therefore used by the ceramist to control the final desired properties of the ceramic body during manufacture. The addition of ball clay will increase the green strength after pressing but the resultant increase of thermal expansion may be undesirable depending on the expansion of the glaze materials being used. Any well designed tile body contains both of these clays to provide a basis for control during production.

Illite – A second major classification of clays is the illite group, classified as to their similarity in sheet structure to the mica family. The name illite derives from the State of Illinois, where the mineral was first identified. The major illite clay ores used in the ceramics industry are those found in the Emilian Appennines in Italy. Other major deposits exist in Sarospatak in Hungary, Karlovy Vary in Czechoslovakia, and Illinois in the United States. The Italian deposits are by far the most significant, as their presence gave rise to the major ceramic tile business in which Italy has dominated and influenced the rest of the world. The clay tiles produced from illites exhibit good green strength and low water absorption (1 to 5%). Higher absorption characteristics sometimes desired in wall tile bodies are obtained by adding calcium carbonate in the form of limestone.

Smectite – The smectites are water swellable clays having a sheet or platelet structure. Smectite is the mineralogical term for this class of clays, which includes montmorillonite, hectorite, and saponite. Montmorillonite clays derive their name from the Montmorillon section of France where this material was first observed and later classified. Most smectites are more commonly known under the geological term bentonite. By convention, a bentonite is understood to be an ore or product with a substantial smectite content. The name bentonite derives from Fort Benton, Wyoming, the site of an important deposit. Lattice substitutions within the smectite clays creates a charge imbalance which is compensated by exchangeable alkali and alkaline earth cations. This contributes to the ability of these clay to swell and impart considerable plasticity in ceramic formulations. When the exchangeable cations are predominately sodium, the individual platelets can separate to produce a colloidal structure in water.

Halloysite – Halloysite is the name of another significant tile making material named after its original observer, Omalius d" Halloy. Halloysite is a hydrated kaolin-like material. The mineral is generally whitish in color with yellow or pink shades. Major deposits exist in Utah, Nevada, Idaho, North Carolina and Georgia in the US, and Dordogne in France, with minor deposits in the former Republics of Czechoslovakia and Yugoslavia.

⌘⌘⌘

Generally employed at lower percentages than most other materials in ceramic tile bodies, Halloysite is usually used to lower thermal expansion and provide a more refractory component. The hydrated nature of this material is prone to cause high shrinkage in clay/tile bodies and is therefore to be avoided in precise size control areas.

Chlorites – The chlorites from the Emilian Appennines in Italy are important to that country's tile industry. Similar in structure to the micas, these minerals are significant for their use in a wide variety of tile bodies, depending on their color and refractoriness. The grey-blue chlorites from these deposits are generally used in the traditional high porosity glazed wall tile industry (Majolica), the grey clays of the Antognola series are used in bodies for both glazed and unglazed (Tuscan) floor tiles. The high iron content red chlorites of this region are significant in their use in high quality glazed floor tiles and in the roofing tile business.

Talc – One of the most significant minerals used in the ceramic tile industry is talc, the English spelling of which arises from the Arabic "talq". Talc ores may contain various impurities, such as calcite, tremolite and chlorite, which remarkably influence their properties. Major significant deposits are located in Texas, Georgia and New York in the US, The Piedmont, Lombardy and Sardinia in Italy, and the Luzenac region of France.

Talc may be used both as a major component of tile bodies or as a network modifier to control thermal expansion. At lower firing temperatures it is characterized by a high coefficient of expansion which enables the glaze to "fit" well to the body, while at higher firing temperatures the material greatly increases the resistance to thermal shock. When these properties are combined with the low release of gases on firing, talc becomes an ideal material for tile bodies, especially in the relatively new technology of rapid firing under one hour.

Where economically viable deposits of talc are available this material usually repaces limestone or calcite in the tile body. The absence of the large gas evolution from the carbonate structure allows a more rapid rise in temperature in the early part of the firing curve and allows much faster firing times with a correspondingly lower expectation of subsequent glaze defects such as pinholes due to gas entrapment in the glaze layer.The use of talc in ceramic tile making is more widespread in the U.S.A. due to better deposits than in the rest of the world. This material also reduces the possibility of high moisture expansion in finished pieces and subsequent glaze crazing resulting from that expansion.

⌘⌘⌘

Pyrophyllite – A mineral with significant parallels to talc is pyrophyllite. The name arises from two Greek words: "piros" (fire) and "phyllon" (leaf), the term probably refers to the tendency of pyrophyllite to "exfoliate" on heating. Significant deposits exist in the US (North Carolina), Japan, The Ural Mountains of Russia, Newfoundland, Brazil, South Africa and Indonesia. The use of pyrophyllite in glazed white body wall tile is significant in North America, Brazil and Indonesia. Its use results in an extremely white body after firing which can be said to have zero shrinkage under controlled conditions due to the material's characteristic expansion and subsequent shrinkage during the firing cycle. The use of this mineral in ceramic tile manufacture generally results in an extremely white body with exceptional size control during manufacture. The smooth surface of the pressed body lends itself to all types of glaze application techniques with resultant very appealing high gloss surfaces.

Feldspathics – Feldspathic minerals, throughout the world, play a significant part in the fluxing of ceramic tile bodies at high temperatures. The major feldspathics are the feldspars, nepheline syenite, pegmatite and aplite. In areas where good base materials such as talc or pyrophyllite are unavailable, feldspars are used in combinations with white burning clays, limestone and silica sand to create good quality white body tiles. The "white" is actually generally ivory or buff due to iron contamination in the feldspar. Some areas produce a white body with a pinkish cast from the contribution of the feldspar. Potash-containing feldspars are used to lower the thermal expansion of tile bodies. Low iron soda feldspars are generally used in the manufacture of frits for glazes, while nepheline syenite, due to its more refractory nature, is used in high density, low absorption bodies for mosaics or paver tiles.

Wollastonite – Wollastonite, a calcium silicate having an acicular structure, is significant in increasing physical strength, decreasing shrinkage, and lowering firing temperatures in ceramic tile bodies. The name was given to this mineral in honor of the English chemist and mineralogist W.H. Wollaston.

Carbonates – The carbonate minerals – limestone, $CaCO_3$, dolomite, $CaMg(CO_3)_2$ and magnesite, $MgCO_3$ – serve various functions in ceramic tiles. They are used to control shrinkage during firing, add strength to the fired body, occasionally to change thermal expansion for a better glaze "fit" and for fluxing or filler purposes.

⌘⌘⌘

Silica Sand – Silica sand, in finer grinds, is an important material for controlling the thermal expansion of tile bodies in areas where the available clays and/or feldspars do not serve this function adequately. Its use is generally avoided, if possible, by ceramists due to the difficulty in fine grinding and the fact that oversize silica particles in the fired body are often point sources for pinholes, since the silica creates a localized high-silica glass which has low solubility for dissolved gases.

Typical Ceramic Tile Bodies

Wall Tile –	Pyrophyllite	65%
	Ball Clay	20%
	Whiting	10%
	Kaolinite	5%
	Talc	60%
	Ball Clay A	15%
	Ball Clay B	15 %
	Wollastonite	10%
Floor Tile –	Illitic Red Clay	30%
	Ball Clay	15%
	Feldspar	30%
	Limestone	20%
	Silica Sand	5%
	Red Shale	45%
	Ball Clay	20%
	Limestone	10%
	Feldspar	25%
Ceramic Mosaic/ – Paver Tiles	Kaolin	35%
	Nepheline Syenite	35%
	Fine Powdered Silica	30%

CERAMIC TILE/WHITEWARE PROCESSING

A description of the basic processing and equipment used in ceramic tile manufacture is included below to familiarize the non-ceramist with the basics of this technology. With few exceptions, the equipment and processes

⌘⌘⌘

used in dry pressing and/or extrusion of ceramic tile are similar to those used in all parts of the ceramic industry and will not be repeated for other areas.

Materials Handling

Most bulk raw materials are conveyed to the production site via truck or bottom discharge rail cars. For some materials, such as air-floated clays, transport is via pneumatic discharge container trucks. Some raw material suppliers have begun to extol the virtues of shipping pre-blended slurries to tile producers. In this case, transport is via tanker trucks and the off-loading is accomplished by piping systems as opposed to conventional dry materials handling systems. The use of pre-blended slurries necessitates economic decisions comparing the capital costs of grinding/weighing/mixing equipment on-site versus the loss of some process flexibility and the cost of shipping water in the slurry. For simple tile factories with limited product lines and tight capital this may be a viable solution.

Minor body ingredients and most glaze frits and/or prepared glaze materials are delivered in paper or plastic sacks in 50 to 80 lb. quantities on wood skids. Some manufacturers receive glaze or engobe materials in “Super Sacks” of up to 1000 pound. capacity. Material suppliers will also deliver prepared glaze formulae in the correctly weighed quantities to glaze preparation plants. The tile manufacturer then adds this pre-weighed quantity to his mills with the proper dilution water. New factories, or those in “Third World” environments can use this service advantageously to simplify startups.

Grinding & Classifying

While most tile producers in the United States begin with beneficiated and size classified raw materials, there are parts of the world where the major bulk materials are delivered in the true “raw” state and must be beneficiated in some way.

Hard raw materials, such as calcium carbonate rock or feldspars, are reduced to desired size by successively grinding through one or more jaw crushers and then to a ring mill for final sizing. Oversize particles are screened after ring milling and returned back through the process in a closed loop system.

Most clay materials are not beneficiated at the plant site prior to blending or milling, but in some cases, especially where continuous ball milling is subsequently used, the clay portion of the body may be pre blended in high speed blungers to insure dispersion or “slaking” of the clay particles prior to final blending or milling.

⌘⌘⌘

Milling & Blending

The milling/blending portion of the process consists of two steps: weighing or proportioning, and milling, mixing, or blending.

Weighing, in most modern ceramic processing factories, is by continuous belt systems, while minor ingredients may be weighed by hand on separate balances and added to the mix at the appropriate time or sequence in the process. Continuous belt systems consist of a hopper into which each of the basic raw materials is loaded, either by front end loader vehicles from bulk storage piles or by conveyor belts from storage silos. The materials are fed along a belt which weighs the materials and records the cumulative weight of each component, usually in a computer controlled system. These weighed materials are then fed to a collector belt which transfers the composite to a large ball mill. Water is automatically fed to the mill through a metering system also controlled by the computer. This portion of the process is by far the most crucial, since the old saying of "garbage in, garbage out" is never more true than in the correct weighing and /or proportioning of raw materials in the first step of a ceramic process. The control of all raw materials in the process is extremely important for continuous, smooth running of a plant.

Milling of clay slurries is in ceramic- or rubber-lined ball mills with capacities from 1000 to 3000 kilograms. The grinding media are silica pebbles from 20 mm to 60 mm in diameter. Silica pebbles are generally used, especially in rubber-lined mills, since the alternative alumina pebbles are more prone to promote excessive wear on the rubber lining. The intimate mixing of the raw materials is extremely important to the uniform completion of the solid state reactions which take place during firing.

The size distribution of the pebbles is important to promote grinding; a usual distribution being:

45-50% pebbles from 20-30 mm diameter
25-30% pebbles from 40-50 mm diameter
20-25% pebbles from 50-60 mm diameter

As the pebbles wear with continued use, the normal practice is to add pebbles of the largest size to replenish the charge, since the wear of the media will fill in the voids of the charge in the mill.

Grinding speeds may be increased with the use of alumina pebbles, but the increased speed must be weighed against the increased cost of the alumina media.

The theory of milling speeds versus mill charge versus media composition versus materials charge can take up many pages of specific description, too numerous to describe in detail here. There are several publications which describe this process in detail, the most complete being "From Technology Through Machinery to Kilns for SACMI Tile" published by the SACMI company, a well known Italian supplier of turnkey ceramics facilities.

The important control parameters for the milling process are:

Media charge and size distribution.
Raw material charge.
Mill rotation speed.
Type and amount of deflocculant.
Water content.

The process of defloculation involves the use of one or more agents whose purpose is to impart a charge to the surface of the clay particles, thereby allowing those particles to repel each other in the slurry, allowing less water (which must subsequently be removed by spray drying or some other process) in the slurry and reducing the viscosity of the slurry to facilitate pumping.

Common deflocculating agents used in the ceramics industry include: trisodium phosphate, sodium silicate, and sodiumhexametaphosphate. Recent work with acrylate defloculants has shown considerable promise.

The total milling time for the slurry (or slip) is usually governed by the percentage of residue in the slurry which will not pass through a sieve of some known dimension. A representative residue content would be 8 to 9% on a 100 mesh sieve. Generally, this would be the result obtained by milling a typical tile formula of 1700 kg. total weight for 8 hours.

After milling, the slip is stored for up to 24 hours, generally in underground storage tanks, to allow mixing with slip from other mills and to allow the formula to come to equilibrium.

Body preparation

The most common method today of preparing the powdered body material for the forming process is spray drying. Older methods included filter pressing and/or drum drying, both with subsequent granulation processes to size the powder for the forming process. These processes, except for special cases, however, have been outmoded by the spray drying process.

A spray dryer is a large, cylindrical, vertically oriented chamber into which the slurry is pumped under pressure through nozzles of controlled orifice size. Heated air (500 to 600°C) is forced upward into the chamber. Since the natural state of least free energy of the droplets of slip is a sphere, the resultant product of the spray dryer is spherical powder particles of controlled, generally uniform, grain size, with 5-6% residual moisture, ready (after an aging step) for pressing. This aging step, usually twenty four hours, is necessary to allow the residual moisture in the particles to equilibrate, since the outside of the particle is dryer than the inside immediately after spray drying.

⌘⌘⌘

Pressing

Tile shapes are pressed on hydraulic presses , ranging from 500 to 2500 tons of capacity, using steel- or rubber-lined steel dies. Spray dried powder lends itself ideally to hydraulic, rapid, automatic pressing. The characteristic free flowing nature and uniform particle size of these powders are ideal for rapid, uniform filling of die cavities, even for somewhat complex trim shapes.

The pressing cycle, as illustrated in Figure 3 comprises several steps:

1) Die filling. Several cavities on one press platen are filled at one time by distribution equipment designed to spread the powder uniformly over the several voids.
2) First pressing. The first press stroke is designed to remove the air entrapped between the powder particles.
3) Deaeration. Die pressure is relaxed to vent the liberated air.
4) Pressing to final thickness and strength. This step determines the final properties of the formed pieces.
5) Ejection from the cavity and removal from the press.

Figure 3. The Tile Pressing Cycle

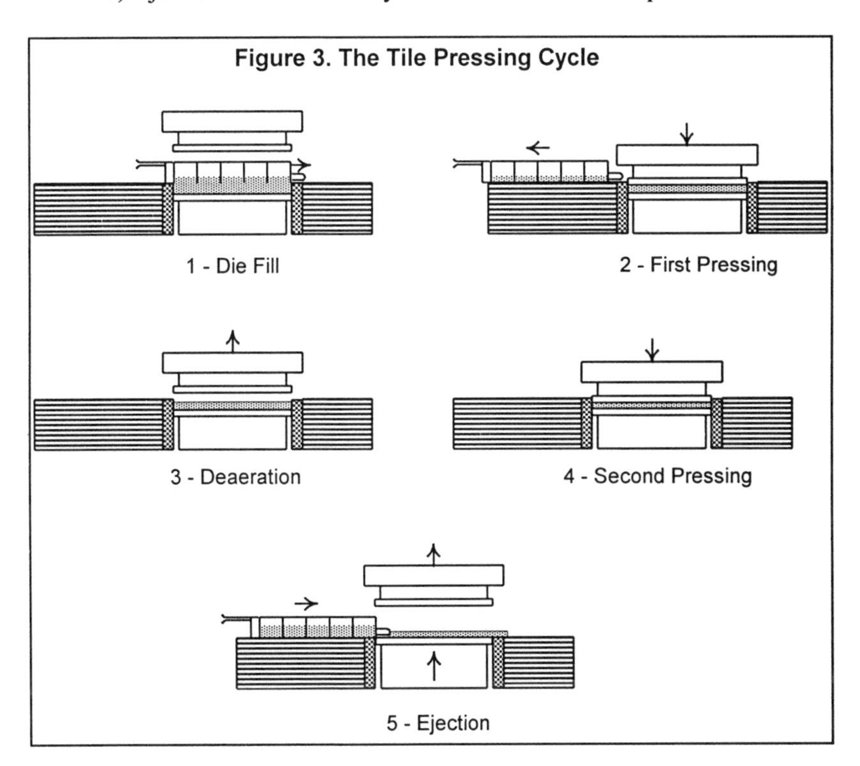

⌘⌘⌘

While there are a great many old friction presses still in operation in various parts of the tile industry, and isostatic pressing is being used in other areas , such as sparkplug blank forming, hydraulic presses are the backbone of the industry due to their consistent pressing pressures during the pressing cycle and their ability to run at consistent cadence for long periods.

Drying

The 5-6% residual moisture in the pressed piece must be removed prior to firing of the tile to prevent the tile from exploding due to a sudden formation of steam inside a sealed surface. Dryers are either vertical or horizontal depending on floor space available, although horizontal radiant dryers are being used more frequently because they require less handling of the "green" tile and result in less breakage and/or chipping of the tile prior to glazing or firing.

At this point, the tile process may take several directions. Glazed floor tile, usually referred to as monocottura (or single firing) is next glazed, decorated and subsequently fired to form the finished product. Most wall tile products are first fired at this point to prepare a "Bisquit" tile , not glazed or decorated. This provides a stronger tile to be subjected to the multiple glazing and decorating processes associated with the normally more highly decorated wall tiles. Unglazed "paver" tiles, in which the finished tile product color has been incorporated into the body of the tile, will go direcly to the kiln at this point. This text will assume the manufacture of single fired glazed floor tile at this point, and proceed with the process steps with that in mind. The general outline of these specific processes, however, holds for all types of tile and whiteware maufacture.

Glazing

The types and styles of glaze surfaces and glazing techniques are as varied as the imaginations of the equipment manufacturer, the glaze suppliers, and the tile manufacturers. The one general constant, however, is the preparation of the glazes themselves prior to application. Almost all glazes in the ceramic tile industry are ball milled prior to application on the tile. In cases where older tunnel kilns are used for firing, raw glazes are the norm. Raw glazes are those compounded directly from the minerals used as raw materials with no intermediate step of smelting (or "fritting") the glaze. Fritted glazes (or frits) are commonly used in the fast firing tile industry because of the high amount of pre-reaction obtained during the fritting process. With firing times from 35 to 55 minutes, the tile manufacturer can ill afford the time necessary for raw glazes to react during the firing cycle.

Nearly all tile glazes are milled in alumina-lined mills using alumina grinding media. The same general rules for body milling apply in glaze

⌘⌘⌘

preparation. Important control parameters are the viscosity and specific gravity of the glaze slip. These parameters, together with the speed of the glazing line, determine whether or not a uniform coating of glaze is applied. Non-uniform glaze thickness will contribute to shading problems on the finished product.

The actual glaze application may be by one of several methods. The oldest of these is spraying with compressed air nozzles. Spraying results in uniform coatings with little or no glaze on the tile edges due to the "shadow" effect of a point glazing source. Sprayed glazes are most commonly used in glazing of complex trim shapes. The coating applied by spraying may not be dense and uniform enoughfor fast firing, however, due to the discontinuous nature of the applied glaze. Other variations of spraying include disco applications whereby the glaze is pumped through the hollow shaft of a high speed motor and centrifugally applied through rapidly spinning discs mounted on the shaft.

Waterfall glazing, using a denser glaze slip, is the preferred method for fast fire glazing, especially for high gloss wall tiles. There are two basic types of waterfall apparatus, the veil method and the "bell" waterfall. Both are similar in final effect, but the bell is easier to maintain because of the lack of a small orifice which may clog in operation. Dry glazing, the application of dry fritted glazes to the surface of the tile, is gaining more acceptance, especially for rustic styled floor tiles.

Decoration is generally by screen printing of colors suspended in a medium (usually called " pasta") and applied through masked screens (one color per screen) to the glazed, but not fired tile. Roll printing, rotary screens, pad printing, and dry glazed screen printing may also be used, depending on the desired end effect.

Firing

Firing is the main step in any ceramic process. Firing is characterized by a number of complex physical changes and chemical reactions which determine the final characteristics of the finished tile product.

– Up to 100°C, the elimination of hygroscopic water takes place. This moisture may be present due to incomplete drying prior to glazing, or as a result of moisture reabsorbed during the glazing process.
– Up to 200°C, the zeolitic water, molecular water bound to the crystalline lattice by absorption, is eliminated.
– From 350 to 650°C, the combustion of organic substances usually present in clays occurs. This is followed by the dissociation of sulfides and sulfates with the resultant emission of sulfurous acid.

⌘⌘⌘

– Combined water, with the resultant alteration of crystal lattices, is expelled between 450 to 650°C.
– At 573°C, the transformation of alpha to beta quartz occurs with an abrupt increase in volume.
– CO_2 is volatilized between 800 and 950°C due to the decomposition of lime and dolomite.
– Beginning at 700°C, new crystalline phases begin to form. These phase consist of silicates and complex aluminosilicates. These phases form the final crystalline structure of the tile.
– Glass phases holding the crystalline matrix together begin to form from 1000° up to 1200°C depending on the body formulation. Glaze melting begins in this region as well.

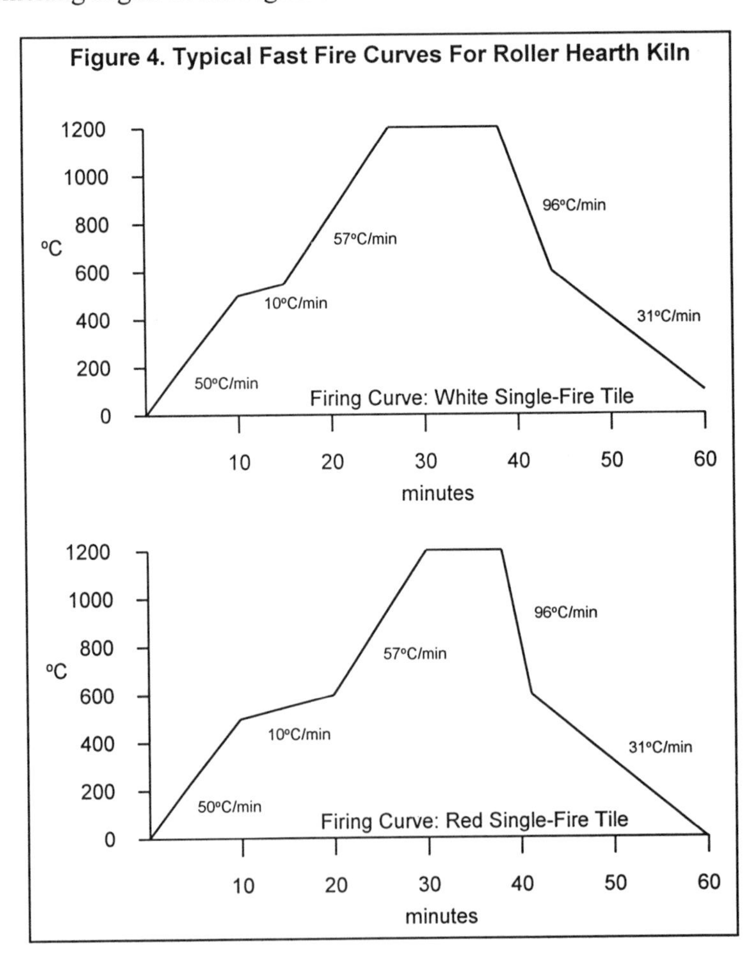

⌘⌘⌘

Figure 4 illustrates a typical tile fast firing curve for a roller hearth kiln. Two basic types of kilns remain in service today with the older tunnel kilns being gradually replaced or outnumbered by roller hearth rapid fire kilns, illustrated in Figure 5. Firing in tunnel kilns is done on refractory setters, either manually or automatically loaded. Tunnel kilns remain in wide use today for smaller tile sizes such as 41/4 x 41/4 inch. Roller hearth kilns fire the tile on ceramic rollers, almost always automatically loaded.

Figure 5. Electrically Heated Roller Kiln

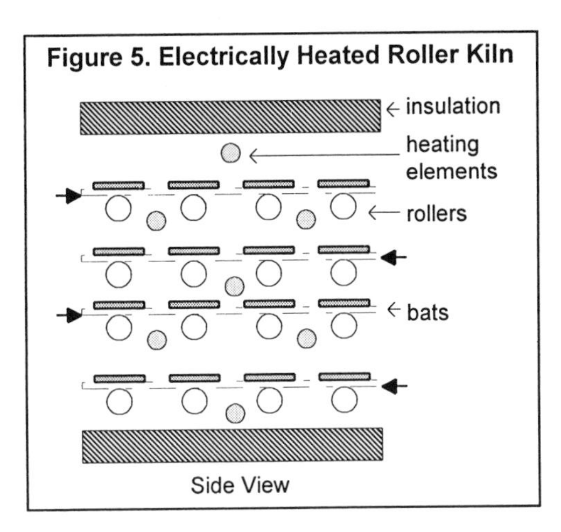

Sorting and packaging

Sorting and packaging of ceramic tiles is an automatic operation throughout the world with few exceptions. Physical dimensions, breaking strength, and color can be automatically inspected for and the product sorted accordingly. Only visual defects such as chips and dirt spots require visual inspection. Packaging is automated, generally utilizing wrap-around cartons of corrugated paper. Palletizing and transport to warehouse storage are also highly automated involving conveyors or automatically guided robot vehicles.

Other Whiteware Industries

The probability that materials other than those common in tiles will be used in the production of sanitaryware and/or dinnerware is slim. The degree to which the various material choices are made is determined by the individual manufacturer. In general, the sanitaryware manufacturer uses cleaner materials, due to the fact that he must produce vitrified porcelain bodies with higher purity materials to meet sanitary codes. Generally, porcelain bodies are similar in composition to those of ceramic mosaic bodies. A higher percentage of white burning clays is used, however, to give the required

⌘⌘⌘

casting properties, since all sanitaryware bodies are slip cast, either in the traditional bench casting methods or the newer pressure casting techniques.

The artware industry is generally composed of lower firing bodies similar to tile compositions, but not noted for significant final strength due to the " art" usage rather than the functional usage of the pieces produced. Some different uses of materials exist, however. Salt or rock salt is used in the firing process of some artware pieces to impart a luster to the finished piece not easily obtained by other methods. This process is normally avoided in conventional manufacture due to the wear and tear produced in kilns from the corrosive nature of the salt.

Artware glazes may incorporate some minerals directly on the surface of the piece for a desired effect. These include chalcedony, pyrolusite and garnet among others.

REFRACTORIES

Refractories are materials, generally ceramics, which are employed where resistance to very high temperature is required, as for furnace linings and metal-melting pots. Materials with a melting point above Seger cone 26, or 1580°C, are called refractory, with those above cone 36, or 1790°C, being highly refractory. Temperature resistance alone, however, does not determine the choice of materials as refractories. In addition to the ability to resist softening and deformation at operating temperatures, other factors such as load bearing capacity, resistance to slag attack, and spalling resistance must be considered. In some cases heat transfer characteristics and electrical resistivity are important.

Many refractories are derived directly from natural minerals, but synthetic materials are also widely use. Clays, especially those having inherent temperature resistance, are the oldest and most common of the naturally occurring refractory minerals. Major natural refractory materials are kaolin, chromite, bauxite, zirconia, and magnesite. These are often marketed under specific trade names. Refractory materials may be acid, such as silica, or basic, such as magnesite or bauxite, for use in acid- or basic-process steel furnaces. Graphite and chromite are generally considered neutral refractories.

Magnesia is insoluble in the slag of open-hearth furnaces and is generally used for linings. The magnesite source may be economically distilled from seawater and further processed. Refractories made with 60 to 80% alumina will withstand temperatures from 1810 to 1835°C.

The chief artificial refractory materials are silicon carbide and aluminum oxide. Ninety eight percent pure silicon carbide can be used up to 2300°C

⌘⌘⌘

before decomposition occurs, while the melting point of 99.5% pure alumina is given as 2050°C.

BIBLIOGRAPHY

Anonymous, 1995, *From Technology Through Machinery to Kilns for SACMI Tile*, Vol. 1 & 2, SACMI, Imola, Italy

Brady, G.S., and Clauser, H.R., 1977, *Materials Handbook*, 11th Ed., McGraw-Hill, New York

SITI, 1993, *La Technologia Ceramica*, Vol. 1-7, Societa Impianti Termoelettrici Industrialli, Marano Tcino, Italy

⌘⌘⌘

NOTES

⌘⌘⌘

TWELVE

FORMULARY

The following formulas are provided as examples of the use of industrial minerals as performance fillers. They are not specifically intended as commercially preparable compositions, but as a means of illustrating the utilization of these minerals within the context of an entire preparation. Each industry has its own conventions in the presentation of suggested formulas, with more or less information provided on sources of supply, compounding procedures, and evaluation properties. Those interested in detailed information on any of the formulas which follow should consult the source noted under the formula title.

PAINT FORMULAS

Acrylic Latex Semi-Gloss Enamel

(American Colloid Co.)

	Pounds
Water	326.5
Bentonite[1]	9.2
Disperse at high speed for 10 minutes, then add	
Tamol® 731	9.2
Igepal® CO-630	3.3
Nopco® NXZ	1.0
Titanium Dioxide[2]	200.0
Calcium Carbonate[3]	50.0
Disperse at high speed for 15 minutes, let down	
UCAR® 5053	37.0
Texanol®	13.9
Nopco NXZ	2.8
Propylene Glycol	43.2
Triton® GR-7M	1.1
PMA 18	1.0
NH_4OH (28%)	1.8
Total Weight	1000.0

[1]No. 353 Bentonite
[2]KRONOX® 2020 or equivalent
[3]Vicron® 15-15 or equivalent

⌘⌘⌘

PVA Interior Latex Flat White

(Union Carbide)

	Pounds
Water	180.00
Ethylene Glycol	15.00
KTPP	2.00
Tamol® 731	6.00
Cosan® PMA-10	0.12
Defoamer	1.00
Titanium Dioxide[1]	150.00
Clay[2]	150.00
Calcium Carbonate[3]	150.00
Disperse at high speed for 10 minutes, then add at slow speed	
Water	259.00
Triton® N-101	2.00
Texanol®	8.00
Natrasol® 250 HBR	5.00
UCAR® 367	225.00
Defoamer	1.00
Total	1154.12

[1]KRONOS® 2131 or equivalent
[2]Huber 70C or equivalent
[3]Vicron® 25-11 or equivalent

⌘⌘⌘

PVA Exterior Latex House Paint
(Union Carbide)

	Pounds
Water	133.30
Ethylene Glycol	28.00
Tamol® 850	14.00
KTPP	2.00
Triton® N-101	2.00
Cosan® PMA-100	1.50
Colloid® Defoamer 691	1.00
Zinc Oxide[1]	50.00
Titanium Dioxide[2]	250.00
Clay[3]	150.00
Disperse at high speed for 15 minutes, then add at slow speed.	
Water	164.50
Natrasol® 250 HBR	4.50
Texanol®	12.00
UCAR® 367	379.00
Colloid 691	2.00
Total	1193.80

[1]Zinc Oxide, No. 503
[2]KRONOS® 2102 or equivalent
[3]Huber 70C or equivalent

⌘⌘⌘

Water-Based Red Epoxy Primer

(Rhone-Poulenc)

	Pounds
Epi-Rez® WJ-5522	330.0
Diacetone Alcohol	7.0
Defoamer	3.5
Water	60.0
Red Iron Oxide[1]	71.1
Barium Sulfate[2]	71.1
Wollastokup® 10 ES	106.7
Halox® SW-111	100.0
Zeeosphere® 400	71.1
Mica	7.5
Disperse at high speed to a Grind of 6. Then slowly add:	
Epi-Rez WJ-5522	147.0
Water	14.2
Then add to above a pre-mixed solution of:	
CMD J60-8290 Curing Agent	60.0
Water	110.0

[1]RO-4097 Red Iron Oxide or equivalent
[2]No. 106 Lo-micron® White Barytes or equivalent

⌘⌘⌘

Acrylic Latex White House Paint

(NYCO)

	Pounds
Water	125.0
Ethylene Glycol	23.2
Cosan® PMA-100	1.5
KTPP	2.0
Tamol® 850	6.0
Triton® N-101	2.0
Defoamer	1.0
Zinc Oxide[1]	50.0
Titanium Dioxide[2]	250.0
NYAD® 400 Wollastonite	50.0
Calcium Carbonate[3]	100.0
Disperse at high speed for 15 minutes	
Water	21.0
Attagel® 50	5.0
Disperse at high speed for 5 minutes, then add at slow speed:	
Water	197.84
Natrasol® 250 HBR 4.0	
Texanol®	11.88
Rhoplex® AC-64 336.0	
Defoamer	1.00
Total	1186.42

[1]Zinc Oxide, No. 503
[2]KRONOX 2102 or equivalent
[3]Vicron 25-11 or equivalent

⌘⌘⌘

Interior Egg Shell Tint Base

(Rohm & Haas)

	Pounds
Propylene Glycol	52.0
Tamol® 731	10.0
Dowicil® 75	1.0
Natrasol® 250 GR	50.0
Triton® CF-10	2.0
PAG-188	2.0
Add the following while mixing at low speed:	
Titanium Dioxide[2]	250.0
Silica[3]	75.0
Disperse in a high speed mill for about 20 minutes and let down at low speed with:	
Water	50.0
Rhoplex® AC-417	422.0
Texanol®	15.0
Propylene Glycol	11.0
PAG-188	4.0
Triton® N-57	4.0
Premix next 3:	
Acrysol RM-4	14.0
Water	28.6
Ammonium Hydroxide (28%)	1.2
Water/Natrasol 250 GR (4%)	100.0
Water/Ammonium Hydroxide	8.3
Total	1100.1

[1]To adjust pH to 8.6 to 9.0
[2]KRONOS® 2020 or equivalent.
[3]No. 20 Silica

⌘⌘⌘

White Metal Primer

(National Starch & Chemical)

	Pounds
Natrasol® 250 HR (1.5% soln.)	89.0
Potassium Tripolyphosphate	1.0
Tamol® 731	4.0
Triton® CF-10	5.0
Raybo® 60	8.0
Nopco® NXZ	1.0
Ethylene Glycol	20.0
Titanium Dioxide[1]	125.0
Calcium Carbonate[2]	50.0
Clay[3]	75.0
Silica[4]	50.0
Mica[5]	25.0
Cosan® 635 W	1.0
Disperse in a high speed mill, then add slowly while mixing at low speed:	
Premix next 3:	
Arolon® 585	100.0
Cobalt Drier	1.0
Zirconium Drier (6%)	1.5
Nacrylic® 8403	468.0
Nopco NXZ	1.0
Ammonium Hydroxide (28%)	1.0
Dowanol® DPM 28.0	
Water	52.0
Total	1106.5

[1]KRONOS® 2101 or equivalent
[2]Vicron® 41-8 or equivalent
[3]Huber 70C or equivalent
[4]No. 31 Silica
[5]No. 262 Mica

⌘⌘⌘

Concrete Floor Paint

(Union Oil)

	Pounds
Water	249.9
Ethylene Glycol	27.9
Natrasol® 250 MR	5.0
Polywet® ND-2	5.0
Igepal® CO-630	2.0
2-amino-2-methyl-1-propanol	1.0
PMA-30	1.0
Colloid® 681-F	2.0
Titanium Dioxide[1]	200.0
Lo-micron® Barytes[2]	50.0
Mica[3]	25.0
Disperse at high speed to a 5 Grind, then add at low speed:	
Water	166.6
Colloid® 681-F	2.0
*CEE-5 Emulsion	73.3
AMSCO RES® 6190	249.0
Total	1059.7

*Drier treated with 0.2% zirconium, 0.1% cobalt, 1% calcium.

[1]KRONOX® 2020 or equivalent
[2]Barytes, No. 1
[3]No. 262 Mica

⌘⌘⌘

White Traffic Marking Paint
(Diamond Shamrock)

	Pounds
Titanium Dioxide[1]	150.0
Calcium Carbonate[2]	325.0
Talc[3]	100.0
Mica[4]	50.0
Metasap® V	3.3
Alkyd Solution (50% solids)	443.8
Lead Drier (24% lead)	8.7
Cobalt Drier (12% cobalt)	1.2
Toluene	85.5
Anti-Skinning Agent	0.7
Total	1168.2

[1]KRONOS® 2010 or equivalent
[2]Vicron® 41-8 or equivalent
[3]Talc IT-3X® or equivalent
[4]No. 262 Mica or equivalent

⌘⌘⌘

PVA White Masonry Paint

(Cargill)

	Pounds	Gallons
Premix:		
Water	262.0	31.50
STRODEX® PK-90 Dispersant	2.0	0.21
Ammonium Hydroxide (28%)	4.0	0.50
Triton® X-100 Surfactant	5.0	0.60
Potassium Tripolyphosphate	1.0	------
Increase Speed and Add Slowly:		
Chemacoil® TA-10	78.0	10.0
Disperse Pigments:		
Tri-Pure® R-901 Titanium Dioxide	200.0	5.72
Nytal® 300 Talc	200.0	8.40
Duramite® Calcium Carbonate	150.0	6.65
Slow Mixing Speed and Add Slowly:		
Colloid® 581-B Defoamer	1.0	0.13
Premix the next 2:		
Ethylene Glycol	18.6	2.00
Nuodex PMA-18 Preservative	1.5	0.15
Cellosize QP-15000 (2%) Thickener	126.0	15.00
Everflex BG Vinyl Acetate Copolymer Emulsion (52%)	188.0	20.80
Super-Cobalt Drier	0.5	0.06
	1237.6	101.72

Viscosity - Fresh	87 Krebs Units
Overnight	90 Krebs Units
4 Months	90 Krebs Units
Pigment Volume Concentration	51.0%
Chemacoil TA-100 on Vehicle Solids	44%

Excellent durability over a 10-year period of exposure in the Chicago area.

⌘⌘⌘

Quality White House Paint
(Union Carbide)

	Pounds	Gallons
Pigment Grind:		
Water	271.9	32.64
CELLOSIZE® Hydroxyethyl Cellulose ER-30M	4.0	0.36
Dispersant[1]	10.5	1.00
Potassium Tripolyphosphate	2.0	0.10
Nonionic Surfactant[2]	2.2	0.25
2-Amino-2-methyl-1-propanol[3]	1.0	0.13
Propylene Glycol	25.9	3.00
Antifoam[4]	1.9	0.25
Titanium Dioxide[5]	225.0	6.54
Zinc Oxide[6]	25.0	0.54
Mica[7]	25.0	1.07
Talc[8]	175.0	7.37
Letdown:		
Opaque Polymer[9]	---	---
UCAR® Latex 367	408.3	45.12
UCAR Filmer IBT	8.9	1.13
Mildewcide[10]	2.2	0.25
Antifoam[4]	1.9	0.25
Total	1190.7	100.00

[1]Dispersant - Tamol® 960 Rohm & Haas
[2]Nonionic Surfactant - Triton® N-101 Rohm & Haas
[3]2-Amino-2-methyl-1-propanol - AMP-95 Angus
[4]Antifoam - Colloid® 640 Colloids
[5]Titanium Dioxide - TiPure® R-900 DuPont
[6]Zinc Oxide - Pasco® 311 Pacific Smelting
[7]Mica - Mica 325WG KMG
[8]Talc - IT-325, Vanderbilt
[9]Opaque Polymer - Ropaque® OP-62 Rohm & Haas
[10]Mildewcide - Skane® M8 Rohm & Haas

⌘⌘⌘

Paint Properties:

PVC, %	40.1
Solids, %	
by Volume	38.7
by Weight	57.7
Viscosity (initial)	
Stormer, KU	85
ICI, poise	0.8
pH (initial)	8.5
Weight per Gallon, lb.	11.91
Freeze-Thaw Stability, 3 cycles	Pass (+4 KU)
Heat Stability, 2 weeks at 120°F	Pass (+3 KU)

Film Properties:

Contrast Ratio, 3 mil film	0.964
Reflectance, %	91.9
Sheen, 85°	5.9

⌘⌘⌘

Quality White House Paint (Reduced TiO_2)
(Union Carbide)

	Pounds	Gallons
Pigment Grind:		
Water	234.2	28.12
Cellosize® Hydroxyethyl Cellulose ER-30M	4.2	0.39
Dispersant[1]	10.5	1.00
Potassium Tripolyphosphate	2.0	0.10
Nonionic Surfactant[2]	2.2	0.25
2-Amino-2-methyl-1-propanol[3]	1.0	0.13
Propylene Glycol	25.9	3.00
Antifoam[4]	1.9	0.25
Titanium Dioxide[5]	175.0	5.09
Zinc Oxide[6]	25.0	0.54
Mica[7]	25.0	1.07
Talc[8]	100.0	4.21
Letdown:		
Opaque Polymer[(9)]	101.9	11.85
UCAR® Latex 367	382.4	42.25
UCAR® Filmer IBT	9.9	1.25
Mildewcide[(10)]	2.2	0.25
Antifoam[(4)]	1.9	0.25
Total	1105.2	100.00

[1]Dispersant - Tamol® 960 (Rohm & Haas)
[2]Nonionic Surfactant - Triton® N-101 (Rohm & Haas)
[3]2-Amino-2-methyl-1-propanol - AMP-95 (Angus)
[4]Antifoam - Colloid® 640 (Colloids)
[5]Titanium Dioxide - TiPure® R-900 (DuPont)
[6]Zinc Oxide - Pasco® 311 (Pacific Smelting)
[7]Mica - Mica 325WG (KMG)
[8]Talc - IT-325® (Vanderbilt)
[9]Opaque Polymer - Ropaque® OP-62 (Rohm & Haas)
[10]Mildewcide - Skane® M8 (Rohm & Haas)

⌘⌘⌘

Paint Properties:

PVC, %	44.0
Solids, %	
by Volume	38.8
by Weight	53.0
Viscosity (initial)	
Stormer, KU	83
ICI, poise	0.8
pH (initial)	8.7
Weight per Gallon, lb.	11.05
Freeze-Thaw Stability, 3 cycles	Pass (+4 KU)
Heat Stability, 2 weeks at 120°F	Pass (+4 KU)

Film Properties:

Contrast Ratio, 3 mil film	0.962
Reflectance, %	92.6
Sheen, 85°	16.2

⌘⌘⌘

Weather Barrier Formulation

(Georgia Marble)

	Pounds	**Gallons**
Water	143.3	17.3
Cellosize® ER-52M[1]	4.0	0.3
Colloid® 226/35[2]	8.0	0.8
KTPP	2.0	---
Ethylene Glycol	20.0	2.2
Surfynol® 104[3]	1.2	0.2
Colloid 643[2]	1.5	0.2
Ti-Pure® R-960[5]	103.0	3.2
AZO®-55[6]	40.6	0.9
Gamaco®[4]	385.0	17.1
Nopcocide® N-96[7]	5.0	0.5
Grind at high speed, then let down with:		
Texanol®[8]	6.0	0.8
UCAR® Vehicle 123[1]	496.6	55.9
Colloid 643[2]	2.0	0.3
28% Ammonium Hydroxide	2.5	0.3
Total	1220.7	100.0

Physical Properties

Weight per Gallon	12.2lbs.
% Solids by Weight	68.5
% Solids by Volume	53.8
PVC	40.0%
Viscosity, KU	110

[1]Union Carbide
[2]Rhone-Poulenc
[3]Air Products
[4]Georgia Marble
[5]DuPont
[6]ASARCO
[7]Diamond Shamrock
[8]Eastman Chemicals

⌘⌘⌘

TT-P-1975 Federal Specification Metal Primer

(Georgia Marble)

	Pounds	Gallons
Water	266.4	32.0
Cellosize® ER-15M[4]	1.0	0.1
Colloid® 111M[1]	15.2	1.7
Triton® CF-10[2]	2.7	0.3
Ethylene Glycol	26.8	2.9
AMP 95[3]	2.0	0.3
Nuosept®-95[5]	2.7	0.3
Colloid 770[1]	2.4	0.3
Ti-Pure® R-900[6]	133.9	3.9
Nalzin® 2[7]	89.2	2.7
325 WG Mica[8]	3.7	1.5
Calwhite® II[9]	46.5	2.1
Grind at high speed, then let down with:		
(Premix next 2)		
Butyl Carbitol®[4]	7.2	0.9
Acrysol® SCT-275[4]	8.2	1.0
UCAR® Vehicle 163[4]	423.6	47.6
Troysan Polyphase® AF-1[10]	5.8	0.6
Colloid 770[1]	4.8	0.6
28% Ammonium Hydroxide	9.0	1.2
Totals	1083.1	100.0

Physical Properties

Weight per Gallon	10.8lbs.
% Solids by Volume 3	6.5
PVC	27.9%

[1]Rhone-Poulenc
[2]Rohm & Haas
[3]Angus
[4]Union Carbide
[5]Huls
[6]DuPont
[7]NL Chemicals
[8]KMG Minerals
[9]Georgia Marble
[10]Troy

⌘⌘⌘

Coarse Finish Texture Coating

Designed for trowel application

(Georgia Marble)

	Pounds	Gallons
76 RES 1018	272.0	32.00
2% Sol'n of Natrosol® 250 HBR	35.0	4.18
50% Sol'n KTPP	8.0	0.67
MichemLube® 511	8.0	1.00
Super Ad-It®	7.0	0.79
Mineral Spirits	33.3	5.00
Texanol®	31.7	4.00
Tronox® CR-800	50.0	1.45
Pyrax® WA	100.0	4.29
Georgia Marble #8	475.0	21.11
Georgia Marble 40-200	200.0	8.89
Georgia Marble OZ	375.0	16.67
Nopco® 8034	7.0	0.93
Totals	1602.0	100.98

Physical Properties

Weight per Gallon	15.9lbs.
% Solids	
By Weight	84.4
By Volume	69.0
PVC	77.0
% Binder	8.5
Pigment:Binder Ratio	8.82:1

⌘⌘⌘

Economy Flat Wall Paint

(Georgia Marble)

	Pounds	Gallons
Water	391.7	47.02
Cellosize® ER-30M[1]	5.5	0.50
AMP-95[2]	1.0	0.13
Tamol® 731[3]	4.6	0.50
Nuosept® 95[4]	2.3	0.25
PA-454[5]	1.0	0.13
KTPP	1.0	0.05
Triton® N-101[1]	2.2	0.25
Propylene Glycol[6]	13.0	1.50
Tronox® CR-813[7]	75.0	2.34
ASP-NC2[8]	200.0	9.30
#10 White[9]	150.0	6.65
Celite® 281[10]	25.0	1.30
Grind at high speed; slow mill, then add:		
PA-454	1.9	0.25
UCAR® Filmer IBT[1]	8.1	0.54
Ultrafiltrate Latex 50-872-UF[1]	220.0	24.80
Total	1,102.3	95.51

Physical Properties:

Stormer, KU	100
ICI, poise	0.9
pH	9.0
Weight/Gallon:	11.38lbs.
% Solids, By Volume:	29.4%
PVC:	68.0%

[1]Union Carbide
[2]Angus
[3]Rohm & Haas
[4]Huls
[5]U.S. Movidyn
[6]Union Chemical Div.
[7]Kerr-McGee
[8]Engelhard Corp.
[9]Georgia Marble
[10]Manville

⌘⌘⌘

High Quality Flat Wall Paint
(Georgia Marble)

	Pounds	Gallons
Water	330.6	39.69
Cellosize® ER-30M[1]	4.5	0.41
Nuosept® 145 [2]	2.3	0.25
Tamol® 731[3]	5.7	0.62
Triton® N-101[1]	2.2	0.25
Tergitol® NP-40[1]	2.2	0.25
AMP-95[4]	1.0	0.13
Propylene Glycol	21.6	2.50
Colloid® 640[5]	1.9	0.25
Tronox® CR-813[6]	175.0	5.47
Satintone® W[7]	175.0	7.99
#10 White[8]	100.0	4.43
Minex®[9]	475.0	3.40
Grind at high speed; slow mill, then add:		
Latex 379[1]	282.7	31.20
Butyl Carbitol® Solvent[1]	13.8	1.75
Colloid 640[5]	1.9	0.25
Propylene Glycol	10.0	1.16
Total	1,205.4	100.00

Physical Properties

Stormer, KU	92
ICI, poise	1.0
pH	8.4
Weight/Gallon	12.05#
% Solids, By Volume	37.8%
% Solids, By Weight	56.9%
PVC	57.0%

[1]Union Carbide
[2]Huls
[3]Rohm & Haas
[4]Angus
[5]Rhone-Poulenc
[6]Kerr-McGee
[7]Engelhard Corp.
[8]Georgia Marble
[9]Unimin

⌘⌘⌘

Interior Flat Wall Paint
(Georgia Marble)

	Pounds	Gallons
Water	125.0	15.00
Nuosept® 95[1]	2.5	0.21
Tamol® 731[2]	9.2	1.00
Igepal® CO-630[3]	3.5	0.40
AMP-95[4]	2.0	0.25
Ti-Pure® R-901[5]	190.0	5.67
Satintone® W[6]	125.0	5.71
Calwhite®[7]	60.0	2.66
Attagel® 50[8]	3.0	0.15
Colloid® 643[3]	1.5	0.21
Grind at high speed; slow mill, then add:		
Water	288.5	34.63
Fulatex® PD-124[9]	243.0	27.00
Texanol®[10]	10.0	1.25
Premix next 2:		
Ethylene Glycol[11]	6.5	5.00
Natrosol® Plus[12]	6.5	0.58
Colloid 643[3]	2.0	0.28
Total	1118.2	100.00

Physical Properties

Weight/Gallon	11.18lbs.
% Solids, by volume	30.1%
% Solids, by weight	47.5%
PVC	50.2%

[1]Huls
[2]Rohm & Haas
[3]Rhone-Poulenc
[4]Angus
[5]DuPont
[6]Engelhard
[7]Georgia Marble
[8]Engelhard
[9]H.B. Fuller Co.
[10]Eastman Chemical
[11]Unocal Chemical
[12]Aqualon

⌘⌘⌘

Vinyl Acrylic Interior Flat Paint
(Georgia Marble)

	Pounds	Gallons
Premix first 6 ingredients:		
Water	250.00	30.01
Natrosol® H4BR[1]	3.75	0.35
AMP-95[9]	2.00	0.25
Nopcosperse® 44[2]	2.60	0.29
Igepal® CO-640[3]	2.00	0.25
Drewplus® L-475[4]	2.00	0.28
Ti-Pure® R-931[5]	200.00	6.06
Gamaco®[6]	100.00	4.43
Glomax® LL[7]	100.00	4.56
Attagel® 40[8]	2.00	0.10
Grind at high speed; slow mill, then add:		
Water	190.50	22.87
Ethylene Glycol	27.90	3.00
Canguard® 327[9]	1.50	0.18
Texanol®[10]	12.60	1.60
Drewplus® L-475[4]	1.00	0.14
UCAR® Latex 351[11]	234.60	25.50
AMP-95	1.00	0.13
Total	1133.45	100.00

Physical Properties

Stormer, KU	80-85
Weight/Gallon	11.3lbs.
% Solids, By Volume	31.5%
% Solids, By Weight	50.0%
PVC	48.0%

[1]Aqualon
[2]Henkel Corp.
[3]Rhone-Poulenc
[4]Drew Ind. Div.
[5]DuPont
[6]Georgia Marble
[7]Dry Branch Kaolin Co.
[8]Engelhard Corp.
[9]Angus Chemical
[10]Eastman Chemical
[11]Union Carbide Corp.

⌘⌘⌘

Premium Exterior Semi-Gloss
(Georgia Marble)

	Pounds	Gallons
Water	66.6	8.00
Colloid® 226/35[1]	6.2	0.60
Nuosept® 95[2]	1.5	0.17
AMP-95[3]	1.5	0.19
Polyphase® AF-1[4]	4.0	0.42
Igepal® CO-630[5]	3.5	0.40
Tiona® RCL-2[6]	250.0	7.50
Gamaco®[7]	20.0	0.89
ASP®-170[8]	30.0	1.40
Drewplus® L-464[9]	1.8	0.25
Grind at high speed; slow mill, then add:		
Fulatex® PD-449[10]	400.5	45.00
Water	218.4	26.22
Premix the next 4:		
Propylene Glycol[11]	47.6	5.50
Texanol®[12]	15.1	1.90
Rheolate® 278[13]	7.0	0.82
Natrosol® 250 MHBR[14]	2.7	0.24
Drewplus® L-464[8]	3.7	0.50
Total	1080.1	100.00

Physical Properties

Stormer, KU	84
Weight/Gallon	10.9#
% Solids, by volume 3	7.5%
% Solids, by weight	51.9%
PVC	27.3%

[1]Rhone-Poulenc
[2]Huls
[3]Angus Chemical
[4]Troy Chemicals
[5]Rhone-Poulenc
[6]SCM Chemicals
[7]Georgia Marble
[8]Engelhard Corp.
[9]Drew Ind. Div.
[10]H.B. Fuller Co.
[11]Dow Chemical
[12]Eastman Chemical
[13]Rhoex, Inc.
[14]Aqualon Co.

⌘⌘⌘

Semi-Gloss Powder Coating

(U.S. Silica)

	phr
Cargill® 3000[1] (saturated OH functional polyester)	100.0
Cargill 2400[1] (blocked urethane curing agent)	22.0
Modaflow® Powder II[2]	1.1
R900 TiO_2[3]	49.0
MIN-U-SIL® 5[4]	52.0

Processing Parameters

–Weigh all components into a mixer
–Premix 60 seconds at 100o rpm and 60 seconds at 2000 rpm in the high intensity mixer (Welex or equivalent)
–Extrude in Buss PR-46 (or equivalent) zone 1 at 80°C and zone 2 at 150°C
–Collect extruded material through cooled pinch rolls
–Grind in a 10 analytical mill (Tekmar or equivalent)
–Sieve through a 100 mesh U.S. Standard screen

Performance Properties (Bake 30 min. at 360°F)

Film Thickness 1.8 mils
Pencil Hardness F-H
Gel Time at 400°F, 190 Seconds
Excellent Impact Resistance
Excellent Exterior Durability

[1]Cargill
[2]Monsanto
[3]DuPont
[4]U.S. Silica

⌘⌘⌘

Latex Interior Flat

(Georgia Marble)

	lbs./100 gal.
Water	350.0
QP-15,000[1]	3.5
Tamol® 731[2]	4.0
Triton® N-101[2]	4.0
KTPP	1.0
Stir until smooth, add:	
R-900[3]	90.0
SNOW*TEX® 45[4]	164.0
MIN-U-SIL® 30[4]	102.0
MIN-U-GEL® 400[5]	6.0
Drew Plus® L-475[6]	0.6
Dowcil® 75[7]	1.0
Exxate® 1300[8]	8.2
High speed disperse to 5+ Hegman	
Drew Plus L-475[6]	0.4
Water	80.6
Ammonia	0.8
Wallpol® 40-136[9]	275.0
	1091.1

60°C Gloss -	2.0	Contrast Ratio -	98.6
Viscosity -	82 KU at 25°C	Reflectance -	88.3%
PVC -	47.9%	lbs./gal. -	10.9
Pigment-	33.2%	N.V.V. -	31.8%
		N.V.W. -	48.0%

Gloss, contrast ratio, and reflectance measured at 1.9 mil dry film thickness.

[1]Union Carbide
[2]Rohm & Haas
[3]DuPont
[4]U.S. Silica
[5]Floridin
[6]Drew Chemical
[7]Dow
[8]Exxon
[9]Reichhold Chemical

⌘⌘⌘

Fire Resistant Interior/Exterior Flat White Latex Paint

(Georgia Marble)

Water	350.0
KTPP	1.0
Ethylene Glycol	25.0
Natrosol® 250 MR[1]	5.0
Nopco® NDW[2]	2.0
Tamol® 731[3] 25%	7.0
Texanol®[4]	5.0
PMA-30[5]	0.5
Titanox® 2030[6]	250.0
SNOW*TEX® 45[7]	100.0
MIN-U-SIL® 10[7]	25.0
Camel Carb®[8]	75.0
Microcel® T-38[9]	10.0
Zinc Borate 2335 Fine1[10]	20.0
Disperse above on a Cowles Mixer	
Igepal® CO-630[11]	3.0
Nopco NDW[2]	2.0
Wallpol® 40-136[12]	290.0
Water	30.0
Kloro® 7000	58.0
	1258.5

PVC	45.6%	Viscosity	102 KU at 25°C
N.V.W.	55.8%	Constrast Ratio (.003" wet)	.961
N.V.V.	37.4%	Reflectance (.003" wet)	.892
pH	7.8		

[1]Aqualon
[2]Henkel
[3]Rohm & Haas
[4]Eastman Chemical
[5]Troy
[6]Kronox
[7]U.S.Silica
[8]Genstar
[9]Johns Manville
[10]U.S. Borax
[11]GAF
[12]Reichhold Chemicals

⌘⌘⌘

Heavy Metal Free Water-Borne Barrier Primer

(Georgia Marble)

	lbs./100 gal.
Becksol® 13-402 (Reichhold Chemicals)	194.0
Triethylamine	11.9
Surfynol® 104 B.C. (Air Products)	8.5
Butyl Cellosolve®	36.6
Secondary Butyl Alcohol	36.6
Mix10 minutes	
MIN-U-SIL® (U.S. Silica)	50.2
Vicron® 1515 (Pfizer)	51.3
YLO 3288D (Pfizer)	149.8
Talcron® MP40-27 (Pfizer)	12.0
Black Tint Base (See Below)	76.9
High speed disperse to a 5 n.s.	
Nuocure® 10% Cobalt (Huls)	0.5
Nuocure 6% Manganese (Huls)	1.8
Exkin® #2 (Huls)	2.6
Add slowly:	
Deionized Water	392.2
	1024.9

Black Tint Base	**lbs./100 gal.**
Ball Mill To A 7 + N.S.	
Beckosol® 13-402	490.8
Dimethyl Ethanolamine	27.9
Surfynol® 104 B.C.	2.8
Butyl Cellosolve	245.4
Special Black #5 (Degussa)	92.2
	859.1

Viscosity -	66 KU at 25°C	P/B -	1.52/1
Wt./gal. -	10.2	PVC -	0.35
N.V.W. -	44.6%	pH -	8.4
N.V.V. -	29.6%	VOC, gm/l - (calculated)	380

⌘⌘⌘

Film Properties

The following typical film properties were obtained with a spray applied film of 1.0 mil over CRS and Bonderite 1000. Panels were force dried 10 minutes at 250°F then air dried 7 days.

	Polished CRS	**Bonderite 1000**
Gloss 85°/60°/20°	45/28/4	
Pencil Hardness	HB-F	HB-F
Direct Impact, Inch lbs.	120	140
Reverse Impact, Inch lbs.	80	80
500 Hour Salt Fog		
Scribe Creep	1/16-1/8"	0-1/16"
Scribe Blisters	Medium 4	Few 4
Field Blisters	None	None
500 Hour QCT (Humidity)		
Blisters	None	None

⌘⌘⌘

Exterior Acrylic Flat Formulation

(Rheox)

	Trade Names	% By Weight
Water	---	20.0
pH Stabilizer	NH_4OH	0.17
Mix 10 Min. Under Shear		
Defoamer	Nopco® NXZ	0.25
Dispersants	KTPP	0.08
	Tamol® 850	0.50
Biocide	Nuosept® 95	0.08
Wetting Agent	Igepal® CO-610	0.25
Propylene Glycol	---	1.68
Coalescent	Texanol®	1.02
Titanium Dioxide	Kronos® 2101	20.96
Talc	Nytal® 300	10.47
Zinc Oxide	Kadox® 515	4.19
Thickener	Benaqua® 4000	0.72
Mix Under Shear for 15 Min.		
Water	---	5.83
Fingicide	Skane® M-8	0.17
Acrylic Resin	R & H AC-235	33.63
		100.00

PVC	37.2
Solids - Wt.	56
Solids - Vol.	39
W.P.G.	11.80

⌘⌘⌘

Corrosion Inhibiting, Lead And Chromate Free, VOC Compliant Primer

(Cargill)

	Pounds	Gallons
Cargill® High Solids Phenolic Modified Alkyd 57-5754	246.2	28.24
Synthetic Red Iron Oxide RO-3097 (Harcros Pigments, Inc.)	310.6	7.61
Zinc Phosphate JO852 (Mineral Pigments Corporation)	61.1	2.29
Sicorin® RZ (BASF Corporation)	6.5	0.31
Nytal® 400 (R.T. Vanderbilt)	223.5	9.43
Sparmite® (Harcros Pigments, Inc.)	56.1	1.53
Methyl Isoamyl Ketone	104.2	15.38
Grind to a 5 Hegman		
Cargill® High Solids Phenolic Modified Alkyd 57-5754	242.4	27.80
Manosec® Cobalt 18% (Manchem, Inc.)	1.0	0.11
10% Calcium Cem-All (Mooney Chemicals, Inc.)	3.6	0.45
Manosec Zirconium 18% (Manchem, Inc.)	3.0	0.33
Activ-8® (R. T. Vanderbilt Co.)	1.2	0.15
Exkin® No. 2 (Huls America, Inc.)	1.2	0.16
Methyl Isoamyl Ketone	42.1	6.21
	1302.7	100.00

Paint Properties

% Nonvolatile (By Weight)	78.98
(By Volume)	59.56
Pigment to Binder Ratio	1.78
Pigment Volume Concentration	35.74
Weight Per Gallon	13.03
Theoretical VOC at 4 to 1 Dilution (Grams/Liter)	420
% Pigment Solids (By Wt. of Primer)	50.58
% Vehicle Solids (By Wt. of Primer)	28.12
Theoretical VOC (Pounds/Gallon)	2.74
(Grams/Liter)	328
Viscosity Krebs Stormer (KU)	100
(Thinned 4 to 1 with TT-T306 Type II)	

⌘⌘⌘

Typical Cured Film Properties

Gloss (60°)	5-15
Dry Times (Hours)	
Dry Hard	3/4
Dry Through	4
Salt Spray	Pass 366 hours

⌘⌘⌘

Fast Dry Semi-Gloss Mint Green Enamel

(Cargill)

	Pounds	Gallons
Cargill® High Solids		
Short Tofa Alkyd 57-5720	206.6	23.89
Hitox® Titanium Dioxide (Hitox Corp.)	82.6	2.41
Heliogen® Green L 8690 (BASF Corporation)	1.6	0.09
Supercoat® (E.C.C. America)	292.5	12.96
Nytal® 400 (R.T. Vanderbilt Co.)	18.0	0.76
Aerosil® R972 (Degussa Corporation)	3.5	0.19
Bykumen® (Byk-Chemie USA)	7.0	0.95
Methyl Isoamyl Ketone	65.9	0.72
Grind to a 6 Hegman		
Cargill High Solids		
Short Tofa Alkyd 57-5720	206.6	23.88
Methyl Isoamyl Ketone	27.1	3.99
Butyl Carbitol® (Union Carbide)	2.1	0.26
Exkin No. 2 (Huls America, Inc.)	0.8	0.11
Methyl Isoamyl Ketone	1.2	0.17
Manosec® Cobalt 18% (Manchem, Inc.)	1.0	0.11
Manosec Zirconium 18% (Manchem)	3.8	0.42
Activ-8® (R.T. Vanderbilt Co.)	1.4	0.17
Methyl Isobutyl Ketone	133.1	19.92
	1054.8	100.00

Paint Properties

% Nonvolatile (By Weight)	67.76
(By Volume)	49.42
Pigment to Binder ratio	1.26
Pigment Volume Concentration	33.40
Weight Per Gallon	10.55
Theoretical VOC (Pounds/Gallon)	3.40
(Grams/Liter)	407
Viscosity #4 Ford Cup at 77°F (Seconds)	68

Typical Cured Film Properties

Cure Schedule	7 Days Air Dry
Substrate	Cold Rolled Steel

⌘⌘⌘

Gloss (60°)	60
Pencil Hardness	Pass 3B
% Crosshatch Adhesion	100
Impact (In. Lbs.)	
Direct	105
Reverse	< 5

⌘⌘⌘

Water Reducible Air Dry Stain

(Cargill)

	Pounds	Gallons
Cargill® Water Reducible 7350	156.5	18.93
Ammonium Hydroxide - 28%	8.7	1.00
AMP-95 (Commercial Solvents Corp.)	2.1	0.27
Deionized Water	238.7	28.66
Butyl Cellosolve®	10.1	1.35
Mapico Red #347 (Cities Service)	38.6	0.89
Nytal® 300 (R.T. Vanderbilt)	77.1	3.24
Patcote® 577 (C. J. Patterson)	2.3	0.34
BYK®-301 (Byk-Mallinckrodt)	0.8	0.10
Disperse, then add:		
Cargill® Water Reducible 7350	156.5	18.93
Ammonium Hydroxide - 28%	10.8	1.24
(Premix next 4)		
Butyl Cellosolve	34.2	4.55
Activ-8® (R.T. Vanderbilt)	0.8	0.10
12% Cobalt (Interstab)	0.6	0.07
6% Manganese Intercar (Interstab)	2.0	0.26
Butyl Cellosolve	7.2	0.96
Deionized Water	159.2	19.11
	906.4	100.00

% Non-Volatiles (By Weight)	42.39
(By Volume)	36.02
Pigment/Binder Weight Ratio	0.43/1
Solvent VOC	2.10 lbs./gal.
252 gms/liter	
Viscosity	75-95 K.U.

⌘⌘⌘

RUBBER COMPOUNDS

Natural Rubber For Tear Strength, Low Heat Build-Up & Heat Resistance

(PPG)

Banbury 1		Banbury 2	
Natural Rubber	100	Zinc Oxide	4
Hi-Sil® 233	50	Sulfur	0.5
HPPD	2	TBBS	2
TMQ	2	MDB	2
100°C Aromatic resin	5		
Hydrogenated rosin	6		
Tall oil	1		
PEG 3350	2		
Stearic acid	2		

Properties:

Cure Rate 138°C, T_{90}	25 minutes
Mooney Scorch 130C, T_5	10
Mooney Viscosity, ML100	66

Cure: 30' / 138°C	Original	Aged 70hrs/121°C
Durometer	57	70
M300, Mpa	3.9	
Tensile, Mpa	29	12
Elongation, %	69	440

MG Trouser Tear	51 kN/m
PICO Abrasion Index	82
Compression Set, 70hrs/100°C	53
DeMattia Cut Growth, 500% growth	40 KC
Goodrich Flexometer: 100°C; 22.5%; 1mpa	
Set	7%
Heat Build-up	17°C

⌘⌘⌘

SBR Rice Husking Roll Cover

(PPG)

Banbury 1		Banbury 2	
SBR1502	80	Zinc oxide	3
SBR1900	20	Insoluble sulfur 80%	9
NR	10	TBBS	2
Hi-Sil® 233s	90	DTDM	1
Mercaptosilane A189	2.4		
100C Resin	10		
ZMTI	1		
Stearic acid	4		
Zinc octoate	2		
PEG3350	2		

Properties:

Specific gravity		1.26
Cure Rate 150°C, T_{90} minutes		26
Mooney Scorch 121C, T_5 minutes		19
Viscosity, ML100		66
Garvey Extrusion;	Die Swell	2%
	Edge - Surface rating	8A
Cure: 40'/150°C		
Durometer	At 23°C	91
	At 100°C	82
M100, Mpa		11.0
Tensile		20.8
Elongation		230%
Pendulum Rebound Z	At 23°C	23
	At 100°C	60
PICO Abrasion, mg loss	At 70°C	25
	At 100°C	45

⌘⌘⌘

Neoprene Compound

(R.T. Vanderbilt)

Neoprene® WHV	100
Octylated Diphenylamine[1]	2
Mixed Diaryl-p-Phenylenediamine[2]	2
Magnesium Oxide[3]	4
Stearic Acid	1
Hard Clay[4]	120
Paraffinic Wax	1
Petrolatum	1
Vulcanized Vegetable Oil[5]	10
Butyl Oleate	10
Aromatic Oil ASTM D-2226 type 102[6]	85
Zinc Oxide	5
Ethylene Thiourea, 75% in binder[7]	1
	342

[1]Agerite® Stalite® S
[2]Vanox® 2-AZ
[3]Maglite® D
[4]Dixie Clay®
[5]Neophax® A
[6]Sundex® 790
[7]END-75

Specific Gravity	1.38

Stock Properties:

Mooney Scorch (MS121°C)	
Minimum Viscosity, units	3
Time to Ten Point Rise, Minutes	27

Vulcanizate Properties:

Cure: Press - 15 minutes at 160°C

Stress-Strain and Hardness:	*Original*	*Oven Aged/70 hrs/100°C*
Tensile Strength, MPa (psi)	7.6 (1100)	9.1 (1320)
Elongation, %	900	860
100% Modulus, Mpa (psi)	0.55 (80)	
Hardness, Duro A	27	36
Die C Tear, kN/m (pli)	15.8 (90)	

⌘⌘⌘

EPDM For Textile Adhesion

(PPG)

Banbury 1		Banbury 2	
EPDM501	100	HMT	1.5
SILENE® 732d	120	Sulfur	1.5
Paraffinic oil	55	TBBS	1
Zinc oxide	5	ZBDC	2
Stearic acid	2	TMTD	0.8
Amide wax	0.5		
Resorcinol	2.5		

Properties:

Specific gravity	1.18
Cure Rate 154°C, T_{90} minutes	28
Mooney Scorch 121°C, T_5 minutes	19
Viscosity ML100	25

Cure : 30'/154°C	Original	Aged 70 hrs/150°C
Durometer	55	68
M300, Mpa	2.7	
Tensile	7.8	9.5
Elongation, %	740	320

Compression Set, 70 hrs /100°C 79%

Adhesion to Untreated Fabrics, kN/m

Fabric	Cure	Peel Strength
Nylon	60'/138°C	5.0
	60'/154°C	11
	30'/168°C	10
Rayon	60'/160°C	8.0
Activated Polyester	60'/160°C	5.0

⌘⌘⌘

EPDM For Super Heat Resistance

(PPG)

Banbury 1		Banbury 2	
EPDM 5206	100	Sulfur	0.4
Hi-Sil® 532 EP	60	ZBDC	2
Mercapto silane A189	1.2	MDB	2
Paraffinic oil	20	MBTS	1.2
ZMTI	1.5	PEG3350	1
TMQ	1		
NBC	3.5		
Stearic acid	2		
Zinc oxide	10		

Properties:

Specific gravity	1.16
Cure Rate 154°C, T_{90} minutes	21
Mooney Scorch 121, T_5 minutes	21
Viscosity, ML100	78

Cure: 30 /154C	Original	Aged 28days/165°C	Aged 7days/190°C
Durometer	62	65	64
Tensile, Mpa	14.8	15.6	4.4
Elongation, %	700	475	150

Compression Set:	
70 hours/100°C	54.0
70 hours/150°C	74.3

Monsanto Fatigue to failure, 100%, KC	215
DeMattia Cut Growth, 500% growth, KC	100+

⌘⌘⌘

Roll Cover
(R.T. Vanderbilt)

Nordel® 1070	95.0
Hypalon® 40	5.0
Zinc Oxide	5.0
Stearic Acid	1.0
Hard Clay[1]	100.0
Paraffinic Oil ASTM D-2226 Type 104B[2]	40.0
Sulfur	2.0
2-Mercaptobenzothiazole[3]	1.5
Tetramethylthiuram Disulfide[4]	0.8
Dipentamethylene Thiuram Tetrasulfide[5]	0.8
	251.1

[1]Dixie Clay®
[2]Sunpar® 150
[3]Captax®
[4]Methyl Tuads®
[5]Sulfads®

Specific Gravity	1.22

Vulcanizate Properties:
Slab Cure: 20 min. at 160°C
Stress-Strain and hardness:

Tensile Strength, Mpa (psi)	11.37 (1650)
Elongation, %	650
Hardness, Duro A	70

⌘⌘⌘

Neoprene For Textile Adhesion

(PPG)

Banbury 1		Banbury 2	
Nroprene GNA	60	Zinc oxide	5
Neoprene GRT	40	Resorcinol resin	1
BR1220	5		
Hi-Sil 233	45		
Magnesium oxide	3		
ODPA	1		
Aromatic oil	10		
Hydrog. castor oil	5		
Amide wax	1		
Stearic acid	2		
HMMM	2		

Properties:

Specific gravity	1.35
Cure Rate 154°C, T_{75} minutes	16
Mooney Scorch 121°C, T_5 minutes	17
Viscosity, ML100	57

Cure: 30'/154°C	Original	Aged 170 hours/100°C
Durometer	61	71
M300, Mpa	3.9	
Tensile	19	17
Elongation, %	850	550

MG Trouser tear, kN/m	100
DeMattia Cut Growth, 500% growth	13 KC

Adhesion to untreated fabric, kN/m	
Nylon	10.0
Activated polyester 811	12.7
Rayon	7.0
Cotton	8+
RFL Nylon	18+

Low temperature stiffness ASTM D1053:		
	T10	-43°C
	T100	-45°C

⌘⌘⌘

NBR Colored Hose Cover

(PPG)

NBR 1042	100
Silene® 732D	140
DOP	35
25°C CI resin	5
Stearic acid	1
Zinc oxide	3
Sulfur	0.2
TBBS	1
TMETD	3

Properties:

Cure Rate 150°C, T_{90} minutes		8
Mooney Scorch 121°C, T_5 minutes		30
Viscosity, ML100		68
Garvey Die Extrusion:	Swell	12%
	Rating	10A (glossy surface)

Cure: 20'/150°C	Original	Aged 70 hrs/121°C
Durometer	72	85
M300, Mpa	5.1	9.3
Tensile	5.9	9.3
Elongation, %	620	300
Die C Tear, kN/m	37	37

Compression Set, 70 hrs/100°C 46%

Fluid Immersions	#3 Oil/70 hrs/100°C	Fuel B/170 hrs/70°C
% Volume change	-2.8	7.3
Durometer	81	55
Tensile, Mpa	6.6	5.9
Elongation, %	450	280
Die C Tear, kN/m	37	18

⌘⌘⌘

Air Hose Cover

(R.T. Vanderbilt)

Nordel® 1145	95.0
Hypalon® 40	5.0
Zinc Stearate	1.0
Magnesium Silicate[1]	100.0
Precipitated Hydrated Amorphous Silica	150.0
Paraffinic Oil ASTM D-2226 Type 104B[2]	90.0
Red Iron Oxide	10.0
Sulfur	2.5
Zinc Dibutyldithiocarbamate[3]	2.0
Tetramethylthiuram Disulfide[4]	0.8
2-Mercaptobenzothiazole[5]	1.0
	457.3

[1]Vantalc® 6H
[2]Sunpar® 150
[3]Butyl Zimate®
[4]Methyl Tuads®
[5]Captax®

Specific Gravity	1.35
Stock Properties:	
Mooney Scorch (MS 121°C)	
Minimum Viscosity, units	18
Time to Ten Point Rise, Minutes	28

Vulcanizate Properties:
Cure: 30 min. at 153°C
Stress-Strain and Hardness

Tensile Strength, Mpa (psi)	8.79 (1275)
Elongation, %	600
100% Modulus, Mpa (psi)	2.34 (340)
Hardness, Duro A	59

⌘⌘⌘

Polybutadiene / NR Athletic Footwear Sole

(PPG)

Banbury 1		Banbury 2	
BR1220	70	Sulfur	2
Natural rubber	30	TBBS	2
Hi-Sil® 233	50	HMT	2
100°C Resin	10	Zinc oxide	1
ODPA	1		
Titanium dioxide	3		

Properties:

Specific gravity 1.14	
Cure Rate 160°C, T_{90} minutes	4.3
Mooney Scorch 130°C, T_5 minutes	15
Viscosity, ML100	108

Cure: 10'/150°C	Original	Aged 170 hrs 110°C
Durometer	72	82
M300, Mpa	4.8	
Tensile	22	21
Elongation, %	825	555

MG Trouser Tear, kN/m		18 (knotty)
PICO Abrasion index		135
Compression Set, 70 hrs 100°C		88%
DeMattia Cut Growth, 500%, KC		40
Pendulum Rebound (Z) , %	23°C	56
	100°C	61

⌘⌘⌘

High Tear Strength / Low Heat Build-up Compound

(PPG)

Banbury 1		Banbury 2	
Natrual Rubber	100	Sulfur	1
Hi-Sil® 233	48	MBS	2
TMQ	1	MDB	1
HPPD	2	TBTD	0.3
Process aid MS40	3		
Stearic acid	2		
PEG3350	1		
Zinc oxide	4		

Properties:

Specific gravity	1.13	
Cure rate 138°C, T_{90}	16 minutes	
Mooney scorch 130°C, T_5	13 minutes	
Mooney viscosity ML100	75	
Cure; 20'/138°C	Original	Aged 170 hrs 100°C
Durometer	62	77
M300, Mpa	5.3	
Tensile	32	21
Elongation , %	715	445
MG Trouser tear, kN / m		
At 23°C	22	
At 100°C	27	
PICO Abrasion Index	77	
Pendulum rebound, At 23°C	70.1	
At 100°C	79.3	
Goodrich Flexometer:		
100°C; 25%; 1.6 Mpa (blow-out conditions)		
Permanent set	20%	
Heat build-up	55°C (no blow-out)	
Dynamic Mechanical Analysis:		
1 Hz; 20% flexural strain		
E′ @ 30°C	14.9	
E′ @ 90°C	11.3	
E″ @ 0°C	2.4	
Tandelta @ 80°C	0.118	
DeMattia Cut growth, 500%	20 KC	

⌘⌘⌘

Vacuum Control Tubing

(R.T. Vanderbilt)

Nordel® 1145	100.0
Zinc Oxide	5.0
Stearic Acid	1.0
Magnesium Silicate[1]	100.0
Precipitated Hydrated Amorphous Silica	150.0
Paraffinic Oil ASTM D-2226 Type 104B[2]	75.0
2-Mercaptobenzothiazole[3]	1.0
Tetramethylthiuram Monosulfide[4]	1.5
Zinc Dimethyldithiocarbamate[5]	1.5
Sulfur	2.5
	437.5

[1]Vantalc® 6H
[2]Sunpar® 150
[3]Captax®
[4]Unads®
[5]Methyl Zimate®

Specific Gravity	1.35

Stock Properties:

Mooney Scorch (MS 121°C)	
Minimum Viscosity, units	48
Time to Ten Point Rise, Minutes	23

Vulcanizate Properties:

Steam Cured: 20 min. at 80 psi	
Stress-Strain and Hardness	
Tensile Strength, Mpa (psi)	8.27 (1200)
Elongation, %	760
100% Modulus, Mpa (psi)	2.75 (400)
Hardness, Duro A	66

⌘⌘⌘

White Tread
(PPG)

Banbury 1		Banbury 2	
Natural rubber	70	Zinc oxide	3
SBR 1502	30	PEG3350	1.5
Mercapto silane A189	0.8	Sulfur	2.8
Hi-Sil® 210	50	MBS	2.0
Titanium dioxide	5		
ODPA	1		
100C resin	10		
Stearic acid	2		
MC wax	2		

Properties:

Specific gravity	1.17
Cure Rate 150°C, T_{90}	14 minutes
Mooney scorch 130°C, T_5	20 minutes
Mooney viscosity ML100	60

Cure: 20'/150°C	Original	Aged 70 hrs/100C
Durometer	67	71
M300, Mpa	6.2	
Tensile	23	16
Elongation, %	610	430
MG Trouser tear , kN/m	115	
PICO Abrasion index	85	

Pendulum Rebound: @ 23°C	56
@ 100°C	73
Goodrich Flexometer:	
100°C;22.5%;1PMa	21°C Heat Build-up
DeMattia Cut Growth 500%	40 KC

⌘⌘⌘

Low Rolling Resistance Tread
(PPG)

Banbury 1		Banbury 2	
Natural rubber	100	Sulfur	2.8
Hi-Sil® 255	40	TBBS	3
100C Resin	5		
ODPA	1		
Stearic acid	2		
PEG 3350	1		
Zinc oxide	4		

Properties:

Cure rate 150C, T_{90}	20 minutes
Mooney scorch 121°C, T_5	30+ minutes
Mooney viscosity, ML100	58

Cure: 30'/150°C	Original	Aged 700 hrs/85°C
Durometer	64	72
M300, Mpa	5.8	
Tensile	28	15
Elongation, %	660	430

MG Trouser tear, kN/m	18
PICO Abrasion index	90

Pendulum Rebound, %	@ 23°C		@ 100°C
Zwick	63.1		78.8
Goodyear	78.1		84.7
Compression Set, 70 hrs/100°C		56%	
Goodrich Flexometer:			
100°C;22.5%;1MPa		15 C HBU	
Dynamic Mechanical Analysis :			
1 Hz;20% strain; 30°C			
E' , Mpa		7.0	
E"		0.38	
Tan delta		0.055	
DeMattia Cut Growth, 500%		15 KC	

⌘⌘⌘

Traffic Cones Yellow

(R.T. Vanderbilt)

Nordel® 1145	95.0
Hypalon® 40	5.0
Zinc Oxide	5.0
Stearic Acid	1.0
Hard Clay[1]	50.0
Precipitated Hydrated Amorphous Silica	100.0
Paraffinic Process Oil[2]	60.0
Yellow Iron Oxide	10.0
Titanium Dioxide[3]	5.0
Sulfur	2.0
2-Mercaptobenzothiazole[4]	1.0
Zinc Dibutyldithiocarbamate[5]	2.0
Tetramethylthiuram Disulfide[6]	0.8
	336.8

[1]Dixie Clay®
[2]Sunpar® 150
[3]Ti-Pure® R-960
[4]Captax®
[5]Butyl Zimate®
[6]Methyl Tuads®

Specific Gravity	1.27

Stock Properties:

Mooney Scorch (MS 121 C/250°F)	
Minimum Viscosity, units	28
Time to Ten Point Rise, Minutes	12

Vulcanizate Properties:

Cure: 8 min. at 160°C	
Stress-Strain and Hardness	
Tensile Strength, Mpa (psi)	10.0 (1450)
Elongation, %	650
Hardness, Duro A	69

⌘⌘⌘

Translucent Footwear Soling

(PPG)

Banbury 1		Banbury 2	
SBR 8107	60	Zinc oxide	3
IR 2200	40	sulfur	2.5
Hi-Sil® 233	60	MBS	0.6
Naphthenic oil	30	TMTM	0.6
ODPA	1	DOTG	0.6
Stearic acid	1		
PEG3350	3		

Properties:

Specific gravity	1.13
Cure rate 150°C, T_{90}	6.0 minutes
Mooney scorch 121°C, T_5	15 minutes
Mooney Viscosity ML100	40

Cure: 6'/150°C	
Durometer	55
M300, Mpa	1.7
Tensile	12
Elongation	760%
NBS Abrasion index	80

⌘⌘⌘

SBR/NR For Heat Resistance

(PPG)

Banbury 1		Banbury 2	
SBR 1215	60	Sulfur	0.4
Natural rubber	40	Vultac 7	1.5
Ciptane® 1	31	TBBS	3.0
TMQ	2	OTOS (TCS)	0.5
HPPD	2	Zinc oxide	5
Stearic acid	2		
PEG 3350	1		

Properties:

Specific gravity	1.10
Cure rate 155C, T_{90}	8.9 minutes
Mooney scorch 121°C, T_5	29 minutes
Mooney Viscosity ML100	30

Cure: 10'/155°C	Original	Aged 700 hrs/100°C	Aged 170 hrs/125°C
Durometer	56	64	65
M300, Mpa	3.8		
Tensile	23	11	13
Elongation, %	765	345	360

Compression Set, 70hrs 100°C	56.8%
PICO Abrasion index	50
Pendulum Rebound, % At 23°C	55.4
At 100°C	68.8
Goodrich Flexometer: 100°C;22.5%;1MPa	
Permanent Set	3.3%
Heat Build-up	14°C

Dynamic Mechanical Analysis, 1 Hz; 10% flexural strain

	Original	Aged 340 hrs/100°C
E' -20C, MPa	46	32
E' +10C	7.0	8.4
E' + 60C	4.6	6.0
Tan delta -20C	0.78	0.79
Tan delta +30C	0.158	0.148

⌘⌘⌘

Nordel Compound
(R.T. Vanderbilt)

Nordel® 2744	70.0
Nordel® 2760	25.0
Hypalon® 40	5.0
Zinc Oxide	5.0
Stearic Acid	1.0
Precipitated Hydrated Amorphous Silica	50.0
Soft Clay[1]	230.0
Paraffinic Oil ASTM D-2226 Type 104B[2]	115.0
Red Iron Oxide	10.0
Sulfur	2.0
Di Benzothiazyl Disulfide[3]	1.0
Tetramethylthiuram Disulfide[4]	0.5
Zinc Dibutyldithiocarbamate[5]	2.0
	516.5

[1]McNamee® Clay
[2]Sunpar® 150
[3]Altax®
[4]Methyl Tuads®
[5]Butyl Zimate®

Specific Gravity 1.40

Stock Properties:
Mooney Scorch (MS 121 C/250°F)

Minimum Viscosity, units	15
Time to Ten Point Rise, Minutes	18

Vulcanizate Properties:
Mold Cure: 20 min. at 160°C
Stress-Strain and Hardness

Tensile Strength, MPa (psi)	7.93 (1150)
Elongation, %	560
100% Modulus, MPa (psi)	2.96 (430)
Hardness, Duro A	63
Die C Tear, kN/m (pli)	30.64 (175)

Compression Set, %, Method B, Pellets Cured 25 Minutes at 160°C
22 Hours at 70°C (153°F) 42

⌘⌘⌘

SBR/NR For Heat Resistance

(PPG)

Banbury 1		Banbury 2	
SBR 1215	60	Sulfur	0.4
Natural rubber	40	Vultac 7	1.5
Ciptane® 1	31	TBBS	3.0
TMQ	2	OTOS (TCS)	0.5
HPPD	2	Zinc oxide	5
Stearic acid	2		
PEG 3350	1		

Properties:

Specific gravity	1.10
Cure rate 155°C, T_{90}	8.9 minutes
Mooney scorch 121°C, T_5	29 minutes
Mooney Viscosity ML100	30

Cure: 10'/155°C	Original	Aged 700 hrs/100°C	Aged 170 hrs/125°C
Durometer	56	64	65
M300, Mpa	3.8		
Tensile	23	11	13
Elongation, %	765	345	360

Compression Set, 70hrs/100°C		56.8%
PICO Abrasion index		50
Pendulum Rebound, %	At 23°C	55.4
	At 100°C	68.8

Goodrich Flexometer:
100°C;22.5%;1MPa

Permanent Set	3.3%
Heat Build-up	14°C

Dynamic Mechanical Analysis:
1 Hz; 10% flexural strain

	Original	Aged 340 hrs 100°C
E' -20C, MPa	46	32
E' +10C	7.0	8.4
E' + 60C	4.6	6.0
Tan delta -20°C	0.78	0.79
Tan delta +30C	0.158	0.148

⌘⌘⌘

Electric Connector Insulation

(R.T. Vanderbilt)

Nordel® 1320	100.0
Zinc Stearate	1.0
Zinc Oxide	5.0
Magnesium Silicate[1]	75.0
Surface-modified Calcined Kaolin[2]	75.0
Paraffinic Oil ASTM D-2226 Type 104B[3]	30.0
Paraffin Wax[4]	5.0
Vinyl Functional Silane[5]	1.0
Zinc Dibutyldithiocarbamate[6]	2.0
DiBenzothiazyl Disulfide[7]	1.0
Tetramethylthiuram Disulfide[8]	0.8
Sulfur	1.5
	297.3

[1]Vantalc® 6H
[2]Translink® 37
[3]Sunpar® 2280
[4]Vanwax® H Special
[5]Vinyl Silane A-172
[6]Butyl Zimate®
[7]Altax®
[8]Methyl Tuads®

Specific Gravity	1.3608

Vulcanizate Properties:

Cure: 20 minutes at 160°C (320°F)

Stress-Strain and Hardness

Tensile Strength, Mpa (psi)	9.31 (1350)
Elongation, %	750
100% Modulus, Mpa (psi)	3.44 (500)
Hardness, Duro A	60

⌘⌘⌘

Tarpaulin Strap

(R.T. Vanderbilt)

Nordel® 1145	90.0
Nordel® 1440	10.0
Zinc Oxide	5.0
Stearic Acid	1.0
Hard Clay[1]	100.0
Paraffinic Oil[2]	50.0
Titanium Dioxide[3]	20.0
Tetramethylthiuram Disulfide[4]	0.8
2-Mercaptobenzothiazole[5]	1.0
Zinc Dibutyldithiocarbamate[6]	2.0
Sulfur	2.0
	281.8

[1]Dixie Clay® [4]Methyl Tuads®
[2] Sunpar® 150 [5]Captax®
[3]Ti-Pure® R-960 [6]Butyl Zimate®

Specific Gravity 1.26

Vulcanizate Properties:

Cure: 8 min. at 160°C

Stresss-Strain and Hardness	
Tensile Strength, MPa (psi)	12.41 (1800)
Elongation, %	780
100% Modulus, MPa (psi)	0.68 (100)
Hardness, Duro A	48

⌘⌘⌘

ADHESIVES & SEALANTS

Dextrin Carton Sealing Adhesive

(Washabaugh)

	Weight%
Dextrin	30.0
Borax	6.0
4.8 µm Kaolin (Water Washed)	6.0
Water	58.0

Packaging Adhesive

(Washabaugh)

	Weight%
Polyvinyl Alcohol	9.0
4.8 µm Kaolin (Water Washed)	4.5
+325 Mesh Silica	4.5
Water	82.0

Solid Fiber Laminating Adhesive

Low Viscosity, High Wet Tack

(Washabaugh)

	Weight%
Water	82.0
ASP® 600	10.8
Airvol® MM81	7.2

Viscosity: 600-800 Mpa.S @ 25 C Brookfield RVF, 20 rpm
Solids Content: 18%
pH: 4.5 - 4.8
Preparation:
Add the clay to the water and mix thoroughly. Add the Airvol and mix thoroughly. Adjust the pH, if necessary, to 4.5 - 4.8 using phosphoric acid. Heat with mixing to 190 - 195 °F. Mix at this temperature for 30 to 45 minutes. Cool to 100 - 125°F and draw off or use.

ASP 600 - 0.6µm kaolin (Engelhard Corp.)
Airvol MM81 - polyvinyl alcohol resin (Air Products and Chemicals Inc.)

⌘⌘⌘

Spiral Tube Winding Adhesive

Moderate Viscosity, Moderate Wet Tack

(Washabaugh)

	Weight%
Water	70.0
ASP 600	23.0
Airvol MM-51	7.0

Viscosity:1700-1900 Mpa.S @ 25 C Brookfield Rvf, 20 Rpm
Solids Content:20%
pH: 4.5 - 4.8

Preparation:
Add the clay to the water and mix thoroughly. Add the Airvol and mix thoroughly. Adjust the pH, if necessary, to 4.5 - 4.8 using phosphoric acid. Heat with mixing to 190 - 195°F. Mix at this temperature for 30 to 45 minutes. Cool to 100 - 125°F and draw off or use.

ASP 600 - 0.6μm kaolin (Engelhard Corp.)
Airvol MM51 - polyvinyl alcohol resin (Air Products and Chemicals Inc.)

⌘⌘⌘

Vinyl Wall Covering Adhesive
(Union Carbide)

	Weight%
Ucar® Latex 160	52.23
Cellosize® 100m, 4% Solution	8.82
Daxad®30	0.98
Butyl Carbitol®	1.76
ASP® 900	19.60
Nytal® 300	4.90
Attagel® 50, Thickener	1.47
PMO-36, Preservative	0.04
Balab® 748	0.02

Solids Content: 61%
Viscosity: Spreadable Paste

Ucar Latex 160 - acrylic latex (Union Carbide Corp.)
Nytal 300 - talc (R.T.Vanderbilt Co. Inc.).
ASP 900 - kaolin clay (Engelhard Corp.)
Attagel 50 - colloidal attapulgite (Engelhard Corp.)
Daxad 30 - dispersant (W.R.Grace & Co.)
Cellosize 100M - hydroxyethylcellulose (Union Carbide)

⌘⌘⌘

Polychloroprene Adhesive

(Degussa)

	Parts By Weight
Baypren® C320	100
Magnesium Oxide, Extra Light	4
Zinc Oxide, Activated	5
Alresen® PA 565	33
Aerosil® 200, Fumed Silica	11
Ethyl Acetate	160
Gasoline 80/110	160
Toluene	80

Baypren C320 - polychloroprene, highly crystalline (Miles Inc.)
Zinc Oxide, Activated - (Miles Inc.)
Alresen Pa 565 - alkylphenolic resin, base reactive (Hoechst Celanese)
Aerosil 200 - fumed silica (Degussa Corporation)

Polyvinyl Acetate Solvent Adhesive

(Degussa)

	Parts By Weight
Mowilith® 50, Polyvinyl Acetate, Solid	100
Methyl Acetate	120
Methyl Alcohol	65
Aerosil® 200, Fumed Silica	1.4 to 5.7

Mowilith 50 - polvinyl acetate (Hoechst Celanese)
Aerosil 200 - fumed silica (Degussa)

⌘⌘⌘

Polyvinyl Acetate Dispersion Adhesive

(Degussa)

	Parts By Weight
Mowilith® DS5, Polvinyl Acetate	100.0
Dioctyl Phthalate	10.0
WW Resin	15.0
Quartz Powder	50.0
Water	25.0
Aerosil®200, Fumed Silica	0.2 to 0.6

This adhesive is quick setting at room temperature and produces rigid joints.

Mowilith DS5 - polyvinyl acetate (Hoechst Celanese)
WW Resin - (W. Biesterfeld)
Quartz Powder - (Quartzwerke)
Aerosil 200 - fumed silica (Degussa).

Room Temperature Fast Cure Epoxy Resin Adhesive

(Degussa)

	Parts By Weight
Epon® 828	100
Talc	50
Aerosil® 200 or 300	1 to 3
Diethylenetriamine	11

This adhesive is quick setting at room temperature and produces rigid joints.

Epon 828 - liquid epoxy resin (Shell Chemical Co.)
Aerosil 200/300 - fumed silica (Degussa)

⌘⌘⌘

Room Temperature Curing Epoxy Resin Adhesive

(Degussa)

Parts By Weight	
Epon® 828	100
Phenol (As Accelerator)	0 to 5
Dibutyl Phthalate	5
Aerosil® 200	5
Polyaminoamide Resin (Medium Viscosity)	80

Epon 828 - liquid epoxy resin (Shell Chemical Co.)
Aerosil 200/300 - fumed silica (Degussa)

Flexible Room Temperature Cure Epoxy Resin

(Degussa)

	Parts By Weight
Epon® 828	100
Asbestos Powder	30
Aerosil® 300	1 to 3
Thiokol® LP-3	50
Diethylenetriamine	11

This adhesive is quick setting at room temperature but should be post-cured for 90 minutes at 90°C. Produces flexible glued joints.

Epon 828 - liquid epoxy resin (Shell Chemical Co.)
Aerosil 300 - fumed silica (Degussa)
Thiokol LP-3 - liquid polysulfide (Morton International)

⌘⌘⌘

High Temperature Cure Epoxy Resin Adhesive

(Degussa)

	Parts By Weight
Epon® 828	100.0
Asbestos Powder	25.0
Zinc Sulfide	4.0
M-Phenylene Diamine	14.0
Aerosil® 300, Fumed Silica	1.0 to 3.0
Glycerine	0.3 to 1.0

Heat-curing adhesive with higher resistance to chemicals and heat.

Cure conditions: 30 minutes at 115°C, followed by 30 minutes at 165°C.

Epon 828 - liquid epoxy resin (Shell Chemical Co.)
Aerosil 300 - fumed silica (Degussa)

Oil Based - Gun Grade Caulking Compound

(Kilchesty)

	Parts By Weight
Linseed or Soy Oil	25 to 35
Asbestos	6 to 12
Calcium Carbonate (Limestone)	40 to 60
Tinting Pigment, Tio2	0.05 to 3.0
Gelling Agent	0 to 2
Drier (Lead, Cobalt, Etc)	0 to 1

⌘⌘⌘

High Quality Polyvinyl Acetate Consumer Caulk

(Newman)

	Parts By Weight
Airflex® 600BP	560.0
Triton® X405	4.0
Tamol® 850	4.0
AMP 95	4.0
Tipure® R-901	15.0
Snowflake® White	675.0
Kathon® LX	0.3
Isopar® K	20.0
Z-6040 Silane	0.7

Airflex 600 BP - vinyl acetate-ethylene copolymer (Air Products)
Snowflake White - 5.5 µm calcium carbonate (ECC International)
Tamol 850 - Dispersant, Rohm And Haas Co., Philadelphia, Pa 19105.
Amp 95 - Dispersant, Angus Chemical Co., Buffalo Grove, Il 60089.
Kathon Lx - Preservative, Rohm And Haas Co., Philadelphia, Pa 19105.
Tipure - Titanium Dioxide, Du Pont, Chemicals And Pigments Dept., Wilmington, De 19898.
Triton X405 - Surfactant, Union Carbide, Danbury, Cn 06817-0001.
Isopar K - Isoparaffin Solvent, Exxon Chemical Americas, Houston, Tx 77079.
Z6040 - Silane, Dow Corning Corp., Midland, Mi 48686-0994.

⌘⌘⌘

Vinyl Acrylic Caulking Compound

(Washabaugh)

	Weight (%)
Vinyl Acrylic Emulsion	40.2
Urea Formaldehyde Resin	5.3
Ethylene Glycol	2.2
Dispersants	1.0
Dibutyl Phthalate	4.1
3 µm $CaCO_3$	44.8
TiO_2	1.0
Colloidal Attapulgite	1.4

Solids: 81%

Properties:
Good Water Resistance And Adhesion
Excellent Flexibility
Non Sagging And Moderate Cost

PVC Plastisol

(Statt)

	Parts By Weight
E-PVC K-74 , Polymer	100.0
Dioctyl Phthalate	100.0
Calcium Carbonate	20.0
Lead Stabilizer	3.0
Ppt. Silica	6.9

⌘⌘⌘

Butyl-Based Caulk

(Newton)

	Parts By Weight
Exxon® Butyl 065 (50% In Mineral Spirits)	200.0
Vistanex® LM-MS, Polyisobutylene	20.0
Iso-Stearic Acid	5.0
IT 3X, Industrial Talc	300.0
Atomite® Whiting	200.0
Rutile Titanium Dioxide	25.0
Terpene Phenolic Resin	35.0
Indopol® H-300, Polybutene	100.0
Z3, Blown Soya Oil	15.0
Cobalt Napthenate, 6%	0.5
Cab-O-Sil®, Fumed Silica	20.0
Mineral Spirits	55.0

Vistanex LM-MS - polyisobutylene, MW = 8,700-10,000 (Exxon)
Atomite Whiting - 3μm calcium carbonate (ECC International)
IT-3X - talc, (R.T. Vanderbilt Co.)
Cab-O-Sil, fumed silica (Cabot Corp.)

⌘⌘⌘

Polybutene, Non-Drying Tape Sealant
(Washabaugh)

	Weight %
Amorphous Polypropylene	5.05
Amoco® H-300 Polybutene	24.00
Bucar® 5214 Butyl Rubber	1.70
Snobrite® Kaolin	16.40
Atomite® Calcium Carbonate	44.05
Celite® 281 Diatomaceous Silica	4.00
Filfloc® S60-900 Cotton Fiber	4.80

Viscosity at 190°C = 200-1800 mpa.s

Amoco H-300 - polybutene (Amoco Chemical Co.).
Bucar 5214 - butyl rubber (Occidental Chemical Corp.)
Snobrite - 0.8μm air-float kaolin (Evans Clay Co.)
Atomite - 3.0μm calcium carbonate (ECC America Inc.)
Celite 281 - diatomaceous silica (Celite Corp.)
Filfloc S60-900 - cotton fiber

Butyl Semi-Resilient Sealing Tape
(Newton, 1986)

	Parts By Weight
Vistanex® MM L-80, Polyisobutylene	100
Vistanex® LM-MS, Polisobutylene	50
Hi-Sil® 422, Ppt. Silica	50
Calcene® TM, Calcium Silicate	225
Indopol® H-300, Polybutene	100

Vistanex MM L-80 - polyisobutylene, Mw = 64,000 - 81,000 (Exxon)
Vistanex LM-MS - polyisobutylene, Mw = 8,700 - 10,000 (Exxon)
Hi-Sil 422 - precipitated silica (PPG Industries)
Calcene TM - treated calcium silicate (PPG Industries)
Indopol H-300 - polybutene (Amoco Chemical Co.)

⌘⌘⌘

Hot Melt Insulated Glass Window Sealant

(Chu)

	Parts By Weight
Kraton® G 1650	30
Kraton Gx 1726	70
Regalrez® 1018	250
Endex® 160	50
Polypropylene	30
Calcium Carbonate	130
Carbon Black	10
UV Stabilizer, Antioxidant	5
Adhesion Promoter	5

Kraton G 1650 - block copolymer (Shell Chemical Co.)
Kraton GX 1726 - block copolymer (Shell Chemical Co.)
Regalrez 1018 - mid block resin (Hercules, Inc.)
Endex 160 - end block resin (Hercules, Inc.)

One Part Polysulfide Sealant

(Degussa)

	Weight %
Lp-32, Liquid Polysulfide Polymer	37.5
Plasticizers	18.8
Adhesive Resin	3.7
Chalk	15.0
Kaolin	5.6
Titanium Dioxide, Rutile	11.2
Aerosil 130, Fumed Silica	1.5
Na-Chromate Paste	6.7

LP-32 - liquid polysulfide polymer (Morton International)
Aerosil 130 - fumed silica (Degussa)

⌘⌘⌘

Filled Architectural Sealant

(Chu)

	Parts By Weight	
Kraton® G 1726	70	
Kraton® G 1652	30	
Regalrez® 1018	400	
Endex® 160	25	
$CaCO_3$	350	
Sylox® TX	75	
TiO_2	30	
A-189, Methoxy Silane	4	
Lica® 09, Coupling Agent	3	
UV Antioxidant	3	
Total	990	> 90% Solids
Solvent		
Tolu-Sol 5, Solvent	19	
Toluene	22	
Propyl Acetate	69	
Total	110	> 10% Solvent

Kraton GX 1726 and G 1652 - block copolymers (Shell Chemical Co.)
Regalrez 1018 - mid block resin (Hercules, Inc.)
Endex 160 - end block resin (Hercules, Inc.)
A-189 - Mercaptopropyltrimethoxysilane (Osi Specialties,Inc.)
Sylox TX - silica gel (Davison Chemical)

⌘⌘⌘

Two Part Polysulphide Joint Sealing Compound

(Degussa)

Part A	**Parts By Weight**
Lp-32, Liquid Polysulfide	100.00
Plasticizers	35.00
Chalk	25.00
Kaolin	30.00
Titanium Dioxide, Rutile	10.00
Adhesive Resin	5.00
Flowers Of Sulphur	0.10
Stearic Acid	1.00
Aerosil 200, Fumed Silica	3.09
Total	209.19

Part B	
Lead Dioxide	7.50
Plasticizers	6.75
Stearic Acid	0.75
Total	15.00

LP-32 - liquid polysulfide polymer (Morton International)
Aerosil 200 - fumed silica (Degussa)

⌘⌘⌘

Two Part Polysulfide Insulating Glass Sealant

(Tapper)

Part A	**Parts By Weight**
Liquid Polysulfide	100.0
Plasticizer	25.0
A-187 Silane	2.0
Titanium Dioxide	5.0
Vicron® 15-15	75.0
Ultra-Pflex®	30.0
Toluene	5.0
Total	242.0

Part B	
Curing Agent 50% Paste	2-15

Vicron 15-15 - ground limestone (Specialty Minerals Inc.)
A-187 - Silane (OSI Specialties,Inc.)
Ultra-Pflex - 0.07μm surface treated ppt. calcium carbonate (Specialty Minerals)

One Part Polyurethane Joint Sealing Compound

(Degussa)

	Weight %
Polyurethane Polymer	29.8
Plasticizers	18.3
Drying Agent Paste	3.8
Isocyanate	1.3
Pigment Paste	2.4
Chalk	37.5
Aerosil 150 Or R 972 Fumed Silica	3.8
Amine Hardener	3.1

Polyurethane Polymer - (Miles Inc.)
Aerosil 150 or R972 - fumed silicas (Degussa)

⌘⌘⌘

Two Part Polyurethane Joint Sealing Compound

(Degussa)

Part A	**Parts By Weight**
Desmophen® 250 U	10.0
Plasticizers	180.0
Age Resister	3.5
Calcium Octoate, (4% Calcium)	5.0
Aerosil® 130, Fumed Silica	30.0
Chalk	430.0
Titanium Dioxide, Rutile	40.0
Lead Octoate, (24% Lead)	5.0
Total	708.5

Part B	
Desmodur®	170.0
Black Paste	1.7
Plasticizers	19.0
Aerosil 130, Fumed Silica	21.0
Total	211.7

Mixing Ratio : 100 Parts A + 30-35 Parts B

Desmophen 250 U - polyol (Miles Inc.)
Age Resister - (Miles Inc.)
Desmodur - isocyanate, (Miles Inc.)
Aerosil 130 - fumed silica (Degussa)

⌘⌘⌘

One Part High Modulus Silicone Sealant

(Degussa)

	Weight %
Siloprene® E-50, Polymer	61.5
Baysilon® Oil M 1000	24.0
Aerosil® 130 Or R972 Fumed Silica	9.3
Polymerizer® VP-1-4077	5.2
Dibutyl Tin Diacetate	(1 Drop)

Siloprene E-50 - 50,000 mpa.s hydroxyl terminated silicone polymer (Miles)
Baysilon M-1000 - 1000 mpa.s silicone fluid (Miles)
Polymerizer VP-F1-4077 - silane crosslinker (Miles)
Aerosil 130 Or R972 - fumed silicas (Degussa)

One Part Medium Modulus Oxime Cure Silicone Sealant

(Tapper)

	Weight %
Silanol Polymer (80,000 mpa.s)	60 - 80
Silicone Plasticizer	5 - 20
Fumed Silica	2 - 6
Calcium Carbonate	20 - 30
Oxime Crosslinker	5 - 7
Tin Catalyst	0.05 - 0.10

One Part Low Modulus Silicone Sealant

(Tapper)

	Weight %
Silanol Polymer (4000 mpa.s)	46.0
Calcium Carbonate	50.3
Methyl Vinyl Di (N-Methyl Acetamido) Silane	3.0
Aminoxy Siloxane Copolymer	0.7

⌘⌘⌘

Resilient Tape - Partially Cured

(Newton)

	Parts By Weight
Exxon® Butyl 065	60.0
Chlorobutyl 1066	40.0
Purecal® U	50.0
N-327 Carbon Black	100.0
Indopol® H-100 Polybutene	15.0
Zinc Oxide	2.0
Magnesium Oxide	0.4
Stearic Acid	0.4

Butyl 065 - butyl rubber (Exxon)
Chlorobutyl 1066 - chlorobutyl rubber (Exxon)
Indopol H-100 - viscous polybutene (Amoco Chemical Co.)
Purecal U - precipitated calcium carbonate (BASF Wyandotte)

Polyisobutylene Hot Flow Sealant

(Newton)

	Weight %
Vistanex® MM L-80	10
Vistanex LM-Ms	15
A-Fax 500-HL-O App	20
Escorez® 2101 Resin	30
N-327 Carbon Black	10
Exxon® LD 600 LDPE	15

Vistanex MM L-80 - polyisobutylene, mw = 64,000-81,000 (Exxon)
Vistanex LM-MS - olyisobutylene, mw= 8,700-10,000 (Exxon)
Exxon LD 600 - low density polyethylene (Exxon)
Escorez 2101 - hydrocarbon resin (Exxon)

⌘⌘⌘

One Part Polysulfide Sealant

(Zeneca)

	Weight %
Thiokol® LP32C, Polymer Resin	35.4
Winnofil® SPT, Thixotrope	17.7
Supercoat®, Extender/Filler	15.9
Nevoxy® EPX-L5, Adhesion Promoter	1.8
Santicizer® 63l, Plasticizer	7.1
Cereclor® 63l, Plasticizer	12.4
Charge disperse in a planetary mixer under vacuum.	
Barium Oxide, Moisture Scavenger	3.3
Add and continue mixing under vacuum.	
Allow for moisture to be removed and cool if necessary.	
Calcium Peroxide, Curative	6.4

Add and continue mixing under vacuum. Cartridge immediately after manufacture.

Winnofil SPT - surface treated ppt. $CaCO_3$ (Zeneca Resins)
Supercoat - surface treated ground $CaCO_3$ (ECC International)
Nevoxy EPXL5 - hydroxy modified resins (Neville Chemical Co.)
Santicizer 160 - plasticizer (Monsanto Co.)
Cerecolor 631 - plasticizer (ICI Americas)

⌘⌘⌘

Two Part Terminated Polybutadiene Sealant

(Zeneca)

	Weight %
Part A	
PolyBD® R45HT, OH Resin Vehicle	26.5
Winnofil® SPT Premium, Thixotrope	39.8
Omya® BLR3, Filler	13.3
Colanyl® Black GRL30, Pigment	0.5
Molecular Sieve 3A, Moisture Scavenger	2.7
Disperse Using A Planetary Mixer.	
Santicizer® 261, Plasticizer	13.3
Silane® A187, Adhesion Promoter	0.5
Tinuvin® 327, Uv Stabilizer	0.13
Irganox® 565, Antioxidant	0.005
Dibutyltin Dilaurate, Catalyst	0.001

Add And Continue To Disperse Until Homogenous. Dearate Under Vacuum. Store In A Tin With A Secure Lid.

Part B	
Suprasec 2020, Curing Agent	3.2

Two Part - Polysulfide Rubber

(Statt)

Part A	**Parts By Weight**
Thiokol LP 977	100.0
Ppt Silica	13 - 27.5
Total	113-127.5
Part B	
MnO^2 - FA	100.0
Dibutyl Phthalate	100.0
Total	200.0

Mixing Ratio : 6 Parts A + 1 Part B.

Thiokol LP 977 - liquid polysulfide polymer (Morton International)

⌘⌘⌘

Chu,S.G., 1986, "Sealants Based on Block Copolymers", *Caulks and Sealants Short Course*, Adhesive and Sealant Council, Washington, DC, pp 269-339.

Kilchesty, A.A., 1986, "Oleo Resinous and Oil-Base Caulking Compounds", ibid, pp 13-23.

Newton, M.V. et al, 1986, "Butyl Sealants: Formulating, Developing, Processing", ibid, pp 25-75.

Statt, B.K. et al, 1992, "Precipitated Silicas in Adhesives and Sealants", *Supplier Short Course on Additives,* , Adhesive and Sealant Council, Washington, DC, April, pp IV, 1-30.

Tapper, M., Mathur, K., 1992, " Use of Calcium Carbonates in Sealants", ibid, pp II, 1-19.

Washabaugh, F.J., 1992, "Kaolin, Talc and Attapulgite", *Supplier Short Course on Additives,* , Adhesive and Sealant Council, Washington, DC, April, pp III, 1-19.

⌘⌘⌘

PLASTICS

Typical BMC Formulation

(Georgia Marble)

	Wt. %
Polyester resin	26.32
t-Butyl perbenzoate (catalyst)	0.26
Calcium carbonate (filler)	52.63
Zinc stearate (lubricant)	0.79
Glass fiber, 1/4 in. (reinforcement)	20.00

Carbonate-Filled BMC

(U.S. Silica)

	Pounds
14071 Polyester (Aristech)	38.4
Styrene monomer	0.4
29B75 Peroxide	0.4
CM2015 Plasticolor pigment	0.8
ASP® 400 Clay (Engelhard)	0.8
Zinc Stearate	0.8
Cal White® II $CaCO_3$	57.8
Modifier M Thickener	0.6
ADD: Glass fiber, 1/2 in.	42.8

Viscosity (cp)	25,500
Flexural strength (1000 psi)	22
Flexural modulus (1000 psi)	1500
Tensile strength (psi)	6800
Notched Izod (ft. lbs.\in.)	9
Barcol hardness	70

⌘⌘⌘

BMC, Premix
(Koppers)

	Wt. %
3102-5 Polyester resin (Koppers)	26
BSH $CaCO_3$ (Omya)	58
Glass fibers, 1/16 - 1/2 in.	15
Lupersol® t-butyl perbenzoate (Lucidol)	0.5
Coad® 20 zinc stearate (Norac)	0.5

Silica-Filled BMC
(U.S. Silica)

	Pounds
14071 Polyester (Aristech)	41.9
Styrene monomer	0.4
29B75 Peroxide	0.4
CM2015 Plasticolor pigment	0.8
ASP® 400 Clay (Engelhard)	0.8
Zinc Stearate	0.8
MIN-U-SIL® 5 Silica (U.S. Silica)	13.6
MIN-U-SIL 30 Silica (U.S. Silica)	40.7
Modifier M Thickener	0.6
ADD: Glass fiber, 1/2 in.	42.8

Viscosity (cp)	46,000
Flexural strength (1000 psi)	18.7
Flexural modulus (1000 psi)	1800
Tensile strength (psi)	9000
Notched Izod (ft. lbs.\in.)	2.6
Barcol hardness	68

⌘⌘⌘

BMC - Durable Microwave Ovenware

(U.S. Silica)

	Wt. %
MR-14029 Polyester (Aristech)	18.95
MR-63004 Polystyrene (Aristech)	8.12
TiO_2, 70% dispersion in MR-14029	1.73
Ultrafine XM Zinc stearate (Mallinkrodt)	0.54
t-Butyl peroxyisopropyl carbonate (Akzo)	0.27
Treated silica blend*	63.28
ADD: Glass fiber	7.11

Treated silica blend:	**Pounds**
MIN-U-SIL® 30 silica (U.S. Silica)	70
MIN-U-SIL 5 silica (U.S. Silica)	30
Prosil® 248 silane (PCR)	0.5

Typical Automotive SMC, Acrylic Modified

(Georgia Marble)

	Wt. %
Polyester resin	23.43
Acrylic additive	12.89
Styrene	2.73
t-Butyl perbenzoate	0.40
Gama-Sperse® 6451 $CaCO_3$ (Georgia Marble)	58.59
Zinc stearate	1.56
Magnesium oxide	0.40

SMC

(Koppers)

	Wt. %
3102-5 Polyester resin (Koppers)	23.9
BSH $CaCO_3$	40.7
Lupersol® t-Butyl perbenzoate (Lucidol)	0.25
Coad® 20 Zinc stearate	1.0
Magox® 900, magnesium oxide thickener	0.7
Low shrink additive	3.45

⌘⌘⌘

SMC - Grill Opening Panel

(Georgia Marble)

	phr
MR-13006 Polyester (Aristech)	65
Neulon® T+	35
t-Butyl perbenzoate	0.75
PDO	1
Zinc stearate	1
VR-3 Viscosity reducer	2
Calwhite® II $CaCO_3$ (Georgia Marble)	160
CM-2015 Black pigment	0.25
"A" Paste free moisture	0.15 - 0.19
Calcium oxide thickener	0.15
PPG 5509 glass fiber, 1 in. chop	27 - 29

Silica -Filled Vinyl Ester SMC

(U.S. Silica)

	Pounds
Derakane® 790 Vinyl ester	43.38
AF70 Antifoam (Dow)	0.02
29B75 Peroxide	0.40
Zinc stearate	1.70
CM2015 Plasticolor pigment	0.90
MIN-U-SIL® 40 silica (U.S. Silica)	51.90
PG9033 Thickener	1.70
ADD: Glas fiber, 1/2 in.	32.60

Viscosity (cp)	24,500
Flexural strength (1000 psi)	21.3
Flexural modulus (1000 psi)	1200
Tensile strength (psi)	10,000
Notched Izod (ft. lbs.\in.)	10.4
Barcol hardness	50

⌘⌘⌘

Carbonate -Filled Vinyl Ester SMC

(U.S. Silica)

	Pounds
Derakane® 790 Vinyl ester	37.48
AF70 Antifoam (Dow)	0.02
29B75 Peroxide	0.30
Zinc stearate	1.30
CM2015 Plasticolor pigment	0.70
Cal White® II $CaCO3$ (Georgia Marble)	58.90
PG9033 Thickener	1.30
ADD: Glas fiber, 1/2 in.	34.90

Viscosity (cp)	30,000
Flexural strength (1000 psi)	33.8
Flexural modulus (1000 psi)	1400
Tensile strength (psi)	16,400
Notched Izod (ft. lbs.\in.)	17.9
Barcol hardness	65

White SMC - Shower Floors/Laundry Tubs

(Georgia Marble)

	phr
MR 14011 Polyester (Aristech)	100
t-Butyl perbenzoate	1
Zinc stearate	6
Cal White® II $CaCO3$ (Georgia Marble)	160
PDI 1100 White pigment	11.2
Modifier ME Thickener	2.4
PPG 5528 Glass fiber, 1 in. chop	20 - 22

⌘⌘⌘

Non-White SMC - Shower Floors/Laundry Tubs
(Georgia Marble)

	phr
MR 13017 Polyester (Aristech)	70
MR 63004 Polyester (Aristech)	30
t-Butyl perbenzoate	1
Zinc stearate	6
Cal White® II CaCO3 (Georgia Marble)	160
PDI 1100 White pigment	11.2
Modifier ME Thickener	2.4
PPG 5528 Glass fiber, 1 in. chop	20 - 22

Typical Cultured Marble
(Georgia Marble)

	Wt. %
Polyester resin	25
MEK-P Catalyst	0.5
Calcium carbonate	74.5

Cultured Marble
(Georgia Marble)

	Wt. %
Polylite® 32-162-10 Polyester (Reichold)	15 - 30
Marblend® $CaCO_3$	60 - 75
Superox® 46744 colored veining mix (Reichold)	0 - 10
Benzoyl peroxide, catalyst	0.3 - 0.6

Typical Properties:

Density, g/cc	2.2
Barcol hardness	65
Tensile strength, psi	2000
Compressive strength, psi	18,000
Gardner Impact, in. lb.	28

⌘⌘⌘

Cultured Onyx I

(Georgia Marble)

	Wt. %
Polylite® 32-162-10 Polyester (Reichold)	36
Glass frit	60
Superox® 46744 colored veining mix (Reichold)	1 - 4
Benzoyl peroxide, catalyst	0.75

Cultured Onyx II

(Georgia Marble)

	Wt. %
Polylite® 32-162-10 Polyester (Reichold)	36
C 31 Alumina hydrate (Alcoa)	54
Superox® 46744 colored veining mix (Reichold)	1 - 4
Benzoyl peroxide, catalyst	0.75

Cultured Granite

(Georgia Marble)

	Wt. %
Polylite® 32-162-10 Polyester (Reichold)	36
Glass frit	60
Colored granules*	1 - 4
Benzoyl peroxide, catalyst	0.75

*May be natural crushed stones or crushed cured pigmented poyester.

Typical Rigid PVC

(Georgia Marble)

	phr
PVC resin	100
Acrylic process aid	3
Tin stabilizer	2
Lubricant	1
Calcium carbonate	5 -20

⌘⌘⌘

Typical Flexible PVC
(Georgia Marble)

	phr
PVC resin	100
Plasticizer	50
Stabilizer	2
Lubricant	0.25
Calcium carbonate	10 - 40

Typical Wire Insulation PVC
(Georgia Marble)

	phr
PVC resin	100
Phthalate plasticizer	39
Epoxy plasticizer	10
Stabilizer	7.25
Lubricant	0.25
Calcined clay	10
Calcium carbonate	10

PVC Wire Insulation
(U.S. Silica)

	phr
Geon® 30 PVC (Goodrich)	100
Dioctyl phthalate	40
Ba\Zn stabilizer	10
Snow Tex 45 calcined clay (U.S. Silica)	8

Volume Resistivity (ohm-cm):	
Average of 10 samples	5.8×10^{11}
Standard deviation	1.7×10^{11}

⌘⌘⌘

Rigid Epoxy I
(U.S. Silica)

	phr
ERL-2774 Epoxy (Union Carbide)	100
HHPA hardener (Allied)	80
MIN-U-SIL® silica blend*	540
2MI imidazole (BASF)	0.15

*70/ 30 blend of MIN-U-SIL 30 & MIN-U-SIL 5

Physicals (25°C/100°C):

Tensile strength (1000 psi)	10.58/5.8
Tensile modulus (10^6 psi)	3.53/2.58
Elongation, %	0.27/0.51
Flexural strength (1000 psi)	12.67
Compressive strength (1000 psi)	36.90
Viscosity, cp (Brookfield No.6 @ 99°C)	
10 rpm	1370
20 rpm	2200
50 rpm	6400

Rigid Epoxy II
(U.S. Silica)

	phr
ERL-2774 Epoxy (Union Carbide)	100
HHPA hardener (Allied)	80
325 mesh silica	540
2MI imidazole (BASF)	0.15

Physicals (25°C/100°C):

Tensile strength (1000 psi)	8.95/7.58
Tensile modulus (10^6 psi)	2.62/2.62
Elongation, %	0.37/0.48
Flexural strength (1000 psi)	10.32
Compressive strength (1000 psi)	30.30
Viscosity, cp (Brookfield No.6 @ 99°C)	
10 rpm	19,000
20 rpm	15,700
50 rpm	12,600

⌘⌘⌘

Flexible Epoxy I

(U.S. Silica)

	pbw
ERL-2774 Epoxy (Union Carbide)	60
HHPA hardener (Allied)	58
Epi-Rez® 505 (Celanese)	40
MIN-U-SIL® silica blend*	475
2MI imidazole (BASF)	0.15

*70/ 30 blend of MIN-U-SIL 30 & MIN-U-SIL 5

Physicals (25°C/100°C):

Tensile strength (1000 psi)	10.10/1.13
Tensile modulus (10^6 psi)	2.51/0.07
Elongation, %	0.26/1.7
Flexural strength (1000 psi)	12.78
Compressive strength (1000 psi)	30.20
Viscosity, cp (Brookfield No.6 @ 99°C)	
10 rpm	1500
20 rpm	1750
50 rpm	4000

Flexible Epoxy II

(U.S. Silica)

	phr
ERL-2774 Epoxy (Union Carbide)	60
HHPA hardener (Allied)	58
Epi-Rez® 505 (Celanese)	40
325 mesh silica	475
2MI imidazole (BASF)	0.15

Physicals (25°C/100°C):

Tensile strength (1000 psi)	9.50/0.82
Tensile modulus (10^6 psi)	2.50/0.07
Elongation, %	0.48/1.23
Flexural strength (1000 psi)	13.29
Compressive strength (1000 psi)	23.20
Viscosity, cp (Brookfield No.6 @ 99°C)	
10 rpm	8400
20 rpm	9250
50 rpm	10,000

⌘⌘⌘

FORMULA SOURCES

American Colloid Co.
1500 West Shure Dr.
Arlington Heights, IL 60004

Cargill/McWhorter
400 East Cottage Place
Carpentersville, IL 60110

Diamond Shamrock
350 Mt. Kemble Ave.
Morristown, NJ

Degussa Corp.
150 Springside Dr.
Akron, OH 44333

Georgia Marble Co.
1201 Roberts Blvd.
Kennesaw, GA 30144

Koppers Industries, Inc.
1600 Koppers Bld., 436 7th Ave.
Pittsburgh, PA 15219

National Starch & Chemical Co.
10 Finderne Ave.
Bridgewater, NJ 08807

NYCO Minerals, Inc.
124 Mountain View Dr.
Willsboro, NY 12996

PPG Industries, Inc.
440 College Park Dr.
Monroeville, PA 15146

Rheox
Wyckoffs Mill Rd.
Hightstown, NJ 08520

Rhone-Poulenc
Prospect Plains Rd.
Cranbury, NJ 08512

Rohm & Haas
Independence Mall West
Philadelphia, PA 19105

Union Carbide Corp.
39 Old Ridgebury Rd.
Danbury, CT 06817

U.S. Silica Co.
P.O. Box 187
Berkeley Springs, WV 25411

R.T. Vanderbilt Co., Inc.
30 Winfield St.
Norwalk, CT 06855

Zeneca Resins
730 Main St.
Wilmington, MA 01887

⌘⌘⌘

NOTES

⌘⌘⌘

THIRTEEN

COMMERCIAL MINERAL PRODUCTS

The following pages present basic technical data kindly provided by a number of suppliers of industrial minerals. This is just a small sampling of the many producers and products available to the formulator and compounder. This information does not represent a recommendation for or against any product, but is designed to confer a sense of the range of quality and quantity among mineral products currently marketed. Particulars regarding specifications and intended application areas should be solicited from the individual producers.

Commercial Mineral Products

Ball Clay
Kentucky-Tennessee Clay Co.

Product	<325 Mesh %[1]	pH	Surf. Area m^2/g	Al_2O_3 %	Fe_2O_3 %	CaO %	Na_2O %	K_2O %	L.O.I. %
Bell Dark	97.9	3.8	24.0	27.7	1.0	0.3	0.1	0.4	10.5
Dresden M	97.1	4.6	23.9	26.2	1.2	0.3	0.1	1.2	12.5
HTP	99.2	4.8	30.9	32.0	1.1	0.3	0.1	0.7	10.5
J-2	97.7	4.5	21.3	28.4	1.1	0.4	0.1	0.5	10.8
Jackson	97.4	4.2	25.0	29.5	1.0	0.4	0.1	0.3	11.8
Kentucky 4 & 12	97.0	4.0	24.5	27.0	1.0	0.3	0.2	0.9	13.2
Kentucky #5 Bond	99.1	4.9	21.8	29.7	1.0	0.4	0.2	1.2	9.7
Kentucky Stone	97.0	3.8	19.9	20.8	1.3	0.3	0.1	1.3	7.1
Kentucky Tile Blend	98.0	4.0	20.5	25.4	1.1	0.3	0.1	1.1	8.7
KT # 1-4	98.5	4.5	18.7	27.7	0.9	0.3	0.2	1.1	10.6
KT-556	97.8	4.2	20.0	27.8	1.0	0.04	0.2	0.9	9.3
KTS-2	97.7	5.2	18.0	24.3	1.0	0.3	0.3	1.3	9.9
KTS-Classic	97.5	5.5	18.0	24.6	1.0	0.3	0.2	1.4	9.5
L-1	97.7	4.0	18.7	21.6	1.0	0.5	0.2	1.4	9.2
M&D	99.2	7.1	43.0	28.7	2.3	0.7	0.1	0.6	9.6
Martin #5	98.0	4.4	14.0	24.6	0.9	0.3	0.2	1.6	9.3
MT Light	98.6	5.8	27.0	26.3	1.3	0.4	0.2	0.6	9.2
Old Mine #4	98.8	4.4	24.4	27.2	1.1	0.4	0.2	1.1	12.5
Tennessee #1 SGP	98.5	4.5	16.0	28.1	0.8	0.3	0.2	1.2	10.2
Tennessee #5	98.1	4.2	17.0	27.0	0.9	0.4	0.2	1.2	13.2
Tennessee #6 Bond	97.7	4.2	21.9	27.2	0.8	0.3	0.2	0.5	9.3
Todd Dark	91.9	4.0	24.1	26.2	1.3	0.7	0.1	1.0	17.7
Todd Light	98.8	4.3	27.6	27.4	1.1	0.5	0.1	1.2	10.5

[1]minimum

⌘⌘⌘

Ball Clay
United Clays

Product	pH	Surf. Area m^2/g	Al_2O_3 %	Fe_2O_3 %	CaO %	Na_2O %	K_2O %	L.O.I. %
101 Ceramic	8.2	13	17.1	6.7	0.3	0.6	8.2	5.2
401 Ceramic	8.7	15	27.6	1.5	0.3	0.5	2.6	8.5
Colton	4.5	18	28.3	1.2	0.1	0.1	1.8	10.3
HB-11	4.9	19	32.1	1.0	0.2	0.1	0.4	11.4
HR-11	4.7	16	33.6	0.9	0.1	0.1	1	11.9
Huntingdon	4.8	20	38.3	0.8	0.1	0.1	0.1	13.6
Imperial	4.4	25	27.0	1.3	0.2	0.1	1	11.2
OMR	7.5	46	27.4	2.6	0.5	0.1	0.7	9.5
OMR-2	6.0	33	28.0	1.9	0.4	0.1	0.5	10.0
OMR-4	5.0	27	28.8	1.4	0.3	0.1	0.5	10.2
Regent	4.5	24	27.8	1.0	0.2	0.1	0.7	9.9
Rex	4.4	16	29.3	0.8	0.1	0.1	1.6	9.5
Royal	5.0	19	27.2	1.1	0.1	0.1	1.3	11.0
SB Blend	5.2	18	28.1	1.1	0.2	0.1	1.3	10.3
Slurry #1 Blend	5.2	20	27.9	1.0	0.2	0.2	1.5	10.0
Slurry #6 Blend	4.6	23	26.5	1.2	0.2	0.1	0.6	10.1
Slurry #7 Blend	4.8	18	30.5	0.9	0.1	0.2	0.9	11.1
Starcast	5.3	20	29.2	0.9	0.2	0.1	0.9	10.7
Sterling	4.4	22	26.8	1.2	0.1	0.1	0.6	11.2
Stratton	4.5	24	26.9	1.0	0.2	0.1	0.4	9.4
TBC-1	4.4	26	24.0	1.5	0.3	0.1	0.8	8.5
TBC-2	4.6	25	26.0	1.3	0.1	0.1	0.5	9.0
Unicast	4.5	15	38.1	0.6	0.1	0.1	0.1	13.6
Victoria	5.3	24	28.3	1.1	0.2	0.1	0.5	10.7
Weldon	5.1	25	27.2	1.3	0.2	0.1	0.8	10.5

⌘⌘⌘

Barite
J.M. Huber Corp.

Product	Brightness %	<325 Mesh, %	Median Particle Size, μm	<10μm %	Moisture %
Huberbrite® 1	84	Trace	1	100	0.20
Huberbrite 3	94	Trace	3	90	0.10
Huberbrite 7	94	Trace	6.5	71	0.10
Huberbrite 10	93	99.5	8.5	57	0.10
Huberbrite 12	92	98.6	11	46	0.10

Barite
Mountain Minerals Company Ltd.

Product	Brightness %	Median Particle Size, μm	pH	Oil Absorption %	Moisture %
Sparwite® W-10	85	1.9	7.0	14	0.10
Sparwite W-10HB	92	1.9	7.0	14	0.10
Sparwite W-20	85	4.4	7.0	12	0.10
Sparwite W-20HB	90	4.4	7.0	12	0.10
Sparwite W-44	85	10.5	7.0	10	0.10
Sparwite W-44HB	90	10.5	7.0	10	0.10
Sparwite W-SHB	95+	1.6	7.0	16	0.12

Barite
Polar Minerals

Product	Brightness %	Median Particle Size, μm	Oil Absorption %	Moisture %
1040	90min.	6-8	10	<0.25
1065	90min.	3-4.5	10	<0.25
1075	90min.	2-3	11-13	<0.25
1090P	95	0.9	8	<0.25
2010	78	7-10	9-10	<0.25
2040	78	6-8	10-10.5	<0.25
2065	80	3-5	10-11	<0.25
2075	80	2-3	12-14	<0.25

⌘⌘⌘

Calcium Carbonate
ECC International

Product	Brightness[1] %	<325 Mesh, %	Median P.S. μm	Oil Abs. %	$CaCO_3$ %
Ground Grades					
Atomite®	95		3.0	14-16	97.6
CC-103™					
CP Filler™		93.00	13.5		96.5
Drikalite®	94	99.65	7.0	8-10	97.4
Duramite®	93	99.80	11.0	7-9	97.4
Marble Dust™	90	82.00	21.0		96.6
Marblemite™		45<200	80.0		97.1
Micro-White® 15	95		2.0	16-18	97.6
Micro-White 25			3.0	14-16	
Micro-White 40			4.0	12-14	
Micro-White 100	92	93.00	13.5		95.0
No. 1 White™	92	99.50	13.0	7-9	97.0
SC-53™					
Snowflake P.E.™	94	99.90	5.5	8-10	97.6
Snowflake White®	94		5.5	10-12	97.6
Supermite®	95		1.0	18-21	97.6

[1] Hunter

Product	<325 Mesh, %	Median P.S., μm	Moisture %	$CaCO_3$ %
Slurry Grades				
Micro-White 07		0.7	75% Solids	97.6
Micro-White 10		1.0		
Micro-White 15		2.0		
Micro-White 25		3.0		
Food Grades				
Micro-White 10 Codex		1.0	0.10	97.0
Micro-White 25 Codex		3.0	0.10	96.8
Micro-White 50 Codex	99.65	7.0	0.10	95.5
Micro-White 100 Codex	93.00	17.0	0.10	95.0

⌘⌘⌘

Product	Median P.S., µm	Oil Abs. %	Surf. Area m^2/g	Hegman	$CaCO_3$ %
Surface Modified Grades[2]					
Kotamite®	3.0	13	2.8	6	97.6
Micro-White 15 SAM	2.0	14	5.0	6	97.6
Opacicote™	1.1			6-7	97.6
Supercoat®	1.0	16	7.2	7	97.6

[2] Properties Before Surface Modification

Calcium Carbonate
Genstar Stone Products Co.

Product	Brightness[1] %	Median P.S., µm	<10µm %	Oil Abs. %	Solids %
Ground Grades					
Camel-CAL	96	0.7		28	
Camel-CARB	93	7.0	62	13	
Camel-FINE	95	2.0		22-24	
Camel-FIL	93	5.5-6.0	65	17	
Camel-TEX	93	5.0	78	14	
Camel-WITE	95	3.0	95	15	
Surface Treated Grades[2]					
Camel-CAL ST	96	0.7		28	
Camel-FINE ST	95	2.0		22-24	
Camel-WITE ST	95	3.0	95	15	
Slurry Grades					
Camel-CAL	96	0.7			74.5-75.5
Camel-FINE	95	2.0			71.5-72.5
Camel-WITE	95	3.0	95		71.5-72.5

[1] Hunter

[2] 1% stearate

⌘⌘⌘

Calcium Carbonate
Georgia Marble Co.

Product	Brightness[1] %	<325 Mesh, %	Median P.S. μm	Hegman	Oil Abs. %
Fine Ground Grades					
General Use					
Calwhite®	94	99.992	6.0	5	10
Gamaco®	94	99.994	3.8	6	16
Gama-Sperse® 80	94	99.994	2.3	7	19
Gama-Sperse 255	94	99.980	12.0	4	9
Gama-Sperse 6451	94	99.992	6.0	5	10
Gama-Sperse 6532	94	99.994	3.8	6	16
Wingdale White	94	99.800	6.5	3	10
Fillers For Plastics					
Calwhite II	94	99.950	7.0		
Gamaco II	94	99.994	3.0		
Gama-Plas™	94	99.800	7.0		
Gama-Sperse 6532NSF	94	99.994	3.0		
Stearate Surface Modified[2]					
Gama-Fil™ D-2T		99.994	2.0		
Gama-Sperse CS-11		99.994	1.0		

[1] Hunter

[2] 1% stearate

Product	Brightness[1] %	<325 Mesh, %	Median P.S.μm	<2μm %	<1μm %	Oil Abs., %	Hegman
Ultra Fine Ground							
Gama-Sperse 2	94	99.994	2.0			21	7+
Gama-Fil D2	94	99.994	2.0			21	6.0
Fine And Ultra Fine Slurry[2]							
Gamaco Slurry	95		2.8				
Gama-Fil 40	95		2.4				
Gama-Fil 55	95		1.8	55	28		
Gama-Fil 90	95		0.8	92	58		

[1] Hunter

[2] 73-75% solids

⌘⌘⌘

Product	Brightness[1] %	<325 Mesh, %	pH
Regular Coarse and Medium Ground Grades			
No Color Specified			
350	88	67-77<200	9.0-9.5
RO-40	88	70-80<200	9.0-9.5
Rock Dust	88	76-84<200	9.0-9.5
No. 9 NCS	88	88-94	9.0-9.5
Color Controlled Grades			
No. 8 White	90	78-80	9.0-9.5
No. 9 White	92	88-94	9.0-9.5
No. 10 White	92	1max.	9.0-9.5
15M®	92	12max.	9.0-9.5
Screen Controlled Grades			
30-50		20<50	9.0-9.5
40-200		20<200	9.0-9.5
Mar'Blend®		42<200	9.0-9.5
OZ		5<14	9.0-9.5
XO		15<40	9.0-9.5
Z		10<20	9.0-9.5

[1] Hunter

Calcium Carbonate
J.M. Huber Corp.

Product	Brightness %	<325 Mesh, %	Median P.S., µm	Oil Abs. %	Hegman	$CaCO_3$ %
F SERIES						
Granular Products						
Hubercarb® F 6-20		2<20				98.3
Hubercarb F 12-40		80<20	700			98.3
Hubercarb F 16-200		83<20	420			98.3
Hubercarb F 30-200		99<20	280			98.3
Fine And Medium Fine						
Hubercarb F 3	82	100	3	17		98.3
Hubercarb F 3T[1]	82	100	3	15		98.3
Hubercarb F 60	74	46	43	9		98.3
Hubercarb F 325	80	96	8	16		98.3

[1] Stearate Treated

⌘⌘⌘

Product	Brightness %	<325 Mesh, %	Median P.S., μm	Oil Abs. %	Hegman	$CaCO_3$ %
M SERIES						
Medium Fine						
Hubercarb M 70	80	60	30	12		92.0
Hubercarb M 200	81	82	12	12		92.0
Hubercarb M-300	85	97	6.5	13		92.0
Fine Products						
Hubercarb M 6	86	99.99	6	16	4	92.0
Hubercarb M 4	87	99.99	4	17	5.5	92.0
Hubercarb M 3	87	99.99	3	17	6	92.0
Hubercarb M 3T[1]	87	99.99	3	15	6	92.0
Q SERIES						
Granular And Medium Fine						
Hubercarb Q 6-20		15<20				96.5
Hubercarb Q 20-60		30<20				96.5
Hubercarb Q 40-200		98<20				96.5
Hubercarb Q 60	80	20	20	12		96.5
Hubercarb Q 100	83	70	24	12		96.5
Hubercarb Q 200	84	82	19	12		96.5
Hubercarb Q 325	86	99.9	13	14	2.0	96.5
Fine And Ultra Fine						
Hubercarb Q 6	87	99.99	6	16	4.0	96.5
Hubercarb Q 4	88	99.99	4	17	6.0	96.5
Hubercarb Q 3	89	99.99	3	18	6.0	96.5
Hubercarb Q 2	90	99.99	2	19	6.5	96.5
Hubercarb Q 1	90	99.99	1	20	5.5	96.5
Surface Modified[1]						
Hubercarb Q 1T				21		96.5
Hubercarb Q 2T				16		96.5
Hubercarb Q 3T				15		96.5
Hubercarb Q 200T				12		96.5
S SERIES						
Fine And Medium Fine						
Hubercarb S200	88	81	24	13		97.7
Hubercarb S325	89	99.5	14	15		97.7
Hubercarb S6	90	99.99	8	17	4	97.7
Hubercarb S4	91	99.99	5	18	6	97.7

[1] Stearate Treated

⌘⌘⌘

Product	Brightness %	<325 Mesh, %	Median P.S., µm	Oil Abs. %	Hegman	$CaCO_3$ %
FILLERS & EXTENDERS						
Hubercarb Optifil	93	Trace	2	17	6.5	99.3
Hubercarb Optifil T[1]	92	Trace	2	16	6.5	99.3
Hubercarb W 3N	92	Trace	3	15	6	99.3
Hubercarb W 3	92	Trace	4	15.0	6	99.3
Hubercarb W 4	91	99.5	5	13.0		99.3

[1] Stearate Treated

Calcium Carbonate
Polar Minerals, Inc.

Product	Brightness %	Median P.S., µm	Oil Abs. %	Hegman	$CaCO_3$ %
8101	93	0.9-1.4	42	7+	99
8101C[1]	93	0.9-1.4	28	7+	99
8103	92	2.5-3.5	22	6.5	99
8103C[1]	92	2.5-3.5	12	6.5	99
8105	92	3.5-4.5	18	4.5	99
8105C[1]	92	3.5-4.5	10	4.5	99
8107	92	4.5-6.0	15	3+	99

[1] Stearic Acid Treated

Calcium Carbonate
Specialty Minerals, Inc.

Product	Brightness[1] %	<325 Mesh, %	Median P.S., µm	Surface Area m^2/g	$CaCO_3$ %
Surface Treated Grades					
Hi-Pflex® 100	95	99.996	3.5	3.0	97
Pfinyl® 402	95	99.996	5.5	2.0	97
Super-Fil®	95	99.999	2.7	3.5	98.5
Super-Pflex® 100	97	99.97	0.7	7.0	98
Super-Pflex 200	97	99.97	0.7	7.0	98
Ultra-Pflex®	98	99.9	0.07	19.0	98

[1] Hunter

⌘⌘⌘

Product	Brightness[1] %	<325 Mesh, %	Median P.S., μm	Surface Area m^2/g	$CaCO_3$ %
Untreated Grades					
Vicron® 15-15	96	99.996	3.5	3.1	97
Vicron 25-11	96	99.996	5.5	1.8	97
Vicron 31-6	95	99.9	6.5	1.5	97
Vicron 41-5	95	99.9	8.0	1.3	97

[1] Hunter

Calcium Carbonate
United Clays, Inc.

Product	<325 Mesh, %	Median P.S., μm	Oil Abs. %	Hegman	$CaCO_3$ %
Aerogem™ #15	65.0	30-35	10.5	2.5-3.0	95.0
Aerogem #16	75.0	25-30	11.5	3.0-4.0	95.0
Aerogem #17	85.0	12.0	12.0	3.0-4.0	95.0
Aerogem#18	99.5	10.0	13.5	4.0-4.5	95.0
Aerogem #19	99.9	8.0	14.0	5.0-5.5	95.0
Aerogem #20	99.99	5.0	15.0	5.5-6.5	95.0

Calcium Carbonate
Zeneca Resins

Product	Ultimate Particle Size, μm	Surface Area m^2/g	Oil Abs. %	Surface Coating, %
Stearate Coated Grades				
Winnofil® S	0.065-0.090			2.6
Winnofil SPM	0.075	21		2.6
Winnofil SPT-L	0.075	21	31	2.6
Winnofil SPT Premium	0.075	21	31	2.6
Carboxylated Polybutadiene Coated				
Winnofil FX	0.075			

⌘⌘⌘

Feldspar
Kentucky-Tennessee Clay Co.

Product	Bright-ness %	<325 Mesh %	Median P.S μm	pH	Surf. Area m^2/g	Al_2O_3 %	CaO %	Na_2O %	K_2O %
Minsilspar	91	88	19	8.2	0.6-0.7	14.00	1.10	4.90	2.80
Minspar 1	91	6	175	8.0	0.1-0.2	18.50	1.50	6.50	4.10
Minspar 200	94	96	12	8.7	0.8-0.9	18.50	1.50	6.50	4.10

Feldspar
Pacer Corp.

Product	<325 Mesh, %	Al_2O_3 %	Fe_2O_3 %	CaO %	MgO %	Na_2O %	K_2O %
Custer (200 Mesh)	88.0	17.00	0.15	0.30	Trace	3.00	10.00
Custer (325 Mesh)	94.7	17.00	0.15	0.30	Trace	3.00	10.00

⌘⌘⌘

Kaolin Clay
Albion Kaolin Co.

Product	<325 Mesh %	Median P.S µm	pH	Max. Moist %	Al_2O_3 %	Fe_2O_3 %	L.O.I. %
Airfloat							
Albion Form® 100	98.5	1.3		1.0	38.60	0.40	13.7
Albion Sperse® 100	98-99	1.8	4.5-5.5	1.0	38.50	0.27	13.5
Albion Sperse A Slurry[1]	98-99	1.8			38.50	0.27	13.5
Albion Sperse B Slurry[2]	98-99	1.8			38.50	0.27	13.5

[1] 71.5% Solids - Sodium Silicate dispersed
[2] 71.5% Solids - Sodium Polyacrylate dispersed

Kaolin Clay
ECC America, Inc.

Product	Bright-ness %	<325 Mesh Min. %	Median P.S µm	pH	Oil Abs. %	Max. Moist. %
Ecca-Tex® 10R[1]	88	99.99	0.45	7.0	43	1.0
Ecca-Tex 20R[1]	90	99.99	0.35	7.0	43	1.0
Ecca-Tex 32R[1]	86	99.99	0.25	7.0	39	1.0
Ecca-Tex 90*[1]	88	99.99	0.45	7.0	43	1.0
Ecca-Tex 180[1]	90	99.99	0.50	7.0	41	1.0
Ecca-Tex 190[1]	90	99.99	0.35	7.0	43	1.0
Ecca-Tex 195[1]	90	99.99	0.25	7.0	45	1.0
Ecca-Tex 295[1]	86	99.99	0.25	7.0	38	1.0
Ecca-Tex 360[1]	84	99.85	0.45	7.0	38	1.0
Ecca-Tex 460	81-84	99.85	0.45	6.5-7.5	38	1.0
Polarlink® 5R[2]	88	99.99	0.45	7.0	43	1.0
Polarlink 15R[2]	86	99.99	0.25	7.0	39	1.0
Polarlink 25R[2]	90	99.99	0.25	7.0	43	1.0
Polarlink 35R[2]	86	99.99	0.25	7.0	39	1.0
Polarlink 45R[2]	90	99.99	0.25	7.0	43	1.0
Polestar® 400[3]	92	99.99	0.8	5.0	100	0.5

[1] Dispersant Treated

[2] Mercaptosilane Treated

[3] Calcined

*Delaminated

⌘⌘⌘

Kaolin Clay
J.M.Huber Corp.

Product	Brightness %	<325 Mesh Min. %	Median P.S μm	pH	Surf. Area m^2/g	Max. Moist. %
Waterwashed						
Nucap® 100G[1]	87.5-88.5	99.99	0.2	6.5-7.5	20-24	1.0
Nucap 100W[1]	87.0-88.5	99.99	0.2	6.5-8.5	20-24	1.0
Nucap 290[1]	90-92	99.99	0.2	5.0-6.5	20-24	1.0
Polyfil® HG	87.5-89.0	99.99	0.3	6.0-7.5	20-24	1.0
Polyfil HG-90	90-92	99.99	0.2	6.0-7.5	20-24	1.0
Polyfil OMH	87.0-88.5	99.995	1.4-2.1	6.0-7.5	16-20	1.0
Calcined						
Nulok® 170[2]	90-92	99.98	1.4	8.0-10.0	7-9	0.5
Polyfil WC[3]	90-93	99.99	1.4	5.0-6.0	7-9	0.5
Delaminated						
Polyfil DL	87.5-89.0	99.99	1..0	6.0-8.0	11-15	1.0

[1]Sulfur-functional

[2] Amino-functional

[3] Organofunctional

Kaolin Clay
Kentucky-Tennessee Clay Co.

Product	Brightness %	<325 Mesh %	Median P.S μm	pH	Surf. Area m^2/g	Moisture %	Al_2O_3 %	Fe_2O_3 %	L.O.I. %
Afton		99.0	0.3	4.0	22-24	1-3	38.8	1.20	14.0
Allen	74min.	98.0	1.2	4.5-6.5		1-3	38.1	0.55	13.5
Allen G		99.5		5.5	13.5		38.5	0.60	13.6
Diamond[2]		99.5		4.8	18.0		37.9	0.60	13.5
Franklin R	78min.	99.5	1.2	4.5-6.5		1.0max.	38.1	0.55	13.6
Hamilton		98.0		4.0	24.0		38.8	1.20	14.0
Hillman		97.5		4.2	16.0		38.1	0.60	13.5
Kaolin Slurry[1]		95.0		5.5	13.5		38.2	0.50	13.5
Kingsley		97.5		5.5	13.0		38.4	0.40	13.6
KT-Cast		95.0		5.5	12.0		38.8	0.50	13.6
Rogers[2]		99.0		4.5	24.0		37.5	1.00	13.2
Samson		97.5		4.7	18.0		38.1	0.60	13.4

⌘⌘⌘

Sapphire[2]		99.5	5.0	22.0		38.2	0.70	13.4
Supreme	80	99.5	4.5-6.5			39.2	0.50	13.6
Wilson		97.0	5.5	15.0		38.1	0.60	13.5
Windsor	74min.	99.7	4.0	22-24	1.0max.	38.8	1.20	14.0

[1]70% solids

[2]Extruded 50/50 clay/flint

Kaolin Clay

R.T. Vanderbilt Co.

Product	Brightness %	<325 Mesh %	Median P.S μμ	pH	Oil Abs. %	Surf. Area m^2/g	Al_2O_3 %	Fe_2O_3 %	L.O.I. %
Hard Clay									
Bilt-Plates® 145	66-70	99.9	0.2	4.7-5.2	42	25	38	1.4	14.2
Bilt-Plates 156	76min.	99.94	0.2	4.7-5.2	41	26	38	1.4	14.2
Bilt-Cote® FC		99.8	0.2	4.7-5.2			38	1.4	14.2
Bilt-Cote H		96	0.2	4.7-5.2		27	38	1.4	14.2
Continental®	66-70	99.2<200	0.2	4.7-5.2	36		38	1.4	14.2
Dixie®	70min.	99.8	0.2	4.7-5.2	42	23	38	1.4	14.2
Par®/Par Rg	66-70	99.5	0.2	4.7-5.2	40		38	1.4	14.2
Randall®	70min.	99.8	0.2	4.7-5.2	42		38	1.4	14.2
Vanclay®	76min.	99.94	0.2	4.7-5.2	41	26	38	1.4	14.2
Soft Clay									
Langford	68-70	99	1.3	4.2-4.7	36		40	0.3	14.0
McNamee®	75min.	99.65	1.3	4.2-4.7	35	16	40	0.3	14.0
Peerless® No.1	73min.	99.6<200	1.3	4.2-4.7	30	15	40	0.3	14.0
Peerless No.2	68-70	99.6<200	1.3	4.2-4.7	33	17	40	0.3	14.0
Peerless No.2AF	68--70	98.5		4.2-4.7			40	0.3	14.0
Peerless No.3	65-68	99.2<200	1.3	4.2-4.7	33	16	40	0.3	14.0
Peerless No.3AF	68min.	99.2<200		4.2-4.7			40	0.3	14.0
Peerless No.4	73min.	97<200	1.3	4.2-4.7	31		40	0.3	14.0
Peerless ShS-Dr				4.2-4.7			40	0.3	14.0
Coated Clay[1]									
Bilt-Cote H-1		96.5<200	0.2	4.7-5.2		27			
Bilt-Cote H-2		96.5<200	0.2	4.7-5.2		27			
Bilt-Cote H-3		96.5<200	0.2	4.7-5.2		27			
Bilt-Cote H-4		96.5<200	0.2	4.7-5.2		27			

[1]Coated with polynaphthalenesulphonate for prill coating

⌘⌘⌘

Kaolin Clay
U.S. Silica Co.

Product	Brightness %	<325 Mesh %	Median P.S µm	Oil Abs. %	Hegman	Moisture %	Al_2O_3 %	Fe_2O_3 %	L.O.I. %
Calcined									
Snow*Tex® 45	91	99.95	1.5	60-70	5.0	0.40	39.68	0.93	0.23

Kaolin Clay
Wilkinson Kaolin Associates, Ltd.

Product	Brightness %	<325 Mesh[1] %	Median P.S µm	pH[2]	Oil Abs. %	Al_2O_3 %	Fe_2O_3 %	L.O.I. %
Wilklay™ CR		98.0.	0.78	4.5-7.0		38.4	0.85	13.7
Wilklay FE		98.0	0.7	5.0		38.4	0.85	13.7
Wilklay PF	80min.	99.5		4.5-6.5				
Wilklay RP-2	78	99.5	0.36	5.2	40	38.4	0.88	13.7
Wilklay SA-1	79.5	99.0		5.4	32	38.4	0.40	13.82
Wilklay WC-5		99.0	1.2	6.9		38.4	0.38	14.21

[1]Minimum
[2]At 28% solids

⌘⌘⌘

Muscovite Mica
Aspect Minerals Inc.

Product	Brightness %	<325 Mesh %	Median P.S µm	Oil Abs. %	Moisture %	Al_2O_3 %	Fe_2O_3 %	K_2O %	L.O.I. %
Industrial Grades									
AlbaFlex® 25		75-85	52-58	70	0.50	38.15	2.42	7.53	4.27
AlbaFlex 50		90	42-47	70	0.50	38.15	2.42	7.53	4.27
AlbaFlex 100		95	34-39	75	0.50	38.15	2.42	7.53	4.27
AlbaFlex 200			22-27	75	0.50	38.15	2.42	7.53	4.27
AlbaShield 15		50-69		65	0.50	38.15	2.42	7.53	4.27
AlbaShield® 25		70-89		65	0.50	38.15	2.42	7.53	4.27
AlbaShield 50		90		70	0.50	38.15	2.42	7.53	4.27
AlbaShield 1000		95		65	0.50	38.15	2.42	7.53	4.27
AlbaShield 2000		99		67	0.50	38.15	2.42	7.53	4.27
Cosmetic Grades									
Carolina® Silver115	68-74	100<400	7-10	93	≤1.0	37-39	1.5-2.5	6-9	3-4
Carolina Silver 150	68-74	100<400	24-28	93	≤1.0	37-39	1.5-2.5	6-9	4-5
Carolina Silver 590	68-74	90	42-48	93	≤1.0	37-39	1.5-2.5	6-9	4-5

Muscovite Mica
KMG Minerals

Product	<325 Mesh, %	Moisture %	Al_2O_3 %	Fe_2O_3 %	K_2O %	L.O.I. %
Flake Grades						
10-KS	15.5<100	0.02	33.10	2.50	9.50	4.20
20-K	14.5<200	0.02	33.10	2.50	9.50	4.20
20-KS	21.6<100	0.02	33.10	2.50	9.50	4.20
40-K	3.0<200	0.02	33.10	2.50	9.50	4.20
F-120	5.8<100	0.02	33.10	2.50	9.50	4.20
L-115	4.0<70	0.02	33.80	2.00	9.70	4.20
L-125	6.0<200	0.02	33.80	2.00	9.70	4.20
L-135	7.0<200	0.02	33.80	2.00	9.70	4.20
L-140	48	0.02	33.80	2.00	9.70	4.20
Micawhite 1117	8.0<200	0.12				4.20
U-125	6.0<200	0.02	31.50	3.00	7.00	4.20
Coarse Oil Well	15.5<100	0.02	33.10	2.50	9.50	4.20

⌘⌘⌘

Med. Oil Well	6.0<100	0.02	31.50	3.00	7.00	4.20
Fine Oil Well	21.6<100	0.02	33.10	2.50	9.50	4.20

Product	<325 Mesh, %	Median P.S[1] µm	pH	Al_2O_3 %	Fe_2O_3 %	K_2O %	L.O.I. %
Dry Ground Grades							
1-K	46.9		6.9	33.10	2.50	9.50	4.20
4-K	62		6.9	33.10	2.50	9.50	4.20
37-C	67.5<200		8.2	31.50	3.00	7.00	4.20
37-CK	67.5<200		7.5	33.10	2.50	9.50	4.20
100-K	51<200		6.9	33.10	2.50	9.50	4.20
F-260	75	42.0	6.9	33.10	2.50	9.50	4.20
FS-255	62		6.9	33.10	2.50	9.50	4.20
Mica S	78		6.9				4.20
Micawhite 200	58		6.9				4.20
P-80-K	56		8.2	31.50	3.00	7.00	4.20
Polymica 325	95.5		6.9				4.20
U-230	51<200		8.2	31.50	3.00	7.00	4.20
U-255	62		8.2	31.50	3.00	7.00	4.20
US-250	61	49.0	8.2	32.20	2.80	8.00	4.20
Wet Ground Grades							
H-360	82.0	44.5	6.9	34.30	2.50	9.90	4.20
HAR 160	82.0		6.9	33.10	2.50	9.50	4.20
HIMOD 270	96.8	37.5	6.9	34.30	2.00	9.90	4.20
S-360	82.0	39.0	6.9	33.10	2.50	9.50	4.20
WG 100	21.5		6.9	33.10	2.50	9.50	4.20
WG 160	85.0		6.9	33.10	2.50	9.50	4.20
WG 325	95.2		6.9	33.10	2.50	9.50	4.20
Micronized Grades							
C-500	95.2		8.2	33.10	2.50	9.50	4.20
C-1000	98.7		8.2	33.10	2.50	9.50	4.20
C-3000	99.5		8.2	33.10	2.50	9.50	4.20
C-4000	99.96		8.2	33.10	2.50	9.50	4.20
F-485	99.5		6.9	33.10	2.50	9.50	5.00
FS-477	96.5		6.9	33.10	2.50	9.50	4.20
HIMOD 450	Trace		6.9	33.80	2.00	9.70	4.20
L-477	92.2	37.5	6.9	33.80	2.00	9.70	4.20

[1] Malvern Method

⌘⌘⌘

Muscovite Mica
Pacer Corp.

Product	<325 Mesh, %	pH	Oil Abs., %	Al_2O_3 %	Fe_2O_3 %	K_2O %	L.O.I. %
Brite-X®	28.3<200	8.0	45	22.0	5.0	7.0	
Brite-X 100	42.4	8.0	45	26.1	2.3	7.6	
Custer 24X		6.8		33.2	0.3	8.9	
Custer LC		6.8		33.2	0.3	8.9	
LCM Coarse	1.5	8.0		26.5	7.0	8.8	
Micaflex®	41.8	8.0		25.0	4.0	6.5	1.0
Micaflow®	41.8	8.0		25.0	4.0	6.5	1.0

Phlogopite Mica
Polar Minerals

Product	Brightness, % (Reflectance)	<325 Mesh %	Median P.S., μm
5040	62	90	20
5040S[1]	62	90	20
5100	58	90<170	50
5100S[1]	58	90<170	50
5200	54	90<70	70
5500	48	90<35	300
5900	42	90<20	650

⌘⌘⌘

Pyrophyllite
R.T. Vanderbilt Co.

Product	Brightness %	<325 Mesh %	Median P.S μm	Oil Abs. %	Surf. Area m^2/g	Al_2O_3 %	Fe_2O_3 %	K_2O %	L.O.I. %
Pyrax® A	65	90	10	24	2	15-19	0.2-0.8	0.2-2.3	2.3-3.9
Pyrax B	80	95	10	26	3	15-19	0.2-0.8	0.2-2.3	2.3-3.9
Pyrax WA	78	87	13	24	3	15-19	0.2-0.8	0.2-2.3	2.3-3.9
Pyrax ABB	65	90.7	11	24	2	15-19	0.2-0.8	0.2-2.3	2.3-3.9
Pyrax HS	75[1]	90			2	15-19	0.2-0.8	0.2-2.3	2.3-3.9
Pyrax MP	65	90.7	11	24	2	15-19	0.2-0.8	0.2-2.3	2.3-3.9
Pyrax RG 3/8"	55[1]	8				15-19	0.2-0.8	0.2-2.3	2.3-3.9
Pyrax RG 1/4"	55[1]	8				15-19	0.2-0.8	0.2-2.3	2.3-3.9
Pyrax RG 16	55[1]	18.1			1	15-19	0.2-0.8	0.2-2.3	2.3-3.9
Pyrax RG 16C	55[1]					15-19	0.2-0.8	0.2-2.3	2.3-3.9
Pyrax RG 140	60[1]	78.5			3	15-19	0.2-0.8	0.2-2.3	2.3-3.9
Pyrax RG 140I	70[1]	80				15-19	0.2-0.8	0.2-2.3	2.3-3.9
Pyrax RG 200	60[1]	95.4			5	15-19	0.2-0.8	0.2-2.3	2.3-3.9
Veecote	80	99			3	15-19	0.2-0.8	0.2-2.3	2.3-3.9

[1] Fired Color

⌘⌘⌘

Silica (fumed)
Cabot Corp.

Product	<325 Mesh, %	pH	Surf. Area m^2/g	Moisture %	L.O.I. %
Undensed Grades					
Cab-O-Sil® L-90	99.98	3.7-4.3	80-100	<0.5	<1
Cab-O-Sil LM-130	99.98	3.7-4.3	115-145	<0.5	<1
Cab-O-Sil LM-150	99.98	3.7-4.3	145-175	<0.5	<1
Cab-O-Sil M-5	99.98	3.7-4.3	175-225	<1.5	<2
Cab-O-Sil PTG	99.98	3.7-4.3	175-225	<1.5	<2
Cab-O-Sil MS-55	99.98	3.7-4.3	230-280	<1.5	<2
Cab-O-Sil H-5	99.98	3.7-4.3	275-325	<1.5	<2
Cab-O-Sil HS-5	99.98	3.7-4.3	300-350	<1.5	<2
Cab-O-Sil EH-5	99.98	3.7-4.3	350-410	<1.5	<2.5
Densed Grades		3.7-4.3			
Cab-O-Sil LM-150D	99.98	3.7-4.3	145-175	<0.5	<1
Cab-O-Sil M-7d	99.98	3.7-4.3	175-225	<1.5	<2
Cab-O-Sil MS-75D	99.98	3.7-4.3	230-280	<1.5	<2
Surface Treated Grades					
Cab-O-Sil TS 720[1]			80-120	0.60	<7
Cab-O-Sil TS 610[2]		4-5	100-140	0.50	
Cab-O-Sil TS 530[3]		4.8-7.5	185-245	0.60	

[1] Silicone Fluid Treated

[2] Hexamethyldisilazane Treated

[3] Dimethyl Dichlorosilane Treated

Product	Solids Content %	pH
DISPERSION GRADES		
Ammonia Stabilized		
Cab-O-Sperse® A105	15	5.0-5.5
Cab-O-Sperse A205	12	5.0-5.5
Cab-O-Sperse A1695	17	9.5-10.0
Cab-O-Sperse A2095	17	9.5-10.0
Cab-O-Sperse A3875	12	7.5-7.8
Potassium Hydroxide Stabilized		
Cab-O-Sperse P1010	30	10.0-10.3
Cab-O-Sperse P1075	28	7.5-8.0
Cab-O-Sperse P1685	18	8.5-9.0

⌘⌘⌘

Sodium Hydroxide Stabilized		
Cab-O-Sperse S109	20	9.0-9.5
Cab-O-Sperse S2095	17	9.5-10.0
Cab-O-Sperse S3295	15	9.5-10.0
Cab-O-Sperse S3810	14	10.0-10.5
Cab-O-Sperse II	20	9.0-10.5

Silica (fumed)
Degussa

Product	pH	Surf. Area m^2/g	Moisture %	L.O.I. %
Hydrophilic				
Aerosil® 90	3.6-4.5	90±15	<1.0	<1.0
Aerosil 130	3.6-4.3	130±25	<1.5	<1.0
Aerosil 150	3.6-4.3	150±15	<0.5	<1.0
Aerosil 200	3.6-4.3	200±25	<1.5	<1.0
Aerosil 300	3.6-4.3	300±30	<1.5	<2.0
Aerosil 380	3.6-4.3	380±30	<2.0	<2.5
Aerosil OX 50	3.8-4.8	50±15	<1.5	<1.0
Aerosil TT600	3.6-4.5	200+50	<2.5	<2.5
Aerosil MOX 80	3.6-4.5	80+20	<1.5	<1.0
Aerosil MOX 170	3.6-4.5	170+30	<1.5	<1.0
Aerosil COK 84	3.6-4.3	107+30	<1.5	<1.0
Hydrophobic				
Aerosil R 202	4.0-6.0	100+20	<0.5	4.0-6.0
Aerosil R 805	3.5-5.5	150+25	<0.5	5.0-7.0
Aerosil R 812	5.5-7.5	260+30	<0.5	1.0-2.5
Aerosil R 812S	5.5-7.5	220+25	<0.5	<1.5-3.0
Aerosil R 972	3.6-4.3	110+20	<0.5	<2.0
Aerosil R 974	3.4-4.2	170+20	<0.5	<2.0

⌘⌘⌘

Silica (precipitated)
Degussa

Product	Average* P.S., µm	pH	DBP Oil Abs.%	Surf. Area m^2/g	Moisture %	SiO_2 %
FK 160	7	4.5	250	160	3	99.4
FK 300 DS	4.5	6.5	260	300	6	99
FK 310	5	7	210	650	2.5	99
FK 320	15	6.3	225	170	6	98
FK 500 LS	4.5	7	330	450	3	99
FK 700	35	7	190	700	6	99
Sipermat® D 10	5	10.3	240	90	3	98
Sipermat D 17	10	8	230	100	3	99.5
Sipermat 22	100	6.3	260	190	6	98
Sipermat 22S	7	6.3	270	190	6	98
Sipermat 22 LS	4.5	6.3	270	170	6	98
Sipermat 283 LS	5	8.3	260	170	6	98
Sipermat 50	50	7	340	450	6	99
Sipermat 50 S	8	7	330	450	6	99
Durosil®	15	9	220	60	6	98
Ultrasil® VN 3 SP	-	6.3	-	185	5.5	98
Ultrasil VN 3 SP Beads	-	6.9	-	175	5.5	98
Quso® G 35	2	8.3	280	180	6	98
Quso WR 55	3	10	-	120	1	98

*Avg. agglomerate size

⌘⌘⌘

Silica (precipitated)
PPG Industries Inc.

Product	Median P.S., µm	pH	Oil Abs. %	Surf. Area m^2/g	Moisture %	SiO_2 %
Thickener Grades						
Hi-Sil® T-600	3.5	6.5-7.3	150	150	6.3	97.5
Hi-Sil T-690	1.4	6.5-7.3	150	150	6.3	97.5
Hi-Sil T-700	1.9	7.0	210	210	4-7	
Flatting Agents						
Lo-Vel® 27	2.0	6.5-7.5	210	180	6.3	97.5
Lo-Vel 29	4.8	6.5-7.3	150	150	6.3	97.5
Lo-Vel 39A	6.7	6.5-7.3	150	150	6.3	97.5

Product	Median P.S., µm	pH	Surf. Area m^2/g	Moisture %
Silane Treated Grade				
Ciptane®		7.0	142	6.5
Agricultural Chemical Grades				
Hi-Sil 135	14	7.0	170	5.0
Hi-Sil 212		7.0	150	5.0
Hi-Sil 213		7.0	150	5.0
Hi-Sil 233	15	7.0	150	5.0
Hi-Sil 250	15	7.0	150	5.0
Hi-Sil ABS	22	7.0	150	5.0
Hi-Sil HOA	45	7.0	155	5.0
Rubber Reinforcing Grades				
Hi-Sil 132	10	7.0	200	5.0
Hi-Sil 135	15	7.0	170	5.0
Hi-Sil 210		7.0	150	5.0
Hi-Sil 233	15	7.0	150	5.0
Hi-Sil 243LD		7.0	150	5.0
Hi-Sil 532EP	7	8.0	60	5.0
Hi-Sil 900	7	7.0	152	5.0
Hi-Sil 915	4	7.0	220	5.0
Hi-Sil ABS	22	7.0	150	4.0
Silene® 732D	9	8.5	35	5.0

⌘⌘⌘

Silica (ground)
U.S. Silica Co.

Product	Mine Site[1]	Brightness[2] %	<325 Mesh %	Median P.S μm	pH	Oil Abs. %	Surf. Area m^2/g	Hegman	SiO_2 %
Sil-Co-Sil® 250	E	80.0	51.0	45.0	6.8	20.0	0.24		99.8
Sil-Co-Sil 125	F	84.0	65.0	21.0	6.8	18.5	0.30		99.7
Sil-Co-Sil 125	D	85.0	71.0	21.0	6.8	18.5	0.35		99.7
Sil-Co-Sil 125	A	83.0	74.0	21.0	6.8	20.5	0.32		99.6
Sil-Co-Sil 106	F	84.0	71.0	19.0	7.0	19.0	0.32		99.7
Sil-Co-Sil 106	E	83.0	76.0	19.0	6.8	23.0	0.32		99.8
Sil-Co-Sil 106	D	86.0	78.0	21.0	6.8	19.0	0.37		99.7
Sil-Co-Sil 90	F	85.0	79.0	18.0	7.0	20.0	0.35	0.0	99.7
Sil-Co-Sil 90	C	82.0	82.0	16.0	6.8	21.0	0.32	0.0	99.6
Sil-Co-Sil 90	E	84.0	84.0	16.0	7.0	24.0	0.35	0.0	99.8
Sil-Co-Sil 90	A	83.0	80.0	16.0	6.8	22.5	0.35	0.0	99.6
Sil-Co-Sil 75	C	82.0	88.0	18.0	6.8	23.0	0.34	1.0	99.6
Sil-Co-Sil 75	F	86.0	88.0	17.0	7.2	20.5	0.38	1.0	99.7
Sil-Co-Sil 75	B	84.0	87.0	16.0	6.0	22.0	0.38	1.0	99.5
Sil-Co-Sil 75	D	86.0	87.0	15.0	7.0	20.0	0.38	1.0	99.7
Sil-Co-Sil 75	A	84.0	88.0	16.0	6.8	24.0	0.38	1.0	99.6
Sil-Co-Sil 63	E	84.0	90.0	14.0	7.3	25.0	0.41	2.0	99.8
Sil-Co-Sil 63	C	83.0	90.5	16.0	6.8	24.0	0.34	2.5	99.6
Sil-Co-Sil 63	G	80.0	90.5	16.0	7.0	21.0	0.38	2.0	99.6
Sil-Co-Sil 63	A	84.0	94.0	15.0	6.8	25.0	0.42	2.5	99.6
Sil-Co-Sil 53	D	87.0	93.0	12.0	7.2	21.0	0.42	3.0	99.7
Sil-Co-Sil 53	C	83.0	94.0	15.0	6.8	25.0	0.35	3.0	99.6
Sil-Co-Sil 53	E	85.0	95.5	14.0	7.5	26.0	0.46	3.0	99.8
Sil-Co-Sil 52	A	85.0	96.0	15.0	7.0	26.0	0.45	3.0	99.6
Sil-Co-Sil 52	B	84.0	97.4	13.0	6.6	23.0	0.40	3.0	99.5
Sil-Co-Sil 51	F	87.0	96.0	11.0	7.2	21.0	0.45	3.0	99.7
Sil-Co-Sil 49	B	84.0	97.5	12.0	6.6	24.0	0.41	3.5	99.5
Sil-Co-Sil 49	E	88.0	97.5	11.0	7.2	22.0	0.45	3.5	99.7
Sil-Co-Sil 47	C	84.0	97.8	13.0	7.0	26.0	0.38	3.5	99.6
Sil-Co-Sil 45	E	85.0	98.4	13.0	7.5	27.0	0.57	3.5	99.8
Sil-Co-Sil 40	F	88.0	99.2	10.0	7.4	23.0	0.37	4.0	99.7
Sil-Co-Sil 40	C	84.0	99.0	12.0	7.2	33.0	0.49	4.0	99.6
Sil-Co-Sil 40	A	86.0	99.2	12.0	7.2	26.5	0.50	4.0	99.6
Sil-Co-Sil 40	E	86.0	99.0	10.0	7.8	27.5	0.61	4.0	99.8

[1] A - Berkeley Springs, WV, B- Colombia, SC, C- Mapleton Depot, PA, D - Mill Creek, OK, E - Ottawa, IL, F - Pacific, MO [2] Photovolt (457 μm)

⌘⌘⌘

Product	Brightness[1] %	Median P.S., μm	pH	Oil Abs. %	Hegman	SiO_2 %
Min-U-Sil® 40	88.6	11.5	7.8	22.0	4.50	99.6
Min-U-Sil 30	89.0	8.0	7.8	23.0	5.25	99.6
Min-U-Sil 15	92.0	4.8	8.0	33.0	6.75	99.7
Min-U-Sil 10	91.5	4.1	7.5	33.3	7.00	99.5
Min-U-Sil 5	92.5	1.7	7.0	42.0	7.75	99.2

[1]Hunter-Reflectance (Y)

Silicates (precipitated)
Degussa AG

Product	<325 Mesh %	Median P.S μm	pH	Surf. Area m^2/g	Moist-ure %	SiO_2 %	Al_2O_3 %	CaO %	Na_2O %	L.O.I. %
Aluminum Silicates										
P820	99.9	0.015	10.4	100	6.00	82.0	9.5		8.0	7.0
Transpafill®	99.8	0.015	7.0	110	6.00	66.0	26		7.0	12.0
Calcium Silicate										
Extrusil®	99.8	0.025	10.0	35	6.00	91.0	0.2	6.0	2.0	7.0
Sodium Aluminum Silicate										
Sipermat® 44	99.9		11.8			42.0	36		22.0	20.0

⌘⌘⌘

Smectite Clay
Southern Clay Products, Inc.

	Viscosity cps	pH
Gelwhite® L[1]	525 @ 10%	9
Gelwhite GP[1]	2500 @ 5%	10
Gelwhite H, NF[1]	1950 @ 5%	9.5
Mineral Colloid® BP[2]	750 @ 5%	9
Mineral Colloid MO[2]	2000 @ 3%	9

[1]Brookfield, 20rpm

[2]Brookfield, 50rpm

Smectite Clay
R.T. Vanderbilt Co.

	Viscosity[1] cps	pH	Al/Mg Ratio[2]
Cosmetic/Pharmaceutical Grades			
Veegum®	225-600 @ 5%	9.0-10.0	0.5-1.2
Veegum F	150-400 @ 5%	9.0-10.0	0.5-1.2
Veegum HV	800-2200 @ 5%	9.0-10.0	0.5-1.2
Veegum K	100-300 @ 5%	9.0-10.0	1.4-2.8
Veegum D	100-300 @ 5%	8.5-9.5	
Veegum HS	40-200 @ 5%	9.0-10.0	3.5-5.5
Veegum Ultra	225-425 @ 5%	4.2-5.2	
Veegum PLUS	350-750 @ 3%	9.0-10.0	
Veegum PRO	300-550 @ 1.5%	8.0-9.0	
Industrial Grades			
Van Gel B	300-900 @ 4%	8.5-9.5	
Van Gel ES	40-200 @ 5%	9.0-10.0	
Van Gel O	200-450 @ 8%	8.5-9.5	
Van Gel C	150-350 @ 6%	9.0-10.0	
Veegum T	250-800 @ 4%	9.0-10.0	
Veegum CER	75-175 @ 1%	9.5-10.5	

[1]Brookfield, 60rpm

[2]For products meeting USP/NF monograph standards

⌘⌘⌘

Talc
Luzenac America, Inc.

Product	Brightness %	Median P.S., μm	Hegman Gind	Oil Abs. %	Surf. Area m^2/g
Industrial Grades					
2C		9.0			
Artic Mist®	88	2.2	6.0		
Cimpact® 669	87	1.2	7.0		
Jetfil® 350V	84	7.5	4.0		
Jetfil 575C	87	3.4	5.5		
Jetfil 625C	88	2.2	6.0		
Jetfil 700C	88	1.5	7.0		
Mistron® 100	80				13
Compacted Mistron 100	80		5.5		13
Mistron 400C	88	3.5			
Mistron CB[1]	87	1.7			
Mistron PXL[1]	87	1.7			
Mistron Super Frost	90	1.7	6.0		
Mistron Vapor	85				13
Compacted Mistron Vapor	85				13
Mistron Vapor R	87	1.7	5.5		
Mistron ZSC	89	1.5	6.0		
Nicron® JS-422	80-84[2]			22	
Nicron JS-528	84-88[2]			28	
Sierralite® II		8.5			
Steawhite® 200		8.5			
T-40/C-117	70-78	10-14			
T-50	75	15.0			
Techfil® 7599	77-80	12.0	3.5		
Vertal® 710	86	7.5	4.0		
Vertal 92	72	15.0			
Vertal 97	74-78	9-13			
Vertal C2+	82	10.0			
Yellowstone		10.0			

[1]Surface treated
[2]Hunter (Y)

⌘⌘⌘

Product	Brightness %	<325 Mesh, %	Median P.S., μm
Cosmetic and Personal Care Grades			
Altalc® 400 SMS[1]	87min.	99.6min.	4.0
Brillante	85min.	90<200	19.0
Olympic	85min.	99.5<200	10.5
Supra A	85min.	98<200	12.0
Supra SMM[1]	85min.		12.0
Supra SMS[1]	85min.		12.0
Suprafino A	89	99.6min.	5.0
Suprafino SMD[1]	89min.	99.6min.	5.0
Suprafino SML[1]	89min.	99.6min.	5.0
Supreme HT	85min.	99.5min.	7.0
Top Note 200	85min.	98<200	11.0
Top Note 300	87min.	98min.	6.0
Ultrafino	89min.	99.6min.	4.0
Ultrafino SMD[1]	89min.	99.6min.	4.0
Ultrafino SMS[1]	89min.	99.6min.	4.0
Vertal CO+	84min.	97<200	16.5
Pharmaceutical Grades			
Alphafil 500 USP	89min.	99.9<400	4.0
Altalc 200 USP	85min.	98<200	9.0
Altalc 300 USP	86min.	98min.	6.0
Altalc 400 USP	87min.	99.9min.	4.0
Altalc 500 USP	87min.	99.9<400	4.0

[1] Surface Treated

⌘⌘⌘

Talc
Polar Minerals

Product	Brightness (Hunter)	Median P.S., μm	Hegman	Oil Abs. %
Industrial Grades				
MV305	92-94	4-6	4	35
MV310	92-94	8-10	1-2	32
MV603	90-92	3-4.5	4	41
MV607	90-92	6-8	1-2	38
MV610	90-92	9-11	0	36
9102	92-94	1.5-2.5	6	45
9103	92-94	3-4	4	41
9107	92-94	6-8	3	38
9110	92-94	8-11	0	36
9603	88min.	3-4.5	5	41
9607	86min.	6-8	3	38
9610	84min.	9-11	0	36
Gel Talc	92-94	4-6	3+	35
Body Filler Medium	92-94	9-11	1	32
Body Filler Fine	92-94	4-6	3+	35
Surface Treated Grades				
9603S1	88min.	3-4.5	5	38
9603Z2	88min.	3-4.5	5	38
Cosmetic and Pharmaceutical Grades				
Ultra 2002	88-92	10-12		
Ultra 3002	88-92	6-8		
Ultra 4002	88-92	3-4		

[1]Silane Treated,

[2]Zinc Stearate Treated

⌘⌘⌘

Talc
Specialty Minerals, Inc.

Product	Brightness (Hunter)	<325 Mesh, %	Median P.S., µm	Hegman	Oil Abs. %
ABT®-2500	89	99.95	2.3		
LVT-325	85			3.0	25
LVT-400	86		4.2	4.0	30
LVT-500	88			5.0	37
LVT-600	88		2.4	6.0	40
Microbloc®[1]	89	99.95	2.3		
MicroPflex™ 1200	89.5		1.2	6.0	
Microtalc® MP10-52	91		0.8	6.5	55
Microtalc MP12-50	89.5		1.2	6.0	54
Microtalc MP15-38	90		2.0	5.75	40
Microtalc MP25-38	90		2.4	5.5	39
Microtalc MP30-36	88	99.8	3.1	4.75	35
Microtuff®-F	89.5		1.2	6.0	
Talcron® MP40-27	89	99.8	4.7	3.75	28
TalcronMP44-26	88.5	99.4	5.0	3.5	26
Talcron MP45-26	86.5	99.1	6.2		25
Ultratalc™ 609	88min.		<0.9	6.75min.	

[1] Proprietary Coating

⌘⌘⌘

Talc
R.T. Vanderbilt Co.

Product	Brightness %	<325 Mesh, %	Median P.S., μm	Oil Abs. %
Tremolitic Talc				
Ceramitalc® HDT	90	96<200		
Ceramitalc No.1	90	97.5		
Ceramitalc 10A	90	97.5		
Ceramitalc 10AC	90	96.5<200		
I.T. X	90	97.8	10.2	23
I.T. 3X	90	97.8	9.3	29
I.T. 5X	90	99.2	7.6	30
I.T. F.T.	90	99.2	7	29
I.T. 325	90	99.93	5.5	29
Nytal® 99	90			19
Nytal 100	90	96.5<200	24	21
Nytal 100HR	90	96.5<200		
Nytal 200	90	97.8	12.6	23
Nytal 300	90	99.93	6.3	29
Nytal 300H	90	99.9		39
Nytal 400	90	99.99	4	39
Floated Platy Talc				
Vantalc® 6H	88	99.9	2.2	52

⌘⌘⌘

Wollastonite
NYCO Minerals, Inc.

Product	Bright-ness %	<325 Mesh %	Median P.S µm	<10µm %	Oil Abs. %	Surf. Area m^2/g	Moist-ure %	L.O.I. %
10 Wollastocoat®[1]	95		3	96	24	4.1	0.15	0.70
325 Wollastocoat	92		10	50	20	1.6	0.10	0.45
400 Wollastocoat	94		7	65	22	1.9	0.10	0.45
G-40 Wollastocoat[2]	83	68			35		0.02	0.20
G-RRIM™[2] Wollastocoat	83	68			35		0.02	0.20
Nyad® 200	90	97<200			23			0.25
Nyad 325	94	99			20	1.5	0.20	0.30
Nyad 400	94	99.9			22	1.9	0.20	0.40
Nyad 475	95			85	30	3.3	0.25	0.55
Nyad 1250	95			96	32	4.0	0.25	0.70
Nyad FP[2]	70	2<100				0.8		
Nyad G®[2]	85	65			45	0.5	0.02	0.20
Nyad G Special[2]	85	70			47	0.4	0.02	0.20
Nyad G Wollastocoat[2]	83	68			40		0.02	0.20
Nycast™ U-1[3]	95		3	96	24	4.1	0.15	0.70
Nycor® R[2]	85	75<200			43	0.8	0.02	0.20

[1]Wollastocoat products are surface treated
[2]High aspect ratio acicular
[3]Surface treated

Wollastonite
R.T. Vanderbilt Co.

Product	Brightness %	<325 Mesh, %	Median P.S., µm	Oil Abs. %	Surf. Area m^2/g	L.O.I. %
Vansil® W10	89	96<200	15.6	19	1.6	2.2
Vansil W20	89	97.2	9.6	20	2.4	2.2
Vansil W30	89	99.92	4.5	21	3.7	2.2

⌘⌘⌘

Zeolite
GSA Resources, Inc.

	Pore Diam. angstroms	Surf. Area m^2/g	Ion Exch. Cap., meq/g	Moisture %
Cabsorb® ZS500H[1]	4.3	521	2.5	<10
Cabsorb ZS500A[1]	4.3	521	2.5	<3
Cabsorb ZC500H[2]	4.3	462	2.5	<15
Cabsorb ZC500A[2]	4.3	462	2.5	<3
Cabsorb ZC400H[2]	4.3	354	1.9	<15
Cabsorb ZS403H[3]	4	40	1.65	
Cabsorb ZS406H[3]	4	40	1.65	

[1]Natural herschelite-sodium chabazite

[2]Natural chabazite

[3]Natural clinoptilolite

⌘⌘⌘

MINERAL SUPPLIERS

Albion Kaolin Co.
1 Albion Rd.
Hephzibah, GA 30815

Aspect Minerals Inc.
P.O. Box 277
Spruce Pine, NC 28717

Cabot Corp.
P.O. Box 188
Tuscola, IL 61953

Degussa Corp.
150 Springside Dr.
Akron, OH 44333

ECC International
5775 Peachtree-Dunwoody Rd., NE
Atlanta, GA 30342

Genstar Stone Products Co.
11350 McCormick Rd.
Hunt Valley, MD 21031

Georgia Marble Co.
1201 Roberts Blvd.
Kennesaw, GA 30144

GSA Resources, Inc.
7660 N. Business Park Dr.
Tucson, AZ 85743

J.M. Huber Corp.
1807 Park 270 Dr.
St. Louis, MO 63146

Kentucky-Tennessee Clay Co.
P.O. Box 6002
Mayfield, KY 42066

KMG Minerals
1469 S. Battleground Ave.
Kings Mountain, NC 28086

Luzenac America, Inc.
9000 East Nichols Ave.
Englewood, CO 80112

Mountain Minerals Co. Ltd.
1425 West Pender St.
Vancouver, B.C., Canada

NYCO Minerals, Inc.
124 Mountain View Dr.
Willsboro, NY 12996

Pacer Corp.
35 South 6th St.
Custer, SD 57730

Polar Minerals
1703 Bluff Rd.
Mt. Vernon, IN 47620

PPG Industries Inc.
440 College Park Dr.
Monroeville, PA 15146

Southern Clay Products, Inc.
1212 Church St.
Gonzales, TX 78629

Specialty Minerals, Inc.
640 North 13th St.
Easton, PA 18042

United Clays
7003 Chadwick Dr.
Brentwood, TN 37027

U.S. Silica Co.
P.O. Box 187
Berkeley Springs, WV 25411

R.T. Vanderbilt Co., Inc.
30 Winfield St.
Norwalk, CT 06855

Wilkinson Kaolin Assoc., Ltd.
831 Milledgeville Rd.
Gordon, GA 31031

Zeneca Resins
730 Main St.
Wilmington, MA 01887

⌘⌘⌘

NOTES

⌘⌘⌘

—A—

⌘⌘⌘

⌘⌘⌘

—B—

⌘⌘⌘

—C—

⌘⌘⌘

⌘⌘⌘

⌘⌘⌘

—D—

⌘⌘⌘

—E—

⌘⌘⌘

—F—

⌘⌘⌘

—G—

—H—

⌘⌘⌘

⌘⌘⌘

—L—

—M—

⌘⌘⌘

⌘⌘⌘

—N—

—O—

⌘⌘⌘

—P—

⌘⌘⌘

⌘⌘⌘

⌘⌘⌘

⌘⌘⌘

—Q—

—R—

⌘⌘⌘

⌘⌘⌘

—S—

⌘⌘⌘

⌘⌘⌘

—T—

⌘⌘⌘

⌘⌘⌘

⌘⌘⌘

—X—

—Y—

—Z—

⌘⌘⌘